Electric Power Distribution Engineering

Are you fascinated by the complex web of electrical power that illuminates our modern world? Do you want to understand the intricate systems responsible for delivering electricity to our homes, businesses, and industries? Look no further than *Electric Power Distribution Engineering, Fourth Edition* by renowned author Turan Gönen, revised and updated by Chee-Wooi Ten and Ali Mehrizi-Sani. This captivating book takes you on a journey through the fascinating realm of electric power distribution, offering a comprehensive yet accessible exploration of the engineering principles, technologies, and practices that underpin this vital aspect of our daily lives. Whether you're a curious non-specialist, an avid reader with a thirst for knowledge, or a librarian or bookseller seeking an invaluable resource, Dr. Gönen's masterwork will both enlighten and captivate you.

An early leader in the academic market, this book provides an overview of classical planning for electric power distribution systems, which has been used for many years in designing and analyzing electric power distribution systems. The authors have taken a bold initiative to update the content, incorporating relevant aspects reflecting the advancements of today's evolving smart grid. Within its pages, readers will discover detailed discussions on the principles of power distribution, including the fundamentals of power generation, transmission, and distribution. The authors provide detailed explanations of the various components and equipment used in distribution systems, such as transformers, circuit breakers, switches, and protective devices. As part of the book, planning for the distribution network involves sizing and considering candidate geographical locations/regions in relation to the capacity of existing infrastructure, allowing for new additions to be built. For example, this includes locations either extending another feeder from distribution substations or building new distribution substations, depending on what makes more sense. Many assumptions have been made for non-existing distribution feeders to calculate ballpark figures for determining voltage profile and power losses if they were to be constructed. Readers will gain insights into how these considerations translate into net positive, net negative, or net-zero loads. All of these aspects can be gradually integrated with renewable energy sources, innovative grid technologies, and distribution automation over time.

The authors involved in this book have made significant contributions to the state-of-the-art development by incorporating recent updates from the literature, thereby addressing the latest advancements. One remarkable feature of Turan Gönen's *Electric Power Distribution Engineering* is its strong focus on practical applications and real-world scenarios. In addition to providing theoretical knowledge, the book also offers numerous examples that effectively bridge the gap between theory and practice. This unique approach enables readers to comprehend the intricacies of distribution system engineering and apply their newfound knowledge to solve complex problems in the field. By seamlessly blending theoretical foundations with practical insights, Dr. Gönen's book emerges as an indispensable resource for aspiring engineers, professionals, and researchers, as it offers a comprehensive understanding of electric power distribution systems and their practical implications.

Electric Power Distribution Engineering

Fourth Edition

Originally authored by Turan Gönen
Revised and updated by Chee-Wooi Ten and
Ali Mehrizi-Sani

CRC Press
Taylor & Francis Group
Boca Raton London New York

CRC Press is an imprint of the
Taylor & Francis Group, an **informa** business

[Fourth] edition published [2024]
by CRC Press
2385 NW Executive Center Drive, Suite 320, Boca Raton FL 33431

and by CRC Press
4 Park Square, Milton Park, Abingdon, Oxon, OX14 4RN

CRC Press is an imprint of Taylor & Francis Group, LLC

[First edition published by McGraw-Hill 1986]
[Sixth edition published by CRC Press 2014]

ISBN: 978-0-367-65495-5 (hbk)
ISBN: 978-0-367-65499-3 (pbk)
ISBN: 978-1-003-12972-1 (ebk)

ISBN: 978-1-032-43935-8 (eBook+)

DOI: 10.1201/9781003129721

Typeset in Nimbus font
by KnowledgeWorks Global Ltd.

Publisher's note: This book has been prepared from camera-ready copy provided by the authors.

Dedication

To Suvie Liv Ten, Sow Geit Teh, and Kin-Kok Ten

Contents

PART II Profiling Electrical Loads

PART III Projecting Network Expansion

Preface

Distribution engineering has played a crucial role in the evolution of the smart grid. State-of-the-art research and developments have focused on distribution networks, with a national promotion of networked microgrid implementations. This framework introduces distributed energy resources (DERs) into both the primary and secondary parts of the system, leading to significant changes in the industry. Resilient strategies have emphasized reconfiguration and operational contingencies, enabling systematic power restoration from distribution dispatching centers. Advanced metering infrastructure (AMI) has also been widely promoted, with new applications derived from customer service, particularly in coordinating outage management and restoration efforts among operators and crews. Given the widespread implementation of IP-based communication infrastructure in AMI and supervisory control and data acquisition (SCADA) systems in control centers, cybersecurity has become a vital subject involving researchers from social and computer sciences. It is essential for new textbooks to cover niche areas, providing instructors with a wider range of choices to enhance the learning experience for undergraduate and graduate students.

This book originally authored by Turan Gönen is a dedicated work that provides critical fundamentals for the next generation of power engineering students in both undergraduate and graduate programs. It has been adopted for years by many instructors for developing courseware modules and content on distribution engineering. The book's focus centers around distribution system planning, covering topics such as load prediction, methodologies for assessing voltage profiles, and power loss estimation on hypothetical feeders based on geographical location. This framework can be particularly useful for incremental feeder or lateral expansions, leveraging existing planning tools from geographic information systems (GIS).

During our extensive search for books on distribution systems, we discovered that each available textbook had a distinct focus. Among them, Dr. Gönen's textbook emerged as a classical reference, renowned for its comprehensive coverage of system planning. The book delves into the intricate details, including sensitivity analysis of capacitance and voltage regulators, which directly impact power losses and voltage profiles that can be optimized for efficiency. With the goal of preserving Turan Gönen's legacy, this editorial effort ensures that the content remains current and applicable. Furthermore, the book provides updates on the latest advancements and state-of-the-art practices in specific chapters, offering readers valuable insights into emerging subjects. In order to maintain a sharp focus on contemporary system planning, older materials that have become less relevant over time have been transitioned to an electronic copy, which will be accessible to readers through the CRC Press portal page. This decision ensures that the book remains a reliable and up-to-date resource in the dynamic field of distribution systems.

Great engineers have a profound impact on humanity. We hope that this book will endure for generations as a classical reference for distribution planning, enabling readers to comprehend the progressive evolution of distribution networks and the transformative effects that technological advancements will have on these networks, leading to networked and resilient systems in the years to come.

Editors,
Chee-Wooi Ten and Ali Mehrizi-Sani

Authors

Turan Gönen, a distinguished figure in the field of electrical engineering, held the esteemed position of professor of electrical engineering and director of the Electrical Power Educational Institute at California State University, Sacramento, USA. Gönen's educational background included a BS and MS in electrical engineering from Istanbul Technical College, Turkey, followed by a PhD in electrical engineering from Iowa State University, Iowa. His academic pursuits extended further as he achieved an MS in industrial engineering and a PhD with a co-major in industrial engineering from Iowa State University alongside an MBA from the University of Oklahoma, USA. Dr. Gönen first authored this book in 1986. His contributions to academia and research were substantial, marked by a portfolio of over 100 technical papers. He held the distinction of being an IEEE Senior Member. Dr. Gönen passed away in 2014.

Chee-Wooi Ten, a senior member of IEEE, is a professor of electrical engineering at Michigan Technological University, Michigan, USA. He focuses on investigating rare events using system risk models and data science, holding keen interest in untangling complex interactions between power grids and transportation systems to advance decarbonization. His efforts focus on crafting cyber-aware security strategies for bulk power systems, validated through steady-state and dynamic analyses, emphasizing the need to anticipate disruptions in the context of distributed generation. Dr. Ten earned degrees from Iowa State University and a PhD from University College Dublin. He joined Michigan Technological University in 2010 after his experience as a power application engineer at Siemens. His research covers critical cyberinfrastructure modeling, and SCADA automation applications in power grids, and he's edited for IEEE and Elsevier journals, authoring the textbook *Electric Power: Distribution Emergency Operation.*

Ali Mehrizi-Sani, a senior member of IEEE, received a PhD in electrical engineering from the University of Toronto, Canada. Currently, he is an associate professor at Virginia Tech, Blacksburg. He is also an associate editor for the *IEEE Transactions on Energy Conversion*, *IEEE Power Engineering Letters*, and *IET Generation, Transmission and Distribution.* Dr. Mehrizi-Sani was a recipient of the 2018 IEEE PES Outstanding Young Engineer Award and the 2017 IEEE Mac E. Van Valkenburg Early Career Teaching Award among other recognitions.

Notation

A	component availability
A	levelized annual cost, $ (Chapter 7)
A	weighted average Btu/kWh net generation (Chapter 2)
A, B, C	phase designation
A, B, C, D	general line (circuit) constants (Chapter 5)
A_{FDR}	feeder availability (Chapter 11)
A_i	availability of component i (Chapter 11)
A_n	area served by one of n substation feeders, mi^2
A_{SD}	service-drop conductor size, cmil (Chapter 6)
A_{SL}	secondary-line conductor size, cmil (Chapter 6)
A_{sys}	system availability (Chapter 11)
AEC	annual equivalent of energy cost, $
$\mathrm{AEIC_c}$	annual equivalent of total installed capacitor bank cost, $ (Chapter 8)
AIC	annual equivalent of feeder investment cost, $
B	average fuel cost, $/MBtu (Chapter 2)
BEC	original (base) annual kWh energy consumption
BVR	bus voltage regulator
BW	bandwidth of voltage-regulating relay
C	capacitance, F
C	common winding (Chapter 3)
C_F	installed feeder cost, $/kVA (Chapter 8)
C_G	generation system cost, $/kVA
C_S	distribution substation cost, $/kVA
C_T	transmission system cost, $/kVA
C_T	transmission cost, $/kVA (Equation 8.26)
C_T	total reactive compensation (= cn) (Equation 8.85)
CR	corrective ratio (Chapter 9)
CT_P	primary-side rating of current transformer (Chapter 9)
CTR	current transformer ratio
D	distance or separation, ft
D	load density, kVA/mi^2
D	ratio of kWh losses to net system input (Chapter 2)
D_g	coincident maximum group demand, W
D_i	demand of load i, W
DF_i	demand factor of load group i
DTA(ij, k)	downtime array coefficients (Chapter 11)
E	source emf; voltage
E_i	event that component i operates successfully (Chapter 11)
EC	energy cost, $/kWh
$\mathrm{EC_{off}}$	incremental cost of off-peak electric energy, $/kWh
$\mathrm{EC_{on}}$	incremental cost of on-peak electric energy, $/kWh (Chapter 6)
$\mathrm{ECL_1}$	eddy-current loss at rated fundamental current
E(T)	expected time during which a component will survive
F	fault point
$\mathrm{F'}_{LD}$	reactive load factor (= Q/S)
F_c	coincident factor
F_D	diversity factor
$\mathrm{F_{LD}}$	load factor
$\mathrm{F_{LL}}$	load-location factor (Chapter 7)
$\mathrm{F_{LS}}$	loss factor

F_{LSA}	loss-allowance factor (Chapter 7)
F_{PR}	peak-responsibility factor
F_R	reserve factor
F_u	utilization factor
FCAF	fuel cost adjustment factor, \$/kWh
FDR feeder	(Chapter 11)
F(t)	unreliability function (Chapter 11)
H	transformer higher-voltage-side winding
HF_I	current harmonic factor
HF_V	voltage harmonic factor
I	failed zone-branch array coefficient (Chapter 11)
I	rms phasor current, A
I_{AB}	current in higher-voltage-side winding between phases A and B, A
I_{ab}	current in lower-voltage-side winding between phases a and b, A
$I_{a,3\phi}$	current in phase a due to single-phase load, A
I_B	base current, A
I_C	current in common winding (Chapter 3)
I_c	core-loss component of excitation current (Chapter 3)
I_e	excitation current (Chapter 3)

Part I

Introduction to Distribution System Planning

1 Distribution System Planning and Automation

To fail to plan is to plan to fail.

A.E. Gasgoigne, 1985

Those who know how can always get a job, but those who know why, may be your boss!

Author Unknown

To make an end is to make a beginning. The end is where we start from.

T.S. Eliot

1.1 INTRODUCTION

The electric utility industry traces its origins back to 1882 with the establishment of the Pearl Street Electric Station in New York City, marking the beginning of electric power generation. Since then, this industry has experienced rapid growth, expanding its generation facilities and transmission and distribution networks nationwide. As we look to the future, the increasing demand for energy and the projected shift toward electricity as the preferred form of energy conversion highlights the significance of the electric power system.

Traditionally, the electric power system comprises three key components: generation, transmission, and distribution. In the past, the distribution system accounted for a substantial portion of the capital investment, nearly equal to that of generation facilities, together representing over 80% of the total system investment. However, recent years have witnessed some changes in this distribution. Figure 1.1 illustrates the investment trends in electric utility plants, with data encompassing privately owned class A and class B utilities, which encompass 80% of all electric utilities in the United States. Figure 1.2 demonstrates the percentage of electric plants in the generation, transmission, distribution, and general plant sectors. Notably, there has been a significant investment in the production sector, closely followed by the distribution sector, underscoring the attention and resources dedicated to individual generation facilities and the growing investment in the distribution sector.

The expenses associated with generation play a pivotal role in the overall operation and maintenance (O&M) costs, accounting for approximately two-thirds of the total expenses. This increase can be attributed to the soaring fuel costs that have significantly impacted the industry. Figure 1.3 outlines the trends in the maintenance expenses-to-plant value ratio for each utility sector: generation, transmission, and distribution. Once again, the production sector incurs the largest O&M expenses, followed by the distribution sector. Considering the economic aspects of the distribution system becomes crucial, emphasizing the importance of meticulous planning, design, construction, and operation in future investments.

Overall, the electric utility industry has witnessed remarkable growth and development since its inception, with generation, transmission, and distribution systems forming its backbone. As energy needs evolve and the transition to electricity intensifies, the industry will continue to navigate challenges such as rising costs and the demand for reliable and efficient distribution. By employing careful and strategic approaches in planning, design, construction, and operation, the distribution

DOI: 10.1201/9781003129721-1

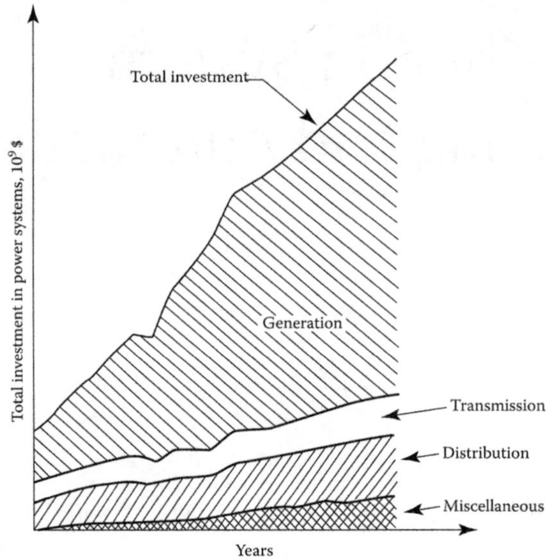

Figure 1.1 Typical investment trends in electric utility plants in service.

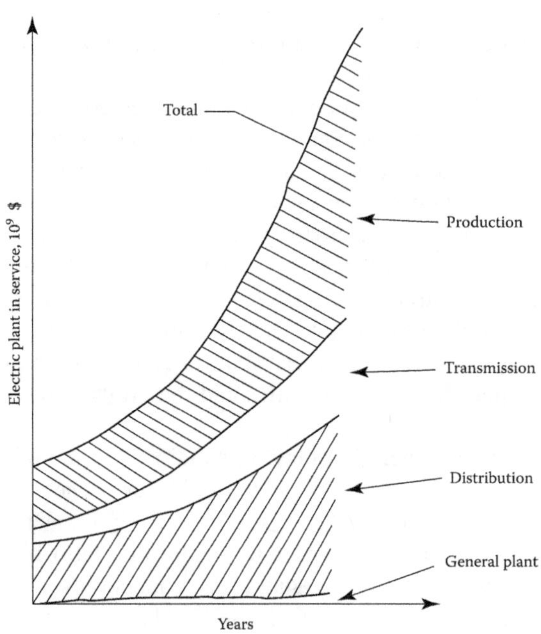

Figure 1.2 Typical trends in electric utility plants in service by percent of sector.

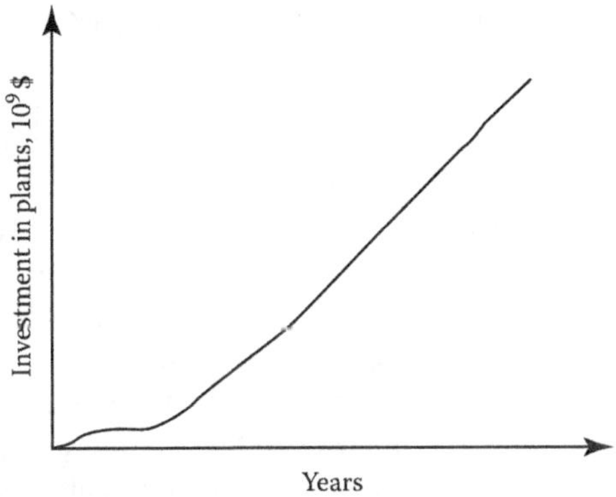

Figure 1.3 Typical ratio of maintenance expenses to plant in service for each utility sector. The data are for privately owned class A and class B electric utilities.

sector can effectively meet these demands and ensure the continued success of the electric power system.

1.2 STATE-OF-THE-ART UPDATE

Over the past decades, distribution networks have primarily operated in a radial configuration, with power flowing from substations to electrical loads. Planning has played a crucial role in the growth and optimization of these networks, with conductor size selection taking into account multi-sections with non-uniform loads to enhance approximation through a combination of economical current density and heuristic approaches [1]. Efforts have been made to address the dimensionality challenges of large mesh networks by treating radial topologies as spanning trees, allowing for incremental expansion through investment and the installation of new branches [2]. However, predicting uncertain dynamics and considering the complexity of other factors in what is known as the stochastic-dynamic problem remains a challenge. To address this complexity, initial conditions should rely on the geographical locations of existing infrastructure, and geographic information systems (GIS) coupled with dynamic programming techniques have been utilized [3].

When expanding the network and considering reconfiguration based on faulted events and potential generation inclusion, studies have explored combinations of open and closed switching simulations to ensure all constraints are satisfied [4]. In investment planning, the consideration of distributed energy resources (DERs) has gained prominence, particularly in the competitive electricity market, where optimal sizing and siting decisions for DER capacities are essential [5]. GIS spatial routing optimization has been extended to the placement of new overhead distribution lines, incorporating factors such as operation, maintenance, and capital investment based on criteria like conductor size and length of new additions [6].

The development of a fuzzy optimal planning model, which considers multiple objectives such as cost minimization and grid reliability maximization, has provided a foundation for comprehensive planning [7]. Extensive surveys have explored the use of genetic algorithms and other heuristic approaches in this context. The promotion of DERs has further driven studies on their locational values in planning, considering aspects such as loss reduction, payback period, and the impact of electrical loads on investment decisions, particularly in relation to the major intake of power supply from substations [8]. Multistage optimization planning has been proposed to accommodate alternative options, including the installation of new line segments, DERs, and incremental changes, with the objective of minimizing the present value of total costs associated with feeders and substations while adhering to generalized constraints [9].

The use of GIS has been instrumental in extending the placement of new transformers geographically, facilitating operational optimization within existing networks [10]. Non-linearity formulations and heuristic algorithms have been introduced to tackle planning challenges, with sensitivity indices established for feeders and substations as related problems [11]. Planning tools have gradually evolved, incorporating high-, medium-, and low-voltage distribution networks concurrently within geographical street maps, enabling automated analysis [12]. Networked microgrids have been introduced within existing distribution networks, where small grids incrementally contribute net generation to distribution feeders. The operational constraints of each microgrid in planning have been analyzed in terms of investment, potential interruptions, loss costs, and operational impacts [13]. The emergence of transportation electrification has introduced unprecedented loading conditions, necessitating a reevaluation of optimization problems, including the location-based placement of charging stations and their potential impacts on the distribution grid [14].

Demand response has emerged as a catalyst for integrating customer choices into distribution network planning. Planning strategies can be statistically driven, utilizing multi-objective functions for system expansion that can alter load patterns both centrally and distributively [15, 16]. Over time, planning approaches have been refined to incorporate contingency and robustness improvement, addressing expansion challenges such as networked microgrids that may involve a significant proportion of thermostatically controlled loads [17–20].

1.3 DISTRIBUTION SYSTEM PLANNING

System planning plays a crucial role in ensuring that the distribution system effectively meets the increasing demand for electricity in a technically reliable and cost-effective manner. Recent efforts have focused on adopting systematic approaches, particularly in the context of networked microgrids within distribution systems. This trend will continue as robust planning tools are continuously developed to evaluate various alternative options and their impacts on the overall system. These evaluations guide the allocation of resources in the annual budget for system maintenance and improvement [21].

The planning process takes into account the rising growth rates and high load densities, which can be efficiently addressed through the addition of distribution systems. All components, ranging from secondary conductors to bulk power substations, are considered in the planning process. However, optimal distribution system planning can present challenges due to factors such as the limited availability of land in urban areas and ecological considerations.

System planners must accurately determine the magnitude and geographic distribution of loads, which are directly correlated with population density and growth patterns. This information enables the estimation of capacity requirements for distribution substations, facilitating their optimal placement and sizing to serve the load at minimal cost while minimizing feeder losses and construction expenses. Service reliability is also a crucial consideration. In the past, planning for different segments of the power supply system, including the distribution system, has often been authorized at the division level without adequate review or coordination with long-term plans. Given the escalating costs of energy, equipment, and labor, effective system planning has become increasingly crucial, necessitating the use of efficient planning methods and techniques.

The distribution system primarily caters to the electrical loads of customers, whether retail or contractual. The proximity to end customers and the substantial investment involved are two key factors that need to be considered. Unlike failures in the transmission and generation systems, which typically do not result in customer service interruptions, failures in the distribution system directly impact customer service. Therefore, distribution system planning begins at the customer level. The demand, load characteristics, load factors, and other customer-related factors dictate the type of distribution system required. Once the customer loads are determined, they are organized for service through secondary lines connected to distribution transformers that step down the primary voltage.

The combination of loads on distribution transformers determines the demand on the primary distribution system, which is then allocated to substations that step down from transmission voltage. The size, location, routing, and capacity of substations, as well as the associated transmission lines, are determined by the distribution system loads. To address the complexity of the distribution system planning problem, planners divide it into sub-problems that can be tackled using available methods and techniques, often ad hoc in nature.

When established planning techniques are lacking, planners may reframe the problem to minimize costs related to subtransmission, substations, feeders, laterals, and losses. However, voltage values, voltage dips, flicker, service continuity, and reliability impose constraints on the planner. Pursuing these objectives significantly influences the addition or modification of the subtransmission network, substations' sizes and locations, service areas, placements of breakers and switches, sizes of feeders and laterals, voltage levels and drops, placements of capacitors and voltage regulators, and the loading of transformers and feeders. Other factors, including transformer impedance, insulation levels, availability of spare transformers and mobile substations, dispatch of generation, and customer rates, must also be considered.

Additionally, there are factors beyond the control of distribution system planners that must be taken into account in long-range planning. These factors encompass the timing and location of energy demands, frequency and duration of outages, costs of equipment, labor, and finances, fluctuations in fuel costs, changing prices of alternative energy sources, socioeconomic conditions and trends, unexpected local population changes, evolving public behavior due to technological ad vancements, energy conservation efforts, environmental concerns, economic conditions such as fluctuations in gross national product (GNP), inflation, recession, and government regulations at the federal, state, and local levels.

In summary, the distribution system planner combines loads on distribution transformers to determine the demand on the primary distribution system, which is then assigned to substations. The planner divides the planning problem into manageable sub-problems, aiming to minimize costs while considering voltage and reliability constraints. The planner's decisions have a significant impact on the subtransmission network, substations, breakers, switches, feeders, voltage levels, capacitors, regulators, transformers, and feeders. Furthermore, external factors such as energy demands, outages, costs, socioeconomic conditions, population changes, public behavior, energy conservation, environmental concerns, economic conditions, and regulations must be taken into consideration [21].

1.4 FACTORS INFLUENCING SYSTEM PLANNING

Initially, the number and complexity of factors influencing system planning may appear daunting. The need for increasing power capacity, higher distribution voltages, automation, and advanced control systems are just a few examples of these factors. Designers also face challenging constraints, including limited availability of urban land, ecological considerations, fuel limitations, resistance to rate increases, and the imperative to minimize investments, carrying charges, and production costs.

In essence, the planning problem aims to minimize the costs associated with subtransmission, substations, feeders, laterals, and losses. This task presents a formidable challenge, as the requirements and constraints involved surpass the capacity of unaided human decision-making.

1.4.1 LOAD FORECASTING

The expansion of the distribution system is primarily motivated by the increasing load demands within the geographic area served by a utility company. Therefore, accurate load forecasting and a comprehensive understanding of the system's response to these load increases are vital for effective planning. Load forecasting involves predicting future electricity demands and is typically conducted on two different time scales: long range, which encompasses a period of 15 to 20 years, and short

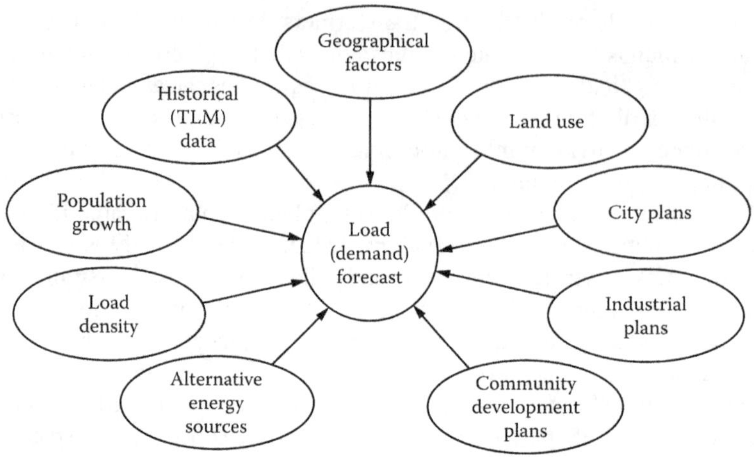

Figure 1.4 Factors affecting load forecast.

range, covering a time horizon of up to 5 years. While it would be ideal to have forecasts that provide detailed predictions at the individual customer level, in practice, forecasts with a coarser level of resolution are often sufficient.

Figure 1.4 shows a visual representation of some of the key factors that influence the load forecast. The growth of electrical loads is closely linked to the development of the communities they serve. Economic indicators, demographic data, and official land use plans are significant inputs that inform the load forecasting process. Long-range forecasts typically yield load densities, measured in kilovolt-amperes (kVA) per unit area, which provide an indication of the expected load distribution across the service territory. On the other hand, short-range forecasts may require more detailed information to capture localized variations in demand.

Load densities, which represent the distribution of electrical load across a specific geographic area, are closely tied to a coordinate grid that encompasses the area of interest. These grid data serve as valuable information for configuration design and aid in the planning process. The master grid, which incorporates load forecasting data, becomes an essential tool for evaluating different geographic locations and determining the necessary actions to accommodate system expansion patterns.

1.4.2 SUBSTATION EXPANSION

Substation expansion is a critical aspect of distribution system planning, and various factors influence this process, as depicted in Figure 1.5. The planner relies on both tangible and intangible information to make informed decisions. For instance, based on the forecasted load, load density, and projected load growth, the planner may determine that a substation expansion or the construction of a new substation is necessary. In the system expansion plan, the current configuration and capacity of the system, in addition to the forecasted loads, play significant roles in determining the appropriate course of action. These considerations ensure that the distribution system can effectively accommodate the expected growth in electricity demand while maintaining reliable and efficient operation.

1.4.3 SUBSTATION SITE SELECTION

Figure 1.6 provides an overview of the factors that influence the selection of suitable sites for substations. One crucial aspect is the proximity to load centers and existing subtransmission lines. The distance from load centers ensures efficient power delivery to customers while the distance from

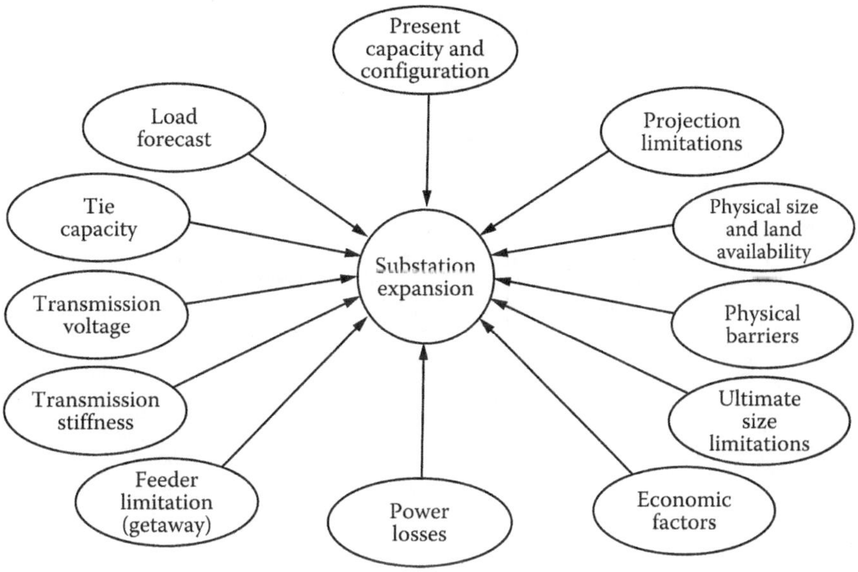

Figure 1.5 Factors affecting load forecast.

existing subtransmission lines affects the feasibility of connecting the substation to the larger power grid. Additionally, various constraints such as land availability, cost, and land use regulations must be considered in the site selection process.

The substation siting process is often described as a screening procedure, as depicted in Figure 1.7, where all potential locations undergo careful evaluation. The service region, typically corresponding to the utility's service territory, serves as the area under evaluation. It represents the geographical area that the utility serves with its distribution system. During the initial screening phase, a set of considerations is applied to assess the suitability of each site. These considerations encompass a range of factors, including safety requirements, engineering feasibility, alignment with system planning objectives, institutional guidelines, economic viability, and aesthetic aspects. The purpose of this stage is to identify areas that are unsuitable for further site development due to factors that may hinder the functionality, efficiency, or compliance of the substation.

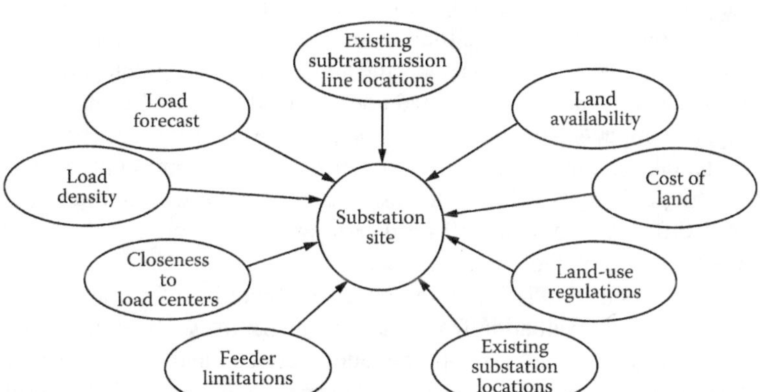

Figure 1.6 Factors affecting substation siting.

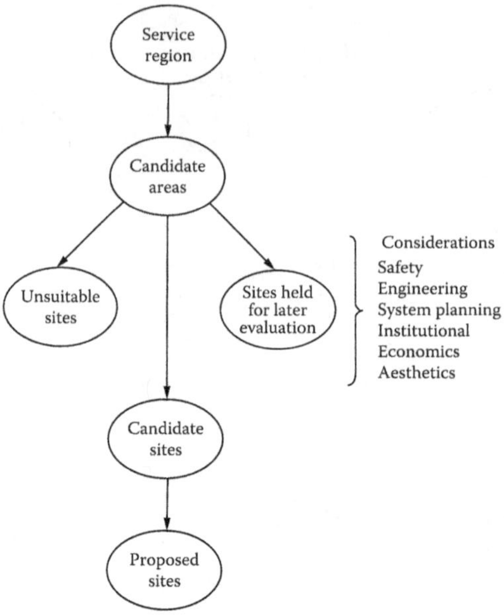

Figure 1.7 Substation site selection procedure.

By conducting this screening procedure, utilities can narrow down the range of potential sites and focus on locations that align with the defined criteria and best meet the operational and regulatory requirements. It is important to note that this initial screening primarily serves as an indicator of areas that are unsuitable for substation development, providing valuable guidance for subsequent stages of the site selection process.

During the substation site selection process, the identified candidate sites are categorized into three groups based on their suitability. The first group consists of sites that are not suitable for immediate development due to inherent limitations or significant challenges. The second group comprises sites that show potential but are not selected for detailed evaluation in the current planning cycle, possibly requiring further assessment in future cycles or specific circumstances. The third group includes candidate sites that are deemed strong contenders for substation development and undergo thorough analysis to assess their feasibility.

The evaluation process may vary among utility companies, with three fundamental approaches utilized. The first approach involves quantitative and qualitative evaluation, where numerical values and quantitative analyses are used to assess the suitability of candidate sites based on objective criteria and performance indicators. The second approach, adverse vs. beneficial effects evaluation, considers both negative impacts (adverse effects) and positive contributions (beneficial effects) associated with each site, taking into account environmental impacts, community concerns, operational efficiency, and system performance. Last, the absolute vs. relative scaling of effects approach compares the effects of different candidate sites either in absolute terms or in relation to each other, helping prioritize sites based on their significance.

To conduct a comprehensive site assessment, a combination of these approaches is recommended, ensuring a thorough evaluation from various perspectives. By employing quantitative and qualitative analyses, considering adverse and beneficial effects, and utilizing absolute and relative scaling, a comprehensive understanding of candidate sites can be achieved. This multifaceted evaluation approach facilitates informed decision-making in selecting the most suitable site for substation construction.

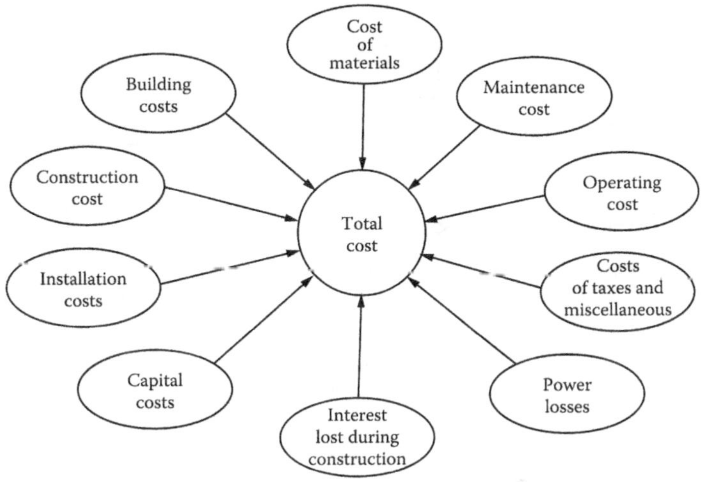

Figure 1.8 Factors affecting the total cost of the distribution system expansion.

1.4.4 OTHER FACTORS

After determining the load assignments to substations, several factors come into play when making decisions about primary voltage selection, feeder route selection, the number of feeders, conductor size selection, and the overall cost. Figure 1.8 illustrates these factors. In general, the voltage levels for subtransmission and distribution systems are governed by company policies and are unlikely to be changed unless the planning engineer can provide compelling evidence of significant benefits through running test cases.

The designer's freedom in choosing the sizes and types of capacity equipment may be limited due to standardization and cost considerations. For example, the designer may need to select a distribution transformer from a predetermined list stocked by the company, which corresponds to the established voltage levels. The decision to add a feeder or expand an existing one depends on the adequacy of the current system within certain limits. Factors such as the size, location, and timing of additional loads that require service play a crucial role in these considerations.

While company policies and existing infrastructure impose certain constraints, the planning engineer can still optimize the design within these limitations. By carefully evaluating the available options and considering factors such as load capacity, equipment availability, and system reliability, the engineer can make informed decisions to enhance the efficiency and effectiveness of the distribution system. It is essential to strike a balance between meeting the growing demand for electricity and minimizing costs while ensuring the reliable and continuous delivery of power to customers.

1.5 PRESENT DISTRIBUTION SYSTEM TECHNIQUES

Electric distribution system planners in the industry rely on computer programs and various tools to aid in their planning activities. These programs include radial or loop power flow analysis, short-circuit and fault-current calculation, voltage drop calculation, and total system impedance calculation. Additionally, tools such as load forecasting, voltage regulation, capacitor planning, reliability analysis, and optimal siting and sizing are employed. However, it is important to note that the practice of using the output of one program as input for the next program is not widely adopted. Instead, computer programs expedite the calculations, allowing distribution engineers to focus on reviewing and analyzing the results.

The use of these tools and their impact on the system design largely depends on the discretion of the planner and the operating policies of the utility company. Figure 1.9 illustrates a functional block

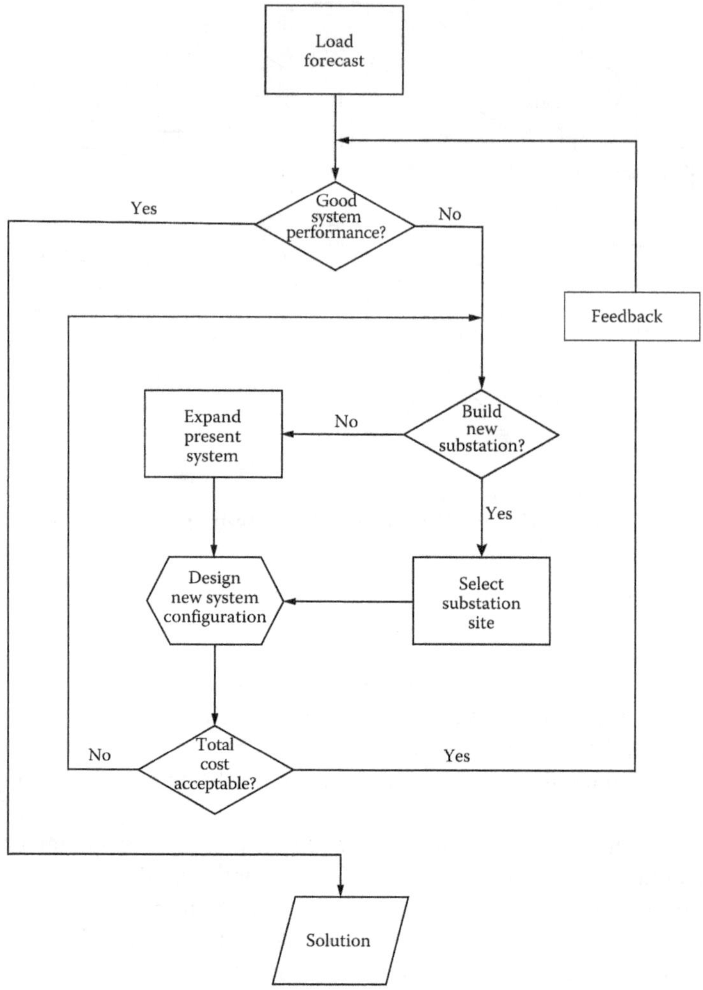

Figure 1.9 A block diagram of a typical distribution system planning process.

diagram that outlines the typical distribution system planning process followed by most utilities. It should be noted that this diagram represents a general planning process and does not specifically reflect the procedures of any particular company. The planning procedure consists of four main activities: load forecasting, distribution system configuration design, substation expansion, and substation site selection.

The configuration design process begins at the customer level, where the specific demand type and load characteristics determine the appropriate distribution system. Secondary lines are then defined to connect the customers to distribution transformers, which step down the voltage from the primary level to the customer-level voltage. The loads of the distribution transformers are aggregated to determine the demands on the primary distribution system. These loads are then assigned to substations, which further step down the voltage from the subtransmission level. The distribution system loads play a crucial role in determining the optimal size, location, and routing of substations, as well as the capacity of associated transmission lines. Each step in the planning process provides important input for the subsequent steps, ensuring a cohesive and coordinated approach to system design.

By utilizing computer programs and tools in the planning process and following a systematic approach, distribution system planners can optimize the design and operation of the distribution system. This leads to improved efficiency, reliability, and cost-effectiveness in meeting the growing demand for electricity while ensuring a high level of service to customers.

The practical implementation of a straightforward procedure in distribution system planning may present challenges. Instead, a more commonly used approach involves the following steps: After receiving load projection data, a system performance analysis is conducted to assess the capability of the existing system to accommodate the projected load increase based on the company's criteria. This analysis constitutes the second stage of the process and utilizes tools such as a distribution power flow program, a voltage profile analysis, and a regulation program. The acceptability criteria used in this analysis reflect the company's policies, obligations to consumers, and other constraints. These criteria can include considerations such as:

1. Service continuity
2. The maximum allowable peak-load voltage drop to the most remote customer on the secondary
3. The maximum allowable voltage dip occasioned by the starting of a motor of specified starting current characteristics at the most remote point on the secondary
4. The maximum allowable peak load
5. Service reliability
6. Power losses

As shown in Figure 1.9, if the current distribution system is insufficient to meet future demand, two options are considered: expanding the existing system through minor additions or constructing a new substation. In the case of expanding the present system, a new network configuration is designed and evaluated for its adequacy. If the configuration proves to be inadequate, alternative configurations are explored until a satisfactory one is identified. The cost of each configuration is calculated, and if the cost is deemed excessive or the performance is unsatisfactory, the original decision to expand or build is reevaluated.

If the decision is to construct a new substation, the next step is to select an appropriate site for the substation placement. However, if the purchase price of the selected site is deemed too high, the expand-or-build decision may require further reconsideration. This iterative process continues until a satisfactory configuration is achieved that effectively addresses existing or future challenges at a reasonable cost. It is worth noting that many of the steps involved in these procedures heavily rely on the utilization of computer programs due to their complexity and the need for accurate analysis and calculations. The use of computer programs significantly aids in achieving optimal solutions and streamlining the planning process.

1.6 DISTRIBUTION SYSTEM PLANNING MODELS

Distribution system planning is a challenging process characterized by numerous variables and the intricate representation of various system requirements and constraints. To tackle these complexities, mathematical models are developed to capture the essence of the system. Distribution system planners can then utilize these models to explore and identify optimal expansion patterns or alternative solutions. These models take into account a range of factors, including but not limited to:

1. Substation locations
2. Substation expansions
3. Substation transformer sizes
4. Load transfers between substations and demand centers
5. Feeder routes and sizes to supply the given loads subject to numerous constraints to minimize the present worth of the total costs involved

In the intricate process of distribution system planning, the optimization of key aspects such as substation locations, expansions, transformer sizes, load transfers, and feeder routes is vital to minimize costs and maximize system performance. Distribution system planners employ a range of powerful operations research techniques to tackle this complex task. Integer programming aids in selecting and placing substations efficiently, considering constraints and load requirements. Network flow analysis ensures optimal load transfers, minimizing energy losses and ensuring effective power distribution. Heuristic algorithms provide practical and near-optimal solutions for determining feasible feeder routes and sizes within reasonable computational time. Sensitivity analysis allows for the evaluation of critical factors and their influence on system design, considering variations in parameters and constraints. Additionally, Monte Carlo simulation enables planners to assess the robustness and reliability of proposed plans by simulating uncertain factors. By utilizing these operations research techniques, distribution system planners can make informed decisions that enhance the efficiency and effectiveness of distribution system planning, optimizing overall system performance while addressing operational and economic constraints. Some of the operations research techniques used in performing this task include

1. The alternative-policy method, by which a few alternative policies are compared and the best one is selected
2. The decomposition method, in which a large problem is subdivided into several small problems and solved separately
3. Programming methods that linearize constraint conditions
4. The quadratic programming method
5. The dynamic-programming method
6. Genetic algorithms method

Each of the aforementioned techniques in distribution system planning brings its own set of advantages and disadvantages. The complexity of long-range planning introduces numerous variables, making the selection of the optimum alternative challenging. This is particularly evident in the distribution system, where a single substation's feeder system can present distribution engineers with an overwhelming number of feasible design options. For instance, a 40-node distribution system alone can have over 15 million possible plans, with approximately 20,000 feasible variations [22].

Given that distribution system costs can account for a substantial portion of an electric utility company's investment budget (up to 60%) and operating costs (around 20%), minimizing these costs becomes a significant challenge [23]. To address this challenge, computer-aided tools have been developed to assist in identifying the most cost-effective distribution configuration. Over the last three decades, extensive research and development have been dedicated to the creation of optimization programs that can analyze numerous possibilities and aid in finding the best design option. While these programs may not consider all aspects of the problem and may incorporate approximations that slightly limit accuracy, they have proven to be valuable in deducing distribution costs. In fact, even with conservative estimates, these tools have demonstrated the potential to reduce distribution costs by 5%–10% [23].

Practical expansion studies of distribution systems have traditionally relied on the expertise and judgment of planning engineers, who base their decisions on factors such as the existing system, power demand forecasts, economic and electrical calculations, and past experience. However, to enhance the effectiveness of these studies and achieve greater cost savings, there has been a shift toward more sophisticated approaches involving mathematical models and computational optimization techniques. By employing these advanced methods, planners can improve upon traditional solutions and generate valuable savings, particularly considering the significant costs associated with system expansion.

In planning the expansion of a given distribution system, the current level of electric power demand is known, and future demand levels can be forecasted over one or several stages, depending on data availability and company policy. The objective is to devise an expansion plan that meets the projected demand while minimizing the associated costs. To address this complex problem, early applications of distribution system planning have divided it into two subproblems, which have been successfully solved:

1. Feeder system planning: This subproblem involves determining the optimal configuration and routing of feeder systems to efficiently deliver power to various load points. Factors such as load distribution, load growth projections, available capacity, and network topology are considered to devise a cost-effective and reliable feeder system design.
2. Substation siting and sizing: This subproblem focuses on selecting suitable locations for substations and determining their optimal sizes to cater to the projected demand. Factors such as load density, geographical considerations, voltage requirements, and cost implications are taken into account to make informed decisions regarding substation siting and sizing.

By addressing these two subproblems in the distribution system planning process, planners can effectively manage the expansion of the system and ensure its ability to meet future power demand while optimizing costs. This approach allows for a systematic consideration of feeder system planning and substation siting and sizing, enabling planners to devise efficient and cost-effective solutions. The integration of mathematical models and computational optimization techniques further enhances the accuracy and efficiency of the planning process, leading to improved outcomes and substantial cost savings.

Nevertheless, more advanced mathematical models have been developed to address the distribution planning problem as a global problem, considering the simultaneous minimization of feeder and substation costs. These complex models have the potential to provide optimal solutions for a single planning stage. However, the complexity of these mathematical problems and the resolution process pose significant challenges. Decisions made in one planning stage regarding the construction of substations and feeders have implications for subsequent stages, and extensive research has been conducted to explore the interdependencies and impacts of such decisions.

By exploring and refining these more sophisticated mathematical models, distribution system planners can gain deeper insights into the global optimization of the system, taking into account the interplay between substations and feeders. This holistic approach can lead to more comprehensive and effective planning decisions, maximizing the system's performance while minimizing costs. The continuous advancement in research and the utilization of cutting-edge optimization techniques contribute to the ongoing improvement of the distribution system planning process, enabling planners to achieve optimal outcomes and meet the challenges of an evolving energy landscape.

1.6.1 ENHANCING DISTRIBUTION SYSTEM PLANNING THROUGH OPTIMIZATION ALGORITHMS

In the realm of distribution system planning, numerous algorithms rooted in optimization programs have been developed, building upon the foundational operations research techniques. One prominent example is a distribution design optimization program currently utilized by more than 25 utilities across the United States. This program operates within an integrated Unix or Windows platform, boasting a user-friendly interface and visual representation of one-line diagrams. It leverages a single open SQL circuit database, facilitating circuit analysis, and supports various equipment selection optimization routes, such as capacitor-regulator sizing and locating. Additionally, it incorporates a constrained linear optimization algorithm to determine multiple feeder configurations.

The program's key features encompass a purpose-built database, editor, display, and user structure explicitly tailored to support optimization applications in augmentation planning and switching studies. To enhance its capabilities, the program employs a linear trans-shipment algorithm in conjunction with a post-optimization radialization technique. While nonlinear methods may excel in certain scenarios, a linear algorithm methodology was chosen for this program due to its suitability for augmentation planning and switching studies. From an application perspective, distribution system planning can be broadly classified into three categories: (1) new system expansion, (2) augmentation of existing systems, and (3) operational planning.

The development and implementation of optimization-based algorithms in distribution system planning signify a significant advancement in the field. By harnessing the power of mathematical optimization, these algorithms offer utilities the ability to efficiently analyze and optimize their distribution systems. The integration of user-friendly interfaces, visual representations, and comprehensive databases empowers planners to make informed decisions regarding equipment selection, feeder configurations, and system expansion. As distribution system planning continues to evolve, these algorithms contribute to more effective and cost-efficient solutions, enhancing the performance and reliability of distribution networks.

1.6.2 NEW EXPANSION PLANNING

Among the three categories of distribution system planning, new system expansion stands out as the most straightforward to optimize. It has garnered significant attention in the technical literature, primarily due to its substantial capital and land requirements. This category can be best described as the planning process for the expanding outskirts of a thriving city. Referred to as greenfield planning by Willis et al. [23], it involves starting with a blank canvas, or "greenfield," and designing a completely new distribution system based on the anticipated development of a particular region. In this planning scenario, the possibilities for the new system design are virtually limitless, presenting both opportunities and challenges for planners.

Optimization algorithms employ efficient linearization techniques to reduce computational time and solve large-scale problems while introducing only a minimal approximation error. This linearization approach represents each segment in the potential system using two key values: a linear cost vs. kVA slope based on segment length and a capacity limit that restricts its maximum loading. This methodology has consistently delivered highly satisfactory outcomes since the 1970s. According to Willis et al. [23], more than 60 utilities in the country alone routinely employ this method for the design of major new distribution additions. Notably, IEEE and Electric Power Research Institute (EPRI) publications have reported significant economic savings of up to 19% compared to traditional manual design practices.

1.6.3 AUGMENTATION AND UPGRADES

Unlike greenfield planning, which involves designing a new distribution system from scratch, distribution planners often face the challenge of economically upgrading an existing system. Although this type of planning may seem easier, it is not necessarily the case, and two reasons contribute to this misconception. First, the availability of new routes, equipment locations, and permissible upgrades for existing equipment is severely limited due to practical, operational, aesthetic, environmental, or community-related factors. This necessitates the delicate balance of various unique constraints and local variations in available options. Second, when dealing with an already established system, linearizing the options for upgrading existing lines is generally not feasible. However, optimization programs have been traditionally employed for augmentation planning due to the lack of better alternatives. These applications have demonstrated the potential to reduce costs in augmentation planning by approximately 5% [23].

As discussed in Section 7.5, comprehensive studies of this nature should include both fixed and variable costs associated with each circuit element. For instance, when evaluating feeder sizes, the cost analysis should encompass two aspects: (1) the investment cost of each installed feeder and (2) the cost of energy losses resulting from I^2R losses in the feeder conductors. It is also possible to incorporate the cost of demand lost, which refers to the expenses incurred to compensate for the reduction in useful system capacity caused by I^2R losses in feeder conductors. Including these factors in the calculations provides a more accurate representation of the overall costs involved.

1.6.4 OPERATIONAL PLANNING

The planning of an existing distribution system is influenced by the actual switching pattern required for its operation. The primary objective is to meet voltage drop criteria, ensure optimal loading, and minimize losses. Unlike other planning approaches, the only available option in this case is switching. The optimization process revolves around minimizing I^2R losses while adhering to loading and operational restrictions. Numerous applications have successfully utilized piecewise linearization-based approximations, yielding favorable outcomes in various optimization scenarios.

However, it is important to note that determining switching patterns has minimal impact on the initial investment decisions related to enhancing feeder routes and/or substation locations. Once these investment decisions are made, the associated costs become fixed investments. Subsequent switching activities during the operational phase primarily influence the minimization of losses and do not significantly affect the initial investment decisions.

1.6.5 BENEFITS OF OPTIMIZATION APPLICATIONS

According to Ramirez-Rosado and Gonen [24], the optimal solution remains the same when considering only the costs of investment and energy losses, which is expected as it results in a lower total cost. They have also demonstrated that the problem can be successfully resolved by considering only investment costs. In one of their studies involving multistage planning, they found that the optimal network structure remained almost unchanged, except for the delay in building a particular feeder until the fourth year when the cost of energy losses was not included. This indicates a minor influence of excluding the cost of energy losses on the optimal network structure.

The primary motivation for applying optimization techniques is to achieve cost reduction. Willis et al. [23] state that a nonlinear optimization algorithm can lead to average savings in augmentation planning comparable to those achieved in greenfield studies. However, this is not the case when it comes to switching. Tests using nonlinear optimization have shown that potential savings in augmentation planning are only a fraction (around one-fourth to one-third) of the savings observed in greenfield studies. On the other hand, linear optimization provides approximately 85% of the savings achievable through nonlinear analysis. Another advantage of optimization efforts is the improved understanding of the system, particularly in terms of cost-performance tradeoffs and interdependencies. In a brief analysis that lasted less than a minute, Willis et al. [23] found that the optimization program identified key cost-saving issues, quantified their interactions with other aspects of the problem, and revealed additional possibilities for cost reduction.

1.7 DISTRIBUTION SYSTEM PLANNING IN THE FUTURE

Having reviewed past and current techniques employed by utility industry planning engineers, it is essential to consider critical factors with a systematic approach for the future. Subsequent sections will explore the potential implications of present-day trends on the future of the planning process.

1.7.1 ECONOMIC FACTORS

Distribution planning will be significantly influenced by various economic factors. One of the key factors is inflation, which will be driven by energy shortages, rising costs of energy source conversion, environmental concerns, and government deficits. Inflation will continue to play a major role in shaping distribution planning. Another important economic factor is the escalating cost of acquiring capital. If inflation continues to erode the value of the dollar, the government may take steps to decrease the money supply, leading to heightened competition for capital needed to expand distribution systems. The third factor to consider is the increasing challenge of raising customer rates. This difficulty in raising rates is partly a result of inflation and also influenced by consumer activist groups, who have made customers more sensitive to rate increases.

1.7.2 DEMOGRAPHIC FACTORS

Important demographic changes impact distribution system planning. One significant trend is the ongoing movement of the population from rural to metropolitan areas, which has been dominant for decades. Although this trend has experienced a reversal during the recent pandemic, economic development in cities typically leads to greater productivity and prosperity for a state.

The driving force behind this trend has been economic factors. The number of single-family farms has been consistently declining, and there are no signs of a reversal in the population shift toward larger urban areas. As people leave the countryside, smaller towns that rely on rural areas for economic activity also experience a decrease in population. This has been a long-standing consideration for distribution planners and does not introduce any new effects that need to be addressed.

However, a new trend attributed to the energy crisis is the migration from suburbs to urban and near-urban areas. This emerging trend is starting to become visible and will result in an increase in multifamily dwellings in areas that already have high population densities.

1.7.3 TECHNOLOGICAL FACTORS

This is a crucial factor as technological advancements are often driven by the urgency of addressing climate change issues. One significant development is the enhancement of fuel cell technology. These devices now have a higher power output, enabling large clusters of fuel cells to meet substantial portions of the total power requirements in densely populated areas. This empowers customers to choose from a wider range of smart home devices that promote long-term energy savings. Among the options available are solar and wind-driven generators. Consumer groups are advocating for utilities to accept any surplus energy from these sources for integration into the overall distribution network. This trend will persist, allowing individual customers to achieve greater energy savings and enhance household tasks through assisted robotics and smart home automation.

1.8 FUTURE ESTABLISHMENT IN DISTRIBUTION PLANNING

Forecasting future distribution planning methods requires the use of statistical techniques to extrapolate. Numerous existing frameworks and algorithms can be enhanced and are anticipated to make significant progress by leveraging new methodologies and computing platforms. The subsequent sections delve into the evolving strategies of distribution planning.

1.8.1 INCREASINGLY CHANGING GRID DYNAMICS

The growing complexity of distribution systems has made their construction, expansion, and modification more challenging. Achieving cost-effectiveness is crucial for efficient multistage planning within a specific time frame. Besides accurate load growth predictions, the addition and removal of components must be carefully coordinated to minimize capital expenditure, achieve performance objectives, and minimize losses. These demands arise at a time when demographic trends are deviating from longstanding norms, and distribution systems are becoming increasingly intricate in design, particularly with the integration of renewable sources.

1.8.2 IMPACTS OF DEMAND-SIDE MANAGEMENT

Distribution companies aim to ensure adequate electric energy supply for all customers. However, various factors such as financial constraints (e.g., high labor and material costs, interest rates), environmental considerations, and fuel shortages or high costs can challenge this principle. As a result, utilities need to reassess their long-term planning strategies and shift toward customer-centric approaches based on consumption rather than capacity expansion.

The implementation of demand-side management offers significant benefits. By optimally adjusting load patterns to align with the available generating capacity, load shaving helps balance and reduce loads on marginal substations and circuits, thereby extending their operational lifespan. Consequently, demand management policies in the future can greatly impact customer loads. To fully leverage the advantages of such policies, distribution systems may require design modifications that align with anticipated loading conditions. However, realizing these benefits relies on system planners having the necessary tools to effectively incorporate demand-side management into the evolving electric energy system. This process of system evolution, driven by changing requirements and constraints, involves inherent uncertainties.

Delgado [25] outlines the key elements for a successful demand-side management program as follows:

1. It must be able to reduce demand during critical system load periods.
2. It must result in a reduction in new generation requirements, purchased power, and/or fuel costs.
3. It must have an acceptable cost/benefit ratio.
4. Its operation must be compatible with system design and operation.
5. It must operate at an acceptable reliability level.
6. It must have an acceptable level of customer convenience.
7. It must provide a benefit to the customer in the form of reduced rates or other incentives.

These elements highlight the fundamental factors that contribute to the success of a demand-side management program. They encompass the capability to reduce demand during critical system load periods, resulting in a decrease in new generation requirements, purchased power, and fuel costs. It is crucial for the program to demonstrate a favorable cost/benefit ratio, align with system design and operation, and maintain an acceptable level of reliability. Prioritizing customer convenience while providing tangible benefits such as reduced rates or incentives is essential. By adhering to these key principles, distribution companies can effectively navigate the dynamic energy landscape, harness the full potential of demand-side management, optimize energy consumption, and enhance overall system efficiency.

1.8.3 COST/BENEFIT RATIO FOR INNOVATION

In the utility industry, the driving force that significantly shapes the future is economics. Consequently, new innovations are unlikely to be adopted solely for their novelty, but rather for their potential to reduce costs or provide economic value that was previously unattainable at comparable expenses. When predicting the replacement of current practices or tools, their acceptance must be evaluated based on these criteria.

One of the anticipated innovations that meet these criteria is the implementation of planning tools on digital computers, specifically designed to handle distribution systems in a network context. It might be tempting to assume that these planning tools would be sufficient for industry-wide use throughout the 1980s. However, this is unlikely to be the case when considering the dominant trends during this period in comparison to those that influenced the development of these tools.

1.8.4 NEW PLANNING TOOLS

There are two categories of network tools to be examined: design and analysis. While the analysis tools are anticipated to become more efficient, they are not expected to undergo substantial changes. However, the environment in which these tools are utilized is projected to undergo significant transformation, which will be explored in the subsequent section. On the other hand, the design tools are expected to undergo significant advancements, as enhanced planning has the potential to greatly influence the utility industry. The outcomes of this development are expected to exhibit the following characteristics:

1. Network design will be optimized with respect to criteria using programming methods of operations research.
2. Network design will be only one facet of distribution system management directed by human engineers using a computer system designed for such management functions.
3. The network editors [26] will be available for designing trial networks; these designs in digital form will be evaluated by extensive simulation programs, which will satisfies performance and load growth criteria.

While improvements in efficiency are expected for analysis tools, significant changes are not foreseen. However, the utilization environment for these tools is projected to undergo substantial transformation, which will be explored further. Conversely, design tools are anticipated to experience notable advancements, driven by the potential impact of enhanced planning on the utility industry. These advancements in design tools are expected to showcase the following characteristics: optimization of network design through operational research programming methods based on specific criteria, integration of network design into distribution system management under the guidance of human engineers utilizing dedicated computer systems, and the availability of network editors for creating trial networks, with designs being evaluated using extensive simulation programs that adhere to performance and load growth criteria.

1.9 EMBRACING EVOLVING TECHNOLOGIES AND REFINED MODELS

Throughout the evolution of planning development, distribution system planners have utilized software systems equipped with established models to conduct the necessary calculations for system analysis. As computing platforms have advanced, algorithms, design techniques, and analyses have improved, benefiting from more refined modeling approaches developed by other researchers. The main proposition of this book is that the continued development of such an approach will engage planners in the future and play a vital role in effectively addressing the challenges discussed earlier.

1.9.1 SYSTEM APPROACH

A mere collection of software systems does not necessarily constitute an efficient problem-solving system, even if the output of one can be used as the input of another. To design a useful tool for engineers, a system approach is adopted, which involves examining the types and sources of information required. The underlying perspective is that this information generates decisions and additional information, which flow from one stage of the design process to the next. At certain points, engineers must assess the generated information and provide their input. Ultimately, the results need to be displayed for use and stored for future reference.

With this understanding of the planning process, the system approach aims to automate as much of the process as possible, ensuring that information transformations occur efficiently. Figure 1.10 illustrates one representation of this information flow, where the outer circle represents the interface between the engineer and the system. The system incorporates analysis programs supported by a

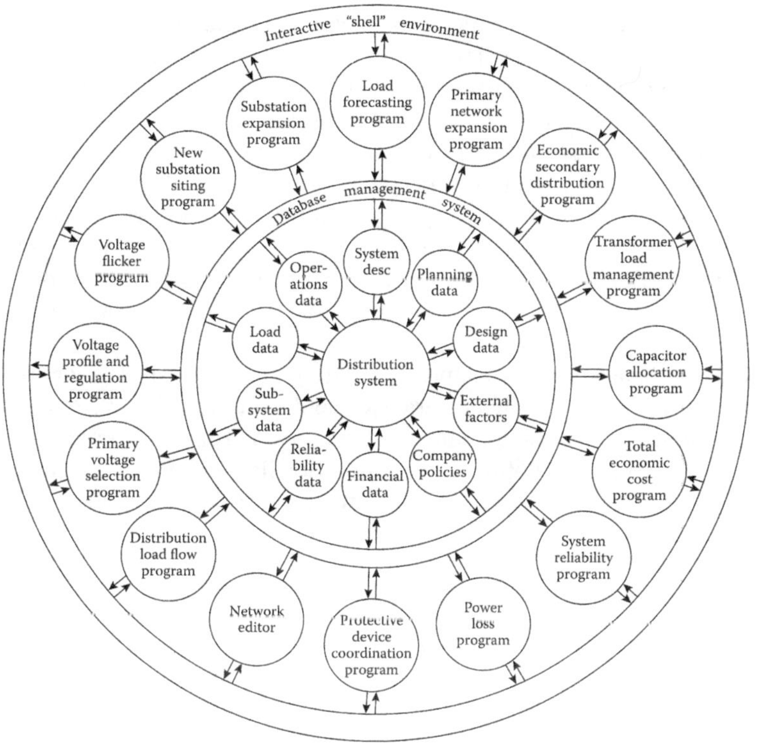

Figure 1.10 A schematic view of a distribution planning system.

relational database management system that stores, retrieves, and modifies various data related to distribution systems [27].

1.9.2 LEVERAGING CLOUD COMPUTING FOR EFFICIENT INFORMATION RETRIEVAL

In today's landscape, cloud computing has revolutionized the rapid retrieval of information, particularly in the storage of electronic archives managed by utilities. As depicted in Figure 1.10, the database assumes a pivotal role in the functioning of such systems. Technological advancements in the past five years have led to remarkable progress in this domain. Not only is it now economically feasible to store vast amounts of data, but desired information can also be retrieved with access times measured in seconds. The relational database serves as the vital interface between the processes requiring data access and the data itself. With the emergence of cloud computing, an organizational structure based on the concept of relations is poised to become the dominant approach in the near future. Relational databases play a crucial role in historical archives derived from daily system operations, ensuring seamless access and management of vital data.

1.9.3 EMERGING PARADIGMS

Alongside the database management program and network analysis programs, it is expected In the realm of distribution system planning, new tools are expected to complement the existing database management and network analysis programs. One notable tool that has gained recognition in the literature is the network editor [26]. Acting as a graphical representation, the network editor forms a graph where vertices correspond to network components (such as transformers and loads) and edges depict the connections between these components.

The network editor encompasses a range of features, including various network objects like feeder line sections, secondary line sections, distribution transformers, and variable or fixed capacitors. Additionally, it incorporates control mechanisms and command functions. A fundamental network object typically comprises a name, an object class description, and a connection list. Notably, the presence of control mechanisms equips planners with intuitive tools for accurate network construction and modification [27]. With the emergence of network editors, designers are empowered to achieve optimal designs for distribution systems that new tools will emerge to support designers in achieving optimal designs. One such tool, identified in the literature as a network editor [26], has garnered attention. The network editor represents a graph in which the vertices represent network components (e.g., transformers and loads) and the edges depict connections between these components.

The network editor boasts various features, including network objects such as feeder line sections, secondary line sections, distribution transformers, and variable or fixed capacitors. It also encompasses control mechanisms and command functions. A basic network object typically comprises a name, an object class description, and a connection list. The presence of control mechanisms equips planners with intuitive tools for accurate network construction and modification [27].

Part II

Profiling Electrical Loads

2 Load Characteristics

Only two things are infinite, the universe and human stupidity.
And I am not so sure about the former.

Albert Einstein

2.1 INTRODUCTION

The interpretation of the term "load" can vary depending on the specific context. In the context of distribution networks, an electrical load refers specifically to the total power consumed by individual customers who are connected to a distribution feeder. This chapter provides essential definitions that play a crucial role in determining usage and contributing to the overall power consumption within a feeder. While transmission systems often represent loads in aggregated quantities, the dynamic nature of loads in distribution systems significantly influences their operation, necessitating the adoption of customer-level models. The practical size of a distribution feeder can vary, accommodating anywhere from a few hundred to over a thousand customers. However, it is important to acknowledge that the examples presented in this chapter represent a critical mass of the feeder, serving as proof-of-concept illustrations, and may not precisely mirror the scale of real-world systems.

As highlighted in Chapter 1, the introduction of metering infrastructure has revolutionized load monitoring in distribution networks. The increasing deployment of smart meters, as part of advanced metering infrastructure (AMI), has substantially enhanced the observability of feeders. Consequently, the acquisition of kilowatt-hour (kWh) information can be obtained at intervals as short as once every hour. This improvement in observability has opened up new opportunities for load analysis and system optimization. However, the integration of smart meters and the AMI has introduced the requirement for regular maintenance of the deployed cyberinfrastructure as a necessary trade-off for these advancements in load monitoring capabilities.

2.2 STATE-OF-THE-ART UPDATE

Metering load would translate the information into a more accurate consumption that is indicated in reliability indices where engineers can anticipate the expected range of annual values in a reliable sense [28]. As smart meters have been ubiqitously deployed in distribution utilities, the more extensive game-theoretic model has refined scheduling for distribution network where load profile may be uniquely deviated based on pricing tariffs [29]. Similarly, the electric vehicle (EV) battery charging will potentially increase over time. The proposed charging characteristics of EV Batteries have articulated the stochastic nature of starting time and initial condition of battery level [30]. As part of the home energy management system, an increased usage of power would result in variation of load patterns. For major example is electronically driven home appliance. As this "load" commitment algorithm can thermostatically determine household consumption based on specific pricing and forecast, the proposed optimization model considers individual household comfort settings [31].

Reconfiguration studies for distribution network would require extensive modeling of the loads how it is characterized with minimum losses [32]. On the demand side (loads), a heuristic approaches can be introduced to aim for reduction of load peak by reshaping the profile, which can improve operational efficiency [33]. EV-incurred load could potentially disrupt operation where overload conditions may trip the circuits. Estimation of household's load growth has accounted for the charging, which can significantly burden existing lower rating distribution transformer [34].

DOI: 10.1201/9781003129721-2

Congestion may occur within a feeder where loads can be determined by the choice of household. Hence, managing the dynamics major appliances and heavy loads can alleviate operating condition for a distribution network [35, 36]. The overall grid reliability can be degraded by a cyber-sabotage where some loads can be intentionally disconnected abruptly through compromised supervisory control and data acquisition (SCADA) [37]. Most importantly, the IP-based meters, known "smart" meters, have been extensively deployed in distribution network with significantly increased number of observation for each load point [38]. By connecting advanced metering infrastructure (AMI) with SCADA network, the data synergy can improve load profile with connectivity verification based on geographical location [39]. Collectively, distribution feeder is connected with loads where each of the validation of lump load (sub-tree) can be summed up, estimated, and validated based on the load characteristics within a medium-voltage (MV) network, especially monitoring power quality in the network, validity of dynamic routing, delay, and communication traffic. [40–42].

The metering point for a smart meter would aggregate a collective point of household consumption. There has been research emphasizing on disaggregating the granular details of appliance consumption as part of the non-intrusive load (NIL) characterization. It is not intrusive when a specific appliance is pre-determined [43]. Understanding aggregated load consumption patterns associated with each distribution transformer (DT) in a context of feeder would allow tremendous opportunities of application. As a backup of contingency if a recloser may be tripped and substation is the sole power supply, an adequate estimation of each subsystem of the associated feeder can be charged with a battery in an MV network [44]. Such planning can account for dynamic power imbalance issues from EVs [45].

From the reverse engineering perspective, line parameters can be fine-tuned with measurements of smart meters since impedance is linearly proportional to line distance [46]. The applications of distribution scheduling can be harnessed centrally to consider a larger scale of consumers' choices and how the overall feeder response to the two-directional behavioral characterization of market participants [47].

2.3 BASIC DEFINITIONS

In a distribution system, most units are measured in kilo (1×10^3) rather than mega (1×10^6) as commonly used in transmission systems, which represents a smaller fraction of the 1000 units in the bulk power system. For example, a 1 MW load in a transmission system can be equivalent to a thousand retail customers. This section presents various ways to define an electrical load in a distribution system.

The ***Demand*** of an installation or system refers to the load at the receiving terminals averaged over a specific interval of time [48]. The load can be expressed in kilowatts (kW), kilovolt-ampere reactive (kVAr), kilovolt-amperes (kVA), kiloamperes (kA), or amperes (A). The ***Demand interval*** represents the duration over which the load is averaged, such as 15 min, 30 min, 1 h, or even longer. It is worth noting that demands averaged over different time intervals may occasionally be identical.

Figure 2.1 shows a plot of daily demand variation curve, or load curve, as a function of time. Note that the selection of both At and total time t is arbitrary. The load is expressed in per unit (pu) of peak load of the system. For example, the maximum of 15-min demands is 0.940 pu, and the maximum of 1-h demands is 0.884, whereas the average daily demand of the system is 0.254. The data given by the curve of Figure 2.1 can also be expressed as shown in Figure 2.2. Here, the time is given in per unit of the total time. The curve is constructed by selecting the maximum peak points and connecting them by a curve. This curve is called the load duration curve. The load duration curves can be daily, weekly, monthly, or annual. For example, if the curve is a plot of all the 8,760 (365×24) hourly loads during the year, it is called an annual load duration curve. In that case, the curve shows the individual hourly loads during the year, but not in the order that they occurred, and the number of hours in the year that load exceeded the value is also shown.

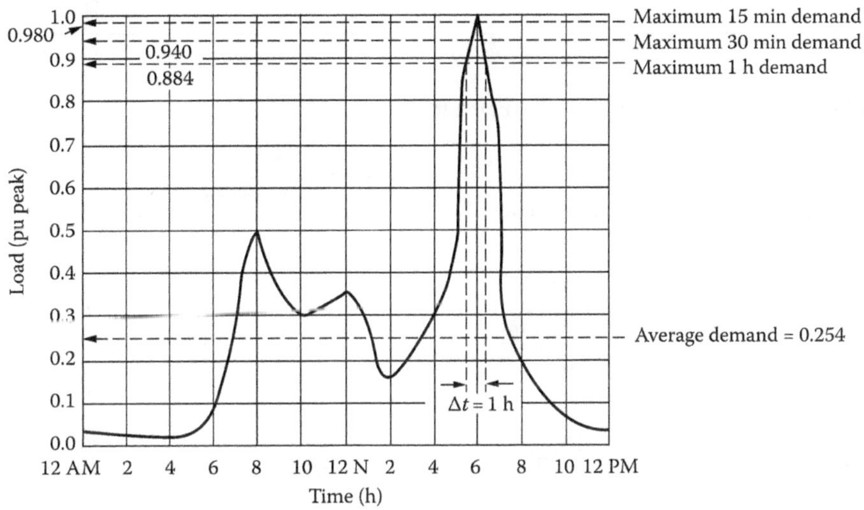

Figure 2.1 A daily demand variation curve.

The hour-to-hour load on a system changes over a wide range. For example, the daytime peak load is typically double the minimum load during the night. Usually, the annual peak load is, due to seasonal variations, about three times the annual minimum. To calculate the average demand, the area under the curve has to be determined. This can easily be achieved by a computer program. ***Maximum demand*** is highest consumption of an observed load, e.g., an installation or system over a specified period of time [48]." The maximum demand statement should also state the demand interval in which it occurs such as daily, weekly, monthly, or annual.

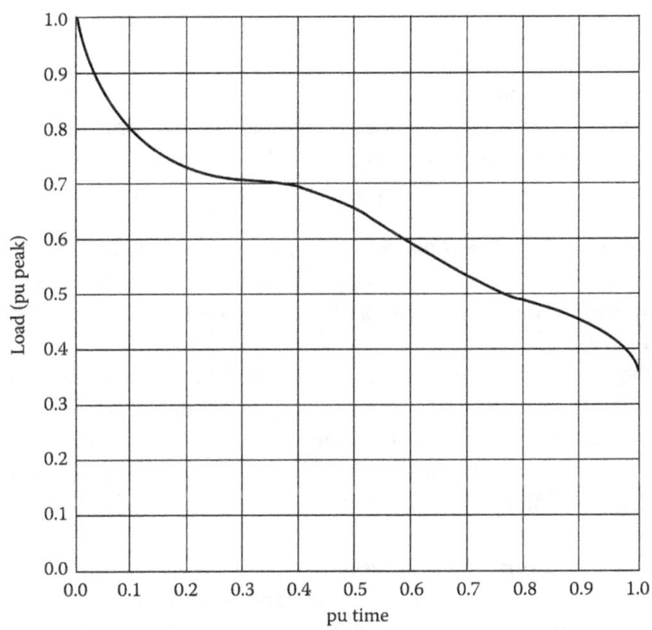

Figure 2.2 A load duration curve.

Table 2.1
Idealized Load Data for the NL & NP's Primary Feeder

	Load, kW		
Time	Street Lighting	Residential	Commercial
12:00 AM	100	200	200
1	100	200	200
2	100	200	200
3	100	200	200
4	100	200	200
5	100	200	200
6	100	200	200
7	100	300	200
8		400	300
9		500	500
10		500	1000
11		500	1000
12 noon		500	1000
1		500	1000
2		500	1200
3		500	1200
4		500	1200
5		600	1200
6	100	700	800
7	100	800	400
8	100	1000	400
9	100	1000	400
10	100	800	200
11	100	600	200
12:00 PM	100	300	200

Example 2.1. Assume that the loading data given in Table 2.1 belongs to one of the primary feeders of the No Light & No Power (NL&NP) Company and that they are for a typical winter day. Develop the idealized daily load curve for the given hypothetical primary feeder.

Solution. Figure 2.3 shows a plot of the load described by 2.1.

Diversified demand (or coincident demand) is referred to the demand of the composite group, as a whole, of somewhat unrelated loads over a specified period of time. The maximum diversified demand is the maximum sum of the contributions of the individual demands to the diversified demand over a specific time interval.

For example, "if the test locations can, in the aggregate, be considered statistically representative of the residential customers as a whole, a load curve for the entire residential class of customers can be prepared. If this same technique is used for other classes of customers, similar load curves can be prepared [49]." As shown in Figure 2.4, if these load curves are aggregated, the system load curve can be developed. The interclass coincidence relationships can be observed by comparing the curves.

Manning [50] defines ***noncoincident demand*** as "the sum of the demands of a group of loads with no restrictions on the interval to which each demand is applicable." Here, again the maximum of the noncoincident demand is the value of some importance. ***Demand factor (DF)*** is the "ratio of the maximum demand of a system to the total connected load of the system [48]." Therefore, the

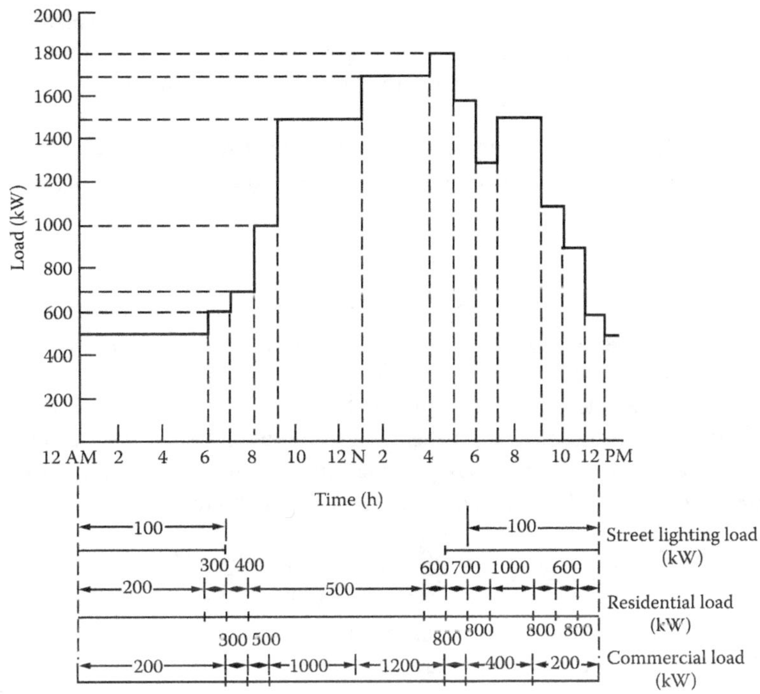

Figure 2.3 The daily load curve for Example 2.1.

DF is

$$\mathrm{DF} \triangleq \frac{\text{Maximum demand}}{\text{Total connected demand}} \tag{2.1}$$

The DF can also be found for a part of the system, for example, an industrial or commercial customer, instead of for the whole system. In either case, the DF is usually less than 1.0. It is an

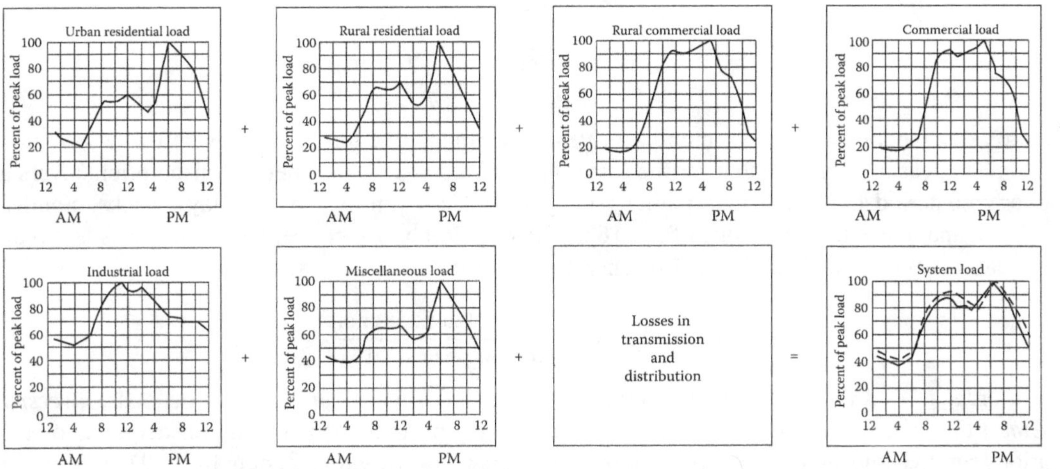

Figure 2.4 Development of aggregate load curves for winter peak period. Miscellaneous load includes street lighting and sales to other agencies. Dashed curve shown on system load diagram is actual system generation sent out. Solid curve is based on group load study data. (From Sarikas, R.H. and Thacker, H.B., AIEE Trans., 31(pt. III), 564, August 1957. Used by permission.)

indicator of the simultaneous operation of the total connected load. ***Connected load*** is "the sum of the continuous ratings of the load-consuming apparatus connected to the system or any part thereof [48]." When the maximum demand and total connected demand have the same units, the DF is dimensionless.

A measure would determine if an equipment is utilized, ***Utilization factor*** (F_u) is introduced. This factor is "the ratio of the maximum demand of a system to the rated capacity of the system [48]" as:

$$F_u \triangleq \frac{\text{Maximum demand}}{\text{Rated system capacity}} \tag{2.2}$$

The utilization factor can also be found for a part of the system. The rated system capacity may be selected to be the smaller of thermal- or voltage-drop capacity [50].

Plant factor is a similar factor that determines the ratio of the total actual energy produced or served over a designated period of time to the energy that would have been produced or served if the plant (or unit) had operated continuously at maximum rating. It is also known as the *capacity factor* or the *use factor*. The following is how the factor is calculated:

$$\text{Plant Factor} \triangleq \frac{\text{Actual energy produced} \times T}{\text{Maximum plant rating} \times T} \tag{2.3}$$

It is mostly used in generation studies. For example, the following annual ratio is defined with actual generation with respect to maximum plant rating.

$$\text{Annual power factor} \triangleq \frac{\text{Actual annual energy generation}}{\text{Maximum plant rating}} \tag{2.4}$$

or

$$\text{Annual power factor} \triangleq \frac{\text{Actual annual energy generation}}{\text{Maximum plant rating} \times 8760} \tag{2.5}$$

The same measure is applied to electrical loads. ***Load factor*** (F_{LD}) is "the ratio of the average load over a designated period of time to the peak load occurring on that period [48]." Load factor F_{LD} is on average load:

$$F_{LD} \triangleq \frac{\text{Average load}}{\text{Peak load}} \tag{2.6}$$

or

$$F_{LD} \triangleq \frac{\text{Average load} \times T}{\text{Peak load} \times T} \tag{2.7}$$

where T is time, in days, weeks, months, or years. The longer the period T, the smaller the resultant factor. The reason for this is that for the same maximum demand, the energy consumption covers a larger time period and results in a smaller average load. Here, time T can be in days, weeks, months, or years, and hence take the value of 24, 168, 730, or 8,760 h, respectively. Load factor is less than or equal to 1.0 with the ideal being 1.0. Therefore,

$$\text{Annual load factor} = \frac{\text{Total annual energy}}{\text{Annual peak load} \times 8,760} \tag{2.8}$$

In order to find the variation of a certain group of loads and a larger group of loads, ***diversity factor*** (F_D) is provided. It is "the ratio of the sum of the individual maximum demands of the various subdivisions of a system to the maximum demand of the whole system [48]." Therefore, the diversity factor is

$$F_D \triangleq \frac{\text{Sum of individual maximum demands}}{\text{Coincident maximum demand}} \tag{2.9}$$

or

$$F_D = \frac{D_1 + D_2 + D_3 + \cdots + D_n}{D_g} \tag{2.10}$$

or

$$F_D = \frac{\sum_{i=1}^{n} D_i}{D_g} \tag{2.11}$$

where

D_i is the maximum demand of load i, disregarding time of occurrence.
$D_g = D_{1+2+3+\cdots+n}$ = coincident maximum demand of group of n loads.

The diversity factor can be equal to or greater than 1.0. From Equation 2.1,

$$\text{DF} = \frac{\text{Maximum demand}}{\text{Total connected demand}}$$

or

$$\text{Maximum demand} = \text{Total connected demand} \times \text{DF} \tag{2.12}$$

Substituting Equation 2.12 into 2.11, the diversity factor can also be given as

$$F_D = \frac{\sum_{i=1}^{n} TCD_i \times DF_i}{D_g} \tag{2.13}$$

where

TCD_i is the total connected demand of group, or class, i load.
DF_i is the demand factor of group, or class, i load.

Coincidence factor is "the ratio of the maximum coincident total demand of a group of consumers to the sum of the maximum power demands of individual consumers comprising the group both taken at the same point of supply for the same time [48]." The ratio of coincidence factor (F_c) is defined by:

$$F_c = \frac{\text{Coincident maximum demand}}{\text{Sum of individual maximum demands}} \tag{2.14}$$

or more specifically,

$$F_c = \frac{D_g}{\sum_{i=1}^{n} D_i} \tag{2.15}$$

Thus, the coincidence factor is the reciprocal of diversity factor, which is inversely related as follows:

$$F_c = \frac{1}{F_D} \tag{2.16}$$

These ideas on diversity and coincidence are the basis for the theory and practice of north-to-south and east-to-west interconnections among the power pools in this country. For example, in the United States during winter, energy comes from south to north, and during summer, just the opposite occurs. Also, east-to-west interconnections help to improve the energy dispatch by means of sunset or sunrise adjustments, that is, the setting of clocks 1 h late or early.

On the contrary, ***load diversity*** is "the difference between the sum of the peaks of two or more individual loads and the peak of the combined load [48]." Load diversity (LD) is determined by summing up individual maximum and subtracted from combined load maximum as follows:

$$\text{LD} \triangleq \sum_{i=1}^{n} D_i - D_g \tag{2.17}$$

As it is assumed that each individual load would have an equal contribution to the coincidence factor (F_c), different weighting factor would contribute to the total sum of the nominator value. The ***contribution factor*** introduced by Manning [50] has extended by defining c_i as "the contribution factor of the i-th load to the group maximum demand." It would give the "weight" of each load and is given in per unit of the individual maximum demand of the i-th load. Contribution factor can then be determined as follows:

$$D_g \triangleq c_1 \times D_1 + c_2 \times D_2 + c_3 \times D_3 + \cdots + c_n \times D_n \tag{2.18}$$

Substituting Equation 2.18 into 2.15,

$$F_c = \frac{c_1 \times D_1 + c_2 \times D_2 + c_3 \times D_3 + \cdots + c_n \times D_n}{\sum_{i=1}^{n} D_i} \tag{2.19}$$

or

$$F_c = \frac{\sum_{i=1}^{n} c_i \times D_i}{\sum_{i=1}^{n} D_i} \tag{2.20}$$

Special cases

Case 1: $D_1 = D_2 = D_3 = \cdots = D_n = D$. From Equation 2.20,

$$F_c = \frac{D \times \sum_{i=1}^{n} c_i}{n \times D} \tag{2.21}$$

or

$$F_c = \frac{\sum_{i=1}^{n} c_i}{n} \tag{2.22}$$

This concludes that the coincidence factor is equal to the average contribution factor.

Case 2: $c_1 = c_2 = c_3 = \cdots = c_n = c$. From Equation 2.20,

$$F_c = \frac{c \times \sum_{i=1}^{n} D_i}{\sum_{i=1}^{n} D_i} \tag{2.23}$$

or

$$F_c = c \tag{2.24}$$

That is, the coincidence factor is equal to the contribution factor.

Last, the ***loss factor*** (F_{LS}) is determined with similar matters as compared to load factor, which is "the ratio of the average power loss to the peak-load power loss during a specified period of time [48]". Therefore, the loss factor

$$F_{LS} \triangleq \frac{\text{Average power loss}}{\text{Power loss at peak load}} \tag{2.25}$$

Equation 2.25 is applicable for the copper losses of the system but not for the iron losses.

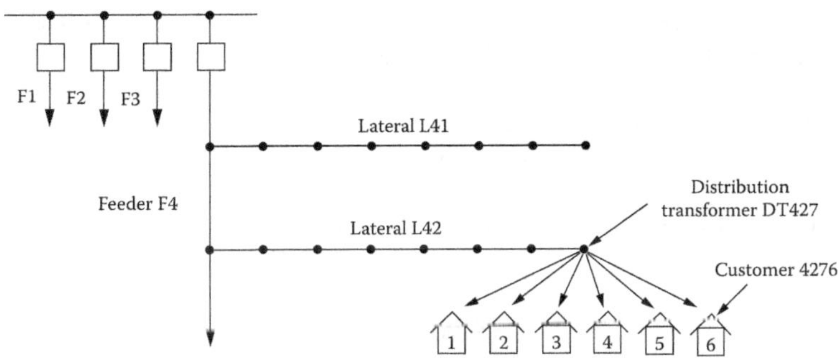

Figure 2.5 Illustration of load connected to a distribution transformer.

Example 2.2. Considering an annual peak load of 2000 kW on a primary feeder, where the total copper loss or $\Sigma I^2 R$ loss is 80 kW per three-phase, and assuming an annual loss factor of 0.15, determine the following:

1. The average annual power loss
2. The total annual energy loss due to the copper losses of the feeder circuits

Solution. 1. From Equation 2.25,

$$\begin{aligned}\text{Average power loss} &= \text{power loss at peak load} \times F_{LS}\\ &= 80\ \text{kW} \times 0.15\\ &= 12\ \text{kW}\end{aligned}$$

2. The total annual energy loss is

$$\begin{aligned}\text{TAEL}_{Cu} &= \text{average power loss} \times 8{,}760\ \text{h/year}\\ &= 12 \times 8{,}760 = 105{,}120\ \text{kWh}\end{aligned}$$

Example 2.3. In Figure 2.5, six residential customers are connected to a distribution transformer (DT). The customer account numbers follow a specific code, such as 4276. Each digit in the code carries significance: the first digit (4) represents feeder F4, the second digit (2) indicates the lateral number connected to feeder F4, the third digit (7) corresponds to the DT on that lateral, and the fourth digit (6) represents the house number connected to that DT.

Assuming a connected load of 9 kW per house and known DF (Demand Factor) and diversity factor values of 0.65 and 1.10, respectively, for the group of six houses, determine the diversified demand of this group on the DT427.

Solution. From Equation 2.13, the diversified demand of the group on the DT is

$$\begin{aligned}D_g &= \frac{\sum_{i=1}^{6} \text{TCD}_i \times \text{DF}}{F_D}\\ &= \frac{(\sum_{i=1}^{6} 9\ \text{kW}) \times 0.65}{1.1}\\ &= \frac{6 \times 9\ \text{kW} \times 0.65}{1.1}\\ &= 31.9\ \text{kW}\end{aligned}$$

Example 2.4. Assume that feeder 4 of Example 2.3 has a system peak of 3000 kVA per phase and a copper loss of 0.5 % at the system peak. Determine the following:

1. The copper loss of the feeder in kW per phase
2. The total copper losses of the feeder in kW per three-phase

Solution. 1. The copper loss of the feeder in kW per phase is

$$\begin{aligned} I^2R &\triangleq 0.5\% \times \text{system peak} \\ &= 0.005 \times 3000 \text{ kVA} \\ &= 15 \text{ kW per phase} \end{aligned}$$

2. The total copper losses of the feeder in kW per three-phase is

$$\begin{aligned} 3I^2R &\triangleq 3 \times 15 \\ &= 45 \text{ kW per three-phase} \end{aligned}$$

Example 2.5. Assume that there are two primary feeders supplied by one of the three transformers located at the NL&NP's Riverside distribution substation, as shown in Figure 2.6. One of the feeders supplies an industrial load that occurs primarily between 8 AM and 11 PM, with a peak of 2000 kW at 5 PM. The other one feeds residential loads that occur mainly between 6 AM and 12 PM, with a peak of 2000 kW at 9 PM, as shown in Figure 2.7. Determine the following:

1. The diversity factor of the load connected to transformer T3
2. The load diversity of the load connected to transformer T3
3. The coincidence factor of the load connected to transformer T3

Solution. 1. From Equation 2.11, the diversity factor of the load is

$$\begin{aligned} F_D &= \frac{\sum_{i=1}^{2} D_i}{D_g} \\ &= 4000 - 3000 = 1000 \text{ kW} \end{aligned}$$

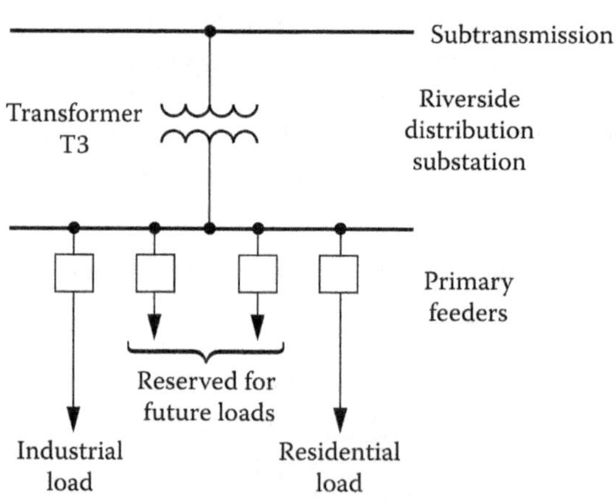

Figure 2.6 NL&NP's riverside distribution substation.

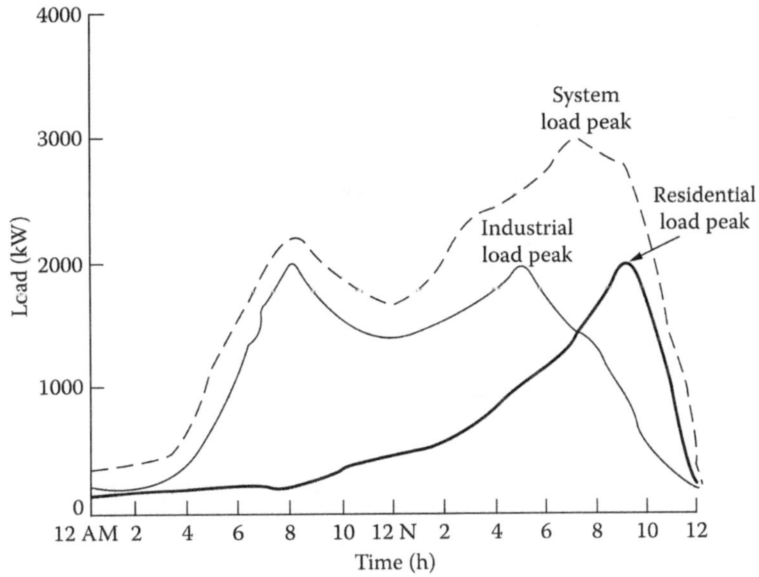

Figure 2.7 Daily load curves of a substation transformer.

2. From Equation 2.17, the load diversity of the load is

$$LD = \sum_{i=1}^{2} D_g - D_g$$
$$= 45 \text{ kW per three-phase}$$

3. From Equation 2.16, the coincidence factor of the load is

$$F_c = \frac{1}{F_D}$$
$$= \frac{1}{1.33}$$
$$\cong 0.752$$

Example 2.6. Using the load curve data provided in Example 2.1 for NL&NP, where the peak load occurs at 4 PM, calculate the following:

1. The class contribution factors for each of the three load classes
2. The diversity factor for the primary feeder
3. The diversified maximum demand of the load group
4. The coincidence factor of the load group

Solution. 1. The class contribution factor is

$$c_i \cong \frac{\text{class demand at time of system (i.e., group) peak}}{\text{class noncoincident maximum demand}}$$

For street lighting, residential, and commercial loads,

$$c_{\text{street}} = \frac{0 \text{ kW}}{100 \text{ kW}} = 0$$

$$c_{\text{residential}} = \frac{600 \text{ kW}}{1000 \text{ kW}} = 0.6$$

$$c_{\text{commercial}} = \frac{1200 \text{ kW}}{1200 \text{ kW}} = 1.0$$

2. From Equation 2.11, the diversity factor is

$$F_D = \frac{\sum_{i-1}^{n} D_i}{D_g}$$

and from Equation 2.18,

$$D_g \triangleq c_1 \times D_1 + c_2 \times D_2 + c_3 \times D_3 + \cdots + c_n \times D_n$$

Substituting Equation 2.18 into 2.11,

$$F_D = \frac{\sum_{i=1}^{n} D_i}{\sum_{i=1}^{n} C_i \times D_i}$$

Therefore, the diversity factor for the primary feeder is

$$\begin{aligned} F_D &= \frac{\sum_{i=1}^{3} D_i}{\sum_{i=1}^{3} C_i \times D_i} \\ &= \frac{100 + 1000 + 1200}{0 \times 100 + 0.6 \times 1000 + 1.0 \times 1200} \\ &= 1.278 \end{aligned}$$

3. The diversified maximum demand is the coincident maximum demand, that is, D_g. Therefore, from Equation 2.13, the diversity factor is

$$F_D = \frac{\sum_{i=1}^{n} \text{TCD}_i \times \text{DF}_i}{D_g}$$

where the maximum demand, from Equation 2.12, is

$$\text{Maximum demand} = \text{Total connected demand} \times \text{DF}$$

Substituting Equation 2.12 into 2.13,

$$F_D = \frac{\sum_{i=1}^{n} D_i}{D_g}$$

or

$$D_g = \frac{\sum_{i=1}^{n} D_i}{F_D}$$

Therefore, the diversified maximum demand of the load group is

$$\begin{aligned} D_g &= \frac{\sum_{i=1}^{3} D_i}{F_D} \\ &= \frac{100 + 1000 + 1200}{1.278} \\ &= 1800 \text{ kW} \end{aligned}$$

4. The coincidence factor of the load group, from Equation 2.15, is

$$F_c = \frac{D_g}{\sum_{i=1}^{n} D_i}$$

or, from Equation 2.16,

$$\begin{aligned} F_c &= \frac{1}{F_D} \\ &= \frac{1}{1.278} \\ &= 0.7825 \end{aligned}$$

2.4 RELATIONSHIP BETWEEN THE LOAD AND LOSS FACTORS

In situations where a feeder is not fully observable (i.e., when smart meters and feeder remote terminal units are not introduced), the metering infrastructure at the substation feeder can be utilized to estimate the relationship between loads and losses. While most instruments are capable of measuring electrical load, accurately calculating power loss often requires the use of estimation models. Fortunately, there is a proportional correlation between load and loss within a feeder. Within a distribution network feeder, the metering point at the feeder head of a substation provides a comprehensive lumped load profile for that feeder through the substation SCADA network. This section focuses on examining the limiting values associated with the relationship between load and loss [50].

Let's consider the primary feeder depicted in Figure 2.8, which is connected to a variable lumped load. Figure 2.9 presents an arbitrary and idealized load curve. However, it does not represent a realistic daily load curve; it is designed for illustrative purposes to explain the variables discussed in this section. Let's assume that the off-peak loss is $P_{LS,1}$ at an off-peak load of P_1, and the peak loss is $P_{LS,2}$ at the peak load of P_2. The load factor can be determined as follows:

$$F_{LD} = \frac{P_{av}}{P_{max}} = \frac{P_{av}}{P_2} \tag{2.26}$$

From Figure 2.9,

$$P_{av} = \frac{P_2 \times t + P_1 \times (T - t)}{T} \tag{2.27}$$

Substituting Equation 2.27 into 2.26,

$$F_{LD} = \frac{P_2 \times t + P_1 \times (T - t)}{P_2 \times T}$$

or

$$F_{LD} = \frac{t}{T} + \frac{P_1}{P_2} \times \frac{T - t}{T} \tag{2.28}$$

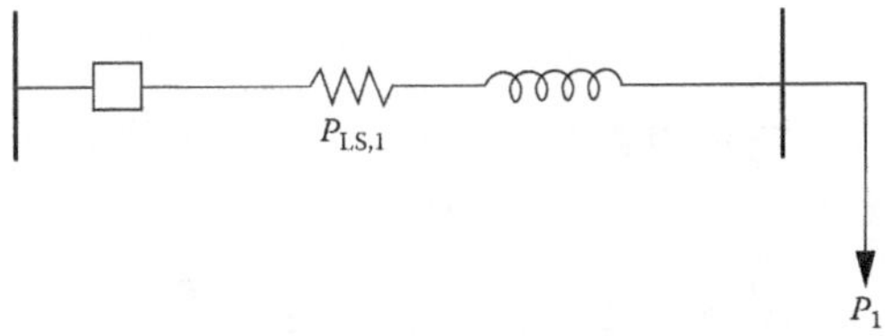

Figure 2.8 A feeder with a variable load.

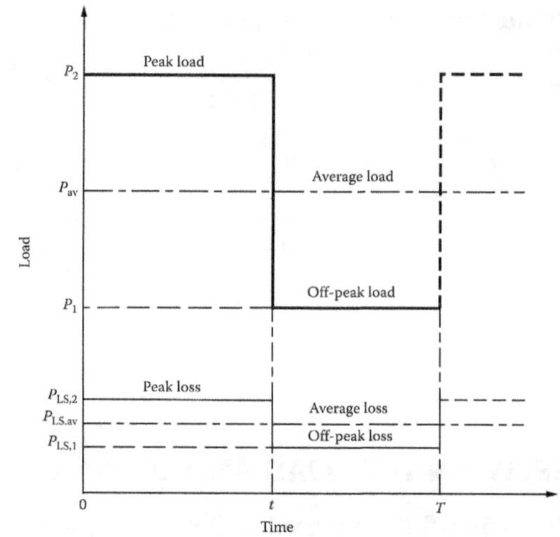

Figure 2.9 An arbitrary and ideal load curve.

The loss factor is

$$F_{LS} = \frac{P_{LS,av}}{P_{LS,max}} = \frac{P_{LS,av}}{P_{LS,2}} \tag{2.29}$$

where
$P_{LS,av}$ is the average power loss
$P_{LS,max}$ is the maximum power loss
$P_{LS,2}$ is the peak loss at peak load.
From Figure 2.9,

$$F_{LD,av} = \frac{P_{LS,2} \times t + P_{LS,1} \times (T - t)}{T} \tag{2.30}$$

Substituting Equation 2.30 into 2.29,

$$F_{LS} = \frac{P_{LS,2} \times t + P_{LS,1} \times (T - t)}{P_{LS,2} \times T} \tag{2.31}$$

where

$P_{LS,1}$ is the off-peak loss at off-peak load
t is the peak-load duration
$T - t$ is the off-peak-load duration

The copper losses are the function of the associated loads. Therefore, the off-peak and peak loads can be expressed, respectively, as

$$F_{LS,1} = k \times P_1^2 \tag{2.32}$$

and

$$F_{LS,2} = k \times P_2^2 \tag{2.33}$$

where k is a constant. Thus, substituting Equations 2.32 and 2.33 into 2.31, the loss factor can be expressed as

$$F_{LS} = \frac{(k \times P_2^2) \times t(k \times P_1^2) \times (T - t)}{(k \times P_2^2) \times T} \tag{2.34}$$

or

$$F_{LS} = \frac{1}{T} + \left(\frac{P_1}{P_2}\right)^2 \times \frac{T-t}{T} \tag{2.35}$$

By using Equations 2.28 and 2.35, the load factor can be related to loss factor for three different cases.

Case 1: Under the assumption that *off-peak load is zero*, i.e., $P_1 = 0$, which is:

$$P_{LS,1} = 0$$

Therefore, from Equations 2.28 through 2.35,

$$F_{LD} = F_{LS} = \frac{1}{T} \tag{2.36}$$

This indicates that the load factor is equal to the loss factor, and they are equal to the t/T constant.

Case 2: Under an assumption that there is a *very short-lasting peak*, which set here,

$$t \rightarrow 0$$

$$\frac{T-t}{T} \rightarrow 1.0$$

Therefore, it simplifies the relationship between loss and load factors:

$$F_{LS} \rightarrow (F_{LD})^2 \tag{2.37}$$

It can be concluded from this finding that the value of the loss factor approaches the value of the load factor squared.

Case 3: This case provides an assumption that the *load is steady*, which set at any given t to be the same as the entire period T:

$$t \rightarrow T$$

In situations where the difference between the peak load and the off-peak load is negligible, such as in the case of a petrochemical plant, Equations 2.28 through 2.35 can be used to determine the following:

$$F_{LS} \rightarrow F_{LD} \tag{2.38}$$

That is, the value of the loss factor approaches the value of the load factor. Therefore, in general, the value of the loss factor is

$$F_{LD}^2 < F_{LS} < F_{LD} \tag{2.39}$$

Hence, the loss factor cannot be directly derived from the load factor. This is due to the fact that the loss factor is determined based on losses as a function of time, which are in turn proportional to the square of the load over time [49–51]. However, Buller and Woodrow [52] proposed an approximate formula that establishes a relationship between the loss factor and the load factor, given by:

$$F_{LS} = 0.3F_{LD} + 0.7F_{LD}^2 \tag{2.40}$$

where

F_{LS} is the loss factor, pu
F_{LD} is the load factor, pu

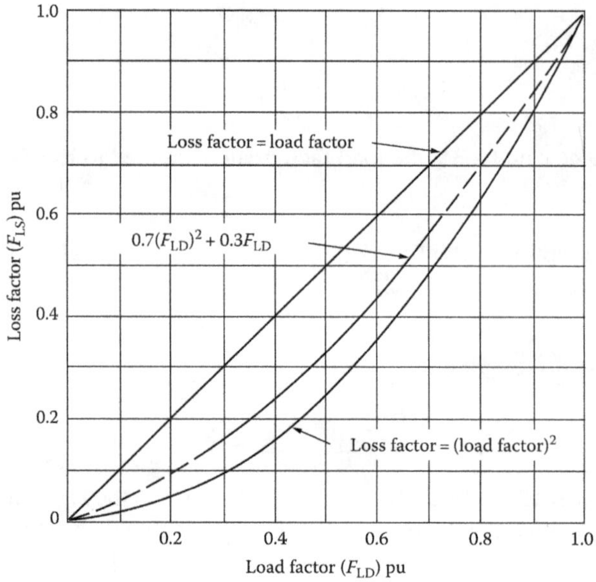

Figure 2.10 Loss factor curves as a function of load factor. (From Westinghouse Electric Corporation, Electric Utility Engineering Reference Book-Distribution Systems, Vol. 3, Westinghouse Electric Corporation, East Pittsburgh, PA, 1965.)

Equation 2.40a provides a reasonably accurate estimation, yielding results that closely align with the actual values. Figure 2.10 illustrates three distinct curves depicting the relationship between the loss factor and the load factor. In more recent studies, researchers have made adjustments to the original formula, specifically tailoring it to better suit rural areas. The modified formula is expressed as follows:

$$F_{LS} = 0.16F_{LD} + 0.84F_{LD}^2 \tag{2.41}$$

Example 2.7. The average load factor of a substation is 0.65. Determine the average loss factor of its feeders, if the substation services

1. An urban area
2. A rural area

Solution. 1. For the urban area

$$\begin{aligned} F_{LS} &= 0.3F_{LD} + 0.7F_{LD}^2 \\ &= 0.3(0.65) + 0.7(0.65)^2 \\ &= 0.49 \end{aligned}$$

2. For the rural area

$$\begin{aligned} F_{LS} &= 0.16F_{LD} + 0.84F_{LD}^2 \\ &= 0.16(0.65) + 0.84(0.65)^2 \\ &= 0.53 \end{aligned}$$

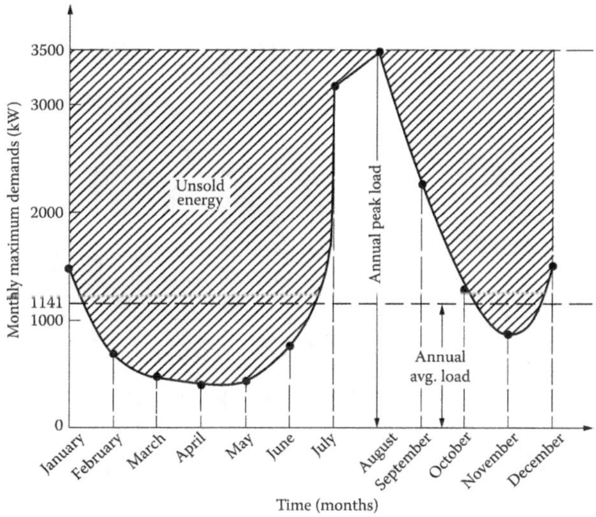

Figure 2.11 A monthly load curve.

Example 2.8. Assume that the Riverside distribution substation of the NL&NP Company supplying Ghost Town, which is a small city, experiences an annual peak load of 3,500 kW. The total annual energy supplied to the primary feeder circuits is 10,000,000 kWh. The peak demand occurs in July or August and is due to air-conditioning load.

1. Find the annual average power demand.
2. Find the annual load factor.

Solution. Assume a monthly load curve as shown in Figure 2.11.

1. The annual average power demand is

$$
\begin{aligned}
\text{Annual } P_{av} &= \frac{\text{Total annual energy}}{\text{Year}} \\
&= \frac{10^7 \text{ kWh/year}}{8760 \text{ h/year}} \\
&= 1141 \text{ kW}
\end{aligned}
$$

2. From Equation 2.6, the annual load factor is

$$
\begin{aligned}
F_{LD} &= \frac{\text{Annual average load}}{\text{Annual peak demand}} \\
&= \frac{1141 \text{ kW}}{3500 \text{ kW}} \\
&= 0.326
\end{aligned}
$$

Or, from Equation 2.8,

$$
\begin{aligned}
\text{Annual load factor} &= \frac{\text{Total annual energy}}{\text{Annual peak load} \times 8,760} \\
&= \frac{10^7 \text{ kWh/year}}{3500 \text{ kW} \times 8760} \\
&= 0.326
\end{aligned}
$$

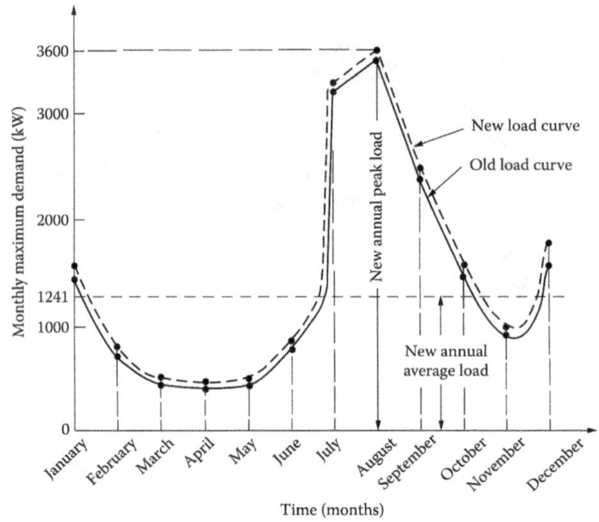

Figure 2.12 The new load curve after the new load addition.

The unsold energy, as shown in Figure 2.11, is a measure of capacity and investment cost. Ideally, it should be kept at a minimum.

Example 2.9. Use the data given in Example 2.8 and suppose that a new load of 100 kW with 100% annual load factor is to be supplied from the Riverside substation. The investment cost, or capacity cost, of the power system upstream, that is, toward the generator, from this substation is \$18.00/kW per month. Assume that the energy delivered to these primary feeders costs the supplier, that is, NL&NP, \$0.06/kWh.

1. Find the new annual load factor on the substation.
2. Find the total annual cost to NL&NP to serve this load.

Solution. Figure 2.12 shows the new load curve after the addition of the new load of 100 kW with 100% load.

1. The new annual load factor on the substation is

$$\text{Annual } F_{LD} = \frac{\text{Annual average load}}{\text{Annual peak demand}}$$

$$= \frac{1141 + 100}{3500 + 100}$$

$$= 0.345$$

2. The total annual and additional cost to NL&NP to serve the additional 100 kW load has two cost components, namely, (1) annual capacity cost and (2) annual energy cost. Therefore,

$$\text{Annual additional capacity cost} = \$18/\text{kW/month} \times 12 \text{ month/year} \times 100 \text{ kW}$$

$$= \$21{,}600$$

and

$$\text{Annual energy cost} = 100 \text{ kW} \times 8760 \text{ h/year} \times \$0.06/\text{kWh}$$

$$= \$52{,}560$$

Therefore, the calculation of total annual additional cost can be itemized as follows:

$$\begin{aligned}\text{Total annual additional cost} &= \text{Annual capacity cost} + \text{Annual energy cost}\\ &= \$21,600 + \$52,560\\ &= \$74,160\end{aligned}$$

Example 2.10. Assume that the annual peak-load input to a primary feeder is 2000 kW. A computer program that calculates voltage drops and I^2R losses shows that the total copper loss at the time of peak load is ΣI^2R = 100 kW. The total annual energy supplied to the sending end of the feeder is 5.61×10^6 kWh.

1. By using Equation 2.40a, determine the annual loss factor.
2. Calculate the total annual copper loss energy and its value at \$0.06/kWh.

Solution. The following provides the two parts of calculation for loss factor, average loss, total annual loss as well as the cost of total loss.

1. From Equation 2.40a, the annual loss factor is

$$F_{LS} = 0.3F_{LD} + 0.7F_{LD}^2$$

where

$$\begin{aligned}F_{LD} &= \frac{5.61 \times 10^6 \text{ kWh}}{2000 \text{ kW} \times 8760 \text{ h/year}}\\ &= 0.32\end{aligned}$$

Therefore,

$$\begin{aligned}F_{LS} &= 0.3 \times 0.32 + 0.7 \times (0.32)^2\\ &\cong 0.1677\end{aligned}$$

2. From Equation 2.25,

$$F_{LS} \cong \frac{\text{Annual power loss}}{\text{power loss at peak load}}$$

or

$$\begin{aligned}\text{Average power loss} &= 0.1677 \times 100 \text{ kW}\\ &= 16.77 \text{ kW}\end{aligned}$$

Therefore,

$$\begin{aligned}\text{Total annual copper loss} &= 16.77\text{kW} \times 8760 \text{ h/year}\\ &= 146,905 \text{ kW}\end{aligned}$$

and

$$\begin{aligned}\text{Cost of total annual copper loss} &= 146,905 \text{ kWh} \times \$0.06/\text{kWh}\\ &= \$8814\end{aligned}$$

Example 2.11. Assume that one of the DTs of the Riverside substation supplies three primary feeders. The 30-min annual maximum demands per feeder are listed in the following table, together with the power factor at the time of annual peak load.

Assume a diversity factor of 1.15 among the three feeders for both real power and reactive power.

1. Calculate the 30-min annual maximum demand on the substation transformer in kW and in kVA.
2. Find the load diversity in kW.
3. Select a suitable substation transformer size if zero load growth is expected and if company policy permits as much as 25% short-time overloads on the distribution substation transformers. Among the standard three-phase (3ϕ) transformer sizes available are the following:
   2,500/3,125 kVA self-cooled/forced-air-cooled
   3,750/4,687 kVA self-cooled/forced-air-cooled
   5,000/6,250 kVA self-cooled/forced-air-cooled
   7,500/9,375 kVA self-cooled/forced-air-cooled
4. Now assume that the substation load will increase at a constant percentage rate per year and will double in 10 years. If the 7500/9375 kVA-rated transformer is installed, in how many years will it be loaded to *its fans-on* rating?

	Demand	
Feeder	kW	PF
1	1800	0.95
2	2000	0.85
3	2200	0.9

Solution. 1. From Equation 2.10,

$$F_{LD} = \frac{1800 + 2000 + 2200}{D_g} = 1.15$$

Therefore,

$$D_g = \frac{6000}{1.15} = 5217 \text{ kW} = \text{P}$$

To find power in kVA, find the PF angles. Therefore,

$$PF_1 = \cos\,\theta_1 = 0.95 \rightarrow \theta_1 = 18.2^\circ$$
$$PF_2 = \cos\,\theta_2 = 0.85 \rightarrow \theta_2 = 31.79^\circ$$
$$PF_3 = \cos\,\theta_3 = 0.90 \rightarrow \theta_3 = 25.84^\circ$$

Thus, the diversified reactive power (Q) is

$$\begin{aligned} Q &= \frac{\sum_{i=1}^{3} P_i \times \tan\theta}{F_D} \\ &= \frac{1800 \times \tan(18.2^\circ) + 2000 \times \tan(31.79^\circ) + 2200 \times \tan(25.84^\circ)}{1.15} \\ &= 2518.95 \text{ kVAr.} \end{aligned}$$

Therefore, the maximum coincidence demand is the apparent power.

$$D_g = \sqrt{P^2 + Q^2} = S$$
$$= \sqrt{5217^2 + 2518.8^2} = 5793.60 \text{ kVA}$$

2. From Equation 2.17, the load diversity is

$$LD = \sum_{i=1}^{3} D_i - D_g$$
$$= 6000 - 5217 = 783 \text{ kW}$$

3. From the given transformer list, it is appropriate to choose the transformer with the 3750/4687-kVA rating since with the 25% short-time overload, it has a capacity of

$$4687 \times 1.25 = 5858.8 \text{ kVA}$$

which is larger than the maximum demand of 5793.60 kVA as found in part (1).
4. Note that the term fans-on rating means the forced-air-cooled rating. To find the increase (g) per year,

$$(1+g)^{10} = 2$$

Hence,

$$(1+g) = 1.07175$$

or

$$g = 7.175 \text{ \%/year}$$

Thus,

$$(1.07175)^n \times 5793.60 = 9375 \text{ kVA}$$

or

$$(1.07175)^n = 1.6182$$

Therefore,

$$n = \frac{\ln 1.6182}{\ln 1.07175}$$
$$= \frac{0.48130}{0.06929} = 7 \text{ years}$$

Therefore, if the 7500/9375 kVA-rated transformer is installed, it will be loaded to its *fans-on* rating in about 7 years.

2.5 MAXIMUM DIVERSIFIED DEMAND

A method for estimating distribution transformer (DT) loads in residential areas was developed by Arvidson [53], known as the diversified-demand method. This approach considers both the diversity between similar loads and the noncoincidence of peak demands among different load types.

To incorporate the noncoincidence of peak demands among different load types, Arvidson introduced the concept of an hourly variation factor. According to Westinghouse [50], this factor is defined as "the ratio of the demand of a particular load type coinciding with the group's maximum demand to the maximum demand of that load type." Table 2.2 provides the hourly variation curves for various household appliances while Figure 2.13 displays multiple curves representing different household appliance types, illustrating the determination of the average maximum diversified demand per customer in kW per load. Each curve in Figure 2.13 represents a 100% saturation level for a specific demand.

To apply Arvidson's method for determining the maximum diversified demand based on a given saturation level and appliance type, the following steps are recommended [50]:

1. Determine the total number of appliances by multiplying the total number of customers by the per-unit saturation.
2. Read the corresponding diversified demand per customer from the curve, in Figure 2.13, for the given number of appliances.
3. Determine the maximum demand, multiplying the demand found in step 2 by the total number of appliances.
4. Finally, determine the contribution of that type load to the group maximum demand by multiplying the resultant value from step 3 by the corresponding hourly variation factor found from Table 2.2.

Example 2.12. Assume a typical DT that serves six residential loads, that is, houses, through six service drops (SDs) and two spans of secondary line (SL). Suppose that there are a total of 150 DTs and 900 residences supplied by this primary feeder. Use Figure 2.13 and Table 2.2. For the sake of illustration, assume that a typical residence contains a clothes dryer, a range, a refrigerator, and some lighting and miscellaneous appliances. Determine the following:

1. The 30-min maximum diversified demand on the DT.
2. The 30-min maximum diversified demand on the entire feeder.
3. Use the typical hourly variation factors given in Table 2.2 and calculate the small portion of the daily demand curve on the DT, that is, the total hourly diversified demands at 4, 5, and 6 PM, on the DT, in kW.

Solution. 1. To determine the 30-min maximum diversified demand on the DT, the average maximum diversified demand per customer is found from Figure 2.13. Therefore, *when the number of loads is six*, the average maximum diversified demands per customer are

$$P_{av,mas} = \begin{cases} 1.6 \text{ kW/house}, & \text{for dryer} \\ 0.8 \text{ kW/house}, & \text{for range} \\ 0.066 \text{ kW/house}, & \text{for refrigerator} \\ 0.61 \text{ kW/house}, & \text{for lightning and misc.appliances} \end{cases}$$

Thus,

$$\begin{aligned} \sum_{i=1}^{4} (P_{av,max})_i &= 1.6 + 0.8 + 0.066 + 0.6 \\ &= 3.076 \text{ kW/house} \end{aligned}$$

Table 2.2
Hourly Variation Factors

							Heat Pump[a]		Water Heater[b] OPWH[c]			
Hour	Lighting and Miscellaneous	Refrigerator	Home Freezer	Range	Air-Conditioning[a]	Cooling Season	Heating Season	House[a] Heating	Both Elements Restricted	Only Bottom Elements Restricted	Uncontrolled	Clothes[d] Dryer
12:00 AM	0.32	0.93	0.92	0.02	0.4	0.42	0.34	0.11	0.41	0.61	0.51	0.03
1	0.12	0.89	0.9	0.01	0.39	0.35	0.49	0.07	0.33	0.46	0.37	0.02
2	0.1	0.8	0.87	0.01	0.36	0.35	0.51	0.09	0.25	0.34	0.3	0
3	0.09	0.76	0.85	0.01	0.35	0.28	0.54	0.08	0.17	0.24	0.22	0
4	0.08	0.79	0.82	0.01	0.35	0.28	0.57	0.13	0.13	0.19	0.15	0
5	0.1	0.72	0.84	0.02	0.33	0.26	0.63	0.15	0.13	0.19	0.14	0
6	0.19	0.75	0.85	0.05	0.3	0.26	0.74	0.17	0.17	0.24	0.16	0
7	0.41	0.75	0.85	0.3	0.41	0.35	1	0.76	0.27	0.37	0.46	0
8	0.35	0.79	0.86	0.47	0.53	0.49	0.91	1	0.47	0.65	0.7	0.08
9	0.31	0.79	0.86	0.28	0.62	0.58	0.83	0.97	0.63	0.87	1	0.2
10	0.31	0.79	0.87	0.22	0.72	0.7	0.74	0.68	0.67	0.93	1	0.65
11	0.3	0.85	0.9	0.22	0.74	0.73	0.6	0.57	0.67	0.93	0.99	1
12 noon	0.28	0.85	0.92	0.33	0.8	0.84	0.57	0.55	0.67	0.93	0.98	0.98
1	0.26	0.87	0.96	0.25	0.86	0.88	0.49	0.51	0.61	0.85	0.86	0.7
2	0.29	0.9	0.98	0.16	0.89	0.95	0.46	0.49	0.55	0.76	0.82	0.65
3	0.3	0.9	0.99	0.17	0.96	1	0.4	0.48	0.49	0.68	0.81	0.63
4	0.32	0.9	1	0.24	0.97	1	0.43	0.44	0.33	0.46	0.79	0.38
5	0.7	0.9	1	0.8	0.99	1	0.43	0.79	0	0.09	0.75	0.3
6	0.92	0.9	0.99	1	1	1	0.49	0.88	0	0.13	0.75	0.22
7	1	0.95	0.98	0.3	0.91	0.88	0.51	0.76	0	0.19	0.8	0.26
8	0.95	1	0.98	0.12	0.79	0.73	0.6	0.54	1	1	0.81	0.2
9	0.85	0.95	0.97	0.09	0.71	0.72	0.54	0.42	0.84	0.98	0.73	0.18
10	0.72	0.88	0.96	0.05	0.64	0.53	0.51	0.27	0.67	0.77	0.67	0.1
11	0.5	0.88	0.95	0.04	0.55	0.49	0.34	0.23	0.54	0.69	0.59	0.04
12:00 PM	0.32	0.93	0.92	0.02	0.4	0.42	0.34	0.11	0.44	0.61	0.51	0.03

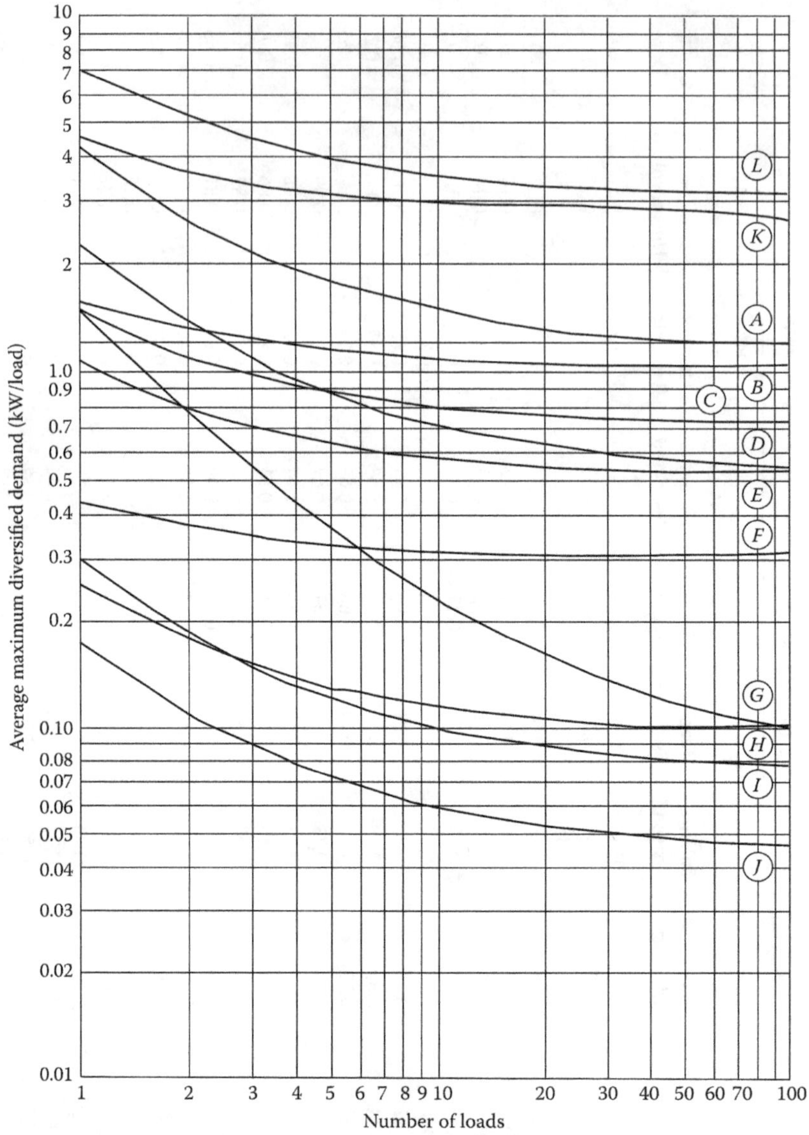

Figure 2.13 Maximum diversified 30-min demand characteristics of various residential loads: A, clothes dryer; B, off-peak water heater, "off-peak" load; C, water heater, uncontrolled, interlocked elements; D, range; E, lighting and miscellaneous appliances; F, 0.5-hp room coolers; G, off-peak water heater, "on-peak" load, upper element uncontrolled; H, oil burner; I, home freezer; J, refrigerator; K, central air-conditioning, including heat-pump cooling, 5-hp heat pump (4-ton air conditioner); L, house heating, including heat-pump-heating-connected load of 15 kW unit-type resistance heating or 5 hp heat pump. (From Westinghouse Electric Corporation, Electric Utility Engineering Reference Book-Distribution Systems, Vol. 3, Westinghouse Electric Corporation, East Pittsburgh, PA, 1965.)

and for six houses

$$(3076 \text{ kW/house})(6 \text{ houses}) = 18.5 \text{ kW}$$

Thus, the contributions of the appliances to the 30-min maximum diversified demand on the DT is approximately 18.5 kW.

2. As in part (1), the average maximum diversified demand per customer is found from Figure 2.13. Therefore, *when the number of loads is 900* (note that, due to the given curve characteristics, the answers would be the same as the ones for the number of loads of 100), then the average maximum diversified demands per customer are

$$P_{av,mas} = \begin{cases} 1.2 \text{ kW/house}, & \text{for dryer} \\ 0.53 \text{ kW/house}, & \text{for range} \\ 0.044 \text{ kW/house}, & \text{for refrigerator} \\ 0.52 \text{ kW/house}, & \text{for lightning and misc.appliances} \end{cases}$$

Hence,

$$\sum_{i=1}^{4} (P_{av,max})_i = 1.2 + 0.53 + 0.044 + 0.52$$
$$= 2.294 \text{ kW/house}$$

Therefore, the 30-min maximum diversified demand on the entire feeder is

$$\sum_{i=1}^{4} (P_{\text{av, max}})_i = 900 \times 2.294$$
$$= 2064 \text{ kW/feeder}$$

However, if the answer for the 30-min maximum diversified demand on one DT found in part (1) is multiplied by 150 to determine the 30-min maximum diversified demand on the entire feeder

$$150 \times 18.5 \cong 2775 \text{ kW}$$

which is greater than the demand 2064.6 kW found previously. This discrepancy is due to the application of the appliance diversities. c. From Table 2.2, the *hourly variation factors* can be found as 0.38, 0.24, 0.90, and 0.32 for dryer, range, refrigerator, and lighting and miscellaneous appliances. Therefore, the total hourly diversified demands on the DT can be calculated as given in the following table in which

(1.6 kW/house)(6 houses)	= 9.6 kW
(0.8 kW/house)(6 houses)	= 4.8 kW
(0.066 kW/house)(6 houses)	= 0.4 kW
(0.61 kW/house)(6 houses)	= 3.7 kW

Time	Dryers, kW	Ranges, kW	Refrigerators, kW	Lighting and Misc. Appliances, kW	Total Hourly Diversified Demand, kW
(1)	(2)	(3)	(4)	(5)	(6)
04:00 PM	9.6 × 0.38	4.8 × 0.24	0.4 × 0.90	3.7 × 0.32	6.344
05:00 PM	9.6 × 0.30	4.8 × 0.80	0.4 × 0.90	3.7 × 0.70	9.67
06:00 PM	9.6 × 0.22	4.8 × 1.00	0.4 × 0.90	3.7 × 0.92	10.674

2.6 RATE STRUCTURE

Despite *deregulation*, most public utilities may still implement their own rate of schemes to sell the energy to their customers. Their rates are subject to government regulation. The total revenue that a utility may be authorized to collect through the sales of its services should be equal to the company's total cost of service. The following is the general revenue calculation based on how the revenue is determined,

$$\begin{aligned}(\text{Revenue requirement}) = {} & (\text{operating expenses}) + (\text{depreciation expenses}) + (\text{taxes}) \\ & + (\text{net valuation}) \times (\text{rate of return})\end{aligned} \tag{2.42}$$

The determination of the revenue requirement is subject to the decision of regulatory commissions. It is the responsibility of management to design rate schedules that can generate the required revenue, which is then reviewed by the commission. However, it is important to note that a regulatory commission cannot guarantee a specific rate of earnings. Instead, it can only provide the opportunity for a public utility to try to earn it.

The rate of return should take into account local conditions and be comparable to the returns earned by similar companies facing similar risks. It should be sufficient to ensure the utility can maintain its creditworthiness and attract the necessary capital to fulfill its obligations. Additionally, rate schedules must comply with legal requirements to avoid unjust and unreasonable discrimination. This means that customers who use the utility's services under similar conditions should be billed at similar prices. Categorizing customers into classes and subclasses is necessary, but within each class, all customers should be treated equally.

Utilities employ various types of rate structures, including but not limited to the following:

1. Flat demand rate structure
2. Straight-line meter rate structure
3. Block meter rate structure
4. Demand rate structure
5. Season rate structure
6. Time-of-day (or peak-load pricing) structure

The flat rate structure establishes a fixed price per kWh that remains constant regardless of the time of use, season, or volume. Metering is not necessary as the rate is negotiated based on the connected load. This type of structure is commonly used for services like parking lot or street lighting.

Similar to the flat structure, the straight-line meter rate structure also provides a single price per kWh without considering customer demand costs.

The block meter rate structure offers tiered pricing, where the price per kWh decreases for higher levels of usage. However, this structure does not actively encourage energy conservation or off-peak usage. As a result, it can lead to excessive peak demand and idle generation capacity, resulting in higher rates to cover the cost of peak-load capacity.

The demand rate structure takes into account the load factor and charges separately for demand and energy usage. It may include a constant price per kWh or decreasing prices for successive blocks of energy used.

The seasonal rate structure sets higher prices per kWh during the on-peak season when the system experiences its peak load and lower prices during the off-peak season when usage is lowest.

The time-of-day rate structure, also known as peak-load pricing, is similar to the seasonal rate structure. It applies higher prices per kWh during the peak period of the day and lower prices during the off-peak period. Both the seasonal and time-of-day rate structures aim to reduce the system's peak load and minimize idle standby capacity.

2.6.1 CUSTOMER BILLING

Customer billing is conducted by measuring the change in meter readings between two consecutive periods, typically occurring monthly. This change reflects the amount of electricity consumed by the customer during that specific interval, measured in kWh. The consumption value is then multiplied by the relevant rate or rates, as well as any applicable adjustment factors, to calculate the customer's bill. Finally, the bill is issued and sent to the customer for payment.

Figure 2.14 shows a typical monthly bill rendered to a residential customer. The monthly bill includes the following items in the indicated spaces:

1. The customer's account number.
2. A code showing which of the rate schedules was applied to the customer's bill.
3. A code showing whether the customer's bill was estimated or adjusted.
4. Date on which the billing period ended.
5. Number of kWh the customer's meter registered when the bill was tabulated.
6. Itemized list of charges. In this case, the only charge shown in box 6 of Figure 2.14 is a figure determined by adding the price of the electricity the customer has used to the routine taxes and surcharges. However, had the customer received some special service during this billing period, a service charge would appear in this space separately.

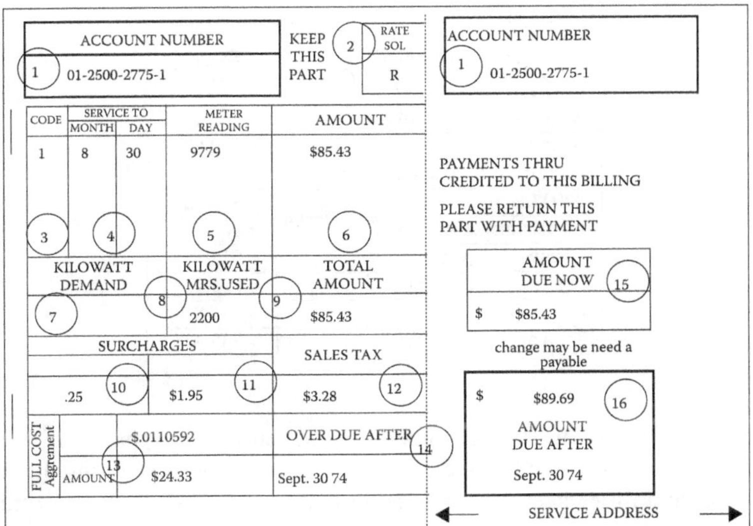

Figure 2.14 A customer's monthly electric bill.

7. Information appears in this box only when the bill is sent to a nonresidential customer using more than 6000 kWh electricity a month.
8. The number of kWh the customer used during the billing period.
9. Total amount that the customer owes.
10. Environmental surcharge.
11. County energy tax.
12. State sales tax.
13. Fuel cost adjustment. Both the total adjustment and the adjustment per kWh are shown.
14. Date on which bill, if unpaid, becomes overdue.
15. Amount due now.
16. Amount that the customer must pay if the bill becomes overdue.

The illustrated example of an electrical bill, depicted in Figure 2.14, is generated according to the following rate structure. It is important to note that a minimum charge is applicable regardless of the customer's electricity consumption, and the initial 20 kWh is covered by a flat rate. The minimum or service charge encompasses the expenses associated with providing service to the customer, including metering, meter reading, billing, and other overhead costs.

Rate Schedule

Minimum Charge (Including First 20 kWh or Fraction Thereof)	$2.25/Month
Next 80 kWh	$0.0355/kWh
Next 100 kWh	$0.0321/kWh
Next 200 kWh	$0.0296/kWh
Next 400 kWh	$0.0265/kWh
Consumption in excess of 800 kWh	$0.0220/kWh

The sample bill shows a consumption of 2200 kWh, which has been billed according to the following schedule:

First 20 kWh @ $2.25 (flat rate)	= $2.25
Next 80 kWh × 0.0355	= $2.84
Next 100 kWh × 0.0321	= $3.21
Next 200 kWh × 0.096	= $5.92
Next 400 kWh × 0.0265	= $10.60
Additional 1400 kWh × 0.0220	= $30.80
2200 kWh	= $55.62
Environmental surcharge	= $0.25
County energy tax	= $1.95
Fuel cost adjustment	= $24.33
State sales tax	= $3.28
Total amount	= $85.43

The utility company applies its rate schedule to bill the customer. Typically, the rates are subject to seasonal variations. In many regions, the demand for electricity rises during the warmer months. Consequently, electric utilities need to utilize additional generators, which are often less efficient and more costly to operate, in order to meet the increased demand. As an illustration, Table 2.3 presents a standard energy rate schedule for commercial users during the on-peak and off-peak seasons.

Table 2.3
Typical Energy Rate Schedule for Commercial Users

On-peak season (June 1–October 31)	
First 50 kWh or less/month for	\$4.09
Next 50 kWh/month	@5.5090/kWh
Next 500 kWh/month	@4.8430/kWh
Next 1400 kWh/month	@4.0490/kWh
Next 3000 kWh/month	@3.8780/kWh
All additional kWh/month	@3.3390/kWh
Off-peak season (November 1–May 31)	
First 50 kWh or less/month for	\$4.09
Next 50 kWh/month	@5.5090/kWh
Next 500 kWh/month	@4.2440/kWh
Next 1400 kWh/month	@3.1220/kWh
Next 3000 kWh/month	@2.7830/kWh
All additional kWh/month	@2.6490/kWh

2.6.2 FUEL COST ADJUSTMENT

The rates mentioned earlier are determined by the average cost, in dollars per million Btu, of the fuel used in NL&NP's thermal power plants. The monthly bill, calculated based on the aforementioned rates, is adjusted up or down for each kWh consumed. The adjustment amount is calculated using the following formula:

$$\text{FCAF} = A \times \frac{B}{10^6} \times C \times \frac{1}{1-D} \tag{2.43}$$

where

FCAF represents the fuel cost adjustment factor, measured in \$/kWh, which is applied to each consumed kWh.

A represents the weighted average Btu per kWh for net generation from NL&NP's thermal power plants during the second calendar month before the end of the billing period.

B represents the difference between the average cost of fuel per million Btu during the second calendar month before the end of the billing period and \$1/million Btu.

C represents the ratio, expressed as a decimal, of the total net generation from all of NL&NP's thermal power plants during the second calendar month before the end of the billing period to the total net generation from all of NL&NP's power plants, including hydro generation owned by NL&NP or purchased by NL&NP from other sources, during the same period.

D represents the loss factor, which is the ratio, expressed as a decimal, of kWh losses (total kWh losses minus losses associated with off-system sales) to net system input for the year ending on December 31 of the previous year. This ratio is updated annually and applied for a period of 12 months.

Example 2.13. Assume that the NL&NP Utility Company has the following, and typical, commercial rate schedule:

1. Monthly billing demand = 30-min monthly maximum kW demand multiplied by the ratio of (0.85/monthly average PF). The PF penalty shall not be applied when the consumer's monthly average PF exceeds 0.85.
2. Monthly demand charge = \$2.00/kW of monthly billing demand.

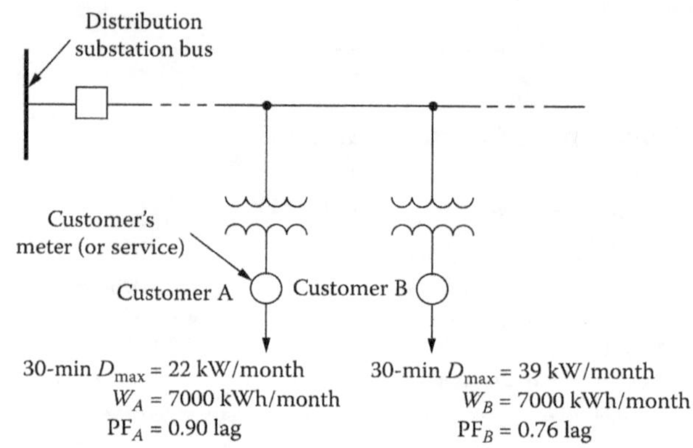

Figure 2.15 Two customers connected to a primary line of the NL&NP.

3. Monthly energy charges shall be as follows:
   2.50 cents/kWh for the first 1000 kWh
   2.00 cents/kWh for the next 3000 kWh
   1.50 cents/kWh for all kWh in excess of 4000
4. The total monthly charge shall be the sum of the monthly demand charge and the monthly energy charge.

Assume that two consumers, as shown in Figure 2.15, each requiring a DT, are supplied from a primary line of the NL&NP.

a. Assume that an average month is 730 h and find the monthly load factor of each consumer.
b. Find a reasonable size, that is, continuous kVA rating, for each DT.
c. Calculate the monthly bill for each consumer.
d. It is not uncommon to measure the average monthly PF on a monthly energy basis, where both kWh and kVAr hours are measured. On this basis, what size capacitor, in kVAr, would raise the PF of customer B to 0.85?
e. Secondary-voltage shunt capacitors, in small sizes, may cost about \$30/kVAr installed with disconnects and short-circuit protection. Consumers sometimes install secondary capacitors to reduce their billings for utility service. Using the 30/kVAr figure, find the number of months required for the PF correction capacitors found in part (d) to pay back for themselves with savings in demand charges.

Solution. a. From Equation 2.7, the monthly load factors for each consumer are the following. For customer A,

$$F_{LD} = \frac{\text{units served}}{\text{Peak load} \times T}$$

$$= \frac{7000\ \text{kWh}}{22\ \text{kW} \times 730\ \text{h}}$$

$$= 0.435$$

and for customer B,

$$\begin{aligned} F_{LD} &= \frac{\text{units served}}{\text{Peak load} \times T} \\ &= \frac{7{,}000 \text{ kWh}}{39 \text{ kW} \times 730 \text{ h}} \\ &= 0.246 \end{aligned}$$

b. The continuous kVA for each DT are the following:

$$\begin{aligned} S_A &= \frac{P_A}{\cos\theta} \\ &= \frac{22 \text{ kW}}{0.90} \\ &= 24.4 \text{ kVA} \end{aligned}$$

and

$$\begin{aligned} S_B &= \frac{P_B}{\cos\theta} \\ &= \frac{39 \text{ kW}}{0.76} \\ &= 51.2 \text{ kVA} \end{aligned}$$

Therefore, the continuous sizes suitable for the DTs *A* and *B* are 25 and 50 kVA ratings, respectively.

c. The monthly bills for each customer are the following:
For customer A

$$\text{Monthly billing demand}^* = 22 \text{ kW} \times \frac{0.85}{0.90} \cong 22 \text{ kW}$$

$$\text{Monthly demand charge} = 22 \text{ kW} \times \$2.00/\text{kW} \cong \$44$$

Monthly energy charge

First 1000 kWh = \$0.025/kWh × 1000 kWh	= \$25
Next 3000 kWh = \$0.02/kWh × 3000 kWh	= \$60
Excess kWh = \$0.015/kWh × 3000 kWh	= \$45
Monthly energy charge	= \$130

Therefore,

$$\begin{aligned} \text{Total monthly bill} &= \text{Monthly demand charge} + \text{Monthly energy charge} \\ &= \$44 + \$130 = \$174 \end{aligned}$$

For customer B

$$\text{Monthly billing demand}^* = 39 \text{ kW} \times \frac{0.85}{0.76} = 43.6 \text{ kW}$$

$$\text{Monthly demand charge} = 43.6 \text{ kW} \times \$2.00/\text{kW} = \$87.20$$

Monthly energy charge

First 20 kWh @ \$2.25 (flat rate)	= \$2.25
Next 80 kWh × 0.0355	= \$2.84
Next 100 kWh × 0.0321	= \$3.21
Next 200 kWh × 0.096	= \$5.92
Next 400 kWh × 0.0265	= \$10.60
Additional 1400 kWh × 0.0220	= \$30.80
2200 kWh	= \$55.62
Environmental surcharge	= \$0.25
County energy tax	= \$1.95
Fuel cost adjustment	= \$24.33
State sales tax	= \$3.28
Total amount	= \$85.43

Therefore,

$$\text{Total monthly bill} = \$87.20 + \$130 \\ = \$217.20$$

d. Currently, customer B at the lagging PF of 0.76 has

$$\frac{7000 \text{ kWh}}{0.76} \times \sin(\cos^{-1} 0.76) = 5986.13 \text{ kVArh}$$

If its PF is raised to 0.85, customer B would have

$$\frac{7000 \text{ kWh}}{0.85} \times \sin(\cos^{-1} 0.85) = 4338 \text{ kVArh}$$

Therefore, the capacitor size required is

$$\frac{5986.3 \text{ KVArh} - 4,338 \text{ KVArh}}{730 \text{ h}} = 2.258 \text{ kVARh} \cong 2.3 \text{ kVAr}$$

e. The new monthly bill for customer B would be as follows:

$$\text{Monthly billing demand} = 39 \text{ kW}$$

$$\text{Monthly demand charge} = 39 \text{ kW} \times \$2.00 = \$78$$

$$\text{Monthly energy charge} = \$130 \text{ as before}$$

Therefore,

$$\text{Total monthly bill} = \$78 + \$130 \\ = \$208$$

Hence, the resultant savings due to the capacitor installation is the difference between the before-and-after total monthly bills. Thus,

$$\begin{aligned}\text{Savings} &= \$217.20 - \$208 \\ &= \$9.20/\text{month}\end{aligned}$$

or

$$\begin{aligned}\text{Savings} &= \$87.20 - \$78 \\ &= \$9.20/\text{month}\end{aligned}$$

The cost of the installed capacitor is

$$\$30/\text{kVAr} \times 2.3\text{kVAr} = \$69$$

Therefore, the number of months required for the capacitors to "payback" for themselves with savings in demand charges can be calculated as

$$\begin{aligned}\text{Payback period} &= \frac{\text{Capacitor cost}}{\text{Savings}} \\ &= \frac{\$69}{\$9.20/\text{month}} \\ &= 7.5 \\ &\cong 8 \text{ months}\end{aligned}$$

However, in practice, the available capacitor size is 3 kVAr instead of 2.3 kVAr. Therefore, the realistic cost of the installed capacitor is

$$\$30/\text{kVAr} \times 3\text{kVAr} = \$90$$

Therefore,

$$\begin{aligned}\text{Payback period} &= \frac{\$90}{\$9.20/\text{month}} \\ &\cong 10 \text{ months}\end{aligned}$$

PROBLEMS

2.1 Use the data given in Example 2.1 and assume that the feeder has a peak loss of 72 kW at peak load and an annual loss factor of 0.14. Determine the following:

a. The daily average load of the feeder
b. The average power loss of the feeder
c. The total annual energy loss of the feeder

2.2 Use the data given in Example 2.1 and the equations given in Section 2.2 and determine the load factor of the feeder.

Table 2.4
Typical Summer-Day Load, in kW

Time	Street Lighting	Residential	Commercial
12:00 AM	100	250	300
1	100	250	300
2	100	250	300
3	100	250	300
4	100	250	300
5	100	250	300
6	100	250	300
7		350	300
8		450	400
9		550	600
10		550	1100
11		550	1100
12 noon		600	1100
1		600	1100
2		600	1300
3		600	1300
4		600	1300
5		650	1300
6		750	900
7		900	500
8	100	1100	500
9	100	1100	500
10	100	900	300
11	100	700	300
12:00 PM	100	350	300

2.3 Use the data given in Example 2.1 and assume that the connected demands for the street lighting load, the residential load, and the commercial load are 100, 200, and 2000 kW, respectively. Determine the following:
  a. The DF of the street lighting load
  b. The DF of the residential load
  c. The DF of the commercial load
  d. The DF of the feeder

2.4 Using the data given in Table 2.4 for a typical summer day, repeat Example 2.1 and compare the results.

2.5 Use the data given in Problem 2.4 and repeat Problem 2.2.

2.6 Use the data given in Problem 2.4 and repeat Problem 2.3.

2.7 Use the result of Problem 2.2 and calculate the associated loss factor.

2.8 Assume that a load of 100 kW is connected at the Riverside substation of the NL&NP Company. The 15-min weekly maximum demand is given as 75 kW, and the weekly energy consumption is 4200 kWh. Assuming a week is 7 days, find the DF and the 15-min weekly load factor of the substation.

2.9 Assume that the total kVA rating of all DTs connected to a feeder is 3000 kVA. Determine the following:
  a. If the average core loss of the transformers is 0.50%, what is the total annual core loss energy on this feeder?
  b. Find the value of the total core loss energy calculated in part (a) at $0.025/kWh.

2.10 Use the data given in Example 2.6 and also consider the following added new load. Suppose that several buildings that have electric air-conditioning are converted from gas-fired heating to electric heating. Let the new electric heating load average 200 kW during 6 months of heating (and off-peak) season. Assume that off-peak energy delivered to these primary feeders costs the NL&NP Company 2 cents/kWh and that the capacity cost of the power system remains at $3.00/kW per month.

a. Find the new annual load factor on the substation.
b. Find the total annual cost to NL&NP to serve this new load.
c. Why is it that the hypothetical but illustrative energy cost is smaller in this problem than the one in Example 2.8?

2.11 The input to a subtransmission system is 87,600,000 kWh annually. On the peak-load day of the year, the peak is 25,000 kW and the energy input that day is 300,000 kWh. Find the load factors for the year and for the peak-load day.

2.12 The electric energy consumption of a residential customer has averaged 1150 kWh/month as follows, starting in January: 1400, 900, 1300, 1200, 800, 700, 1000, 1500, 700, 1500, 1400, and 1400 kWh. The customer is considering purchasing equipment for a hobby shop that he has in his basement. The equipment will consume about 200 kWh each month. Estimate the additional annual electric energy cost for operation of the equipment. Use the electrical rate schedule given in the following table.

Residential

Rate: (net) per month per meter

Energy Charge	
For the first 25 kWh	6.00 ¢/kWh
For the next 125 kWh	3.2 ¢/kWh
For the next 850 kWh	2.00 ¢/kWh
All in excess of 1000 kWh	1.00 ¢/kWh
Minimum: $1.50 per month	

Commercial

A rate available for general, commercial, and miscellaneous power uses where consumption of energy does not exceed 10,000 kWh in any month during any calendar year.

Rate: (net) per month per meter

Energy Charge	
For the first 25 kWh	6.0 ¢/kWh
For the next 375 kWh	4.0 ¢/kWh
For the next 3600 kWh	3.0 ¢/kWh
All in excess of 4000 kWh	1.5 ¢/kWh
Minimum: $1.50 per month	

General power

A rate available for service supplied to any commercial or industrial customer whose consumption in any month during the calendar year exceeds 10,000 kWh. A customer who exceeds 10,000 kWh per month in any 1 month may elect to receive power under this rate.

A customer who exceeds 10,000 kWh in any 3 months or who exceeds 12,000 kWh in any 1 month during a calendar year shall be required to receive power under this rate at the option of the supplier.

A customer who elects at his own option to receive power under this rate may not return to the commercial service rate except at the option of the supplier.
Rate: (net) per month per meter kW is rate of flow. 1 kW for 1 h is 1 kWh.

Demand charge

For the first 30 kW of maximum demand per month	\$2.50/kW
For all maximum demand per month in excess of 30 kW	\$1.25/kW

Energy Charge	
For the first 100 kWh per kW of maximum demand per month	2.00 ¢/kWh
For the next 200 kWh per kW of maximum demand per month	1.2 ¢/kWh
All in excess of 300 kWh per kW of maximum demand per month	0.5 ¢/kWh

Minimum charge: The minimum monthly bill shall be the demand charge for the month.
Determination of maximum demand: The maximum demand shall be either the highest integrated kW load during any 30-min period occurring during the billing month for which the determination is made or 75% of the highest maximum demand that has occurred in the preceding month, whichever is greater.
Water heating: 1.00/kWh with a minimum monthly charge of \$1.00.

2.13 The Zubits International Company, located in Ghost Town, consumed 16,000 kWh of electric energy for Zubit production this month. The company's monthly average energy consumption is also 16,000 kWh due to some unknown reasons. It has a 30-min monthly maximum demand of 200 kW and a connected demand of 580 kW. Use the electrical rate schedule given in Problem 2.12.

a. Find the Zubits International's total monthly electrical bill for this month.
b. Find its 30-min monthly load factor.
c. Find its DF.
d. The company's newly hired plant engineer, who recently completed a load management course at Ghost University, suggested that, by shifting the hours of a certain production from the peak-load hours to off-peak hours, the maximum monthly demand can be reduced to 140 kW at a cost of \$50/month. Do you agree that this will save money? How much?

2.14 Repeat Example 2.12, assuming that there are eight houses connected on each DT and that there are a total of 120 DTs and 960 residences supplied by the primary feeder.

2.15 Assuming that the 30-min monthly maximum demands of customers A and B are 27 and 42 kW, respectively. The new monthly energy consumptions by customers A and B are 8000 and 9000 kWh, respectively. The new lagging load PFs of A and B are 0.90 and 0.70, respectively.

2.16 A customer transformer has 12 residential customers connected to it. Connected load is 20 kW per house, DF is 0.6, and diversity factor is 1.15. Find the diversified demand of the group of 12 houses on the transformer.

2.17 A distribution substation is supplied by total annual energy of 100,000 MWh. If its annual average load factor is 0.6, determine the following:

a. The annual average power demand
b. The maximum monthly demand

2.18 Suppose that one of the transformers of a substation supplies four primary feeders. Among the four feeders, the diversity factor is 1.25 for both real power and reactive power. The 30-min annual demands of per feeder with their PFs at the time of annual peak load are shown as follows:

Feeder	Demand (kW)	PF
1	900	0.85
2	1000	0.9
3	2100	0.95
4	2000	0.9

a. Find the 30-min annual maximum demand on the substation transformer in kW and in kVA.
b. Find the load diversity in kW.
c. Select a suitable substation transformer size if zero load growth is expected and company policy permits as much as 25% short-time overloads on the transformer. Among the standard three-phase transformer sizes available are the following:
2500/3125 kVA self-cooled/forced-air-cooled
3750/4687 kVA self-cooled/forced-air-cooled
5000/6250 kVA self-cooled/forced-air-cooled
7500/9375 kVA self-cooled/forced-air-cooled
d. Now assume that the substation load will increase at a constant percentage rate per year and will double in 10 years. If 7500/9375 kVA-rated transformer is installed, in how many years will it be loaded to its *fans-on* rating?

2.19 Suppose that a primary feeder is supplying power to a variable load. Every day and all year long, the load has a daily constant peak value of 50 MW between 7 PM and 7 AM and a daily constant off-peak value of 5 MW between 7 AM and 7 PM. Derive the necessary equations and calculate the following:

a. The load factor of the feeder
b. The factor of the feeder

2.20 A typical DT serves four residential loads, that is, houses, through six SDs and two spans of SL. There are a total of 200 DTs and 800 residences supplied by this primary feeder. Use Figure 2.13 and Table 2.2 and assume that a typical residence has a clothes dryer, a range, a refrigerator, and some lighting and miscellaneous appliances. Determine the following:

a. The 30-min maximum diversified demand on the transformer
b. The 30-min maximum diversified demand on the entire feeder
c. Use the typical hourly variation factors given in Table 2.2 and calculate the small portion of the daily demand curve on the DT, that is, the total hourly diversified demands at 6 AM, 12 noon, and 7 PM, on the DT, in kW

2.21 Assuming that the monthly demand charge is \$15/kW and that the monthly energy charges are 12 cents/kWh for the first 1000 kWh, 10 cents/kWh for the next 3000 kWh, and 8 cents/kWh for all kWh in excess of 4000. The 30-min maximum diversified demands for customers A and B are 40 kW each. The PFs for customers A and B are 0.95 lagging and 0.50 lagging, respectively.

2.22 The annual peak load of the feeder is 3000 kWh. Total copper loss at peak load is 300 kW. If the total annual energy supplied to the sending end of the feeder is 9000 MWh, determine the following:

a. The annual loss factor for an urban area
b. The annual loss factor for a rural area
c. The total amount of energy lost due to copper losses per year in part (a) and its value at \$0.06/kWh
d. The total amount of energy lost due to copper losses per year in part (b) and its value at \$0.06/kWh.

3 Application of Distribution Transformers

Now that I am almost up the ladder,
I should, no doubt, be feeling gladder,
It is quite fine, the view and such,
If just it did not shake so much.

Richard Armour

3.1 INTRODUCTION

Previous chapter introduces how consumption of load is characterized. This chapter connects the electrical loads with distribution transformers. Each feeder would have hundreds of distribution transformers (DTs), serving as the electrical loads in primary distribution network. All DTs can be connected in any combination of phases, e.g., phase ab, or phase c, which results in unbalanced three-phase circuit. In general, DTs are used to reduce primary system voltages (2.4–34.5 kV) to utilization voltages (120–60 V). Table 3.1 gives standard transformer capacity and voltage ratings according to ANSI Standard C57.12.20-1964 for single-phase DTs. Other voltages are also available, for example, 2400 × 7200, which is used on a 2400 V system that is to be changed later to 7200 V. Secondary symbols used are the letter Y, which indicates that the winding is connected or may be connected in wye, and Gnd Y, which indicates that the winding has one end grounded to the tank or brought out through a reduced insulation bushing. Windings that are delta-connected or may be connected delta are designated by the voltage of the winding only.

In Table 3.1, further information is given by the order in which the voltages are written for low-voltage (LV) windings. To designate a winding with a mid-tap that will provide half the full-winding kVA rating at half the full-winding voltage, the full-winding voltage is written first, followed by a slant, and then the mid-tap voltage. For example, 240/120 is used for a three-wire connection to designate a 120 V mid-tap voltage with a 240 V full-winding voltage. A winding that is appropriate for series, multiple, and three-wire connections will have the designation of multiple voltage rating followed by a slant and the series voltage rating; for example, the notation 120/240 means that the winding is appropriate either for 120 V multiple connection, for 240 V series connection, and for 240/120 three-wire connection. When two voltages are separated by a cross ×, a winding that is appropriate for both multiple and series connections but not for three-wire connection is indicated. The notation 120 × 240 is used to differentiate a winding that can be used for 120 V multiple connection and 240 V series connection, but not a three-wire connection. Examples of all symbols used are given in Table 3.2.

To minimize installation costs, small distribution transformers (DTs) are designed for pole mounting in overhead distribution systems. Preferred oriented steel is commonly used in their construction to reduce size and weight. DTs with a capacity of 100 kVA and below are directly attached to the pole, while those larger than 100 kVA up to 500 kVA are hung on crossbeams or support lugs. If three or more transformers larger than 100 kVA are used, they are installed on a platform supported by two poles. In underground distribution systems, transformers can be installed in street vaults, direct-buried in manholes, placed on pads at ground level, or housed within buildings. The choice of transformer type may depend on factors such as soil content, lot location, public acceptance, or cost considerations. DTs, along with any secondary service junction devices, are typically installed within enclosures placed on either the front or rear lot lines of the customer's premises.

DOI: 10.1201/9781003129721-3

Table 3.1
Standard Transformer kVA and Voltages

kVA		High Voltages		Low Voltages	
Single Phase	**Three Phase**	**Single Phase**	**Three Phase**	**Single Phase**	**Three Phase**
5	30	2,400/4,160 Y	2,400	120/240	208 Y/120
10	45	4,800/8,320 Y	4,160 Y/2,400	240/480	240
15	75	4,800Y/8,320 YX	4,160 Y	2400	480
25	112%	7,200/12,470 Y	4,800	2520	480 Y/277
	150	12,470 Gnd Y/7,200	8,320 Y/4,800	4800	240 × 480
50	225	7,620/13,200 Y	8,320 Y	5040	2,400
75	300	13,200 Gnd Y/7,620	7,200	6900	4,160 Y/2,400
100	500	12,000	12,000	7200	4,800
167		13,200/22,860 Gnd Y	12,470 Y/7,200	7560	12,470 Y/7,200
250		13,200	12,470 Y	7980	13,200 Y/7,620
333		13,800 Gnd Y/7,970	13,200 Y/7,620		
500		13,800/23,900 Gnd Y	13,200 Y		
		13,800	13,200		
		14,400/24,940 Gnd Y	13,800		
		16,340	22,900		
		19,920/34,500 Gnd Y	34,400		
		22,900	43,800		
		34,400	67,000		
		43,800			
		67,000			

Table 3.2
Designation of Voltage Ratings for Single- and Three-Phase Distribution Transformers

Single Phase		Three Phase	
Designation	**Meaning**	**Designation**	**Meaning**
120/240	Series, multiple, or three-wire connection	2,400/4,160 Y	Suitable for delta or wye connection
240/120	Series or three-wire connection only	4,160 Y	Wye connection only (no neutral)
240 × 480	Series or multiple connection only	4,160 Y 2,400	Wye connection only (with neutral available)
120/208 Y	Suitable for delta or wye connection three phase	12,470 Gnd Y/7,200	Wye connection only (with reduced insulation neutral available)
12,470 Gnd Y/7,200	One end of winding grounded to tank or brought out through reduced insulation bushing	4,160	Delta connection only

The specific location for equipment installation may be subject to limitations imposed by customer preference, local ordinances, landscape conditions, and other factors. As a general guideline, transformers should be centrally located with respect to the load they supply in order to optimize cable efficiency, minimize voltage drop, and maintain aesthetic considerations.

For underground distribution systems, secondary service junctions can be in the form of pedestals, hand-holes, or direct-buried splices. If the service cables are directly connected from the DT to the user's apparatus, no junction is required. Secondary or service conductors can be made of either copper or aluminum, although aluminum conductors are commonly used for cost savings. The cables can be single conductor or triplex conductors. Neutrals can be either bare or covered and may be installed separately or assembled with the power conductors. All secondary or service conductors are rated for 600 V, and their sizes range from #6 AWG to 1000 kcmil.

3.2 STATE-OF-THE-ART UPDATE

Transformers modeled in distribution system are often referred to the step-down elements between medium-voltage (MV) and low-voltage (LV) systems, i.e., the distribution transformers (DTs). The input data relies on ambient temperature and loading condition of the individual transformer. This is a balance risk of transformer loading against operational benefits in utilizing the devices [54]. Hence, the power loss reduction is a crucial indicator to establish the parameters and variables through SCADA communication interface [55]. Some databases from the engineering division would provide detailed information for DT to be connected geographically where this can link with SCADA network to determine load pattern and utilization of DTs [56]. The definition of losses has been extended with varying economic impacts, relying on quality of service on those DTs, cost of service, as well as overall performance for the estimation [57]. The grounding connection of connected DTs with the DERs can impact upgrades on the protective relaying of out of step duty or transfer trip due to transient overvoltage conditions [58]. The active power filter associated with a transformer is verified that exhibits close to zero impedance to harmonic current and self-impedance [59], including the modernization by reducing arc hazards [60]. The center-tapped DTs have been modeled for backward-forward exchange within the element, dealing with the non-linearity of the feeder [61]. For the power transformers in a distribution substation, the on-load tap changers can be formulated as a non-linear multi-objective linearized optimization with mixed-integer linear programming (MILP) algorithms, depending with renewable penetration [62]. Another application of power transformers in substation is that is required to compensate voltage changes along the feeder [63]. The impedance information of the feeder circuits can be useful to analyze unbalance as a re-phasing strategy to identify issues of phase relevance [64].

An increased number of EV charging on distribution network could potentially affect the life cycle of DTs that are associated with the households [65]. To model the DTs accurately, a complex theory as stationary reference frame is established to address the unbalanced nature of power flow [66]. From the electronics standpoint, design emphasizes on single-phase line frequency transformer have been applied for photovoltaic systems, including the effects of the solar panels on DTs [67,68]. Extensive research has been brought up on voltage volatility among shared customers on a single DT [69]. Much research in mitigation of transformer loss of life and phase rebalancing incurred by single-phase loads [70, 71]. The emerging technology in solid-state modeling for the dynamic feasibility and deployment challenges has been studied [72, 73].

3.3 TYPES OF DISTRIBUTION TRANSFORMERS

Heat dissipation is a critical consideration when it comes to the loading of transformers, as excessive heat can be detrimental to their operation [74–79]. In the case of liquid-filled transformers, the coils are immersed in an oil-filled tank that has a smooth surface. The oil absorbs the heat generated by the coils and transfers it to the tank's surface, which then facilitates its dissipation into the surrounding air. However, for transformers with a capacity of 25 kVA and larger, the required surface area of the tank to effectively dissipate heat exceeds that needed to enclose the coils alone. As a solution, transformer tanks may be designed with corrugations to increase the surface area, or external tubes can be welded onto the tank. To further enhance heat dissipation, forced-air cooling may be employed by blowing air over the tube surface. This design approach, known as forced-air-cooled, differs from self-cooled types. However, it's worth noting that presently, all distribution transformers are constructed to be self-cooled, eliminating the need for additional forced-air cooling mechanisms.

Hence, distribution transformers can be categorized into two main types: (1) dry-type transformers and (2) liquid-filled transformers. Dry-type transformers are air-cooled and rely on air insulation. On the other hand, liquid-filled transformers can be further divided into two subtypes: (1) oil-filled transformers and (2) inerteen-filled transformers.

The DTs employed in overhead distribution systems can be categorized as:

1. Conventional transformers
2. Completely self-protecting (CSP) transformers
3. Completely self-protecting for secondary banking (CSPB) transformers

Conventional transformers do not come equipped with built-in protection against lightning, faults, or overloads. However, Current-Limiting Self-Protecting (CSP) transformers are specifically designed to offer self-protection from lightning strikes, line surges, overloads, and short circuits. These transformers are equipped with lightning arresters mounted directly on the transformer tank to safeguard the primary winding against lightning and line surges. Overload protection is provided by circuit breakers located inside the transformer tank. Furthermore, internal protective links positioned between the primary winding and primary bushing protect the transformer against internal faults. Single-phase CSP transformers are available for various primary voltages ranging from 2,400 to 34,400 V, with secondary voltages of 120/240 or 240/480/277 V. CSPB (Current-Limiting Self-Protecting Bank) DTs are designed for banked secondary service and resemble CSP transformers, but they feature an additional set of circuit breakers used for sectionalizing the secondary when required.

The transformers used in underground distribution systems can be classified as follows:

1. Subway transformers
2. Low-cost residential transformers
3. Network transformers

Subway transformers are utilized in underground vaults and can be classified into two types: conventional transformers and current-protected transformers. Low-cost residential transformers resemble conventional transformers commonly used in overhead distribution. Network transformers, on the other hand, are specifically designed for secondary networks. These transformers feature an integrated primary disconnecting and grounding switch, as well as a network protector mounted directly on the transformer. They are available in different configurations, including liquid-filled, ventilated dry type, or sealed dry type.

ABB Corporation has developed resibloc dry-type transformers in a wide range of capacities, from 112.5 to 25,000 kVA, with primary voltage levels ranging from 2300 to 34.5 kV and secondary voltage levels from 120 V to 15 kV. These transformers employ a unique construction method, where the windings are hermetically cast in epoxy without the use of molds. The epoxy insulation system is reinforced using a special glass fiber roving technique, resulting in a cohesive winding block structure. As a result, these transformers possess exceptional mechanical strength, and their design has been optimized for capacities up to 25,000 kVA.

The transformers are designed to withstand a temperature rise of 80°C, providing a high level of safety by eliminating the risk of fire and explosion, as well as the potential for liquid leaks. This characteristic significantly reduces the need for maintenance. These transformers can be used in both indoor and outdoor enclosures. The resibloc transformers are engineered to be non-ignitable during electrical arcs of nominal duration. Even if exposed to a direct flame, the resibloc material will self-extinguish upon removal of the flame. These transformers are suitable for use in harsh indoor and outdoor environments. However, it is important to note that while the core and coil technologies have been improved to withstand caustic and humid conditions, resibloc transformers still require proper enclosure protection.

It is crucial that the enclosure for these transformers is designed appropriately to ensure electrical clearances are maintained, even under high wind loading that may cause the enclosure to flex or bend. Failure to maintain adequate clearances can result in transformer failures and electrical safety hazards. Additionally, the design of the enclosure must prevent excess water entry, particularly in coastal areas or regions with heavy rain, to mitigate associated risks. Forced air cooling can be incorporated into the design of these transformers, and temperature sensors located in the low-voltage windings are connected to a three-phase winding temperature monitor, which automatically controls the operation of the forced air cooling system.

Table 3.3 provides electrical characteristics of typical single-phase DTs, while Table 3.4 presents electrical characteristics of typical three-phase pad-mounted transformers. It is important to note that for more precise values, it is recommended to refer to the individual manufacturer's catalogs.

To determine the resistance (R') and reactance (X') of a transformer of the same size and voltage, but with a different impedance value (Z'), the tabulated percentage values of R and X should be multiplied by the ratio of the new impedance value to the tabulated impedance value, which is represented as Z'/Z. This calculation allows for finding the resistance and reactance of the new transformer using the following equations:

$$R' = R \times \frac{Z'}{Z} \% \Omega \tag{3.1}$$

and

$$X' = X \times \frac{Z'}{Z} \% \Omega \tag{3.2}$$

3.4 REGULATION

To determine the transformer regulation for a kVA load with a power factor of cos θ at rated voltage, you can utilize any of the following formulas:

$$\%\text{regulation} = \frac{S_L}{S_T}\left[\%IR\cos\theta + \%IX\sin\theta + \frac{(\%IX\cos\theta - \%IR\sin\theta)^2}{200}\right] \tag{3.3}$$

or

$$\%\text{regulation} = \frac{I_{op}}{I_{ra}}\left[\%R\cos\theta + \%X\sin\theta + \frac{(\%X\cos\theta - \%R\sin\theta)^2}{200}\right] \tag{3.4}$$

or

$$\%\text{regulation} = V_R\cos\theta + V_X\sin\theta + \frac{(V_X\cos\theta - V_R\sin\theta)^2}{200} \tag{3.5}$$

Table 3.3
Electrical Characteristics of Typical Single-Phase Distribution Transformers

			Watts Loss			120/240 V Low Voltage % Regulation				Watts Loss			240/480 and 277/480 Y V Low Voltage % Regulation		
kVA	Percent of Av. Excit. Curr.	No Load	Total	1.0 PF	0.8 PF	% Z	%R	% X	No Load	Total	1.0 PF	0.8 PF	% Z	% R	% X
						2400/4160 Y V high voltage									
5	2.4	34	137	2.06	2.12	2.2	2.1	0.8							
10	1.6	68	197	1.3	1.68	1.7	1.3	1.1	68	202	1.35	1.69	1.7	1.3	1
15	1.4	84	272	1.27	1.59	1.6	1.3	1	84	277	1.3	1.6	1.6	1.3	1.1
25	1.3	118	385	1.1	1.65	1.7	1.1	1.1	118	390	1.11	1.65	1.7	1.1	1.3
38	1.1	166	540	1	1.55	1.6	1	1.3	166	550	1.04	1.54	1.6	1	1.2
50	1	185	615	0.88	1.58	1.7	0.9	1.5	185	625	0.9	1.58	1.7	0.9	1.5
75	1.3	285	910	0.85	1.41	1.5	0.8	1.2	285	925	0.86	1.33	1.4	0.9	1.1
100	1.2	355	1175	0.84	1.55	1.7	0.8	1.5	355	1190	0.85	1.49	1.6	0.8	1.4
167	1	500	2100	0.99	1.75	1.9	1	1.6	500	2000	0.9	1.57	1.7	0.9	1.4
250	1	610	3390	1.16	2.16	2.4	1.1	2.1	610	3280	1.11	2.02	2.2	1.1	1.9
333	1	840	4200	1.08	2.51	3	1	2.8	840	3690	0.88	1.9	2.2	0.9	1.9
500	1	1140	5740	0.97	2.5	3.1	0.9	3	1140	4810	0.95	2	2.3	0.7	2.2
						7,200/12,470 Y V high voltage									
5	2.4	41	144	2.07	2.11	2.2	2.1	0.8							
10	1.6	68	204	1.37	1.8	1.8	1.4	1.2	68	209	1.43	1.8	1.8	1.4	1.1
15	1.4	84	282	1.33	1.69	1.7	1.3	1.2	84	287	1.35	1.7	1.7	1.4	1
25	1.3	118	422	1.22	1.69	1.7	1.2	1.2	118	427	1.24	1.69	1.7	1.2	1.2
38	1.1	166	570	1.1	1.64	1.7	1.1	1.3	166	575	1.1	1.65	1.7	1.1	1.3
50	1	185	720	1.1	1.71	1.8	1.1	1.4	185	725	1.1	1.71	1.8	1.1	1.4
75	1.3	285	985	0.95	1.6	1.7	0.9	1.4	285	1000	0.97	1.52	1.6	1	1.3
100	1.2	355	1275	0.95	1.72	1.9	1.9	1.7	355	1290	0.95	1.6	1.7	1.9	1.4
167	1	500	2100	0.98	1.9	2.1	1	1.9	500	2000	0.91	1.7	1.9	0.9	1.7
250	1	610	3490	1.22	2.45	2.8	1.2	2.6	610	3250	1.17	2.19	2.4	1.1	2.2
333	1	840	4255	1.07	2.5	3	1	2.8	840	3690	0.89	2.03	2.4	0.9	2.2
500	1	1140	5640	0.95	2.55	3.2	0.9	3.1	1140	4810	0.78	1.99	2.4	0.7	2.3

Table 3.3 (Continued)
Electrical Characteristics of Typical Single-Phase Distribution Transformers

kVA	Percent of Av. Excit. Curr.	No Load	Watts Loss Total	Watts Loss 1.0 PF	Watts Loss 0.8 PF	120/240 V Low Voltage % Regulation % Z	%R	% X	No Load	Watts Loss Total	Watts Loss 1.0 PF	Watts Loss 0.8 PF	240/480 and 277/480 Y V Low Voltage % Regulation % Z	% R	% X
13,200/22,860 Gnd Y or 13,800/23,900 Gnd Y or 14,400/24,940 Gnd Y V high voltage															
5	2.4	42	154	2.25	2.3	2.4	2.3	0.9							
10	1.6	73	215	1.45	1.89	1.9	1.4	1.3	73	220	1.49	1.89	1.9	1.5	1.2
15	1.4	84	305	1.48	1.8	1.8	1.5	1	84	310	1.52	1.8	1.8	1.5	1
25	1.3	118	437	1.29	1.79	1.8	1.3	1.3	118	442	1.3	1.78	1.8	1.3	1.2
38	1.1	166	585	1.15	1.72	1.8	1.1	1.4	166	590	1.16	1.72	1.8	1.1	1.4
50	1	185	735	1.14	1.81	1.9	1.1	1.4	185	740	1.15	1.81	1.9	1.1	1.5
75	1.4	285	1050	1.05	1.78	1.8	1	1.5	285	1065	1.06	1.78	1.8	1	1.5
100	1.3	355	1300	0.97	1.81	2	0.9	1.8	355	1310	0.98	1.74	1.9	1	1.6
167	1	500	2160	0.98	1.96	2.2	1	2	500	2060	0.95	1.8	2	0.9	1.8
250	1	610	3490	1.22	2.52	2.9	1.2	2.7	610	3285	1.11	2.16	2.5	1.1	2.3
333	1	840	4300	1.09	2.6	3.1	1	2.9	840	3750	0.91	2.05	2.4	0.9	2.2
500	1	1140	5640	0.95	2.55	3.2	1.1	3	1140	4760	0.76	1.98	2.4	0.7	2.3

Table 3.4
Electrical Characteristics of Typical Three-Phase Pad-Mounted Transformers

			Watts Loss			208 Y/120V Low Voltage % Regulation				Watts Loss			480Y/277V Low Voltage % Regulation		
kVA	Percent of Av. Excit. Curr.	No Load	Total	1.0 PF	0.8 PF	% Z	%R	% X	No Load	Total	1.0 PF	0.8 PF	% Z	% R	% X
4,160 Gnd Y/2,400 X12,470 Gnd Y/7,200 V high voltage															
75	1.5	360	1,350	1.35	2.1	2.1	1.3	1.6	360	1,350	1.35	2.1	2.1	1.3	1.6
112	1	530	1,800	1.15	1.7	1.7	1.1	1.3	530	1,800	1.15	1.7	1.7	1.1	1.3
150	1	560	2,250	1.15	1.9	1.9	1.1	1.6	560	2,250	1.15	1.9	1.9	1.1	1.6
225	1	880	3,300	1.1	1.9	1.9	1.1	1.6	800	3,300	1.1	1.9	1.9	1.1	1.6
300	1	1050	4,300	1.1	1.9	2	1.1	1.7	1050	4,100	1.05	1.8	1.8	1	1.5
500	1	1600	6,800	1.15	2.2	2.3	1	2.1	1600	6,500	1.1	2	2	1	1.7
750	1	1800	10,200	1.28	4.4	5.7	1.1	5.6	1800	9,400	1.18	4.3	5.7	1	5.6
1000	1	2100	12,500	1.2	4.3	5.7	1	5.6	2100	10,900	1.04	4.2	5.7	0.9	5.7
1500	1	2900	19,400	1.26	4.3	5.7	1.1	5.6	3300	16,500	1.04	4.2	5.7	0.9	5.7
2500	1								4800	26,600	1.03	4.2	5.7	0.9	5.7
3750	1								6500	35,500	0.95	4.1	5.7	0.8	5.7
12,470 Gnd Y/7,200 V high voltage															
75	1.5	360	1,350	1.4	1.7	1.7	1.3	1.1	360	1,350	1.4	1.5	1.5	1.3	0.8
112	1	530	1,800	1.2	1.5	1.5	1.1	1	530	1,800	1.2	1.3	1.3	1.1	0.7
150	1	560	2,250	1.2	1.8	1.9	1.1	1.6	560	2,250	1.2	1.7	1.7	1.1	1.3
225	1	880	3,300	1.1	1.8	1.8	1.1	1.4	880	3,300	1.1	1.6	1.6	1.1	1.2
300	1	1050	4,300	1.1	1.6	1.6	1.1	1.2	1050	4,100	1.1	1.4	1.4	1	1
500	1	1600	6,800	1.1	1.7	1.7	1	1.4	1600	6,500	1.1	1.4	1.4	1	1
750	1	1800	10,200	1.3	4.4	5.7	1.1	5.6	1800	9,400	1.2	4.3	5.7	1	5.6
1000	1	2100	12,500	1.2	4.3	5.7	1	5.6	2100	10,900	1	4.2	5.7	0.9	5.7
1500	1	2900	19,400	1.3	4.3	5.7	1.1	5.6	3300	16,500	1	4.2	5.7	0.9	5.7
2500	1							4800	26,600	1	4.2	5.7	0.9	5.7	
3750	1							6500	35,500	0.9	4.1	5.7	0.8	5.7	

Table 3.4 (Continued)
Electrical Characteristics of Typical Three-Phase Pad-Mounted Transformers

	Percent of		Watts Loss			208 Y/120V Low Voltage % Regulation				Watts Loss			480Y/277V Low Voltage % Regulation		
kVA	Av. Excit. Curr.	No Load	Total	1.0 PF	0.8 PF	% Z	%R	% X	No Load	Total	1.0 PF	0.8 PF	% Z	% R	% X
						2400/4160 Y/2400 V low voltage 12,470 Delta V high voltage									
1000	1.38	2443	11,480	1.06	4.09	5.56	0.89	5.49							
1500	1.33	3455	15,716	0.98	4.04	5.56	0.81	5.51							
2500	1.29	4956	23,193	0.92	3.97	5.56	0.73	5.52							
3750	1.37	6775	33,100	0.89	3.97	5.5	0.7	5.45							
5000	1.33	8800	42,125	0.86	3.94	5.5	0.67	5.45							
						24,940 Delta V high voltage									
1000	1.42	2533	11,588	1.07	4.09	5.56	0.91	5.49							
1500	1.37	3625	15,213	0.96	4.03	5.56	0.8	5.5							
2500	1.31	5338	23,213	0.88	3.98	5.56	0.72	5.52							
3750	1.42	7075	33,700	0.9	3.97	5.5	0.71	5.44							
5000	1.33	8725	43,550	0.88	3.96	5.5	0.69	5.44							

where

θ is the power factor angle of load
$V_R = \frac{\text{copper loss}}{\text{output}} \times 100$ is the percent resistance voltage
S_L is the apparent load power
S_T is the rated apparent power of transformer
I_{op} is the operating current
I_{ra} is the rated current
$V_X = \sqrt{(V_Z^2 - V_R^2)}$ is the percent leakage reactance voltage
V_Z is the percent impedance voltage

Note that the percent regulation at unity power factor is

$$\%\text{Regulation} = \frac{\text{Copper loss}}{\text{Output}} \times 100 + \frac{(\%\text{reactance})^2}{200} \tag{3.6}$$

3.5 TRANSFORMER EFFICIENCY

The efficiency of a transformer can be computed using the following equation:

$$\%\text{Efficiency} = \frac{\text{Output in watts}}{\text{Output in watts} + \text{Total losses in watts}} \times 100 \tag{3.7}$$

The total losses of a transformer encompass losses in the electric circuit, magnetic circuit, and dielectric circuit. According to Stigant and Franklin [75], the transformer achieves its maximum efficiency when the iron loss and copper loss are equal. Thus, the load at which the efficiency is maximized can be determined by the following equation:

$$\%\text{Load} = \sqrt{\left(\frac{\text{Iron loss}}{\text{Copper loss}}\right)} \times 100 \tag{3.8}$$

Figures 3.1 and 3.2 depict nomograms that enable quick determination of transformer efficiency (for more accurate values, consult individual manufacturer's catalogs). As the cost of electric energy is presently around 5–6 cents/kWh and expected to double in the coming decades, the focus on transformer efficiency aligns with energy efficiency.

It is important to note that iron losses, also known as core losses, consist of hysteresis loss and eddy-current loss. Hysteresis loss occurs as a result of the power required to sustain the continuous reversals of elementary magnets (or individual molecules) in the iron core due to flux alternations. Eddy-current loss arises from circulating currents induced in the iron core when magnetic flux cuts through the conductive iron material. The eddy-current loss is directly proportional to the frequency and the square of the flux density. To minimize these losses, the core is constructed using thin laminations that are insulated from each other, reducing eddy-current losses. Special grades of steel alloyed with silicon are also employed to decrease both hysteresis and eddy-current losses. It is worth noting that iron losses are practically independent of the load, while copper losses arise from the resistance of the primary and secondary windings.

In general, the costs associated with distribution transformers can be categorized as follows: (1) investment cost, (2) energy loss cost (due to transformer losses), and (3) demand loss cost (related to lost capacity resulting from transformer losses). Of these, the investment cost is typically the largest component, encompassing the cost of the transformer itself, as well as material and labor expenses associated with its installation.

Figure 3.3 illustrates the annual cost per unit load versus the load level. At low-load levels, the relatively high costs are primarily due to the investment cost, while at high-load levels, the costs

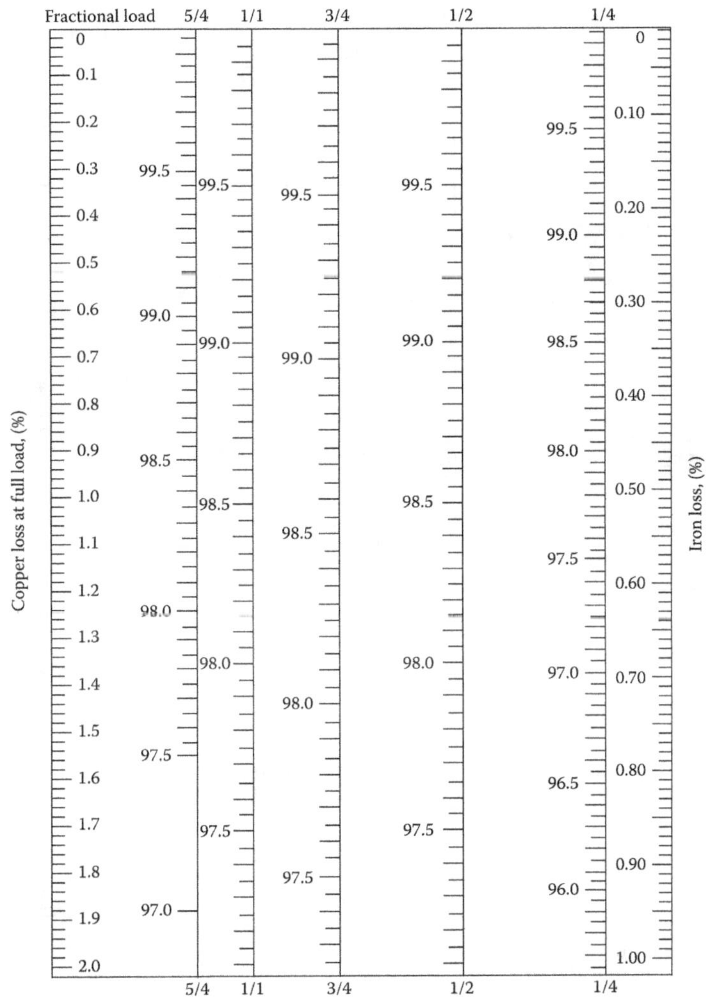

Figure 3.1 Transformer efficiency chart applicable only to the unity PF condition. To obtain the efficiency at a given load, lay a straightedge across the iron and copper loss values and read the efficiency at the point where the straightedge cuts the required load ordinate. (From Stigant, S.A. and Franklin, A.C., *The J&P Transformer Book*, Butterworth, London, U.K., 1973.)

include additional factors such as transformer lifespan reduction, energy losses, and demand loss, in addition to the investment cost. Figure 3.2 indicates an operating range that tends to be closer to the lower end of the curve. It is commonly considered cost-effective to install a transformer at approximately 80% of its nameplate rating and later replace it with a larger capacity transformer at around 180%. However, increasing capital, plant and equipment, and energy costs currently have a tendency to reduce these percentages.

3.6 TERMINAL OR LEAD MARKINGS

The terminals or connections of a transformer serve as the points where external electrical circuits are linked. As per the standards set by NEMA and ASA, the higher-voltage winding is designated as HV or H, while the lower-voltage winding is denoted as LV or x. Transformers equipped with more than two windings have their windings labeled as H, x, y, and z, arranged in descending

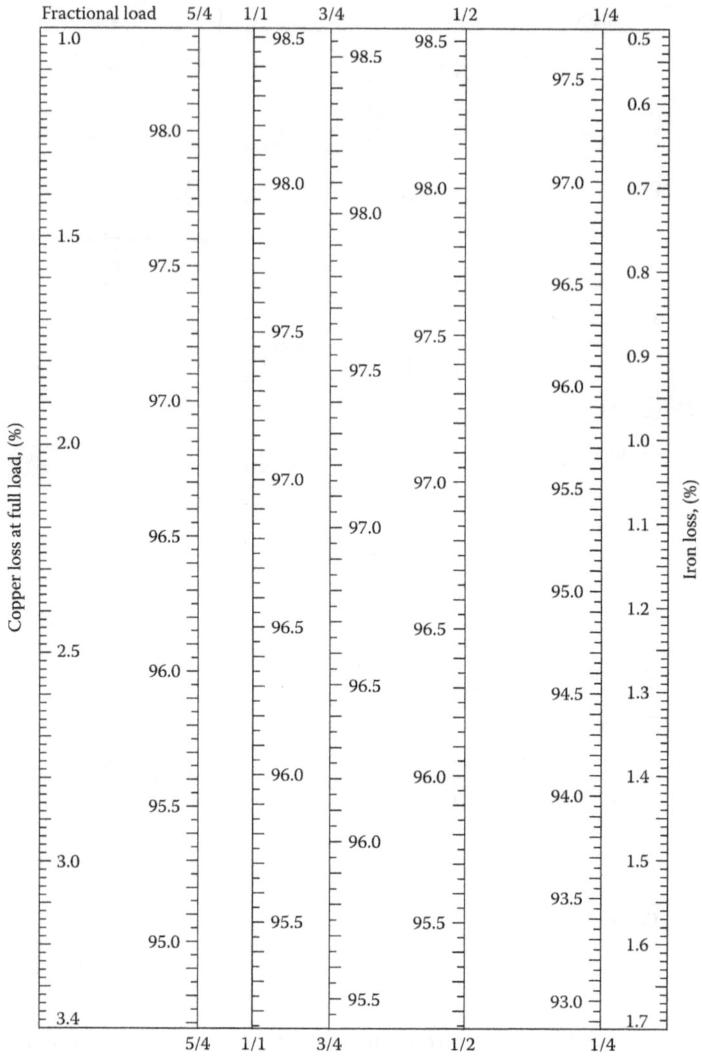

Figure 3.2 Transformer efficiency chart applicable only to the unity PF condition. To obtain the efficiency at a given load, lay a straightedge across the iron and copper loss values and read the efficiency at the point where the straightedge cuts the required load ordinate. (From Stigant, S.A. and Franklin, A.C., emphThe J&P Transformer Book, Butterworth, London, U.K., 1973.)

order of voltage. Terminal H_1 is positioned on the right-hand side when facing the HV side of the transformer. In the case of single-phase transformers, the leads are numbered in a way that ensures the voltage between the highest-numbered H lead and the highest-numbered x lead remains lower than the voltage of the HV winding.

For three-phase transformers, when facing the HV winding, terminal H_1 is situated on the right-hand side, followed by terminals H_2 and H_3 in numerical order from right to left. Conversely, when facing the LV winding, terminal x_1 is located on the left-hand side, and terminals x_2 and x_3 are arranged in numerical order from left to right.

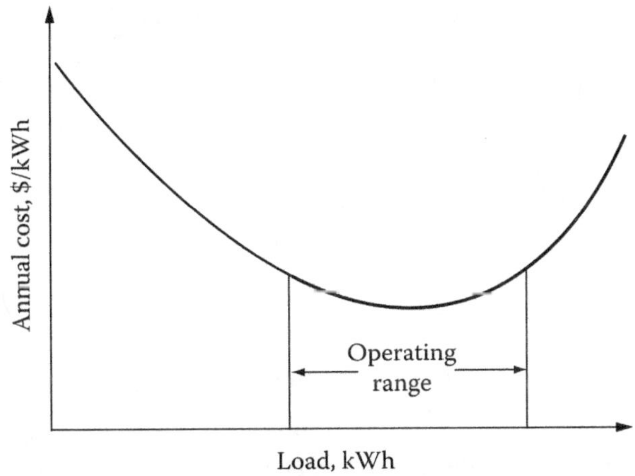

Figure 3.3 Annual cost per unit load vs. load level.

3.7 TRANSFORMER POLARITY

Transformer winding terminals are labeled to indicate the polarity, specifically to distinguish between the HV (high-voltage) and LV (low-voltage) sides. The primary and secondary windings are not explicitly identified since their designation depends on the specific input and output connections.

Transformer polarity refers to the direction of current flow through the HV leads in relation to the current flow through the LV leads at any given moment. It simply signifies the relative orientation of induced voltages between the HV leads and the LV terminals. The polarity of a single-phase DT can be either additive or subtractive. With standard markings, the voltage from H_1 to H_2 always aligns with the voltage from X_1 to X_2 in terms of direction or phase. If the transformer has adjacent terminals H_1 and X_1, as depicted in Figure 3.4a, it is referred to as having *subtractive* polarity. Conversely, if terminals H_1 and X_1 are diagonally opposite, as shown in Figure 3.4b, the transformer is considered to have *additive* polarity.

Transformer polarity can be determined through a simple test involving the connection of two adjacent terminals from the HV and LV winding while applying a moderate voltage to the HV winding, as depicted in Figure 3.5. The voltage between the HV and LV winding terminals that are not connected together is then measured. If the measured voltage is lower than the applied voltage to the HV winding, as shown in Figure 3.5a, the polarity is considered subtractive. Conversely, if

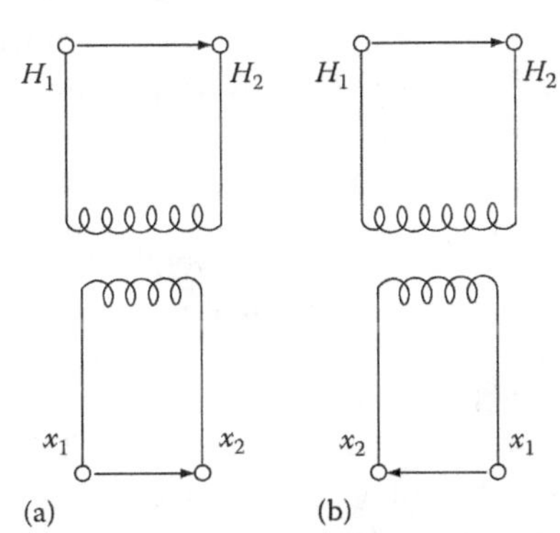

Figure 3.4 Additive and subtractive polarity connections: (a) subtractive polarity and (b) additive polarity.

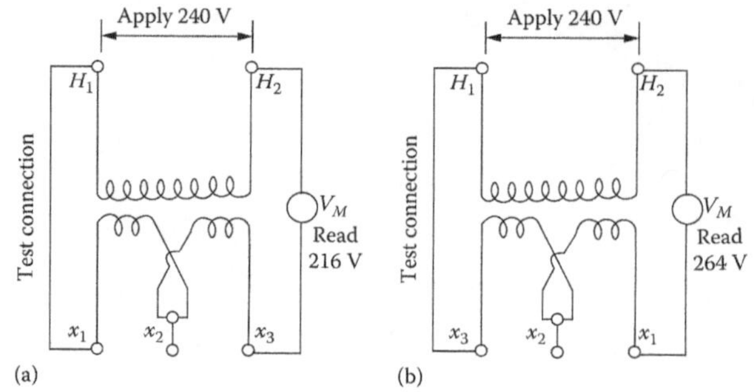

Figure 3.5 Polarity test: (a) subtractive polarity and (b) additive polarity.

the measured voltage is higher than the applied voltage, as illustrated in Figure 3.5b, the polarity is additive.

As per industry standards, all single-phase DTs with a rating of 200 kVA or smaller, and HV winding voltages of 8660 V and below, are designated with additive polarity. On the other hand, all other single-phase transformers have subtractive polarity. Polarity markings are particularly useful when connecting transformers into three-phase banks.

3.8 DISTRIBUTION TRANSFORMER LOADING GUIDES

The rated kVA of a transformer refers to the continuous output that can be obtained at its rated voltage and frequency without exceeding the specified temperature rise. Temperature rise is used as a basis for rating rather than the actual temperature because the ambient temperature can vary significantly during operation. The lifespan of the insulation used in transformers is influenced by the temperature it reaches and the duration it is sustained. Therefore, in order to determine the overload capabilities of a transformer, factors such as ambient temperature, preload conditions, and the duration of peak loads need to be known.

In accordance with Appendix C57.91 titled "The Guide for Loading Mineral Oil-Immersed Overhead-Type Distribution Transformers with 55°C and 65°C Average Winding Rise," which is an appendix to the ANSI Overhead Distribution Standard C57.12, insulation-life curves have been developed for 20 transformer types. These curves indicate a minimum life expectancy of 20 years at hot-spot temperatures of 95°C and 110°C for transformers with temperature rises of 55°C and 65°C respectively. Previous transformer loading guidelines were based on the "8°C insulation-life rule," which states that for transformers with class A insulation (typically oil-filled), the rate of deterioration approximately doubles with every 8°C increase in temperature. In other words, if a class A insulation transformer operates 8°C above its rated temperature, its lifespan would be halved.

3.9 EQUIVALENT CIRCUITS OF A TRANSFORMER

It is possible to represent a given transformer using several equivalent circuits. However, the common practice is to choose the simplest circuit that still provides the desired accuracy in calculations.

Figure 3.6 illustrates an equivalent circuit of a single-phase two-winding transformer. This circuit represents a practical transformer with an iron core and connected to a load (Z_L). When the primary winding is energized, it generates a flux through the iron core. The flux that links both the primary and secondary windings is known as the mutual flux (ϕ_m), and its maximum value is denoted as ϕ_m. Additionally, there are leakage fluxes (ϕ_{l1} and ϕ_{l2}) produced by the primary and secondary windings, respectively. These leakage fluxes result in inductive reactances (x_{l1} and x_{l2}) associated with the primary and secondary windings. Furthermore, the primary and secondary windings have their internal resistances (r_1 and r_2) respectively.

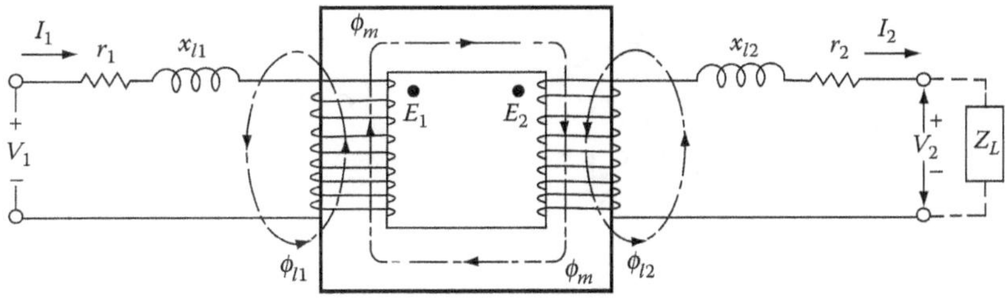

Figure 3.6 Basic circuit of a practical transformer.

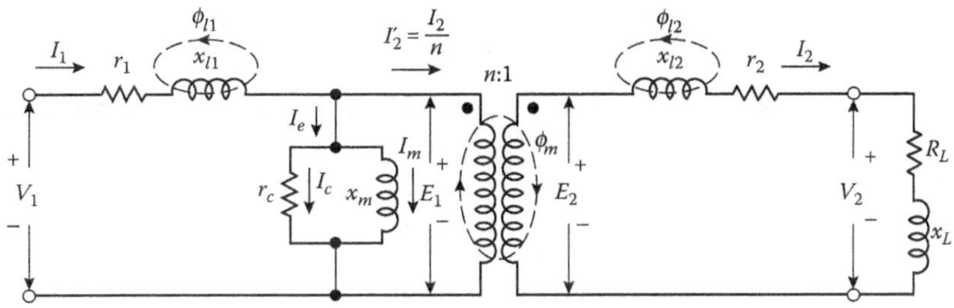

Figure 3.7 Equivalent circuit of a loaded transformer.

Figure 3.7 depicts an equivalent circuit of a loaded transformer. It is important to note that the current I_2' represents the primary (or load) component that corresponds exactly to the secondary current I_2, similar to an ideal transformer. Therefore,

$$I_2' = \frac{n_2}{n_1} \times I_2 \tag{3.9}$$

or

$$I_2' = \frac{I_2}{n} \tag{3.10}$$

where,

I_2 is the secondary current
n_1 is the number of turns in primary winding
n_2 is the number of turns in secondary winding
n is the turns ratio $= \frac{n_1}{n_2}$

The current I_e represents the excitation current component of the primary current I_1, which is necessary to generate the mutual flux. As depicted in Figure 3.8, the excitation current I_e can be divided into two components: (1) the magnetizing current component I_m and (2) the core-loss component I_c. The parameter r_c denotes the equivalent power loss in the transformer core, which includes hysteresis and eddy-current losses, caused by the magnetizing current I_m. The parameter x_m represents the inductive reactance component of the transformer when the secondary winding is open.

Figure 3.9 illustrates an approximate equivalent circuit that combines the primary side, the reflected secondary side, and the load impedances. It is important to note that the primary side perceives the secondary current I_2 as I_2/n, where n is the turns ratio. Moreover, the secondary and load impedances are transformed (or referred) to the primary side as $n^2(r_2 + jx_{l2})$ and $n^2(R_L + jX_L)$, respectively. Additionally, the secondary terminal voltage V_2 is transferred to the primary side as nV_2.

As the excitation current I_e is significantly smaller compared to I_2/n in a loaded transformer, it can be disregarded, as illustrated in Figure 3.10. Consequently, the equivalent impedance of the

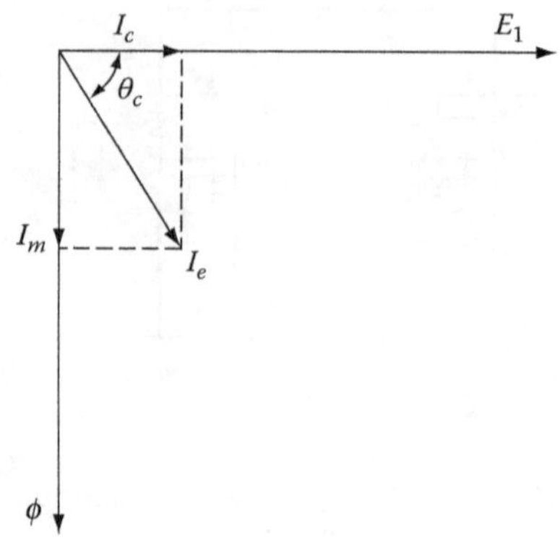

Figure 3.8 Phasor diagram corresponding to the excitation current components at no load.

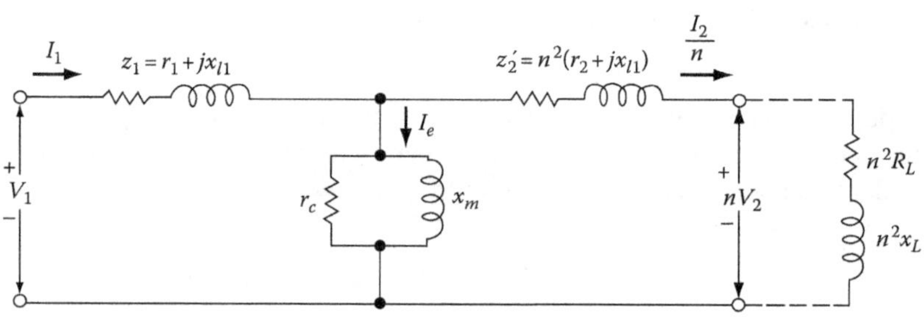

Figure 3.9 Equivalent circuit with the referred secondary values.

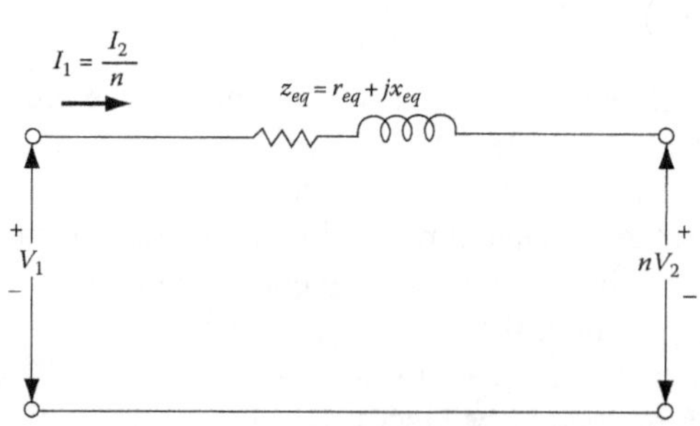

Figure 3.10 Simplified equivalent circuit assuming negligible excitation current.

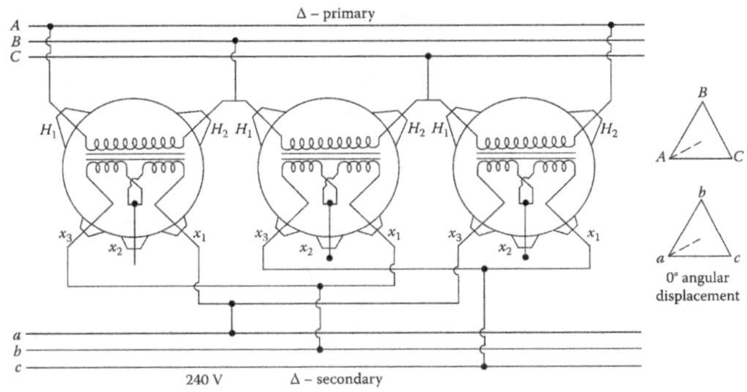

Figure 3.11 Simplified equivalent circuit for a large-sized power transformer.

transformer, when referred to the primary side, is given by

$$\begin{aligned} Z_{eq} &= Z_1 + Z_2' \\ &= Z_1 + n^2 Z_2 \\ &= r_{oq} + jx_{eq} \end{aligned} \tag{3.11}$$

where

$$Z_1 = r_1 + jx_{l2} \tag{3.12}$$

$$Z_2 = r_2 + jx_{l2} \tag{3.13}$$

and therefore the equivalent resistance and the reactance of the transformer referred to the primary are

$$r_{eq} = r_1 + n^2 r_2 \tag{3.14}$$

and

$$x_{eq} = x_{l1} + n^2 x_{l2} \tag{3.15}$$

As before in Figure 3.11, for large-size power transformers,

$$r_{eq} \to 0$$

therefore the equivalent impedance of the transformer becomes

$$Z_{eq} = jx_{eq} \tag{3.16}$$

3.10 SINGLE PHASE TRANSFORMER CONNECTIONS

3.10.1 GENERAL

Single-phase distribution transformers (DTs) outnumber polyphase transformers. This is partially due to the fact that lighting and smaller power loads are typically supplied using single-phase secondary circuits. Even in polyphase secondary systems, single-phase transformers connected as polyphase banks are commonly used.

Single-phase DTs consist of one HV primary winding and two LV secondary windings rated at a nominal 120 V. In older transformers, four insulated secondary leads were brought out of the

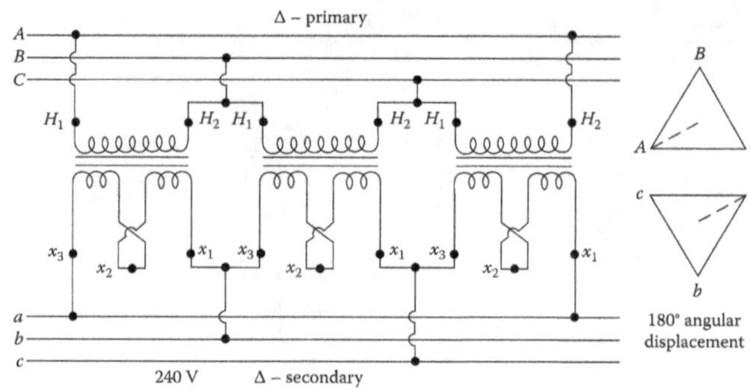

Figure 3.12 Single-phase transformer connections.

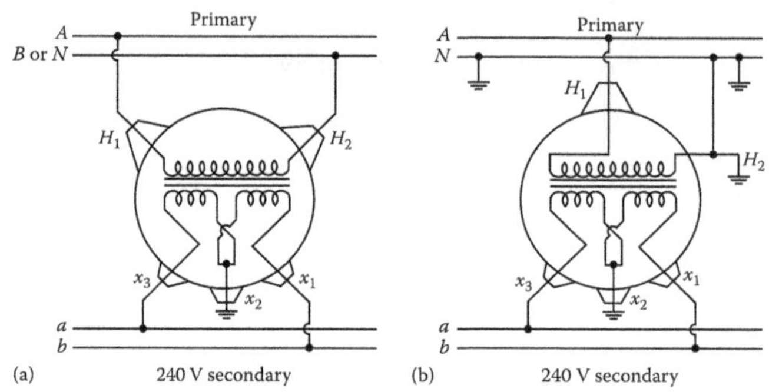

Figure 3.13 Single-phase transformer connections.

transformer tank, with series or parallel connections made outside the tank. However, modern transformers now feature connections made inside the tank, with only three secondary terminals brought out.

Various connection diagrams for single-phase transformers supplying single-phase loads are shown in Figure 3.12. The secondary coils, each rated at a nominal 120 V, can be connected in parallel to provide a two-wire 120 V circuit (Figure 3.12a and b), or they can be connected in series to provide a three-wire 120/240 V single circuit (Figure 3.12c and d). The connections depicted in Figure 3.12a and b are suitable for small loads and short secondary circuit lengths. They are often used for individual customers who require only 120 V single-phase power.

However, this connection is typically not considered sufficient for modern homes. If a mistake is made in polarity when connecting the two secondary coils in parallel (Figure 3.12a), such that LV terminal 1 is connected to terminal 4 and terminal 2 is connected to terminal 3, it will result in a short-circuited secondary. This will cause the fuses on the HV side of the transformer (not shown in the figure) to blow. Conversely, if there is a polarity mistake when connecting the coils in series (Figure 3.12c), the voltage across the outer conductors will be zero instead of 240 V. Voltage adjustment taps, if provided, are located on the HV winding of the transformer. Figures 3.12b and d show single-bushing transformers connected to a multigrounded primary. These connections are used in systems with multigrounded neutral, such as 12,470 Gnd Y/7200, 13,200 Gnd Y/7620, and 24,940 Gnd Y/14,400 V systems. It is essential to maintain good and solid grounds on both the transformer and the system. Figure 3.13 illustrates single-phase transformer connections for single- and two-bushing transformers, serving customers who require only 240 V single-phase power. These connections are commonly used in small industrial applications.

In general, the 120/240 V three-wire connection system is preferred as it offers twice the load capacity of the 120 V system with only 12 times the conductor amount. Each 120 V winding has half

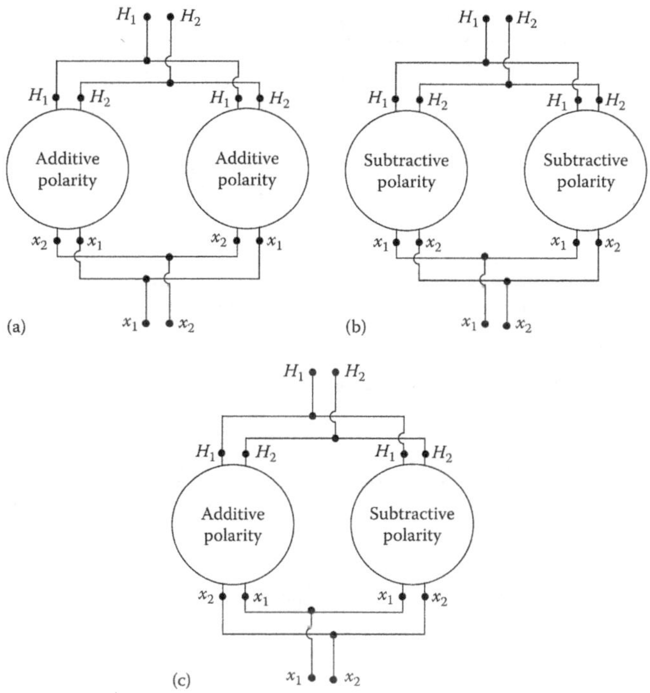

Figure 3.14 Single-phase transformer paralleling.

the total kVA rating of the transformer. When the connected 120 V loads are equal and balanced, no current flows in the neutral conductor. Therefore, it is crucial to maintain balanced loads to optimize transformer capacity usage and minimize regulation. Typically, one leg of the 120 V two-wire system and the middle leg of the 240 V two-wire or 120/240 V three-wire system are grounded to limit the voltage to ground on the secondary circuit.

3.10.2 SINGLE PHASE TRANSFORMER PARALLELING

When there is a need for increased capacity during emergency situations, it is possible to connect two single-phase transformers in parallel. These transformers can have the same or different kVA ratings and can be of either additive or subtractive polarity. However, it is important to ensure the following conditions are observed and the transformers are connected as depicted in Figure 3.14:

1. All transformers have the same turns ratio.
2. All transformers are connected to the same primary phase.
3. All transformers have identical frequency ratings.
4. All transformers have identical voltage ratings.
5. All transformers have identical tap settings.
6. Per unit impedance of one transformer is between 0.925 and 1.075 of the other in order to maximize capability.

However, paralleling two single-phase transformers is not cost-effective because the total cost and losses associated with two small transformers are much higher than that of a single large transformer with the same capacity. Therefore, this method should only be used temporarily to address increased demands for single-phase power during emergency situations. Figure 3.15 illustrates the

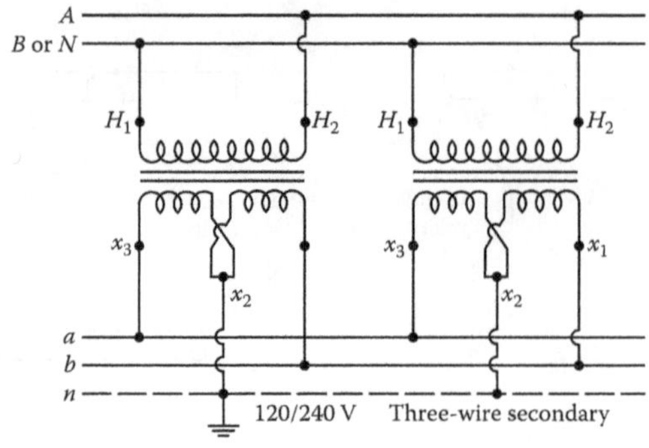

Figure 3.15 Parallel operation of two single-phase transformers.

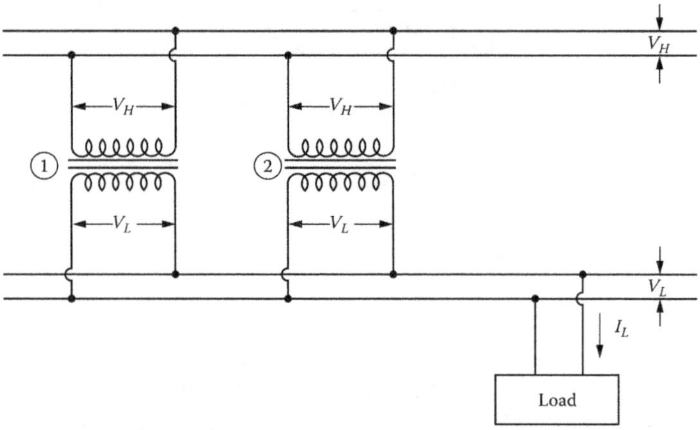

Figure 3.16 Two transformers connected in parallel and feeding a load.

connection of two single-phase transformers, each with two bushings, to a two-conductor primary for supplying 120/240 V single-phase power on a three-wire secondary system.

To understand how the load is divided between the parallel-connected transformers, let's consider the scenario depicted in Figure 3.16. It is assumed that the necessary conditions for paralleling have already been met (as shown in Figure 3.17).

Figure 3.7 shows the corresponding equivalent circuit referred to as the LV side. Since the transformers are connected in parallel, the voltage drop through each transformer must be equal. Therefore,

$$I_1(Z_{eq,T1}) = I_2(Z_{eq,T2}) \tag{3.17}$$

from which

$$\frac{I_1}{I_2} = \frac{Z_{eq,T2}}{Z_{eq,T1}} \tag{3.18}$$

where

I_1 is the secondary current of transformer 1
I_2 is the secondary current of transformer 2
I_L is the load current
$Z_{eq,1}$ is the equivalent impedance of transformer 1
$Z_{eq,2}$ is the equivalent impedance of transformer 2

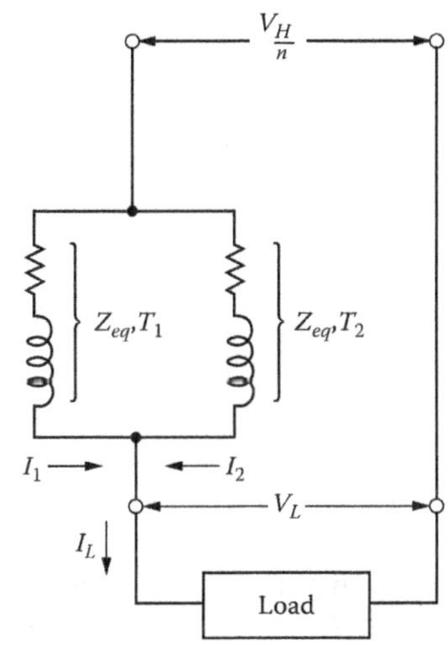

Figure 3.17 Equivalent circuit.

By examining Equation 3.18, we observe that the load division depends solely on the relative percentage impedance of the transformers. If we substitute the ohmic impedances in Equation 3.18 with their equivalents in terms of percentage impedance, we obtain the following equation:

$$\frac{I_1}{I_2} = \frac{(\%z)_{eq,T2}}{(\%z)_{eq,T1}} \frac{S_{T1}}{S_{T2}} \tag{3.19}$$

where
$(\%z)_{T1}$ is the percent impedance of transformer 1
$(\%z)_{T2}$ is the percent impedance of transformer 2
S_{T1} is the kVA rating of transformer 1
S_{T2} is the kVA rating of transformer 2

Equation 3.19 can be expressed in terms of kVA supplied by each transformer since the primary and secondary voltages for each transformer are the same, respectively. Therefore,

$$\frac{S_{L1}}{S_{L2}} = \frac{(\%z)_{T2}}{(\%z)_{T1}} \frac{S_{T1}}{S_{T2}} \tag{3.20}$$

where
S_{L1} is the kVA supplied by transformer 1 to the load
S_{L2} is the kVA supplied by transformer 2 to the load

Example 3.1. Figure 3.18 shows an equivalent circuit of a single-phase transformer with three-wire secondary for three-wire single-phase distribution. The typical DT is rated as 25 kVA, 7200-120/240 V, 60 Hz, and has the following per unit impedance based on the transformer ratings and based on the use of the entire LV winding with zero neutral current:

$$R_T = 0.014 \text{ pu}$$

and

$$X_T = 0.012 \text{ pu}$$

Here, the two halves of the LV may be independently loaded, and, in general, the three-wire secondary load will not be balanced. Therefore, in general, the equivalent circuit needed is that of

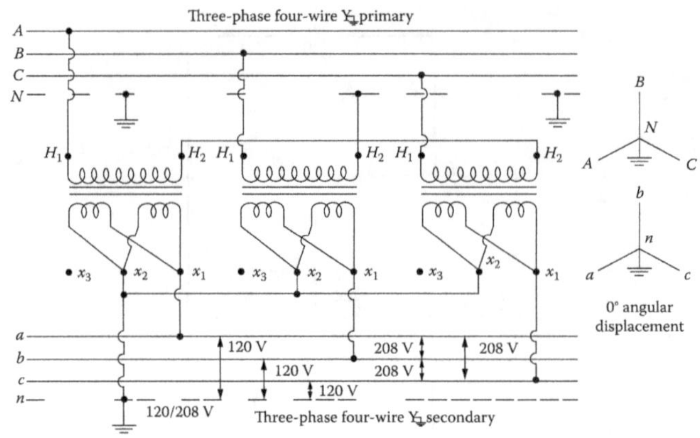

Figure 3.18 An equivalent circuit of a single-phase transformer with three-wire secondary. (From Lloyd, B., *Electric Utility Engineering Reference Book-Distribution Systems*, Vol. 3, Westinghouse Electric Corporation, East Pittsburgh, PA, 1965.)

a three-winding single-phase transformer as shown in Figure 3.18, when voltage drops and/or fault currents are to be computed. Thus, use the meager amount of data (it is all that is usually available) and evaluate numerically all the impedance data shown in Figure 3.18.

Solution. Figure 3.18 is based on the reference by Lloyd [50]. To determine $\overline{Z}_{HX_{1-2}}$ approximately, Lloyd gives the following formula:

$$\overline{Z}_{HX_{1-2}} = 1.5R_T + j1.2X_T \tag{3.21}$$

where $\overline{Z}_{HX_{1-2}}$ is the transformer impedance referred to HV winding when the section of the LV winding between the terminals X_2 and X_3 is short-circuited. From Figure 3.18, the turns ratio of the transformer is

$$n = \frac{V_H}{V_X} = \frac{7200 \text{ V}}{120 \text{ V}} = 60$$

Since the given per-unit impedances of the transformer are based on the use of the entire LV winding,

$$\begin{aligned}\overline{Z}_{HX_{1-3}} &= R_T + jX_T \\ &= 0.014 + j0.012 \text{ pu}\end{aligned}$$

Also, from Equation 3.21,

$$\begin{aligned}\overline{Z}_{HX_{1-2}} &= 1.5R_T + 1.2jX_T \\ &= 1.5 \times 0.014 + j1.2 \times 0.012 \\ &= 0.012 + j0.0144 \text{ pu}\end{aligned}$$

Therefore,

$$\begin{aligned}2\overline{Z}_{HX_{1-3}} - \overline{Z}_{HX_{1-2}} &= 2(0.014 + j0.012) - (0.021 + j0.0144) \\ &= 0.007 + j0.0096 \text{ pu} \\ &= 14.515 + j19.906 = 24.637\angle 53.9^\circ \ \Omega\end{aligned}$$

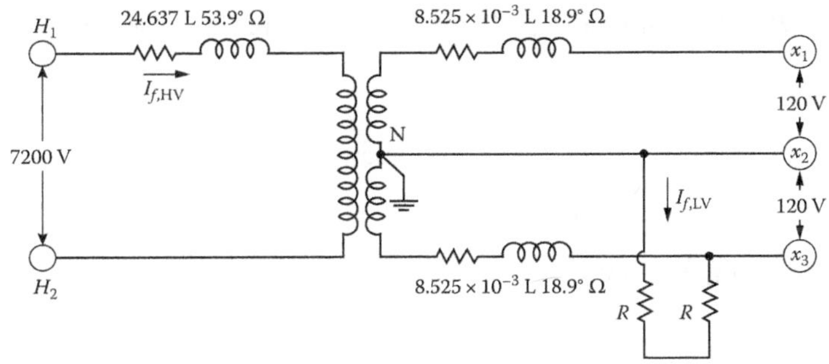

Figure 3.19 Secondary line-to-neutral fault.

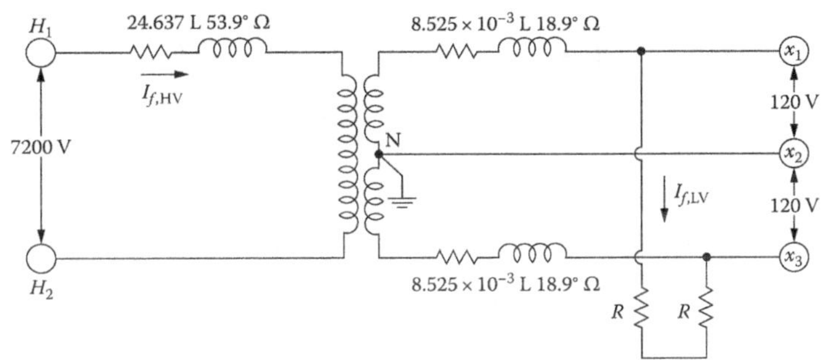

Figure 3.20 Secondary line-to-line fault.

and

$$\frac{2\overline{Z}_{HX_{1-2}} - 2\overline{Z}_{HX_{1-3}}}{n^2} = \frac{2(0.021 + j0.0144) - 2(0.014 + j0.012)}{60^2}$$
$$= 3.89 \times 10^{-6} + j1.334 \times 10^{-6} \text{ pu}$$
$$= 0.008064 + j0.0028 = 8.525 \times 10^{-3}\angle 18.9^\circ \ \Omega$$

Example 3.2. Using the transformer equivalent circuit found in Example 3.1, determine the line-to-neutral (120 V) and line-to-line (240 V) fault currents in three-wire single-phase 120/240 V secondaries shown in Figures 3.19 and 3.20, respectively. In the figures, *R* represents the resistance of service-drop cable per conductor. Usually *R* is much larger than *X* for such cable and therefore *X* may be neglected.

Using the given data, determine the following:

a. Find the symmetrical rms fault currents in the HV and LV circuits for a 120 V fault if *R* of the service-drop cable is zero.
b. Find the symmetrical rms fault currents in the HV and LV circuits for a 240 V fault if *R* of the service-drop cable is zero.
c. If the transformer is a CSPB type, find the minimum allowable interrupting capacity (in symmetrical rms amperes) for a circuit breaker connected to the transformer's LV terminals.

Solution. a. When $R = 0$, from Figure 3.19, the line-to-neutral fault current in the secondary side of the transformer is

$$\bar{I}_{f,LV} = \frac{120}{8.525 \times 10^{-3}\angle 18.9° + (1/60)^2(24.637\angle 53.9°)}$$
$$= 8,181.7\angle -34.4° \text{ A}$$

Thus, the fault current in the HV side is

$$\bar{I}_{f,HV} = \frac{\bar{I}_{f,LV}}{n}$$
$$= \frac{8,181.7}{60} = 136.4 \text{ A}$$

Note that the turns ratio is found as

$$n = \frac{7,200 \text{ V}}{120 \text{ V}} = 60$$

b. When $R = 0$, from Figure 3.20, the line-to-line fault current in the secondary side of the transformer is

$$\bar{I}_{f,LV} = \frac{240}{2(8.525 \times 10^{-3}\angle 18.9°) + (1/30)^2(24.637\angle 53.9°)}$$
$$= 5649\angle -40.6° \text{ A}$$

Thus, the fault current in the HV side is

$$\bar{I}_{f,HV} = \frac{\bar{I}_{f,LV}}{n}$$
$$= \frac{5649}{30} = 188.3 \text{ A}$$

Note that the turns ratio is found as

$$n = \frac{7200 \text{ V}}{240 \text{ V}} = 30$$

c. Therefore, the minimum allowable interrupting capacity for a circuit breaker connected to the transformer LV terminals is 8181.7 A.

Example 3.3. Using the data given in Example 3.2, determine the following:

a. Estimate approximately the value of R, that is, the service-drop cable's resistance, which will produce equal line-to-line and line-to-neutral fault currents.
b. If the conductors of the service-drop cable are aluminum, find the length of the service-drop cable that would correspond to the resistance R found in part (a) in case of (1) #4 AWG conductors with a resistance of 2.58 Ω/mi and (2) #1/0 AWG conductors with a resistance of 1.03 Ω/mi.

Solution. a. Since the line-to-line and the line-to-neutral fault currents are supposed to be equal to each other,

$$\frac{240}{2R+0.032256+j0.02756}=\frac{120}{2R+0.012096+j0.0083}$$

or

$$R\simeq 0.0075\Omega$$

b. The length of the service-drop cable is

1. If #4 AWG aluminum conductors with a resistance of 2.58 Ω/mi or 4.886 × 10^{--4} Ω/ft are used,

$$\begin{aligned}\text{Service-drop length} &= \frac{\text{R}}{4.886\times 10^{-4}}\\ &= \frac{0.0075\Omega}{4.886\times 10^{-4}\Omega/\text{ft}}\\ &\cong 15.35\text{ ft}\end{aligned}$$

2. If #1/0 AWG aluminum conductors with a resistance of 1.03 Ω/mi or 1.9508 × 10^{-4} Ω/ft are used,

$$\begin{aligned}\text{Service-drop length} &= \frac{0.0075\Omega}{1.9508\times 10^{-4}\Omega/\text{ft}}\\ &\cong 38.45\text{ ft}\end{aligned}$$

Example 3.4. Assume that a 250 kVA transformer with 2.4% impedance is paralleled with a 500 kVA transformer with 3.1% impedance. Determine the maximum load that can be carried without overloading either transformer. Assume that the maximum allowable transformer loading is 100% of the rating.

Solution. Designating the 250 and 500 kVA transformers as transformers 1 and 2, respectively, and using Equation 3.20,

$$\begin{aligned}\frac{S_{L1}}{S_{L2}} &= \frac{(\%Z)_{T2}}{\%Z_{T1}}\frac{S_{T1}}{S_{T2}}\\ &= \frac{3.1}{2.4}\times\frac{250}{500}=0.6458\end{aligned}$$

Assume a load of 500 kVA on the 500 kVA transformer. The preceding result shows that the load on the 250 kVA transformer will be (0.6458) × (500 kVA) = 322.9 kVA when the load on the 500 kVA transformer is 500 kVA. Therefore, the 250 kVA transformer becomes overloaded before the 500 kVA transformer. The load on the 500 kVA transformer when the 250 kVA transformer is carrying rated load is

$$\begin{aligned}S_{L2} &= \frac{S_{L1}}{0.6458}\\ &= \frac{250}{0.6458}\\ &= 387.1\text{ kVA}\end{aligned}$$

Thus, the total load is

$$\sum_{i=1}^{2} S_{Li} = S_{L1} + S_{L2}$$
$$= 250 + 387.1$$
$$= 637.1 \text{ kVA}$$

3.11 THREE-PHASE CONNECTIONS

The common approaches for interconnecting three single-phase transformers in three-phase configurations are the delta–delta (Δ–Δ), wye–wye (Y–Y), wye–delta (Y–Δ), and delta–wye (Δ–Y) connections. It is important to note that these connections assume all transformers in the bank have identical kVA ratings.

3.11.1 Δ–Δ TRANSFORMER CONNECTION

Figures 3.21 and 3.22 depict the delta–delta (Δ–Δ) connection, which is achieved by connecting single-phase transformers in a manner that enables the provision of 240 V service with angular displacements of 0° and 180°, respectively.

This connection is commonly employed to supply a simultaneous load of small single-phase lighting and three-phase power. To achieve this service, the mid-tap of the secondary winding in

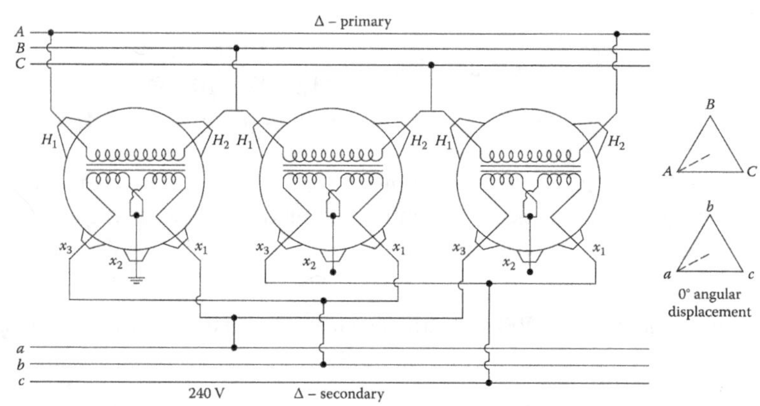

Figure 3.21 Δ–Δ transformer bank connection with 0° angular displacement.

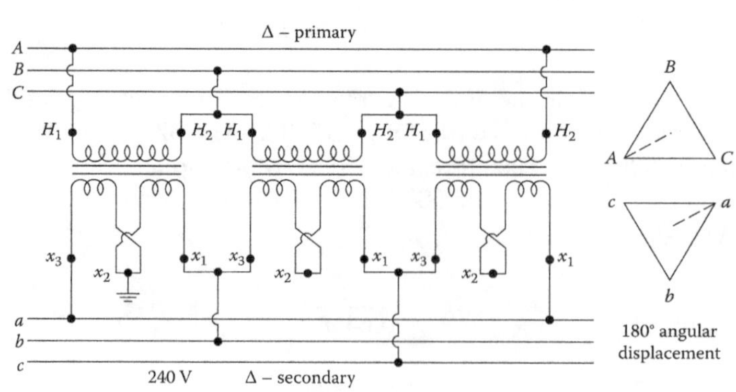

Figure 3.22 Δ–Δ transformer bank connection with 180° angular displacement.

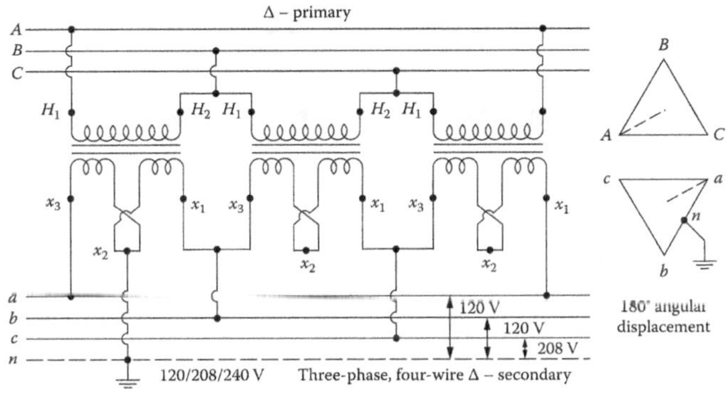

Figure 3.23 Δ–Δ transformer bank connection to provide 120/208/240 V three-phase four-wire service.

one of the transformers is grounded and connected to the secondary neutral conductor, as depicted in Figure 3.23. Consequently, the single-phase loads are connected between the phase and neutral conductors. As a result, the transformer with the mid-tap carries two-thirds of the 120/240 V single-phase load and one-third of the 240 V three-phase load, while the other two units each carry one-third of both the 120/240 V and 240 V loads.

This configuration avoids issues related to third-harmonic overvoltage and telephone interference. However, high circulating currents may occur unless all three single-phase transformers are connected to the same regulating taps and possess identical voltage ratios. Additionally, unless all transformers have identical impedance values, the rating of the transformer bank is reduced. It is important to note that the secondary neutral bushing can only be grounded on one of the three single-phase transformers, as shown in Figure 3.23.

Therefore, to get balanced transformer loading, the conditions include the following:

1. All transformers have identical voltage ratios.
2. All transformers have identical impedance values.
3. All transformers are connected on identical taps.

However, if two of the units have identical impedance values and the third unit has an impedance value within a range of plus or minus 25% of the impedance value of the similar transformers, it is possible to operate the Δ–Δ bank with a small degree of unbalanced transformer loading while still maintaining a reduced bank output capacity. Table 3.5 provides the allowable levels of load unbalance on the odd and similar transformers. It should be noted that Z_1 represents the impedance of the odd transformer unit, while Z_2 represents the impedance of similar transformer units. Therefore, when dealing with unbalanced transformer loading, the load values must be compared against the values in the table to ensure that no individual transformer is overloaded.

Let's assume that Figure 3.10 illustrates the equivalent circuit of a Δ–Δ connected transformer bank referred to the LV side. An equation for voltage drop can be derived for the LV windings as follows:

$$\overline{V}_{ba}+\overline{V}_{ac}+\overline{V}_{cb}=\overline{I}_{ba}\overline{Z}_{ab}+\overline{I}_{ac}\overline{Z}_{ca}+\overline{I}_{cb}\overline{Z}_{bc} \tag{3.22}$$

where

$$\overline{V}_{ba}+\overline{V}_{ac}+\overline{V}_{cb}=0 \tag{3.23}$$

Therefore, Equation 3.22 becomes

$$\overline{I}_{ba}\overline{Z}_{ab}+\overline{I}_{ac}\overline{Z}_{ca}+\overline{I}_{cb}\overline{Z}_{bc}=0 \tag{3.24}$$

Table 3.5
Permissible Percent Loading on Odd and Like Transformers as a Function of the Z_1/Z_2 Ratio

Z_1/Z_2 Ratio	% Load on Odd Unit	% Load on Like Unit
0.75	109	96
0.8	107	96.5
0.85	105.2	97.3
0.9	103.3	98.3
1.1	96.7	102
1.15	95.2	102.2
1.2	93.8	103.1
1.25	92.3	103.9

For the delta-connected secondary,

$$\bar{I}_a = \bar{I}_{ba} - \bar{I}_{ac} \tag{3.25}$$

$$\bar{I}_b = \bar{I}_{cb} - \bar{I}_{ba} \tag{3.26}$$

$$\bar{I}_c = \bar{I}_{ac} - \bar{I}_{cb} \tag{3.27}$$

From Equation 3.24,

$$\bar{I}_{ba}\bar{Z}_{ab} = -\bar{I}_{ac}\bar{I}_{ca} - \bar{I}_{cb} - \bar{Z}_{bc} \tag{3.28}$$

Adding the terms of $\bar{I}_{ba}\bar{Z}_{ab}$ and $\bar{I}_{ba}\bar{Z}_{ca}$ to either side of Equation 3.28 and substituting Equation 3.25 into the resultant equation,

$$\bar{I}_{ba} = \frac{\bar{I}_a\bar{Z}_{ca} - \bar{I}_b\bar{Z}_{bc}}{\bar{Z}_{ab} + \bar{Z}_{bc} + \bar{Z}_{ca}} \tag{3.29}$$

and similarly,

$$\bar{I}_{ac} = \frac{\bar{I}_c\bar{Z}_{bc} - \bar{I}_a\bar{Z}_{ab}}{\bar{Z}_{ab} + \bar{Z}_{bc} + \bar{Z}_{ca}} \tag{3.30}$$

and

$$\bar{I}_{cb} = \frac{\bar{I}_b\bar{Z}_{ab} - \bar{I}_c\bar{Z}_{ca}}{\bar{Z}_{ab} + \bar{Z}_{bc} + \bar{Z}_{ca}} \tag{3.31}$$

If the three transformers shown in Figure 3.24 have equal percent impedance and equal ratios of percent reactance to percent resistance, then Equations 3.29 through 3.31 can be expressed as

$$\bar{I}_{ba} = \frac{(\bar{I}_a/S_{T,ca}) - (\bar{I}_b/S_{T,bc})}{(1/S_{T,ab}) + (1/S_{T,bc}) + (1/S_{T,ca})} \tag{3.32}$$

$$\bar{I}_{ac} = \frac{(\bar{I}_c/S_{T,bc}) - (\bar{I}_a/S_{T,ab})}{(1/S_{T,ab}) + (1/S_{T,bc}) + (1/S_{T,ca})} \tag{3.33}$$

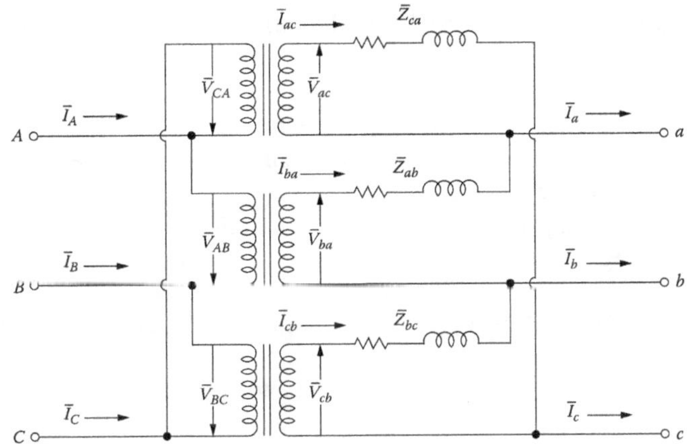

Figure 3.24 Equivalent circuit of a Δ–Δ-connected transformer bank.

$$\bar{I}_{cb} = \frac{(\bar{I}_b/S_{T,ab}) - (\bar{I}_c/S_{T,ab})}{(1/S_{T,ab}) + (1/S_{T,bc}) + (1/S_{T,ca})} \tag{3.34}$$

where

$S_{T,ab}$ is the kVA rating of the single phase between phases a and b
$S_{T,bc}$ is the kVA rating of the single phase between phases b and c
$S_{T,ca}$ is the kVA rating of the single phase between phases c and a

Example 3.5. Three single-phase transformers are connected in Δ–Δ to provide power for a three-phase wye-connected 200 kVA load with a 0.80 lagging power factor and an 80 kVA single-phase light load with a 0.90 lagging power factor, as shown in Figure 3.25.

Assume that the three single-phase transformers have equal percent impedance and equal ratios of percent reactance to percent resistance. The primary-side voltage of the bank is 7620/13,200 V, and the secondary-side voltage is 240 V. Assume that the single-phase transformer connected

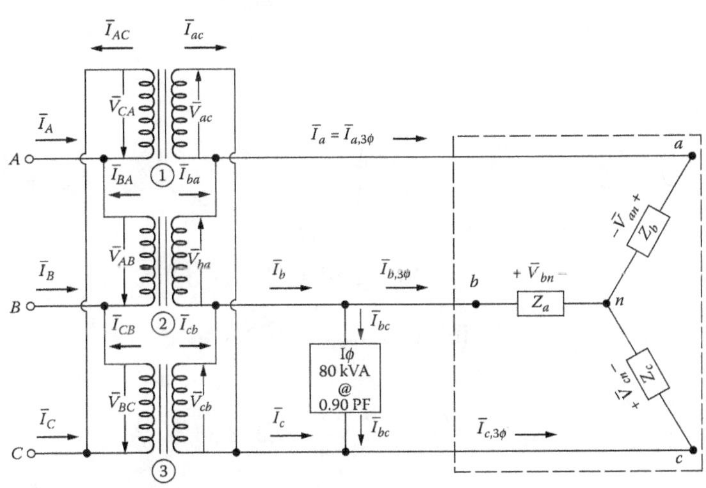

Figure 3.25 For Example 3.5.

between phases b and c is rated at 100 kVA and the other two are rated at 75 kVA. Determine the following:

a. The line current flowing in each secondary-phase wire
b. The current flowing in the secondary winding of each transformer
c. The load on each transformer in kVA
d. The current flowing in each primary winding of each transformer
e. The line current flowing in each primary-phase wire

Solution. a. Using the voltage drop $\overline{V}_{an}$ as the reference, the three-phase components of the line currents can be found as

$$\begin{aligned}|\overline{I}_{a,3\phi}| = |\overline{I}_{b,3\phi}| = |\overline{I}_{c,3\phi}| &= \frac{S_{L,3\phi}}{\sqrt{3}\times V_{L-L}}\\ &= \frac{200}{\sqrt{3}\times 0.240}\\ &= 481.7\text{ A}\end{aligned}$$

Since the three-phase load has a lagging power factor of 0.80,

$$\begin{aligned}|\overline{I}_{a,3\phi}| &= |\overline{I}_{a,3\phi}|(\cos\theta - j\sin\theta)\\ &= 481.7(0.80 - j0.60)\\ &= 385.36 - j289.02\\ &= 481.7\angle -36.9^\circ\text{ A}\end{aligned}$$

$$\begin{aligned}|\overline{I}_{b,3\phi}| &= a\overline{I}_{a,3\phi}\\ &= (1\angle 120^\circ)481.7\angle -36.9^\circ\\ &= -443.08 - j188.89\\ &= 481.7\angle 203.1^\circ\text{ A}\end{aligned}$$

$$\begin{aligned}|\overline{I}_{c,3\phi}| &= a\overline{I}_{a,3\phi}\\ &= (1\angle 120^\circ)481.7\angle -36.9^\circ\\ &= 57.87 + j478.21\\ &= 481.7\angle 83.1^\circ\text{ A}\end{aligned}$$

The single-phase component of the line currents can be found as

$$|\overline{I}_{bc}| = \frac{S_{L,1\phi'}}{V_{L-L}} = \frac{80}{0.240} = 333.33\ textA$$

Since the single-phase load has a lagging power factor of 0.90, the current phasor $\overline{I}_{bc}$ lags its voltage phasor $\overline{V}_{bc}$ by –25.8°. Also, since the voltage phasor $\overline{V}_{bc}$ lags the voltage reference $\overline{V}_{an}$

by 90° (see Figure 3.12), then the current phasor $\bar{I}_{bc}$ will lag the voltage reference $\bar{V}_{an}$ by –115.8° (= –25.8°–90°). Therefore,

$$\begin{aligned}\bar{I}_{bc} &= 333.33\angle -115.8^\circ \\ &= -145.3 - j300 \text{ A}\end{aligned}$$

Hence, the line currents flowing in each secondary-phase wire can be found as

$$\begin{aligned}\bar{I}_a &= \bar{I}_{a,3\phi} \\ &= 481.7\angle -36.9^\circ \text{ A}\end{aligned}$$

$$\begin{aligned}\bar{I}_b &= \bar{I}_{b,3\phi} + \bar{I}_{bc} \\ &= 481.7\angle -203.1^\circ + 333.33\angle -115.8^\circ \\ &= -588.38 - j488.99 \\ &= 765.05\angle 219.7^\circ \text{ A}\end{aligned}$$

$$\begin{aligned}\bar{I}_c &= \bar{I}_{c,3\phi} + \bar{I}_{bc} \\ &= 481.7\angle 83.1^\circ - 333.33\angle -115.8^\circ \\ &= -87.43 - j178.21 \\ &= 198.5\angle -63.8^\circ \text{ A}\end{aligned}$$

b. By using Equation 3.33, the current flowing in the secondary winding of transformer 1 can be found as

$$\begin{aligned}\bar{I}_{ac} &= \frac{(\bar{I}_c/S_{T,bc}) - (\bar{I}_a/S_{T,ab})}{(1/S_{T,ab}) + (1/S_{T,bc}) + (1/S_{T,ca})} \\ &= \frac{((198.5\angle -63.8^\circ)/100) - ((481.7\angle -36.9^\circ)/75)}{(1/75) + (1/100) + (1/75)} \\ &= \frac{1.985\angle -63.8^\circ) - 6.4227\angle -36.9^\circ}{0.0367} \\ &= -116.07 + j56.55 \\ &= 129.11\angle -33.1^\circ \text{ A}\end{aligned}$$

Similarly, by using Equation 3.32,

$$\begin{aligned}\bar{I}_{ba} &= \frac{(\bar{I}_a/S_{T,ca}) - (\bar{I}_b/S_{T,bc})}{(1/S_{T,ab}) + (1/S_{T,bc}) + (1/S_{T,ca})} \\ &= \frac{((481.7\angle -36.9^\circ)/75) - ((765.05\angle 219.7^\circ)/100)}{(1/75) + (1/100) + (1/75)} \\ &= \frac{6.4227\angle -36.9^\circ) - 7.6505\angle 219.7^\circ}{0.0367} \\ &= 300.34 + j28.08 \\ &= 301.65\angle 5.3^\circ \text{ A}\end{aligned}$$

and using Equation 3.34,

$$\begin{aligned}\bar{I}_{cb} &= \frac{(\bar{I}_b/S_{T,ab}) - (\bar{I}_c/S_{T,ca})}{(1/S_{T,ab}) + (1/S_{T,bc}) + (1/S_{T,ca})} \\ &= \frac{((765.05\angle 219.7°)/75) - ((198.5\angle -63.8°)/75)}{0.0367} \\ &= -245.6 - j112.95 \\ &= 270.3\angle 204.7° \text{ A}\end{aligned}$$

c. The kVA load on each transformer can be found as

$$\begin{aligned}\overline{S}_{L,ab} &= V_{ba} \times |\bar{I}_{ba}| \\ &= 0.240 \times 301.65 \\ &= 72.4 \text{ kVA}\end{aligned}$$

$$\begin{aligned}\overline{S}_{L,ab} &= V_{ba} \times |\bar{I}_{ba}| \\ &= 0.240 \times 370.33 \\ &= 64.88 \text{ kVA}\end{aligned}$$

$$\begin{aligned}\overline{S}_{L,ca} &= V_{ac} \times |\bar{I}_{ac}| \\ &= 0.240 \times 129.11 \\ &= 30.99 \text{ kVA}\end{aligned}$$

d. The current flowing in the primary winding of each transformer can be found by dividing the current flow in each secondary winding by the turns ratio. Therefore,

$$n = \frac{7620 \text{ V}}{240 \text{ V}} = 31.75$$

and hence

$$\begin{aligned}\bar{I}_{AC} &= \frac{\bar{I}_{ac}}{n} \\ &= \frac{129.11\angle -33.1°}{31.75} \\ &= 4.07\angle -33.1° \text{ A}\end{aligned}$$

$$\begin{aligned}\bar{I}_{BA} &= \frac{\bar{I}_{ba}}{n} \\ &= \frac{301.65\angle 5.3°}{31.75} \\ &= 9.5\angle 5.3° \text{ A}\end{aligned}$$

$$\bar{I}_{CB} = \frac{\bar{I}_{cb}}{n}$$
$$= \frac{270.3\angle 204.7^\circ}{31.75}$$
$$= 8.51\angle 204.7^\circ \text{ A}$$

e. The line current flowing in each primary-phase wire can be found as

$$\bar{I}_A = \bar{I}_{AC} - \bar{I}_{BA}$$
$$= 4.07\angle -33.1^\circ - 9.5\angle 5.3^\circ$$
$$= -6.05 - j3.1$$
$$= 6.8\angle 270.1^\circ \text{ A}$$

$$\bar{I}_B = \bar{I}_{BA} - \bar{I}_{CB}$$
$$= 9.5\angle 5.3^\circ - 8.51\angle 204.7^\circ$$
$$= 17.19 + j4.44$$
$$= 17.76\angle 14.5^\circ \text{ A}$$

$$\bar{I}_C = \bar{I}_{CB} - \bar{I}_{AC}$$
$$= 8.51\angle 204.7^\circ - 4.07\angle -33.1^\circ$$
$$= -11.14 + j1.34$$
$$= 11.22\angle 186.8^\circ \text{ A}$$

3.11.2 OPEN-Δ OPEN-Δ TRANSFORMER CONNECTION

The Δ–Δ connection offers the greatest flexibility among the various connection configurations. One advantage of this connection is that if one transformer becomes damaged or is taken out of service, the remaining two transformers can still operate in what is referred to as the *open-Δ* or *V connection*, as depicted in Figure 3.26.

Let's consider a scenario where a balanced three-phase load with a unity power factor is supplied by all three transformers in a Δ–Δ bank. Removing one of the transformers from service will cause the currents in the remaining two transformers to increase by a ratio of 1.73, even though the output of the transformer bank remains the same with a unity power factor as before. However, the individual transformers will now operate at a power factor of 0.866. One transformer will deliver a leading load while the other delivers a lagging load. To ensure the safe operation of the remaining portion of the Δ–Δ transformer bank (i.e., the open-Δ open-Δ bank), the connected load needs to be reduced by 57.7%. This reduction can be calculated as follows:

$$S_{\Delta-\Delta} = \frac{\sqrt{3}V_{L-L}I_L}{1000} \text{ kVA} \tag{3.35}$$

and

$$S_{\angle-\angle} = \frac{\sqrt{3}V_{L-L}I_L}{\sqrt{3} \times 1000} \text{ kVA} \tag{3.36}$$

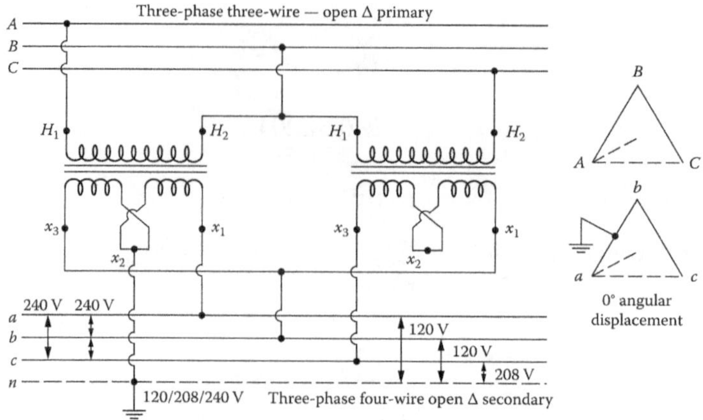

Figure 3.26 Three-phase four-wire open-delta connection. (Note that three-phase four-wire means a three-phase system made up of four wires.)

Therefore, by dividing Equation 3.35 by 3.36, side by side,

$$\frac{S_{\angle-\angle}}{S_{\Delta-\Delta}} = \frac{1}{\sqrt{3}}$$
$$= 0.577, \text{ or } 57.7\% \tag{3.37}$$

where

$S_{\Delta-\Delta}$ is the kVA rating of the $\Delta-\Delta$ bank
$S_{\angle-\angle}$is the kVA rating of the open-Δ bank
V_{L--L} is the line-to-line voltage, V
I_L is the line (or full load) current, A

It is important to note that the two transformers in the open-Δ bank constitute 66.6% of the total capacity of the three transformers in the Δ–Δ bank, but they can only supply 57.7% of the total capacity. This means that the two transformers have a reduced capacity compared to the full bank. The power factor at which these two transformers operate when the load is at unity power factor is 0.866, which can be calculated by taking the ratio of 57.7% to 66.6%. Operating in this manner allows the bank to still provide three-phase currents and voltages in their correct phase relationships. However, the capacity of the bank is decreased to 57.7% of its original rating, as it is based on the 86.6% capacity of the two units comprising the open-Δ bank. Open-Δ banks are commonly used in situations where the load is expected to grow, and when the load does increase, the third transformer can be added to complete a Δ–Δ bank.

Figure 3.27 illustrates an open-Δ connection for a 240 V three-phase three-wire secondary service with a 0° angular displacement. It is important to note that the neutral point "n" shown in the LV phasor diagram is a theoretical point and does not physically exist.

For the purpose of illustration, let's assume that a balanced three-phase load, such as an induction motor shown in the figure, with a lagging power factor, is connected to the secondary. In this case, the phase currents *a*, *b*, and *c* in the secondary can be determined using the following equations:

$$\bar{I}_a = \frac{S_{3\phi}}{\sqrt{3}V_{L-L}}\angle\theta_{\bar{I}a} \tag{3.38}$$

$$\bar{I}_b = \frac{S_{3\phi}}{\sqrt{3}V_{L-L}}\angle\theta_{\bar{I}b} \tag{3.39}$$

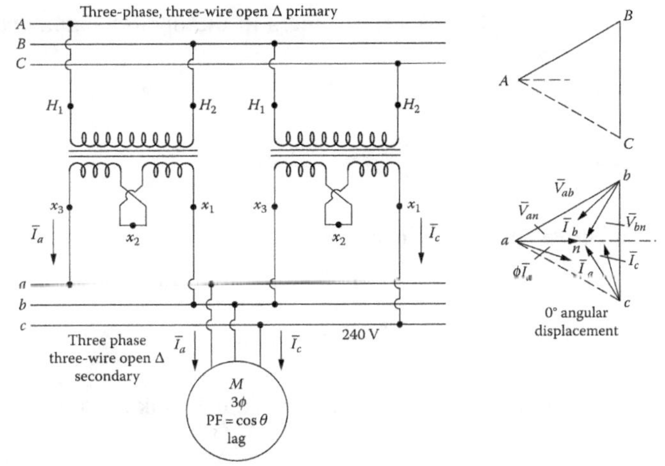

Figure 3.27 Three-phase three-wire open-delta connection.

$$\bar{I}_c = \frac{S_{3\phi}}{\sqrt{3}V_{L-L}} \angle \theta_{\bar{I}c} \tag{3.40}$$

The transformer kVA loads can be calculated as follows. The kVA load on the first transformer is

$$\begin{aligned} S_{T1} &= V_{L-L} \times |\bar{I}_a| \\ &= V_{L-L} \frac{S_{3\phi}}{\sqrt{3}V_{L-L}} \\ &= \frac{S_{3\phi}}{\sqrt{3}} \text{ kVA} \end{aligned} \tag{3.41}$$

and the kVA load on the second transformer is

$$\begin{aligned} S_{T2} &= V_{L-L} \times |\bar{I}_b| \\ &= V_{L-L} \frac{S_{3\phi}}{\sqrt{3}V_{L-L}} \\ &= \frac{S_{3\phi}}{\sqrt{3}} \text{ kVA} \end{aligned} \tag{3.42}$$

Therefore, the total load that the transformer bank can be loaded to (or the total "effective" transformer bank capacity) is

$$\sum_{i=1}^{2} S_{Ti} = \frac{2 \times S_{3\phi}}{\sqrt{3}} \tag{3.43}$$

and hence,

$$S_{3\phi} = \frac{\sqrt{3}}{2} \sum_{i=1}^{2} S_{Ti} \text{ kVA} \tag{3.44}$$

For example, if there are two 50 kVA transformers in the open-Δ bank, even though the total transformer bank capacity appears to be

$$\sum_{i=1}^{2} S_{Ti} = 100\ \text{kVA}$$

in reality the bank's "effective" maximum capacity is

$$S_{3\phi} = \frac{\sqrt{3}}{2} \times 100 = 86.6\ \text{kVA}$$

If there are three 50 kVA transformers in the Δ–Δ bank, the bank's maximum capacity is

$$S_{3\phi} = \sum_{i=1}^{3} S_{Ti} = 150\ \text{kVA}$$

which shows an increase of 73% over the 86.6 kVA load capacity.

Assume that the load power factor is cos θ and its angle can be calculated as

$$\theta = \theta_{\overline{V}_{an}} - \theta_{\overline{I}a} \tag{3.45}$$

or using $\overline{V}_{an}$ as the reference,

$$\theta = 0^\circ - \theta_{\overline{I}a} \tag{3.46}$$

If $\theta_{\overline{I}a}$ is negative, then θ is positive, which means it is the angle of a lagging load power factor. Also, it can be shown that

$$\theta = \theta_{\overline{V}_{bn}} - \theta_{\overline{I}b} \tag{3.47}$$

or

$$\theta = 120^\circ - \theta_{\overline{I}a} \tag{3.48}$$

and

$$\theta = \theta_{\overline{V}_{cn}} - \theta_{\overline{I}b} \tag{3.49}$$

or

$$\theta = +120^\circ - \theta_{\overline{I}c} \tag{3.50}$$

The transformer power factors for transformers 1 and 2 can be calculated as

$$\cos\theta_{T1} = \cos(\theta_{\overline{V}_{ab}} - \theta_{\overline{I}_a}) \tag{3.51}$$

or

$$\text{if } \theta_{\overline{I}_a} = -30^\circ,$$

$$\cos\theta_{T1} = \cos(\theta_{\overline{V}_{ab}} - 30^\circ) \tag{3.52}$$

Table 3.6
Effects of the Load Power Factor on the Transformer Power Factors

Load Power Factor		Transformer Power Factors	
$\cos\vartheta$	ϑ	$\cos\vartheta_{T1} = \cos(\vartheta + 30°)$	$\cos\vartheta_{T2} = \cos(\vartheta - 30°)$
0.866 lag	+30°	0.5 lag	1
1	0°	0.866 lag	0.866 lead

and

$$\cos\theta_{T2} = \cos(\theta_{\overline{V}_{bc}} - \theta_{\overline{I}_c}) \tag{3.53}$$

or

$$\text{if } \theta_{\overline{I}_c} = +30°,$$

$$\cos\theta_{T2} = \cos(\theta_{\overline{V}_{bc}} - 30°) \tag{3.54}$$

Therefore, the total real power output of the bank is

$$\begin{aligned} P_T &= P_{T_1} + P_{T_2} \\ &= V_{L-L}|\overline{I}_a|\cos(\theta + 30°) + V_{L-L} \times |\overline{I}_c|\cos(\theta - 30°) \\ &= \sqrt{3}V_{L-L}I_L\cos\theta \text{ kW} \end{aligned} \tag{3.55}$$

and, similarly, the total reactive power output of the bank is

$$\begin{aligned} Q_T &= Q_{T_1} + Q_{T_2} \\ &= V_{L-L}|\overline{I}_a|\sin(\theta + 30°) + V_{L-L} \times |\overline{I}_c|\sin(\theta - 30°) \\ &= \sqrt{3}V_{L-L}I_L\sin\theta \text{ kVAr} \end{aligned} \tag{3.56}$$

As indicated in Table 3.6, when the connected load of the bank operates with a lagging power factor of 0.866, corresponding to a power factor angle of 30°, transformer 1, as calculated in Equation 3.52, operates with a lagging power factor of 0.5, while transformer 2, as determined in Equation 3.54, operates with a unity power factor. However, when the bank load operates at unity power factor, resulting in a power factor angle of zero, transformer 1 operates with a lagging power factor of 0.866, and transformer 2 operates with a leading power factor of 0.866.

3.11.3 Y–Y TRANSFORMER CONNECTION

Figure 3.28 illustrates the connection of three transformers in a Y–Y configuration on a typical three-phase four-wire multigrounded system, providing 120/208Y-V service at 0° angular displacement. This setup offers a 208 V three-phase power supply for three-phase motors and a 120 V single-phase power supply for lighting and other small single-phase loads. It is advisable to distribute the single-phase loads evenly across the three phases.

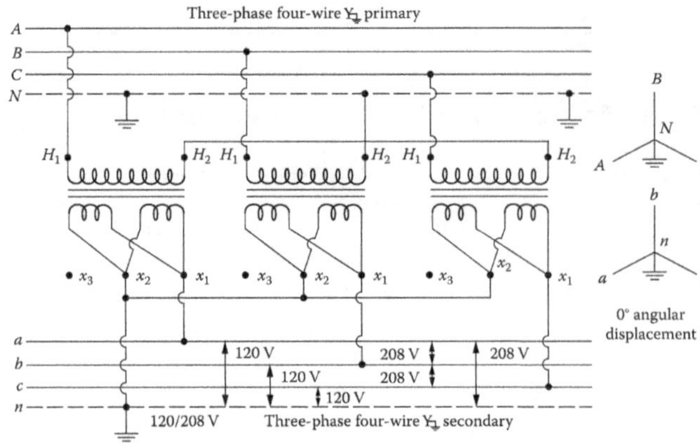

Figure 3.28 Y–Y connection to provide a 120/208 V grounded-wye three-phase four-wire multi-grounded service.

One advantage of the Y–Y connection is its compatibility with systems transitioning from delta to four-wire wye to increase capacity. In such cases, existing transformers can be utilized. For instance, if the previous distribution system operated at 2.4 kV delta, and the new system operates at 2.4/4.16Y kV, the existing 2.4/4.16Y kV transformers can be connected in a wye configuration and employed.

In the Y–Y transformer bank connection, each winding is exposed to only 57.7% (or 1/1.73) of the line voltage, while full line current flows through each transformer winding. It is worth noting that power-distribution circuits supplied from a Y–Y bank may introduce series of disturbances, such as telephone interference, in nearby communication circuits.

Moreover, it is crucial to solidly ground the primary neutral point and securely connect it to the system neutral. Failure to do so may lead to excessive voltages on the secondary side. When the transformer neutral is isolated from the system neutral, an unstable condition arises at the transformer neutral, primarily caused by third-harmonic voltages. On the other hand, grounding the transformer neutral significantly increases the risk of telephone interference and the possibility of resonance between the line capacitance to the ground and the transformer's magnetizing impedance.

3.11.4 Y–Δ TRANSFORMER CONNECTION

Figure 3.29 illustrates three single-phase transformers connected in Y–Δ configuration on a three-phase three-wire ungrounded-wye primary system. This arrangement provides 120/208/240 V three-phase four-wire delta secondary service at a 30° angular displacement.

Figure 3.30 depicts three transformers connected in Y–Δ configuration on a typical three-phase four-wire grounded-wye primary system. This setup offers 240 V three-phase three-wire delta secondary service at a 210° angular displacement.

The Y–Δ connection offers several advantages as the voltage between the outside legs of the wye is 1.73 times the voltage to neutral. This allows for higher distribution voltage by using transformers with primary windings rated for the voltage between any leg and the neutral. For example, 2.4 kV single-phase transformers can be connected in a wye configuration on the primary side to a 4.16 kV three-phase wye circuit.

In the Y–Δ connection, the voltage transformation ratio of the bank is 1.73 times the voltage transformation ratio of the individual transformers. When transformers of different capacities are employed, the maximum safe bank rating is three times the capacity of the smallest transformer.

The primary supply is typically a grounded wye circuit, which can be either three-wire or four-wire, including a neutral wire. The neutral wire, originating from the neutral of the wye-connected

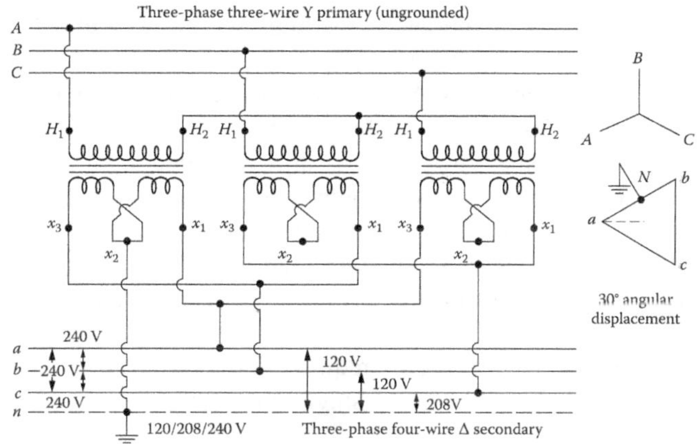

Figure 3.29 Y–Δ connection to provide a 120/208/240 V three-phase four-wire secondary service.

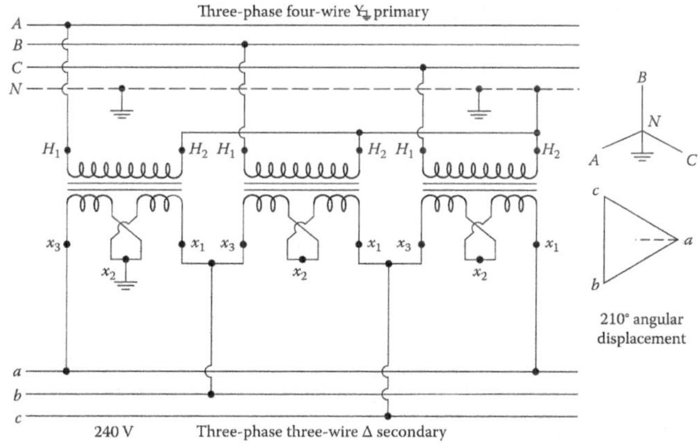

Figure 3.30 Y–Δ connection to provide a 240 V three-phase three-wire secondary service.

substation transformer bank supplying the primary circuit, may be completely independent of the secondary system or joined with the secondary system's neutral. In the case of an independent primary neutral, it serves as an isolated neutral and is grounded only at the substation.

When the same wire serves as both the primary and secondary neutral, it is grounded at multiple points, including each customer's service, creating a multigrounded common neutral. However, in either case, the primary-bank neutral is typically not connected to the primary-circuit neutral, as it is unnecessary and helps prevent transformer winding damage during phase-to-ground faults and excessive fuse blowing throughout the system.

In the case of a Y–Y connection, neglecting the neutral on the primary side can cause voltage waveforms to deviate from their sine-wave form. On the other hand, in the case of a Y–Δ connection, if the primary neutral is not connected, the voltage waveform may also deform. However, this deformation leads to circulating currents in the delta, which act as magnetizing currents and correct the waveform deformation. Hence, there are no objections to neglecting the neutral in this configuration.

However, when a transformer supplies a motor load, neglecting the neutral can result in damaging overcurrents in each three-phase motor circuit. These overcurrents cause equal amounts of current to flow in two wires of the motor branch circuit, while the total of these two currents flows in the third wire. If the unprotected circuit experiences the highest current, it may lead to motor burnout. This applies to ungrounded Y–Δ and Δ–Y banks.

If the transformer bank is used to supply both three-phase and single-phase loads, and if the bank neutral is solidly connected, the disconnection of a large transformer due to fuse operation can cause an even greater overload on the remaining two transformers. Detecting the blowing of a single fuse is challenging since there is no immediate noticeable decrease in service quality, and one of the two remaining transformers may be burnt out due to the overload.

On the contrary, if the bank neutral is not connected to the primary-circuit neutral and is left isolated, the disconnection of one transformer leads to a partial service interruption without the risk of transformer burnout.

To determine the approximate rated capacity required for a Y–Δ-connected bank with an isolated bank neutral to serve combined three-phase and single-phase loads, assuming a unity power factor, the following equation can be used:

$$\frac{2S_{1\phi}+S_{3\phi}}{3}$$

which is equal to rated transformer capacity across lighting phase, where $S_{1\phi}$ is the single-phase load, kVA; $S_{3\phi}$ is the three-phase load, kVA.

To summarize, when the primary-side neutral of the transformer bank is connected to the primary-circuit neutral in a Y–Δ configuration, the transformer bank may be at risk of burnout due to the following factors:

1. The transformer bank may act as a grounding transformer bank for unbalanced primary conditions and may supply fault current to any fault on the circuit to which it is connected, reducing its own capacity for connected load.
2. The transformer bank may be overloaded if one of the protective fuses opens on a line-to-ground fault, leaving the bank with only the capacity of an open-Y open-Δ bank.
3. The transformer bank causes circulating current in the delta in an attempt to balance any unbalanced load connected to the primary line.
4. The transformer bank provides a delta in which triple-harmonic currents circulate.

All the effects mentioned above can lead to additional current flow in the transformer bank beyond its normal load current, which can ultimately result in the burnout of the transformer bank.

3.11.5 OPEN-Y OPEN-Δ TRANSFORMER CONNECTION

As depicted in Figure 3.31, when one phase of the primary supply is opened, the transformer bank is transformed into an open-Y open-Δ configuration and continues to provide power to the three-phase load at a reduced capacity.

Example 3.6. Consider a scenario where two single-phase transformers are connected in an open-Y open-Δ configuration to supply power to a three-phase wye-connected load with a power factor of 0.80 (lagging) and a single-phase load of 50 kVA with a power factor of 0.90 (lagging), as shown in Figure 3.32. The primary-side voltage of the bank is 7620/13,200 V, while the secondary-side voltage is 240 V. Based on the provided information, the following calculations can be performed:

a. The line current flowing in each secondary-phase wire
b. The current flowing in the secondary winding of each transformer
c. The kVA load on each transformer
d. The current flowing in each primary-phase wire and in the primary neutral

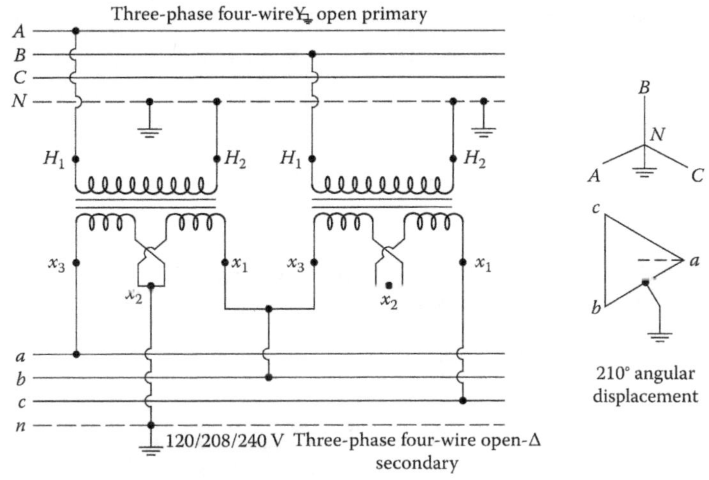

Figure 3.31 Open-wye open-delta connection.

Solution. a. Using the voltage drop $\overline{V}_{an}$ as the reference, the three-phase components of the line currents can be found as

$$\begin{aligned}|\overline{I}_{a,3\phi}| &= |\overline{I}_{b,3\phi}| \\ &= |\overline{I}_{c,3\phi}| \\ &= \frac{S_{L,3\phi}}{\sqrt{3}V_{L-L}} \\ &= \frac{100}{\sqrt{3}\times 0.240} = 240.8\ \text{A}\end{aligned} \tag{3.57}$$

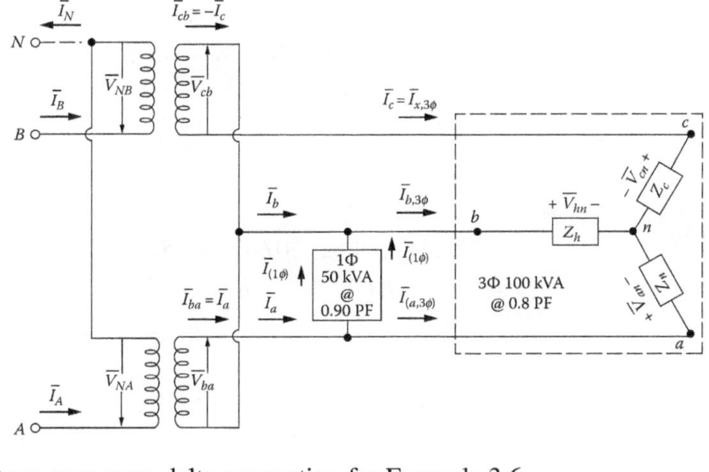

Figure 3.32 Open-wye open-delta connection for Example 3.6.

Since the three-phase load has a lagging power factor of 0.80,

$$\begin{aligned}\bar{I}_{a,3\phi} &= |\bar{I}_{a,3\phi}|(\cos\theta - j\sin\theta)\\ &= 240.8(0.80 - j0.60)\\ &= 192.68 - j144.5\\ &= 240.8\angle -36.9^\circ \text{ A}\end{aligned} \tag{3.58}$$

$$\begin{aligned}\bar{I}_{b,3\phi} &= a^2\bar{I}_{a,3\phi}\\ &= (1\angle 240^\circ)(240.8\angle -36.9^\circ)\\ &= 240.8\angle 203.1^\circ\\ &= -221.5 - j94.5 \text{ A}\end{aligned} \tag{3.59}$$

$$\begin{aligned}\bar{I}_{c,3\phi} &= a\bar{I}_{a,3\phi}\\ &= (1\angle 240^\circ)(240.8\angle -36.9^\circ)\\ &= 240.8\angle 83.1^\circ\\ &= 28.1 - j239 \text{ A}\end{aligned} \tag{3.60}$$

The single-phase component of the line currents can be found as

$$\begin{aligned}\bar{I}_{1\phi} &= \frac{S_{L,1\phi}}{V_{L-L}}\\ &= \frac{50}{0.240} = 208.33 \text{ A}\end{aligned} \tag{3.61}$$

therefore,

$$\begin{aligned}\bar{I}_{1\phi} &= [|\bar{I}_{1\phi}|\cos(30^\circ - \theta_1) j\sin(30^\circ - \theta_1)]\\ &= 208.33[\cos(30^\circ - 25.8^\circ) j\sin(30^\circ - 25.8^\circ)]\\ &= 207.78 + j15.26 \text{ A}\end{aligned} \tag{3.62}$$

Hence, the line currents flowing in each secondary-phase wire can be found as

$$\begin{aligned}\bar{I}_a &= \bar{I}_{a,3\phi} - \bar{I}_{1\phi}\\ &= -192.68 - j144.5 - 207.78 - j15.26\\ &= 400.46 - j129.24\\ &= 420.8\angle -17.9^\circ \text{ A}\end{aligned}$$

$$\begin{aligned}\bar{I}_b &= \bar{I}_{b,3\phi} - \bar{I}_{1\phi}\\ &= -221.5 - j94.5 - 207.78 - j15.26\\ &= 429.28 - j109.76\\ &= 442.8\angle -165.7^\circ \text{ A}\end{aligned}$$

$$\begin{aligned}\bar{I}_b &= \bar{I}_{b,3\phi} - \bar{I}_{1\phi}\\ &= -221.5 - j94.5 - 207.78 - j15.26\\ &= 429.28 - j109.76\\ &= 442.8\angle -165.7^\circ \text{ A}\end{aligned}$$

$$\begin{aligned}\bar{I}_c &= a^2\bar{I}_{c,3\phi}\\ &= 240.8\angle 83.1^\circ \text{ A}\end{aligned}$$

b. The current flowing in the secondary winding of each transformer is

$$\begin{aligned}\bar{I}_{ba} &= \bar{I}_a \\ &= 420.8\angle -17.9^\circ \text{ A}\end{aligned}$$

$$\begin{aligned}\bar{I}_{cb} &= -\bar{I}_c \\ &= -240.8\angle 83.1^\circ \\ &= 240.8\angle 83.1^\circ + 180^\circ \\ &= 240.8\angle 263.1^\circ \text{ A}\end{aligned}$$

c. The kVA load on each transformer can be found as

$$\begin{aligned}S_{L,ba} &= V_{ba} \times |\bar{I}_{ba}| \\ &= 0.240 \times 420.8 \\ &= 101 \text{ kVA}\end{aligned} \tag{3.63a}$$

and the power-factor angle is

$$\begin{aligned}S_{L,cb} &= V_{cb} \times |\bar{I}_{cb}| \\ &= 0.240 \times 240.8 \\ &= 57.8 \text{ kVA}\end{aligned} \tag{3.63b}$$

d. The current flowing in each primary-phase wire can be found by dividing the current flow in each secondary winding by the turns ratio. Therefore,

$$n = \frac{7920 \text{ V}}{240 \text{ V}} = 31.7$$

and hence

$$\begin{aligned}\bar{I}_A &= \frac{\bar{I}_{ba}}{n} \\ &= \frac{420.8\angle -17.9^\circ}{31.75} \\ &= 12.6 - j4.07 \\ &= 13.25\angle -17.9^\circ \text{ A}\end{aligned} \tag{3.64a}$$

$$\begin{aligned}\bar{I}_B &= \frac{\bar{I}_{cb}}{n} \\ &= \frac{420.8\angle -263.1^\circ}{31.75} \\ &= 0.91 - j7.52 \\ &= 7.58\angle -263.1^\circ \text{ A}\end{aligned} \tag{3.64b}$$

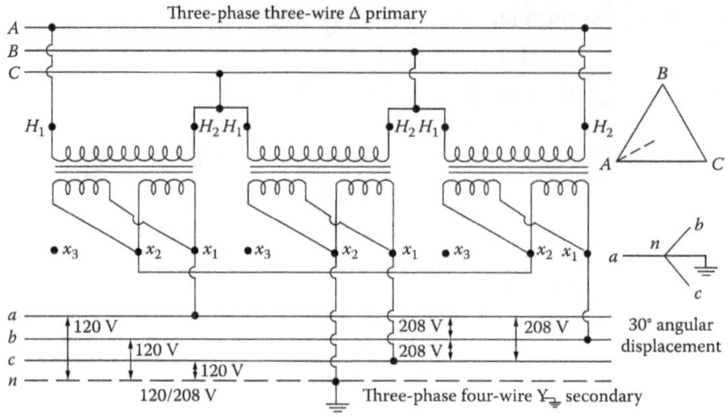

Figure 3.33 Δ–Y connection with 30° angular displacement.

Therefore, the current in the primary neutral is

$$\begin{aligned}\bar{I}_N &= \bar{I}_A + \bar{I}_B \\ &= 13.25\angle -17.9^\circ + 7.58\angle 263.1^\circ \\ &= 11.69 - j11.6 \\ &= 16.47\angle -44.8^\circ \text{ A}\end{aligned} \tag{3.65}$$

3.11.6 Δ–Y TRANSFORMER CONNECTION

Figures 3.33 and 3.34 illustrate the connection of three single-phase transformers in a Δ–Y configuration to deliver 120/208 V three-phase four-wire grounded-wye service, with angular displacements of 30° and 210°, respectively.

To address the issue of unbalanced primary currents in transformer banks with single-phase lighting loads concentrated on one phase, the Δ–Y system is commonly employed. In this system, the neutral of the secondary three-phase system is grounded, and single-phase loads are connected between different phase wires and the neutral, while three-phase loads are connected to the phase wires. This arrangement allows for balanced distribution of single-phase loads across the three phases in each bank, and banks can be paralleled if desired.

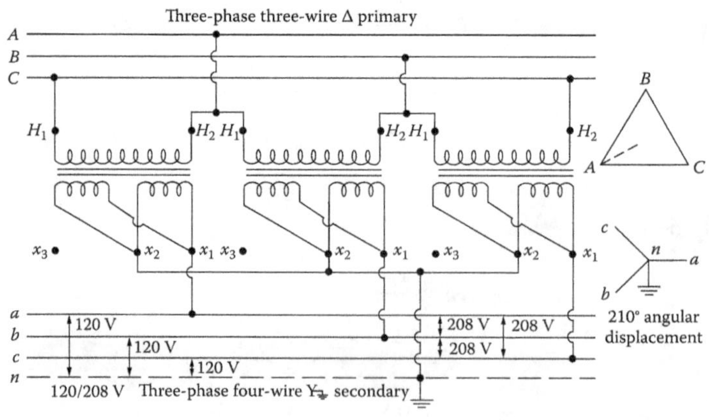

Figure 3.34 Δ–Y connection with 210° angular displacement.

When transformers of different capacities are used, the maximum safe rating of the transformer bank is three times the capacity of the smallest transformer. However, if one transformer becomes damaged or is taken out of service, the entire transformer bank becomes inoperative.

In both the Y–Y and Δ–Δ connections, the line voltages on the secondary side are in phase with the line voltages on the primary side. In contrast, with the Y–Δ or Δ–Y connections, the line voltages on the secondary side are shifted by 30° relative to the line voltages on the primary side. As a result, a Y–Δ or Δ–Y transformer bank cannot be operated in parallel with a Δ–Δ or Y–Y transformer bank. Ensuring identical angular displacements becomes particularly crucial when interconnecting three-phase transformers within the same secondary system or when paralleling three-phase banks of single-phase transformers. Additional conditions for successfully paralleling three-phase distribution transformers (DTs) include the following:

1. All transformers have identical frequency ratings.
2. All transformers have identical voltage ratings.
3. All transformers have identical tap settings.
4. Per unit impedance of one transformer is between 0.925 and 1.075 of the other.

The Δ–Y step-up and Y–Δ step-down connections are particularly suitable for high-voltage (HV) transmission systems. They offer cost-effective solutions and provide a stable neutral point that can be solidly grounded or grounded through resistance to ensure system stability and prevent oscillations.

3.12 THREE PHASE TRANSFORMERS

When it comes to transforming three-phase voltages, three-phase transformers are used. These transformers consist of a core with three legs, with one primary and one secondary winding of each phase placed on separate legs. The core can be designed with only three legs because the fluxes generated by the three windings are 120° out of phase with each other. Two core legs serve as the return path for the flux in the third leg. For instance, if the flux in one leg reaches its maximum value at a certain instant, the flux in the other two legs is half that value and flows in the opposite direction at the same instant.

Compared to three single-phase transformers with the same total capacity rating, a three-phase transformer occupies less space since all three windings can be placed on a single core. Additionally, three-phase transformers are generally more efficient and cost-effective than equivalent single-phase transformer banks, especially at higher ratings. However, if one-phase winding becomes damaged, the entire three-phase transformer must be taken out of service. Three-phase transformers can be connected in any of the aforementioned connection types, with all connections made inside the transformer tank.

Figures 3.35 through 3.39 illustrate various connection diagrams for three-phase transformers. Figure 3.35 shows a Δ–Δ connection, which provides 120/208/240 V three-phase four-wire secondary service at 0° angular displacement. This connection is used to supply 240 V three-phase loads along with small amounts of 120 V single-phase loads. Transformers with capacities up to 150 kVA are typically designed such that when 5% of the rated kVAs is taken from the 120 V tap on the 240 V connection, the three-phase capacity is reduced by 25%.

Figure 3.36 illustrates a three-phase open-Δ connection for 120/240 V service. This connection is used to supply large single-phase loads at both 120 V and 240 V simultaneously, along with small amounts of three-phase load. The two sets of windings in the transformer have different capacity sizes in terms of kVA. The transformer efficiency is relatively low, especially for three-phase loads. The transformer is rated at only 86.6% of the rating of two sets of windings when they are equal in size, and even less when they are unequal.

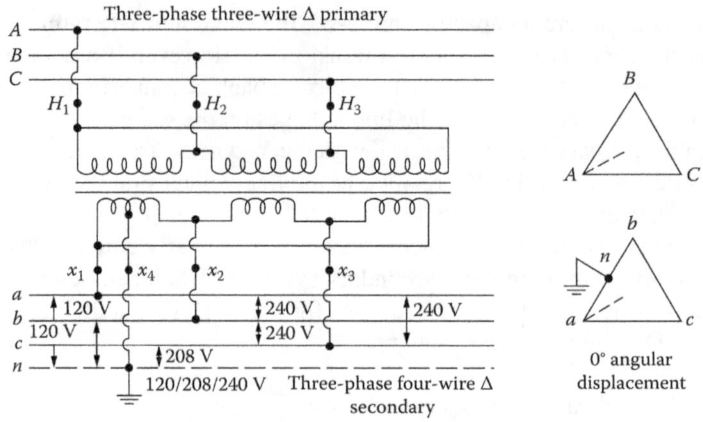

Figure 3.35 Three-phase transformer connected in delta–delta.

Figure 3.37 presents a three-phase Y–Δ connection for 120/240 V service at a 30° angular displacement. This connection is used to supply three-phase 240 V loads along with small amounts of 120 V single-phase loads.

Figure 3.38 showcases a three-phase open-Y open-Δ connection for 120/240 V service at a 30° angular displacement. The statements regarding efficiency and capacity for the three-phase open-Δ connection also apply to this connection.

Finally, Figure 3.39 depicts a three-phase transformer connected in Y–Y for 120/208Y-V service. This connection allows single-phase loads to be balanced across the three phases.

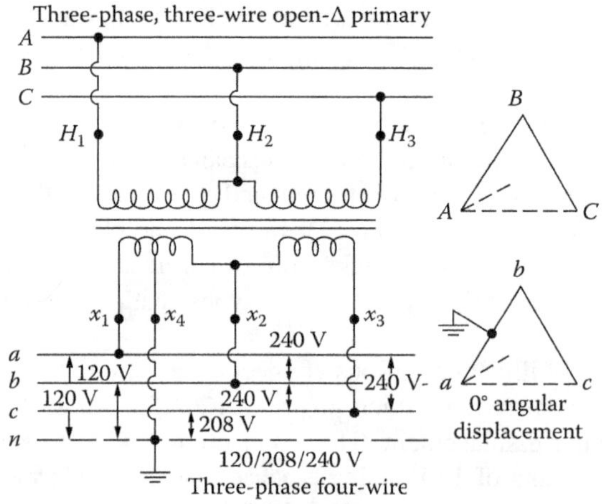

Figure 3.36 Three-phase transformer connected in open-delta.

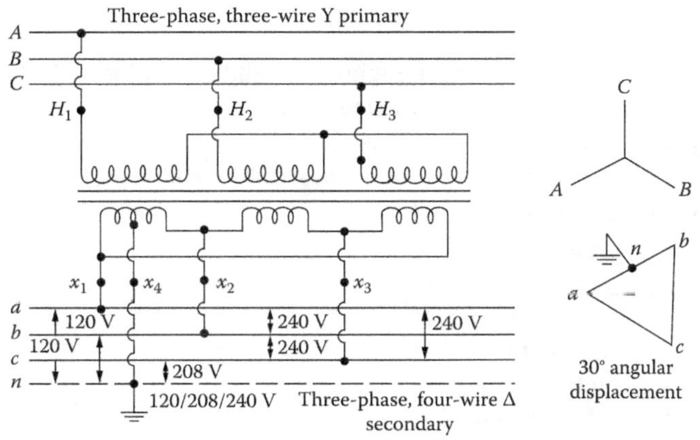

Figure 3.37 Three-phase transformer connected in Y–Δ.

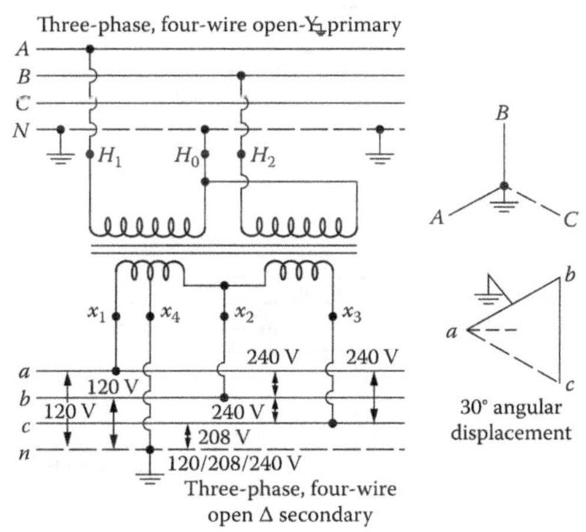

Figure 3.38 Three-phase transformer connected in open-wye open-delta.

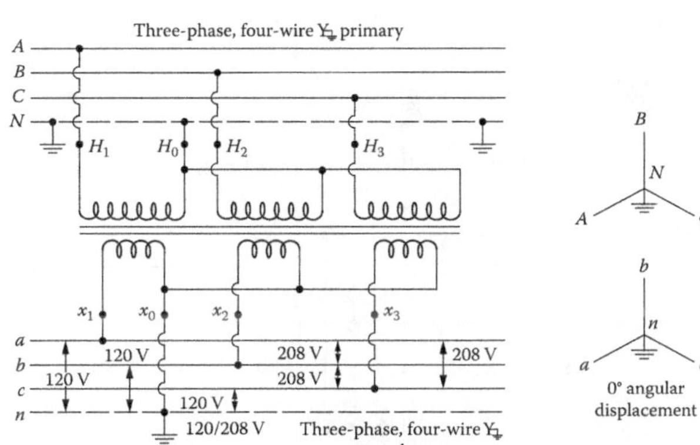

Figure 3.39 Three-phase transformer connected in Y–Y.

PROBLEMS

3.1 Consider the T–T connection given in Figure 3.41 and determine the following:

a. Draw the LV diagram, correctly oriented on the 0° reference shown.
b. Find the value of the $\overline{V}_{ab}$ and $\overline{V}_{an}$ phasors.
c. Find the magnitudes of the following rated winding voltages:

i. The voltage $V_{H_1H_2}$ on transformer 1
ii. The voltage $V_{x_1\phi}$ on transformer 1
iii. The voltage $V_{\phi x_2}$ on transformer 1
iv. The voltage V_{H_10} on transformer 2
v. The voltage V_{0H} on transformer 2
vi. The voltage V_{x_1n} on transformer 2
vii. The voltage V_{nx} on transformer 2

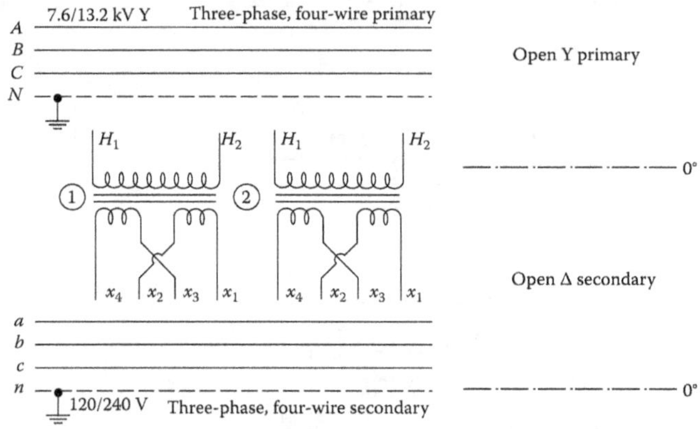

Figure 3.40 For Problem 3.1.

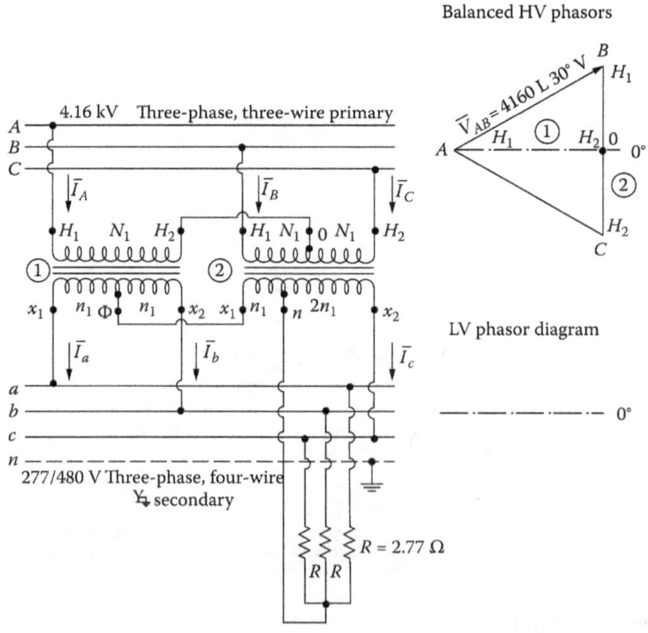

Figure 3.41 A T–T connection.

3.2 Assume that the T–T transformer bank given in Problem 3.3 is loaded with the balanced resistors given. Assume that the secondary voltages are perfectly balanced; the necessary HV applied voltages then are not perfectly balanced. Use secondary voltages of 480 V and neglect magnetizing currents. Determine the following:

a. The LV current phasors
b. The HV current phasors

3.3 Use the results of Problems 3.3 and 3.4 and apply the complex power formula $S = P + jQ = \overline{V}\overline{I}^*$ four times, once for each LV winding, for example, a part of the output of transformer 1 is $\overline{V}_{X_2X_2}\overline{I}_a^*$. Based on these results, find the following:

a. Total complex power output from the T–T bank. (Does your result agree with that which is easily computed as input to the resistors?)
b. The necessary kVA ratings of both LV windings of both the transformers.
c. The ratio of total kVA ratings of all LV windings in the transformer bank to the total kVA output from the bank.

3.4 Consider Figure 3.42 and assume that the motor is rated 25 hp and is mechanically loaded so that it draws 25.0 kVA three-phase input at cos $\theta = 0.866$ lagging power factor:

a. Draw the necessary HV connections so that the LVs shall be as shown, that is, of *abc* phase sequence.
b. Find the power factors cos θ_{T1} and cos θ_{T2} at which each transformer operates.
c. Find the ratio of voltampere load on one transformer to total VA delivered to the load.

3.5 Consider Figure P3.4 and assume that the two-transformer T–T bank delivers 120/208 V three-phase four-wire service from a three-phase three-wire 4160 V primary line. The problem is to determine if this bank can carry unbalanced loads even though the primary neutral terminal N is not connected to the source neutral. (If it can, the T–T performance is quite different from the three-transformer wye-grounded wye bank.) Use the ideal-transformer theory and pursue the question as follows:

a. Load phase a-n with R = 1.20 Ω resistance and then find the following six complex currents numerically: $\overline{I}_a, \overline{I}_b, \overline{I}_c, \overline{I}_A, \overline{I}_B$, and $\overline{I}_C$
b. Find the following complex powers of windings by using the $= P + jQ = \overline{V}\overline{I}^*$ equation numerically:

$$S_{T1_{(x1-n)}} = \text{complex power of } x_1 - n \text{ portion of transformer 1}$$

$$S_{T1_{(H_1-H_2)}} = \text{complex power of } H_1 - H_2 \text{ portion of transformer 1}$$

$$S_{T2_{(H_1-0)}} = \text{complex power of } H_1 - 0 \text{ portion of transformer 2}$$

$$S_{T2_{(H_2-0)}} = \text{complex power of } H_2 - 0 \text{ portion of transformer 2}$$

c. Do your results indicate that this bank will carry unbalanced loads successfully? Why?

3.6 Figure 3.44 shows two single-phase transformers, each with a 7620 V HV winding and two 120 V LV windings. The diagram shows the proposed connections for an open-Y to open-Δ bank and the HV applied phasor-voltage drops. Here, *abc* phase sequence at LV and HV sides and 120/240 V are required.

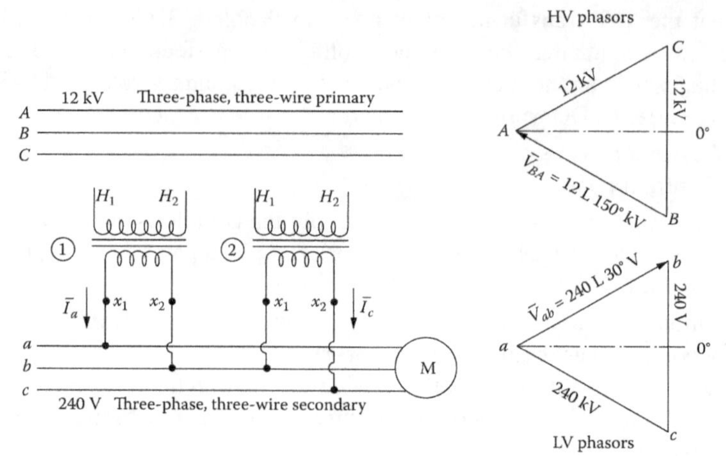

Figure 3.42 For Problem 3.4.

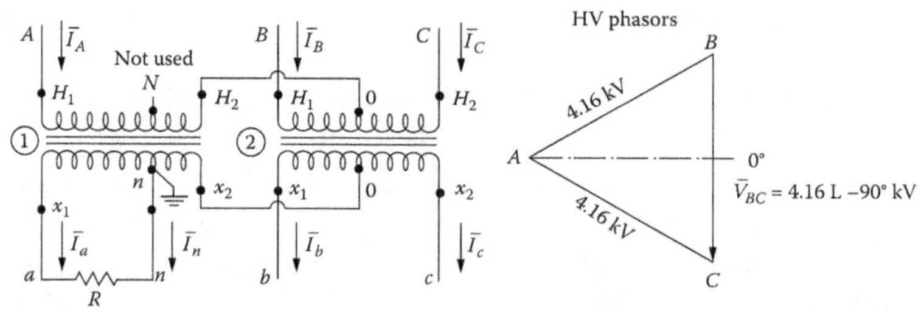

Figure 3.43 For Problem 3.7.

a. Sketch the LV phasor diagram, correctly oriented on the 0° reference line. Label it adequately with x's (1), and (2), etc., to identify.
b. State whether or not the proposed connections will output the required three-phase four-wire 120/240 V delta LV.

3.7 A large number of 25 kVA DTs are to be purchased. Two competitive bids have been received. The bid data are tabulated as follows:
Evaluate the bids on the basis of total annual cost (TAC) and recommend the purchase of the

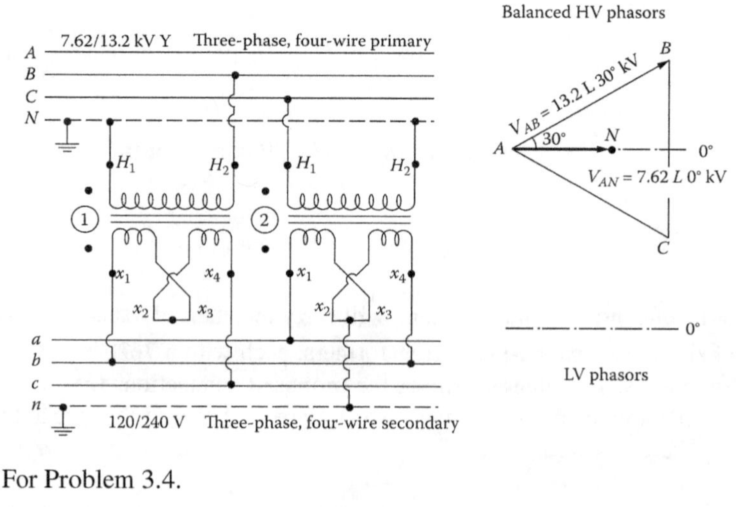

Figure 3.44 For Problem 3.4.

Transformer	Cost of Transformer Delivered to NL&NP's Warehouse	Core Loss at Rated Load (W)	Copper Loss at Rated Voltage and Frequency (W)	Per Unit Exciting Current
A	$355	360	130	0.015
B	$345	380	150	0.02

one having the least TAC. The cost of installing a transformer is not to be included in this study. The following system data are given:

Annual peak load on transformer = 35 kVA
Annual loss factor = 0.15
Per unit annual fixed charge rate = 0.15
Installed cost of shunt capacitors = $10/kVAr
Incremental cost of off-peak energy = $0.01/kWh
Incremental cost of on-peak energy = $0.012/kWh
Investment cost of power system upstream from DTs = $300/kVA.

Calculate the TAC of owning and operating one such transformer and state which transformer should be purchased. (*Hint*: Study the relevant equations in Chapter 6 before starting to calculate.)

3.8 Assume that a 250 kVA DT is used for single-phase pole mounting. The transformer is connected phase-to-neutral 7200 V on the primary and 2520 V phase-to-neutral on the secondary side. The leakage impedance of the transformer is 3.5%. Based on the given information, determine the following:

a. Assume that the transformer has 0.7 pu A in the HV winding. Find the actual current values in the HV and LV windings. What is the value of the current in the LV winding in per units?
b. Find the impedance of the transformer as referred to the HV and LV windings in ohms.
c. Assume that the LV terminals of the transformer are short-circuited and 0.22 per unit voltage is applied to the HV winding. Find the HV and LV winding currents that exist as a result of the short circuit in per units and amps.
d. Determine the internal voltage drop of the transformer, due to its leakage impedance, if a 1.2 per unit current flows in the HV winding. Give the result in per units and volts.

Part III

Projecting Network Expansion

4 Design of Subtransmission Lines and Distribution Substations

A teacher affects eternity.

Author Unknown

Education is the best provision for old age.

Aristotle, 365 BC

Education is ⋯ hanging around until you've caught on.

Will Rogers

4.1 INTRODUCTION

The distribution system, broadly defined, encompasses the components of the electric utility system situated between the bulk power source and the customers' service switches. Within this definition, the distribution system comprises the following elements:

1. Subtransmission system
2. Distribution substations
3. Distribution or primary feeders
4. Distribution transformers
5. Secondary circuits
6. Service drops

However, some engineers specializing in distribution systems prefer to define it as the portion of the electric utility system extending from the distribution substations to the service entrance of consumers. Figure 4.1 illustrates a one-line diagram depicting a typical distribution system. Subtransmission circuits serve the purpose of transmitting power from bulk power sources to the distribution substations. The subtransmission voltage typically ranges between 12.47 and 245 kV. The distribution substation, composed of power transformers and necessary voltage-regulating equipment, buses, and switchgear, then steps down the subtransmission voltage to a lower primary system voltage suitable for local distribution. The three-phase primary feeder, typically operating within the range of 4.16–34.5 kV, distributes power from the substation's low-voltage bus to its load center, where it branches out into three-phase subfeeders and single laterals.

To further facilitate distribution, distribution transformers with ratings ranging from 10 to 500 kVA are typically connected to each primary feeder, subfeeders, and laterals. These transformers reduce the distribution voltage to the desired utilization voltage. The secondary windings provide the path through which energy is distributed from the distribution transformer to consumers via service drops. This chapter provides a brief overview of the design aspects related to subtransmission and distribution substations [77].

DOI: 10.1201/9781003129721-4

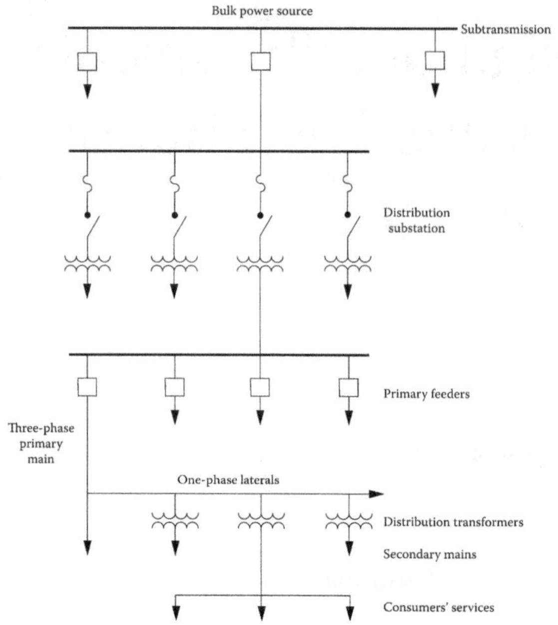

Figure 4.1 One-line diagram of a typical distribution system.

4.2 SUBTRANSMISSION

The subtransmission system is a crucial component of the electric utility system responsible for delivering power from large transmission substations. These subtransmission circuits can be constructed using overhead open-wire configuration on wood poles or underground cables. The voltage levels of these circuits typically range from 12.47 kV to 245 kV, with the most common levels being 69 kV, 115 kV, and 138 kV. There is a growing trend toward using higher primary voltages, leading to an increased adoption of higher voltage levels in subtransmission systems.

The design of subtransmission systems can vary, ranging from simple radial systems to more complex subtransmission networks. The primary factors influencing the design choices are cost and reliability. Figure 4.2 depicts a radial subtransmission system where circuits radiate directly from the bulk power stations to the distribution substations. While this design is straightforward and cost-effective, it suffers from limited service continuity. Consequently, radial systems are generally not preferred. Instead, an enhanced form of the radial-type subtransmission design, as illustrated in Figure 4.3, is typically favored. This design allows for faster service restoration in the event of a fault occurring on one of the subtransmission circuits.

Typically, to ensure higher service reliability, the subtransmission system is designed with loop circuits or multiple circuits forming a subtransmission grid or network. Figure 4.4 illustrates a loop-type subtransmission system. In this design, a single circuit originates from a bulk power bus, passes through several substations, and returns to the same bus.

Figure 4.5 depicts a grid-type subtransmission system, which incorporates multiple circuits. The distribution substations are interconnected, and the design may include more than one bulk power source. This configuration offers the highest level of service reliability but requires sophisticated control of power flow and relaying. The grid-type subtransmission design is the most commonly employed form of subtransmission due to its effectiveness and widespread usage.

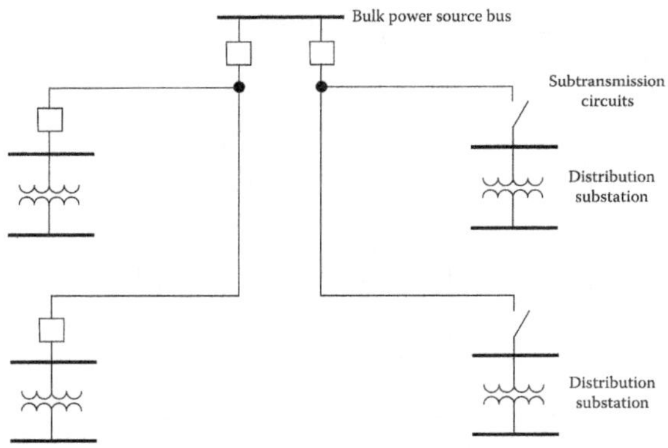

Figure 4.2 Radial-type subtransmission.

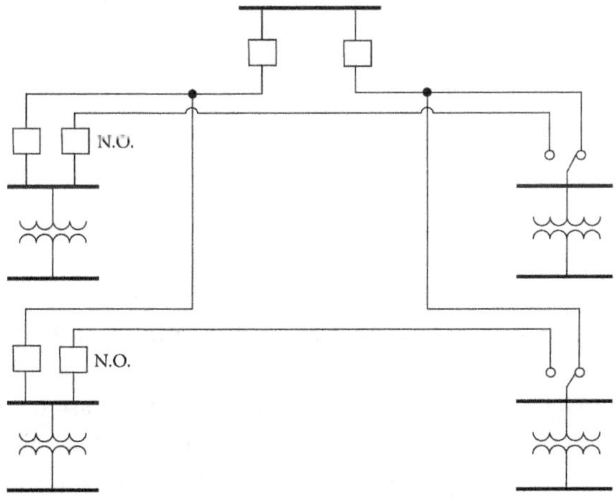

Figure 4.3 Improved form of radial-type subtransmission.

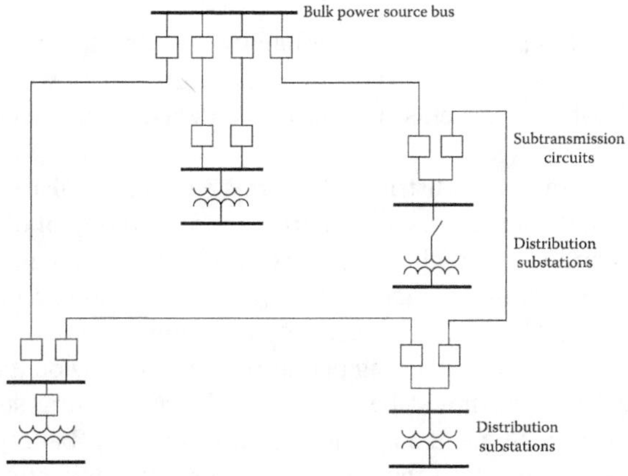

Figure 4.4 Loop-type subtransmission.

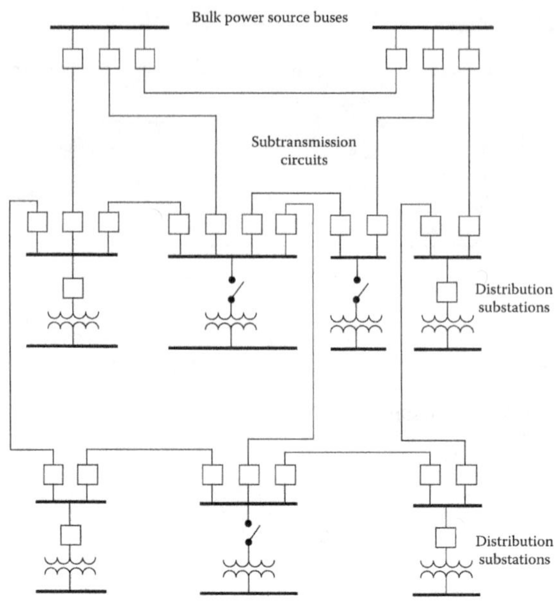

Figure 4.5 Grid- or network-type subtransmission.

4.3 STATE-OF-THE-ART UPDATE

By design, distribution feeders are typically planned to operate within a radial configuration under normal conditions. The design of distribution substations relies on a balance of Volt/VAr control and optimal feeder reconfiguration to coordinate control and management at the feeder level [80]. Enhancements and maintenance at various locations can improve phase balancing of an unbalanced radial feeder, taking into account switch observations in an averaging sense [81]. Alternatively, a heuristic approach utilizing a geographic information system (GIS) can be implemented [82]. In most cases, utilities recognize that deterministic criteria can improve the system by minimizing costs while maintaining satisfactory levels of customer expectations [83]. Further advancements in distribution system planning extend beyond medium-voltage (MV) to include the low-voltage (LV) distribution system, taking into consideration investment and operation costs of the secondary grid [84].

As substation technology has advanced, broader bandwidth communications and more affordable storage mediums have contributed to the establishment of informatics within substations [85]. The development of knowledge-based systems for substation decision support has also improved over time [86].

With the increasing penetration of distributed energy resources (DERs) and networked micro-grids in distribution feeders, voltage issues arising from the integration of distributed generation can impact voltage regulators and load tap changers (LTCs) in today's systems [87]. Addressing these complex planning problems requires a robust strategy to accommodate emerging changes. Optimal sizing and location of distribution substations, considering fixed and variable costs, should be part of the optimization formulation, extending operational constraints on the existing network [88]. Additionally, geographical factors should be considered by associating substation locations with population growth [89]. This includes the consideration of growing battery storage for household solar power generation [90]. Amidst all these considerations, the collective sum of loads must be effectively distributed with protective relaying, such as circuit breakers at the feeder head and re-closers at various points downstream of the feeder [91]. Similarly, the low-voltage (LV) side of the distribution network is accounted for in the design, where sensitivity to detect and trip must be maintained while remaining coordinated with feeder breaker settings [92]. As the communication infrastructure evolves in substation automation, the sophistication of relaying has transformed distribution automation and planning [93]. This technological advancement has expanded to the LV network, incorporating additional metering measurements [94]. Such instrumentation encompasses processes and features that enable automated recovery of distributed resources [95–99].

Subtransmission line costs are determined by a per-mile cost along with a termination cost at the end of the line, which is associated with the substation where it is connected. According to the ABB Guidebook [100], using 1994 prices as a reference, the costs can range from as low as $50,000 per mile for a 46 kV subtransmission line supported by wooden poles, with a capacity of approximately 50 MVA ($1 per kVA-mile), to over $1,000,000 per mile for a 500 kV double circuit construction with a capacity of 2000 MVA ($0.5 per kVA-mile).

4.4 DISTRIBUTION SUBSTATIONS

The design of distribution substations in the electric utility industry has undergone standardization based on previous experiences, but the process of standardization is ongoing. Typical distribution substations are depicted in Figures 4.6 and 4.7, showcasing their appealing aesthetics achieved through the use of underground cables for incoming and outgoing connections, as well as between the transformer secondary and the low-voltage bus structure. Some of these substations incorporate automatic switching for sectionalizing purposes, while others employ preferred-emergency automatic transfer functionality.

Figure 4.8 provides an overview of a modern substation. It showcases two 115 kV, 1200 A vertical-break-style circuit switchers that are responsible for switching and protecting two transformers supplying power to a large tire manufacturing plant. The transformer in the foreground has a rating of 15/20/28 MVA, 115/4.16 kV, and an impedance of 8.8%. The second transformer is rated at 15/20/28 MVA, 115/13.8 kV, with an impedance of 9.1%. For a closer look at a typical modern distribution substation transformer, refer to Figure 4.9. Figure 4.10 showcases a primary unit substation transformer, while Figure 4.11 displays a secondary unit substation transformer. Figure 4.12 highlights a GEAFOL solid dielectric transformer rated at 630 kVA, 10/0.4 kV. Additionally, Figure 4.13 presents an oil distribution transformer, with a cutaway view of a TUMERIC transformer featuring an oil expansion tank in the foreground and TUNORMA with an oil expansion tank in the background. Last, Figure 4.14 depicts another 630 kVA, 10/0.4 kV GEAFOL solid dielectric transformer.

Figure 4.15 showcases a three-phase, core-type transformer with a rating of 40 MVA, 110 kV ± 16%/21 kV. It has dimensions of 5.2 m in height, 9.4 m in length, 3 m in width, and weighs 80 tons. For a different type of transformer, refer to Figure 4.16, which presents a three-phase, shell-type transformer rated at 850/950/1100 MVA, 415 kV ± 11%/27 kV. This transformer stands at a height of 11.3 m, has a length of 14 m, a width of 5.7 m, and weighs 552 tons (excluding cooling oil). Figure 4.17 showcases a fully assembled three-phase, step-up transformer rated at 910 MVA, 20.5/500 kV. This transformer measures 39 ft in height, 36 ft in length, 29 ft in width, and weighs 562 tons. Last, Figure 4.18 provides a view of a typical core and coil assembly of a three-phase, core-type power transformer.

A typical substation consists of various equipment, including power transformers, circuit breakers, disconnecting switches, station buses and insulators, current-limiting reactors, shunt reactors, current transformers, potential transformers, capacitor voltage transformers, coupling capacitors, series capacitors, shunt capacitors, grounding system, lightning arresters and/or gaps, line traps, protective relays, station batteries, and other apparatus [101].

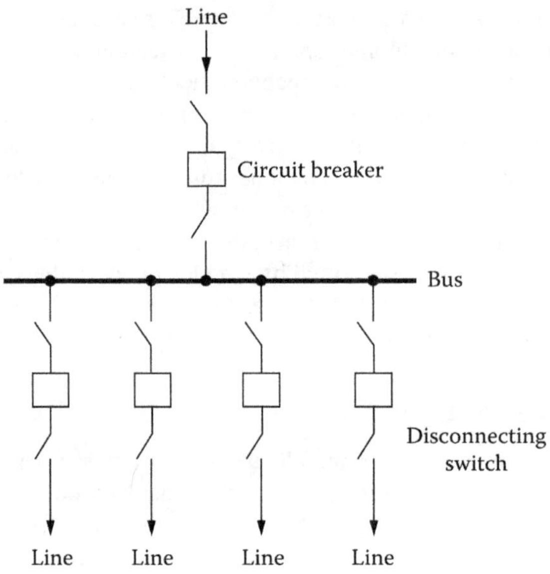

Figure 4.6 A typical single bus scheme.

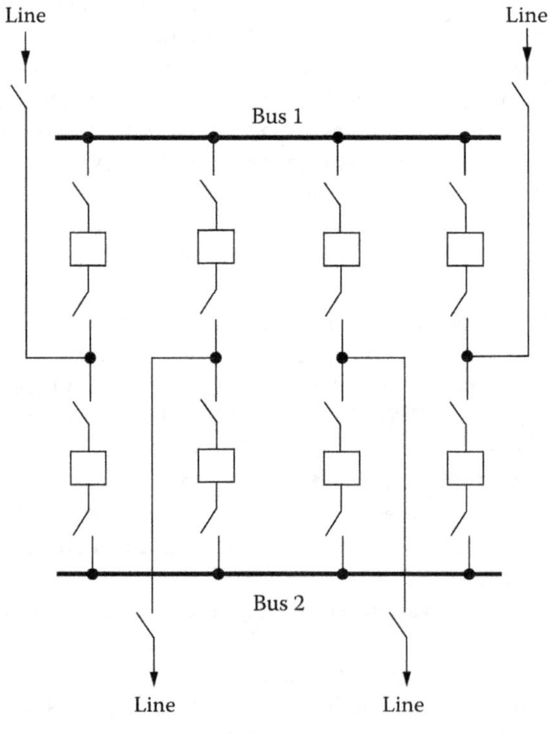

Figure 4.7 A typical double bus–double breaker scheme.

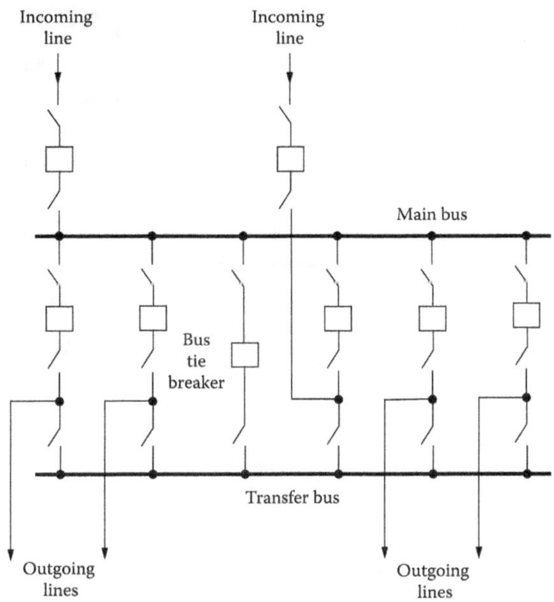

Figure 4.8 A typical main-and-transfer bus scheme.

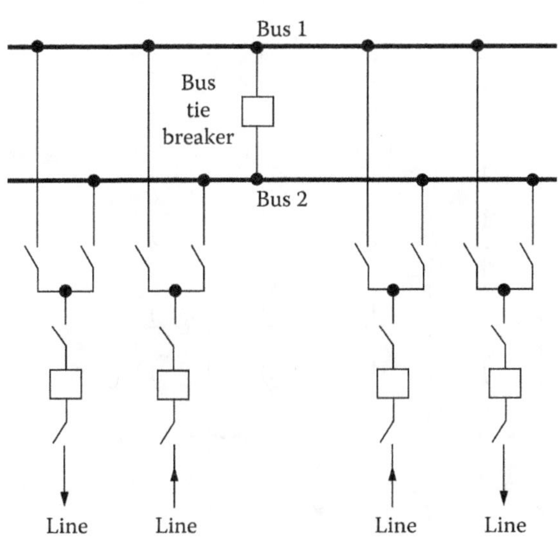

Figure 4.9 A typical double bus—–single breaker scheme.

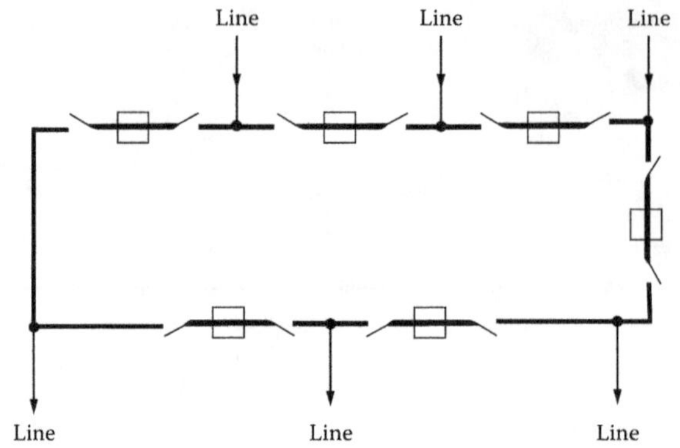

Figure 4.10 A typical ring bus scheme.

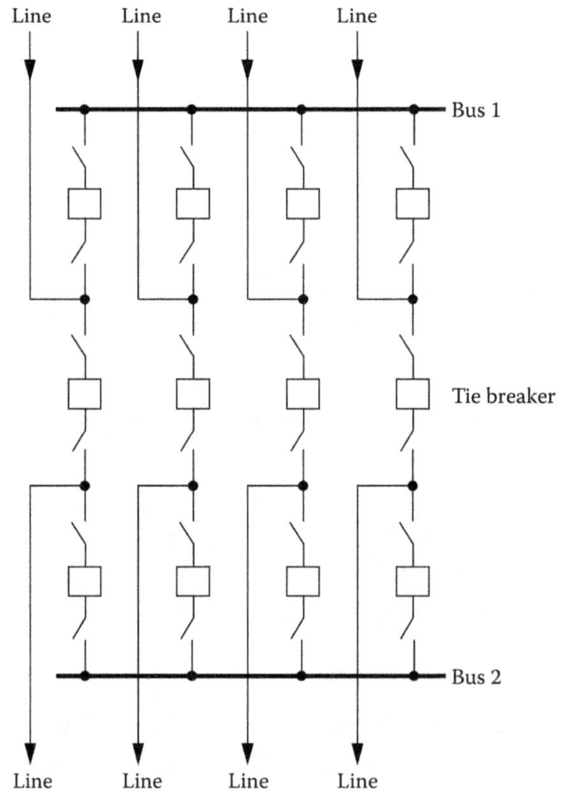

Figure 4.11 A typical breaker-and-a-half scheme.

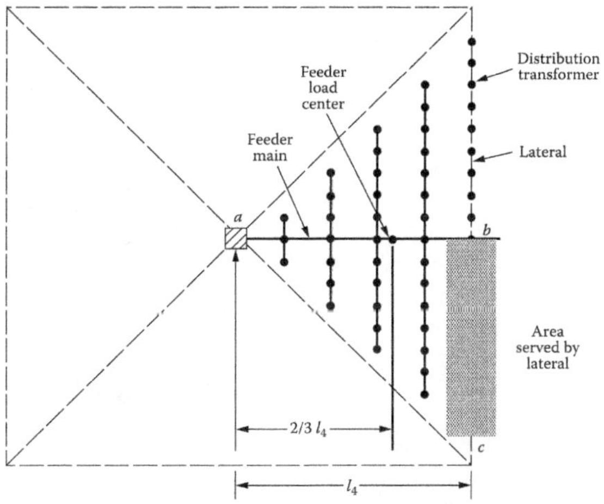

Figure 4.12 Square-shaped distribution substation service area. (Based on Westinghouse Electric Corporation, *Electric Utility Engineering Reference Book-Distribution Systems*, Vol. 3, East Pittsburgh, PA, 1965.)

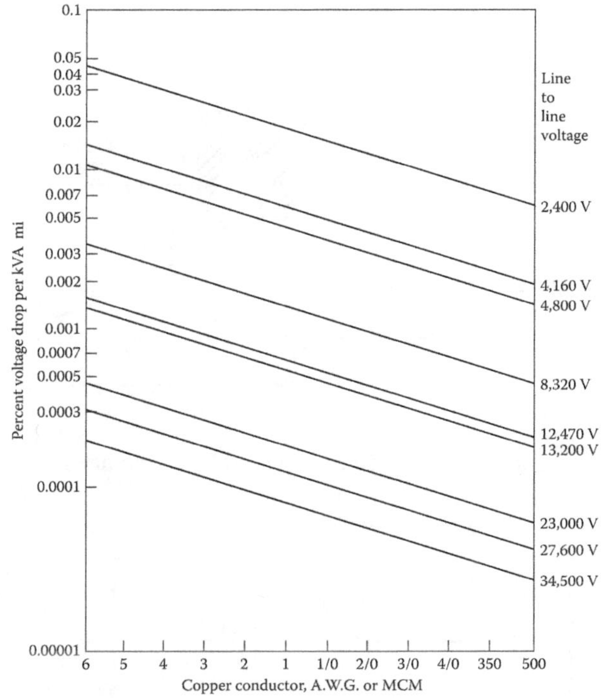

Figure 4.13 The *K* constant for copper conductors, assuming a lagging-load power factor of 0.9.

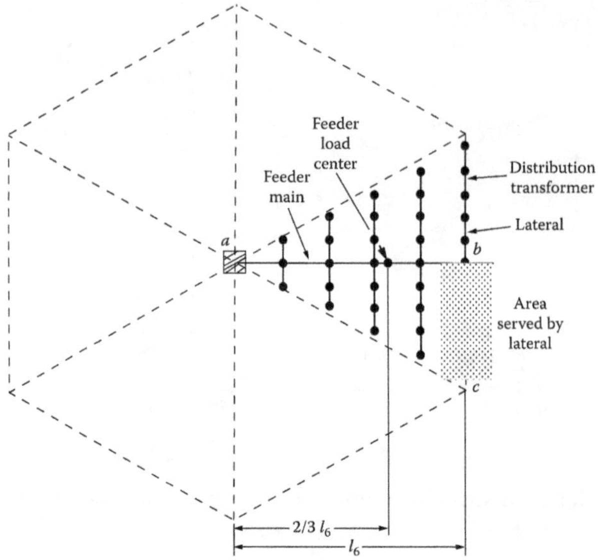

Figure 4.14 Hexagonally shaped distribution substation area. (Based on Westinghouse Electric Corporation, *Electric Utility Engineering Reference Book-Distribution Systems*, Vol. 3, East Pittsburgh, PA, 1965.)

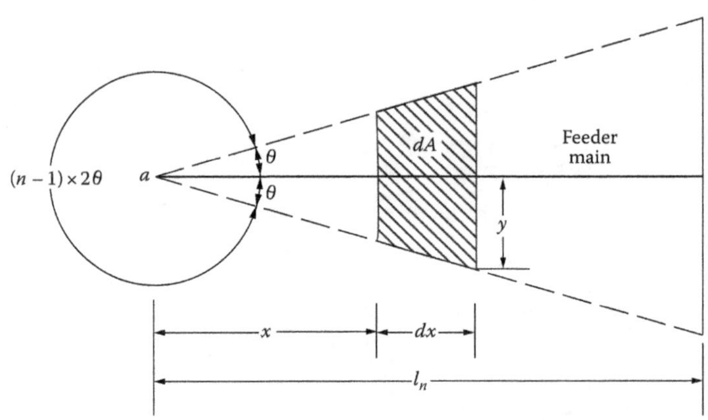

Figure 4.15 Distribution substation service area served by *n* primary feeders.

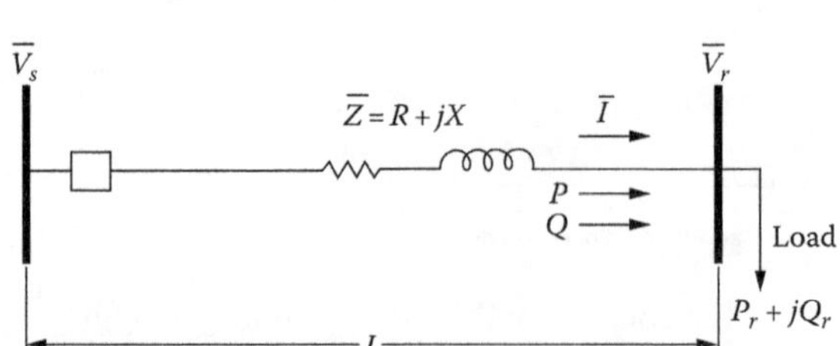

Figure 4.16 An illustration of a primary-feeder main.

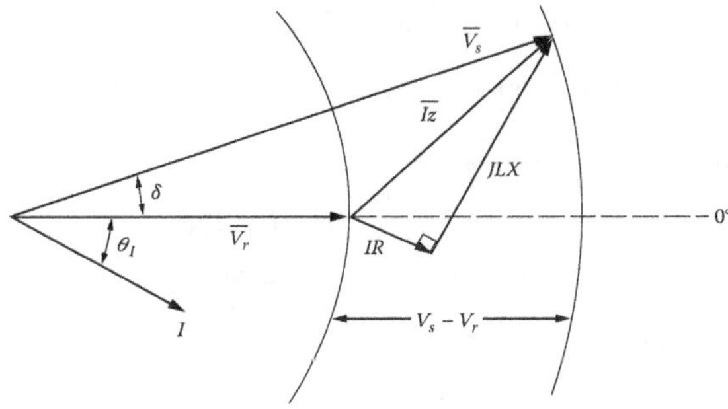

Figure 4.17 Phasor diagram.

Substation costs encompass the expenses associated with the equipment and labor involved in constructing a substation, which also include the cost of land and easements required for rights of way. When considering substation costs for planning purposes, they can be divided into four main categories [102]:

1. *Site costs*: the cost of buying the site and preparing it for a substation
2. *Transmission cost*: the cost of terminating transmission at the site
3. *Transformer cost*: the transformer and all metering, control, oil spill containment, fire prevention, cooling, noise abatement, and other transformer-related equipment, along with typical bus work, switches, metering, relaying, and breakers associated with this type of transformer and their installations
4. *Feeder bus-work/getaway costs*: the cost of beginning distribution at the substation, which includes getting feeders out of the substation

The expenses associated with substation site costs are influenced by local land prices, which are determined by the real estate market. These costs include site preparation activities such as grading, grounding mat installation, foundation construction, buried ductwork, control building establishment, lighting installation, fencing, landscaping, and the creation of access roads. In some cases, the estimated costs of feeder bus work and gateways are included in the overall transformer costs. Substation costs can vary significantly depending on factors such as the type of substation, its capacity, local land prices, and other specific circumstances. According to the ABB Guidebook, substation costs ranged from \$1.8 million to \$5.5 million in 1994 prices. These costs depend on factors such as land costs, labor costs, utility equipment and installation standards, and other relevant circumstances. The total cost of a typical substation can vary between approximately \$36 per kW and \$110 per kW, depending on the specific circumstances.

Example 4.1. Consider a typical substation that might be fed by two incoming 138 kV lines feeding two 32 MVA, 138/12.47 kV transformers, each with a low-voltage bus. Each bus has four outgoing distribution feeders of 9 MVA peak capacity each. The total site cost of the substation is \$600,000. The total transmission cost including high-side bus circuit breakers is estimated to be \$900,000. The total costs of the two transformers and associated equipment is \$1,100,000. The feeder bus-work/getaway cost is \$400,000. Determine the following:

a. The total cost of this substation
b. The utilization factor of the substation, if it is going to be used to serve a peak load of about 50 MVA
c. The total substation cost per kVA based on the previously mentioned utilization rate

Solution. a. The total cost of this substation is

$$\$600,000 + \$900,000 + \$1,100,000 + \$400,000 = \$3,000,000$$

b. The utilization factor of the substation is

$$F_u = \frac{\text{maximum demand}}{\text{rated system capacity}} = \frac{50\text{ MVA}}{2(32\text{ }MVA)} \cong 0.78 \text{ or } 78\%$$

c. The total substation cost per kVA is

$$\frac{\$3,000,000}{50,000\text{ kVA}} = \$60/\text{kVA}$$

4.5 SUBSTATION BUS SCHEMES

The chosen substation scheme (or diagram) dictates the electrical and physical configurations of the switching and busing at the subtransmission voltage level. Conversely, the selection of a specific substation scheme depends on factors such as safety, reliability, cost-effectiveness, simplicity, and other relevant considerations.

The most commonly used substation bus schemes include (1) single bus scheme, (2) double bus–double breaker (or double main) scheme, (3) main-and-transfer bus scheme, (4) double bus–single breaker scheme, (5) ring bus scheme, and (6) breaker-and-a-half scheme.

Figure 4.6 depicts a standard single bus scheme, while Figure 4.7 illustrates a typical double bus–double breaker scheme. Additionally, Figure 4.8 demonstrates a common main-and-transfer bus scheme, and Figure 4.9 presents a typical double bus–single breaker scheme. Furthermore, Figure 4.10 showcases a typical ring bus scheme, and Figure 4.11 provides an illustration of a typical breaker-and-a-half scheme.

Each scheme offers distinct advantages and disadvantages, which vary depending on the economic rationale for a specific level of reliability. Table 4.1 offers a summarized overview of the advantages and disadvantages of different switching schemes.

4.6 SUBSTATION LOCATION

The placement of a substation is determined by several factors, including voltage levels, considerations for voltage regulation, costs associated with subtransmission, substation, primary feeders, mains, and distribution transformers. Additionally, there are other non-technical constraints outlined in Chapter 1 that impose limitations on the substation's location.

Nevertheless, in order to choose an optimal site for a substation, it is essential to adhere to the following guidelines [51]:

1. Place the substation as close as possible to the load center of its service area, aiming to minimize the product of additional load and distance from the substation.
2. Select a substation location that facilitates proper voltage regulation without the need for extensive measures.
3. Choose a substation location that allows convenient access for incoming subtransmission lines and outgoing primary feeders.
4. Ensure that the selected substation location provides sufficient space for future expansion.
5. Ensure that the selected substation location complies with land-use regulations, local ordinances, and the concerns of neighboring properties.
6. Opt for a substation location that minimizes the number of customers affected by any service disruption.
7. Consider other factors such as adaptability and emergency situations.

Table 4.1
Summary of Comparison of Switching Schemes[a]

Switching Scheme	Advantages	Disadvantages
1. Single bus	1. Lowest cost	1. Failure of bus or any circuit breaker results in shutdown of entire substation 2. Difficult to do any maintenance 3. Bus cannot be extended without completely de-energizing substation 4. Can be used only where loads can be interrupted or have other supply arrangements
2. Double bus–double breaker	1. Each circuit has two dedicated breakers 2. Has flexibility in permitting feeder circuits to be connected to either bus 3. Any breaker can be taken out of service for maintenance 4. High reliability	1. Most expensive 2. Would lose half the circuits for breaker failure if circuits are not connected to both buses
3. Main-and-transfer	1. Low initial and ultimate cost 2. Any breaker can be taken out of service for maintenance 3. Potential devices may be used on the main bus for relaying	1. Requires one extra breaker for the bus tie 2. Switching is somewhat complicated when maintaining a breaker 3. Failure of bus or any circuit breaker results in shutdown of entire substation
4. Double bus–single breaker	1. Permits some flexibility with two operating buses 2. Either main bus may be isolated for maintenance 3. Circuit can be transferred readily from one bus to the other by use of bus-tie breaker and bus selector disconnect switches	1. One extra breaker is required for the bus tie 2. Four switches are required per circuit 3. Bus protection scheme may cause loss of substation when it operates if all circuits are connected to that bus 4. High exposure to bus faults 5. Line breaker failure takes all circuits connected to that bus out of service 6. Bus-tie breaker failure takes entire substation out of service
5. Ring bus	1. Low initial and ultimate cost 2. Flexible operation for breaker maintenance 3. Any breaker can be removed for maintenance without interrupting load 4. Requires only one breaker per circuit 5. Does not use main bus 6. Each circuit is fed by two breakers 7. All switching is done with breakers	1. If a fault occurs during a breaker maintenance period, the ring can be separated into two sections 2. Automatic reclosing and protective relaying circuitry rather complex 3. If a single set of relays is used, the circuit must be taken out of service to maintain the relays (common on all schemes) 4. Requires potential devices on all circuits since there is no definite potential reference point. These devices may be required in all cases for synchronizing, live line, or voltage indication 5. Breaker failure during a fault on one of the circuits causes loss of one additional circuit owing to operation of breaker- failure relaying
6. Breaker-and-a-half	1. Most flexible operation 2. High reliability 3. Breaker failure of bus side breakers removes only one circuit from service 4. All switching is done with breakers 5. Simple operation; no disconnect switching required for normal operation 6. Either main bus can be taken out of service at any time for maintenance 7. Bus failure does not remove any feeder circuits from service	1. $1^1/_2$ breakers per circuit 2. Relaying and automatic reclosing are somewhat involved since the middle breaker must be responsive to either of its associated circuits

Source: Fink, D.G. and Beaty, H.W., *Standard Handbook for Electrical Engineers,* 11th edn., McGraw-Hill, New York, 1978.

4.7 RATING OF A DISTRIBUTION SUBSTATION

The growing capacity demands of a system due to increasing load density can be fulfilled by either:

1. Maintaining the service area of an existing substation and enhancing its capacity.
2. Establishing new substations while keeping the rating of the existing substation unchanged.

To aid in distribution planning, it is beneficial to make the following assumptions:

1. For short-term distribution planning, consider system changes at a constant load density.
2. For long-term planning, assume system changes with increasing load density.

In addition, it is customary and advantageous to utilize geometric representations of substation service areas, as suggested by Van Wormer [103], Denton and Reps [104], and Reps [50]. This approach simplifies the comparison of alternative plans that may involve varying sizes of distribution substations, different numbers of primary feeders, and diverse primary-feeder voltages.

Reps [50] conducted an analysis using a square-shaped service area, which represents either a portion or the entirety of a distribution substation's service area. The assumption is that the square area is served by four primary feeders originating from a central feed point, as depicted in Figure 4.12. Each feeder and its respective laterals are three-phase. The dots in the figure symbolize balanced three-phase loads, grouped at specific locations and supplied by distribution transformers.

Here, the percent voltage drop from the feed point a to the end of the last lateral at c is

$$\%VO_{ac} = \%VO_{ab} + \%VO_{bc}$$

Reps [50] simplified the calculation of voltage drop, mentioned earlier, by introducing a constant denoted as "K," which represents the percent voltage drop per kVA-mile. Figure 4.26 provides the values of the K constant for different voltages and copper conductor sizes. This figure specifically applies to three-phase overhead lines with a spacing of 37 inches between phase conductors. The subsequent analysis is based on the research conducted by Denton and Reps [104] and Reps [50].

In Figure 4.12, each feeder serves a total load of

$$S_4 = A_4 \times D \text{ kVA} \tag{4.1}$$

where

S_4 is the kVA load served by one of four feeders emanating from a feed point
A_4 is the area served by one of four feeders emanating from a feed point,mi^2
D is the load density, kVA/mi^2

Equation 4.1 can be rewritten as

$$S_4 = l_4^2 \times D \text{ kVA} \tag{4.2}$$

since

$$A_4 = l_4^2 \tag{4.3}$$

where l_4 is the linear dimension of the primary-feeder service area, mi. Assuming uniformly distributed load, that is, equally loaded and spaced distribution transformers, the voltage drop in the primary-feeder main is

$$\%VD_{4,\text{min}} = \frac{2}{3} \times l_4 \times K \times S_4 \tag{4.4}$$

or substituting Equation 4.2 into Equation 4.4,

$$\%VD_{4,\min} = 0.667 \times K \times D \times k_4^2 \tag{4.5}$$

In Equations 4.4 and 4.5, it is assumed that the total or lumped-sum load is located at a point on the main feeder at a distance of $(2/3) \times l_4$ from the feed point *a*.

Reps [50] extends the discussion to a hexagonally shaped service area supplied by six feeders from the feed point that is located at the center, as shown in Figure 4.14. Assume that each feeder service area is equal to one-sixth of the hexagonally shaped total area, or

$$\begin{aligned} A_6 &= \frac{l_6}{\sqrt{3}} \times l_6 \\ &= 0.578 \times l_6^2 \end{aligned} \tag{4.6}$$

where

A_6 is the area served by one of six feeders emanating from a feed point, mi^2
l_6 is the linear dimension of a primary-feeder service area, mi

Here, each feeder serves a total load of

$$S_6 = A_6 \times D \text{ kVA} \tag{4.7}$$

or substituting Equation 4.6 into Equation 4.7,

$$S_6 = 0.578 \times D \times l_6^2 \tag{4.8}$$

As before, it is assumed that the total or lump sum is located at a point on the main feeder at a distance of $(2/3) \times l_6$ from the feed point. Hence, the percent voltage drop in the main feeder is

$$\%VD_{6,main} = \frac{2}{3} \times l_6 \times K \times S_6 \tag{4.9}$$

or substituting Equation 4.8 into Equation 4.9,

$$\%VD_{6,main} = 0.385 \times K \times D \times l_6^3 \tag{4.10}$$

4.8 GENERAL CASE: SUBSTATION SERVICE AREA WITH *N* PRIMARY FEEDERS

Denton and Reps [104] and Reps [50] expand the discussion to a general scenario where the distribution substation service area is supplied by n primary feeders originating from a specific point, as depicted in Figure 4.15. It is assumed that the load within the service area is uniformly distributed, and each feeder serves a triangular-shaped region. The load served by a feeder in a differential area of *dA* can be described as:

$$dS = D\, dA \text{ kVA} \tag{4.11}$$

where

dS is the differential load served by the feeder in the differential area of dA, kVA
D is the load density, kVA/mil
dA is the differential service area of the feeder, mi^2

In Figure 4.15, the following relationship exists:

$$\tan\theta = \frac{y}{x + dx} \tag{4.12}$$

or

$$\begin{aligned} y &= (x + dx)\tan\theta \\ &\cong x \times \tan\theta \end{aligned} \tag{4.13}$$

The total service area of the feeder can be calculated as

$$\begin{aligned} A_n &= \int_{x=0}^{l_n} dA \\ &= l_n^2 \times \tan\theta \end{aligned} \tag{4.14}$$

The total kVA load served by one of n feeders can be calculated as

$$\begin{aligned} S_n &= \int_{x=0}^{l_n} dS \\ &= D \times l_n^2 \times \tan\theta \end{aligned} \tag{4.15}$$

This total load is located, as a lump-sum load, at a point on the main feeder at a distance of $(2/3) \times l_4$ from the feed point a. Hence, the summation of the percent voltage contributions of all such areas is

$$\%\text{VD}_n = \frac{2}{3} \times l_n \times K \times S_n \tag{4.16}$$

or, substituting Equation 4.15 into Equation 4.16,

$$\%\text{VD}_n = \frac{2}{3} \times K \times D \times l_n^3 \times \tan\theta \tag{4.17}$$

or, since

$$n(2\theta) = 360^\circ \tag{4.18}$$

Equation 4.17 can also be expressed as

$$\%\text{VD}_n = \frac{2}{3} \times K \times D \times l_n^3 \times \tan\theta \frac{360^\circ}{2n} \tag{4.19}$$

Equations 4.18 and 4.19 are only applicable when n $\geq$ 3. Table 4.2 gives the results of the application of Equation 4.17 to square and hexagonal areas.

Table 4.2
Application Results of Equation 4.17

n	θ	$\tan\theta$	$\%\text{VD}_n$
4	45°	1.0	$\frac{2}{3} \times K \times D \times l_4^3$
6	30°	$\frac{1}{\sqrt{3}}$	$\frac{2}{3\sqrt{3}} \times K \times D \times l_6^3$

For $n = 1$, the percent voltage drop in the feeder main is

$$\%\text{VD}_1 = \frac{1}{2} \times K \times D \times l_1^3 \tag{4.20}$$

and for $n = 2$ it is

$$\%\text{VD}_2 = \frac{1}{2} \times K \times D \times l_2^3 \tag{4.21}$$

To calculate the percentage voltage drop in a uniformly loaded lateral, it is recommended to concentrate the total load at a midpoint along its length. Then, multiply the kVA-mile product, which represents the line length and loading, by the corresponding K constant [50].

4.9 COMPARISON OF THE FOUR AND SIX FEEDER PATTERNS

For a square-shaped distribution substation area served by four primary feeders, that is, $n = 4$, the area served by one of the four feeders is

$$A_4 = l_4^2 \text{ mi}^2 \tag{4.22}$$

The total area served by all four feeders is

$$\begin{aligned} TA_4 &= 4A_4 \\ &= 4l_4^2\ mi^2 \end{aligned} \tag{4.23}$$

The kVA load served by one of the feeders is

$$S_4 = D \times l_4^2 \text{ kVA} \tag{4.24}$$

Thus, the total kVA load served by all four feeders is

$$TS_4 = 4D \times l_4^2 \text{ kVA} \tag{4.25}$$

The percent voltage drop in the main feeder is

$$\%\text{VD}_{4,\text{main}} = \frac{2}{3} \times K \times D \times l_4^3 \tag{4.26}$$

The load current in the main feeder at the feed point a is

$$I_4 = \frac{S_4}{\sqrt{3} \times V_{L-L}} \tag{4.27}$$

or

$$I_4 = \frac{D \times l_4^2}{\sqrt{3} \times V_{L-L}} \tag{4.28}$$

The ampacity, that is, the current-carrying capacity, of a conductor selected for the main feeder should be larger than the current values that can be obtained from Equations 4.27 and 4.28.

On the other hand, for a hexagonally shaped distribution substation area served by six primary feeders, that is, $n = 6$, the area served by one of the six feeders is

$$A_6 = \frac{1}{\sqrt{3}} \times l_6^2 \text{ mi}^2 \tag{4.29}$$

The total area served by all six feeders is

$$TA_6 = \frac{6}{\sqrt{3}} \times l_6^2 \text{ mi}^2 \tag{4.30}$$

The kVA load served by one of the feeders is

$$S_6 = \frac{1}{\sqrt{3}} \times D \times l_6^2 \text{ kVA} \tag{4.31}$$

Therefore, the total kVA load served by all six feeders is

$$TS_6 = \frac{6}{\sqrt{3}} \times D \times l_6^2 \text{ kVA} \tag{4.32}$$

The percent voltage drop in the main feeder is

$$\%\text{VD}_{6,\text{main}} = \frac{2}{3\sqrt{3}} \times K \times D \times l_6^3 \tag{4.33}$$

The load current in the main feeder at the feed point *a* is

$$I_6 = \frac{S_6}{\sqrt{3} \times V_{L-L}} \tag{4.34}$$

or

$$I_4 = \frac{D \times l_6^2}{3 \times V_{L-L}} \tag{4.35}$$

The relationship between the service areas of the four- and six-feeder patterns can be found under two assumptions: (1) feeder circuits are thermally limited (TL) and (2) feeder circuits are voltage-drop-limited (VDL).

1. *For TL feeder circuits*: For a given conductor size and neglecting voltage drop,

$$I_4 = I_6 \tag{4.36}$$

Substituting Equations 4.28 and 4.35 into Equation 4.36,

$$\frac{D \times l_4^2}{\sqrt{3} \times V_{L-L}} = \frac{D \times l_6^2}{3 \times V_{L-L}} \tag{4.37}$$

from Equation 4.37,

$$\left(\frac{l_6}{l_4}\right)^2 = \sqrt{3} \tag{4.38}$$

Also, by dividing Equation 4.30 by Equation 4.23,

$$\begin{aligned} \frac{TA_6}{TA_4} &= \frac{6/\sqrt{3} l_6^2}{4 l_4^2} \\ &= \frac{\sqrt{3}}{2} \left(\frac{l_6}{l_4}\right)^2 \end{aligned} \tag{4.39}$$

Substituting Equation 4.38 into Equation 4.39,

$$\frac{TA_6}{TA_4} = \frac{3}{2} \tag{4.40}$$

or

$$TA_6 = 1.50\, TA_4 \tag{4.41}$$

Therefore, the six feeders can carry 1.50 times as much load as the four feeders if they are thermally loaded.

2. *For VDL feeder circuits*: For a given conductor size and assuming equal percent voltage drop,

$$\%\text{VD}_4 = \%\text{VD}_6 \tag{4.42}$$

Substituting Equations 4.26 and 4.33 into Equation 4.42 and simplifying the result,

$$l_A = 0.833 \times l_6 \tag{4.43}$$

From Equation 4.30, the total area served by all six feeders is

$$TA_6 = \frac{6}{\sqrt{3}} \times l_6^2 \tag{4.44}$$

Substituting Equation 4.43 into Equation 4.23, the total area served by all four feeders is

$$TA_4 = 2.78 \times l_6^2 \tag{4.45}$$

Dividing Equation 4.44 by Equation 4.45,

$$\frac{TA_6}{TA_4} = \frac{5}{4} \tag{4.46}$$

or

$$TA_6 = 1.25 TA_4 \tag{4.47}$$

Therefore, the six feeders can carry only 1.25 times as much load as the four feeders if they are VDL.

4.10 DERIVATION OF THE K CONSTANT

Let's consider the primary-feeder main depicted in Figure 4.16. In this case, the effective impedance Z of the three-phase main relies on the characteristics of the load. For instance, when a lumped-sum load is connected at the end of the main, as illustrated in the figure, the effective impedance can be described as follows:

$$\overline{Z} = Z \times l \ \Omega/\text{phase} \tag{4.48}$$

where

z is the impedance of three-phase main line, Ω/(mi phase)
l is the length of the feeder main, mi

When the load is uniformly distributed, the effective impedance is

$$\overline{Z} = \frac{1}{2} \times z \times l \ \Omega/\text{phase} \tag{4.49}$$

When the load has an increasing load density, the effective impedance is

$$\overline{Z} = \frac{2}{3} \times z \times l \ \Omega/\text{phase} \tag{4.50}$$

Taking the receiving-end voltage as the reference phasor,

$$\overline{V}_r = V_r \angle 0^\circ \tag{4.51}$$

and from the phasor diagram given in Figure 4.17, the sending-end voltage is

$$\overline{V}_s = V_s \angle \delta \tag{4.52}$$

The current is

$$\overline{I} = I \angle -\theta \tag{4.53a}$$

and the power-factor angle is

$$\begin{aligned}\theta &= \theta_{\overline{V}_r} - \theta_{\overline{I}} \\ &= 0^\circ - \theta_{\overline{V}_I} - \theta_{\overline{I}}\end{aligned} \tag{4.53b}$$

and the power factor is a lagging one. When the real power P and the reactive power Q flow in opposite directions, the power factor is a *leading* one.

Here, the per unit voltage regulation is defined as

$$\text{VR}_{\text{pu}} \triangleq \frac{V_s - V_r}{V_r} \tag{4.54}$$

and the percent voltage regulation is

$$\text{VR}_{\text{pu}} \triangleq \frac{V_s - V_r}{V_r} \times 100 \tag{4.55}$$

or

$$\%\text{VR} = \text{VR}_{\text{pu}} \times 100 \tag{4.56}$$

whereas the per unit voltage drop is defined as

$$\text{VR}_{\text{pu}} \triangleq \frac{V_s - V_r}{V_B} \tag{4.57}$$

where V_B is normally selected to be V_r.

Hence, the percent voltage drop is

$$\%\text{VD} \triangleq \frac{V_s - V_r}{V_B} \times 100 \tag{4.58}$$

or

$$\%\text{VD} = \text{VD}_{\text{pu}} \times 100 \tag{4.59}$$

where V_B is the arbitrary base voltage. The base secondary voltage is usually selected as 120 V. The base primary voltage is usually selected with respect to the potential transformation (PT) ratio used.

Common PT Ratios	V_B (V)
20	2,400
60	7,200
100	12,000

From Figures 4.16 and 4.17, the sending-end voltage is

$$\overline{V}_r = V_r + \overline{I}\overline{Z} \tag{4.60}$$

or

$$V_s = (\cos\delta + j\sin\delta) = V_r\angle 0^\circ + I(\cos\theta - j\sin\theta)(R + jX) \tag{4.61}$$

The quantities in Equation 4.61 can be either all in per units or in the mks (or SI) system. Use line-to-neutral voltages for single-phase three-wire or three-phase three- or four-wire systems.

In typical *distribution circuits*,

$$R \cong X$$

and the voltage angle δ is closer to zero or typically

$$0^\circ \leq \delta \leq 4^\circ$$

whereas in typical *transmission* circuits,

$$\delta \cong 0^\circ$$

since X is much larger than R.

Therefore, for a typical *distribution* circuit, the sin δ can be neglected in Equation 4.61. Hence,

$$V_s \cong V_s\cos\delta$$

and Equation 4.61 becomes

$$V_s \cong V_r + IR\cos\theta + IX\sin\theta \tag{4.62}$$

Therefore, the per unit voltage drop, for a lagging power factor, is

$$\text{VD}_{\text{pu}} = \frac{IR\cos\theta + IX\sin\theta}{V_B} \tag{4.63}$$

and it is a positive quantity. The VD_{pu} is negative when there is a leading power factor due to shunt capacitors or when there is a negative reactance X due to series capacitors installed in the circuits.

The complex power at the receiving end is

$$P_r + jQ_r = \overline{V}_r\overline{I}^* \tag{4.64}$$

Therefore,

$$\overline{I} = \frac{P_r + jQ_r}{\overline{V}_r} \tag{4.65}$$

since

$$\overline{V}_r = V_r\angle 0^\circ$$

Substituting Equation 4.65 into Equation 4.61, which is the exact equation since the voltage angle δ is not neglected, the sending-end voltage can be written as

$$\overline{V}_s = V_r\angle 0^\circ + \frac{RP_r + XQ_r}{V_r\angle 0^\circ} - j\frac{RQ_r - XP_r}{V_r\angle 0^\circ} \tag{4.66}$$

or approximately,

$$\overline{V}_s = V_r + \frac{RP_r + XQ_r}{V_r} \tag{4.67}$$

Substituting Equation 4.67 into Equation 4.57,

$$V_{\text{pu}} \cong \frac{RP_r + XQ_r}{V_r V_B} \tag{4.68}$$

or

$$\text{VD}_{\text{pu}} \cong \frac{(S_r/V_r)R\cos\theta + (S_r/V_r)X\sin\theta}{V_B} \tag{4.69}$$

or

$$\text{VD}_{\text{pu}} \cong \frac{S_r \times R\cos\theta + S_r \times X\sin\theta}{V_r V_B} \tag{4.70}$$

since

$$P_r = S_r\cos\theta \text{ W} \tag{4.71}$$

and

$$Q_r = S_r \times \sin\theta \text{ var} \tag{4.72}$$

Equations 4.69 and 4.70 can also be derived from Equation 7.63, since

$$S_r = V_r I \text{ VA} \tag{4.73}$$

The quantities in Equations 4.68 and 4.70 can be either all in per units or in the SI system. Use the line-to-neutral voltage values and per phase values for the P_r, Q_r, and S_r.

To determine the K constant, use Equation 4.68,

$$V_{\text{pu}} \cong \frac{RP_r + XQ_r}{V_r V_B}$$

or

$$\text{VD}_{\text{pu}} \cong \frac{(S_{(3\phi)})(S)(r\cos\theta + x\sin)((1/3)\times 1000)}{V_r V_B} \text{ pu V} \tag{4.74}$$

or

$$\text{VD}_{\text{pu}} = S \times K \times S_{3\phi} \text{ pu V} \tag{4.75}$$

or

$$\mathrm{VD_{pu}} = S \times K \times S_n \text{ pu V} \tag{4.76}$$

where

$$K \cong \frac{(r\cos\theta + x\sin\theta)((1/3)\times 1000)}{V_r V_B} \tag{4.77}$$

Therefore, $K = f$ (conductor size, spacing, cos θ, V_B) and it has the unit of

$$\frac{\mathrm{VD_{pu}}}{\text{Arbitrary no. of KVA mi}}$$

To get the percent voltage drop, multiply the right side of Equation 4.77 by 100, so that

$$K \cong \frac{(r\cos\theta + x\sin\theta)((1/3)\times 1000)}{V_r V_B} \times 100 \tag{4.78}$$

which has the unit of

$$\frac{\%\text{VD}}{\text{Arbitrary no. of KVA mi}}$$

In Equations 4.74 through 4.76, s is the effective length of the feeder main that depends upon the nature of the load. For example, *when the load is connected at the end of the main as lumped sum*, the effective feeder length is

$$s = 1 \text{unit length}$$

when the load is uniformly distributed along the main,

$$s = \frac{1}{2} \times l \text{ unit length}$$

when the load has an increasing load density, shown in Figure 4.18,

$$s = \frac{2}{3} \times l \text{ unit length}$$

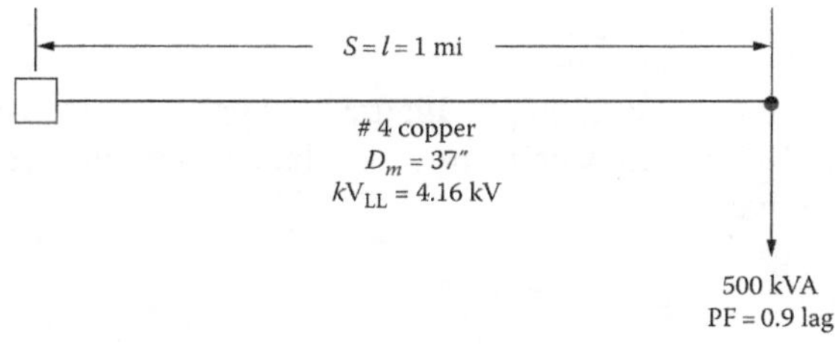

Figure 4.18 The feeder of Example 4.2.

Example 4.2. Assume that a three-phase 4.16 kV wye-grounded feeder main has #4 copper conductors with an equivalent spacing of 37 in. between phase conductors and a lagging-load power factor of 0.9.

a. Determine the K constant of the main by employing Equation 4.77.
b. Determine the K constant of the main by using the precalculated percent voltage drop per kVA-mile curves and compare it with the one found in part (a).

Solution. a. From Equation 4.77,

$$K \cong \frac{(r\cos\theta + x\sin\theta)((1/3)\times 1000)}{V_r V_B}$$

where
$r = 1.503\ \Omega$/mi from Table A.1 for 50°C and 60 Hz
$x_L = x_a + x_d = 0.7456\ \Omega$/mi
$x_a = 0.609\ \Omega$/mi from Table A.1 for 60 Hz
$x_d = 0.1366\ \Omega$/mi from Table A.10 for 60 Hz and 37-in. spacing $\cos\theta = 0.9$, lagging
$V_r = V_B = 2400$ V, line-to-neutral voltage
Therefore, the per unit voltage drop per kVA-mile is

$$K \cong \frac{(1.503\times 0.9 + 0.7456\times 0.4359)((1/3)\times 1000)}{2400^2}$$
$$\cong 0.0001\text{VD}_{\text{pu}}/(\text{KVA}\cdot\text{mi})$$

or

$$K \cong 0.01\%VD/(\text{KVA}\cdot\text{mi})$$

b. From Figure 4.13, the K constant for #4 copper conductors is

$$K \cong 0.01\%VD/(\text{KVA}\cdot\text{mi})$$

which is the same as the one found in part (a).

Example 4.3. Assume that the feeder shown in Figure 4.18 has the same characteristics as the one in Example 4.2, and a lumped-sum load of 500 kVA with a lagging-load power factor of 0.9 is connected at the end of a 1 mi long feeder main. Calculate the percent voltage drop in the main.

Solution. The percent voltage drop in the main is

$$\begin{aligned}\%\text{VD} &= s\times K\times S_n\\ &= 1.0\text{mi}\times 0.01\%\text{VD/(KVA}\times\text{mi)}\times 500\text{ kVA}\\ &= 5.0\%\end{aligned}$$

Example 4.4. Assume that the feeder shown in Figure 4.19 has the same characteristics as the one in Example 4.3, but the 500 kVA load is uniformly distributed along the feeder main. Calculate the percent voltage drop in the main.

Solution. The percent voltage drop in the main is

$$\%\text{VD} = s\times K\times S_n$$

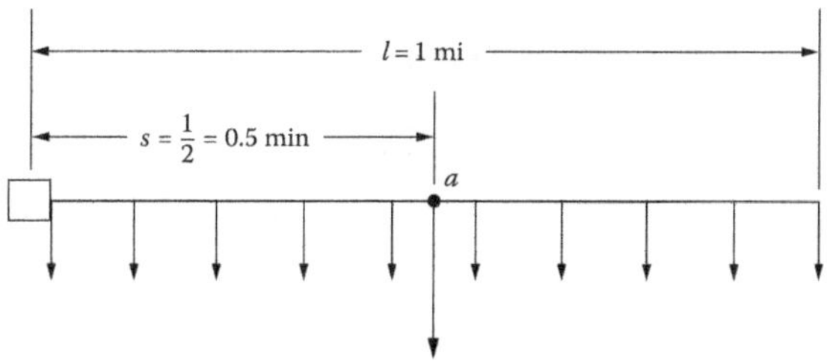

Figure 4.19 The feeder of Example 4.4.

where the effective feeder length s is

$$s = \frac{1}{2} = 0.5\text{mi}$$

Therefore,

$$\begin{aligned}\%\text{VD} &= s \times K \times S_n \\ &= 0.5 \text{ mi} \times 0.01\% \text{ VD/(KVA} \times \text{mi)} \times 500 \text{ kVA} \\ &= 2.5\%\end{aligned}$$

Therefore, it can be seen that the negative effect of the lumped-sum load on the %VD is worse than the one for the uniformly distributed load. Figure 4.19 also shows the conversion of the uniformly distributed load to a lumped-sum load located at point a for the voltage-drop calculation.

Example 4.5. Assume that the feeder shown in Figure 4.20 has the same characteristics as the one in Example 4.3, but the 500 kVA load has an increasing load density. Calculate the percent voltage drop in the main.

Solution. The percent voltage drop in the main is

$$\%\text{VD} = s \times K \times S_n$$

where the effective feeder length s is

$$s = \frac{2}{3}l = 0.6667 \text{ mi}$$

Therefore,

$$\begin{aligned}\%\text{VD} &= \frac{2}{3}l \times K \times S_n \\ &= 0.6667 \text{ mi} \times 0.01\%\text{VD/KVA} \cdot \text{mi)} \times 500 \text{ kVA} \\ &= 3.33\%\end{aligned}$$

Therefore, it becomes apparent that the detrimental impact of an increasing load density is more pronounced compared to a uniformly distributed load, but it is less severe than that of a lumped-sum load. Figure 4.20 also demonstrates the conversion of an increasing load density scenario to a lumped-sum load positioned at point b for the purpose of voltage-drop calculations.

Example 4.6. Use the results of the calculations of Examples 4.3 through 4.5 to calculate and compare the percent voltage drop ratios and reach conclusions.

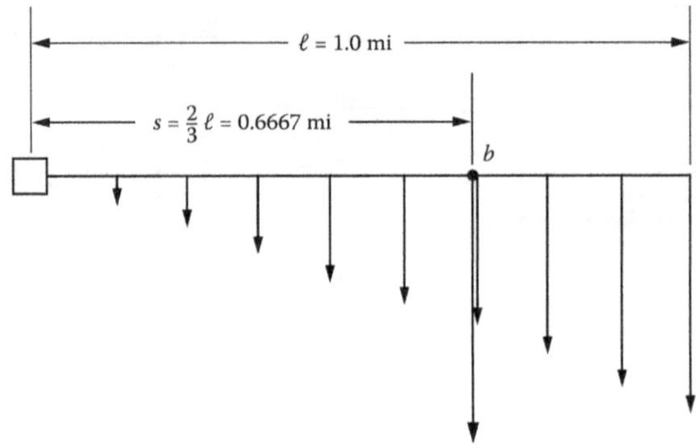

Figure 4.20 The feeder of Example 4.5.

Solution. a. The ratio of the percent voltage drop for the lumped-sum load to the one for the uniformly distributed load is

$$\frac{\%\mathrm{VD}_{\text{lumped}}}{\%\mathrm{VD}_{\text{uniform}}} = \frac{5.0}{2.5} = 2.0 \tag{4.79}$$

Therefore,

$$\%\mathrm{VD}_{\text{lumped}} = 2.0(\%\mathrm{VD}_{\text{uniform}}) \tag{4.80}$$

b. The ratio of the percent voltage drop for the lumped-sum load to the percent voltage drop for the load with increasing load density is

$$\frac{\%\mathrm{VD}_{\text{lumped}}}{\%\mathrm{VD}_{\text{increasing}}} = \frac{5.0}{3.33} = 1.5 \tag{4.81}$$

Therefore,

$$\%\mathrm{VD}_{\text{lumped}} = 1.5(\%\mathrm{VD}_{\text{increasing}}) \tag{4.82}$$

c. The ratio of the percent voltage drop for the load with increasing load density to the one for the uniformly distributed load is

$$\frac{\%\mathrm{VD}_{\text{increasing}}}{\%\mathrm{VD}_{\text{uniform}}} = \frac{2.33}{2.50} = 1.33 \tag{4.83}$$

Therefore,

$$\%\mathrm{VD}_{\text{increasing}} = 1.33(\%\mathrm{VD}_{\text{uniform}}) \tag{4.84}$$

4.11 SUBSTATION APPLICATION CURVES

Reps [50] derived the following formula to relate the application of distribution substations to load areas:

$$\%\mathrm{VD}_n = \frac{((2/3)\times l_n)K(n\times D\times A_n)}{n} \tag{4.85}$$

where

$\%VD_n$ is the percent voltage drop in primary-feeder circuit
$2/3 \times l_n$ is the effective length of primary feeder
K is the $\%VD/(kVA \cdot mi)$ of the feeder
A_n is the area served by one feeder
n is the number of primary feeders
D is the load density

Reps [50] and Denton and Reps [104] developed an alternative form of Equation 4.85 as

$$\%VD_n = \frac{TS_n^{3/2}}{n^{3/2} \times D^{1/2}} \frac{(2/3) \times K}{(\tan\theta)^{1/2}} \tag{4.86}$$

where TS_n is the total kVA supplied from a substation $(= n \times D \times A_n)$.

Based on Equation 4.86, they have developed the distribution substation application curves, as shown in Figures 4.21 and 4.22. These application curves relate the load density, substation load kVAs, primary-feeder voltage, and permissible feeder loading.

The distribution substation application curves are based on the following assumptions [50]:

1. #4/0 AWG copper conductors are used for the three-phase primary-feeder mains.
2. #4 AWG copper conductors are used for the three-phase primary-feeder laterals.
3. The equivalent spacing between phase conductors is 37 in.
4. A lagging-load power factor of 0.9.

The plotted curves represent the relationship between the number of primary feeders (n) and the load density (D) for various values of TS_n, which denotes the total kVA loading of all n primary feeders, including those serving the load area of a substation or feed point. Figures 4.21 and 4.22 display the curves for n versus D, considering both constant TS_n and constant TA_n, where TA_n represents the total area served by all n feeders emanating from the feed point or substation. The curves are illustrated for five primary-feeder voltage levels and two different percent voltage drops, namely 3% and 6%. The percent voltage drop, as stated in [50], is measured from the feed point or distribution substation bus to the last distribution transformer on the farthest lateral of a feeder.

The configuration of distribution substations and primary feeders in a particular system is typically planned to achieve a predetermined percent voltage drop or a specified kVA loading in the primary feeders. In regions with low load density and long distances to be covered by primary feeders, the maximum allowable percent voltage drop in a primary feeder usually sets the limit for its kVA loading. Conversely, in areas with relatively high load density and shorter primary feeders, the maximum allowable loading on a primary feeder is usually determined by its current-carrying capacity. This limit may be reached as a feeder becomes more heavily loaded, even before voltage drop becomes a concern.

The utilization of application curves allows for a clear determination of whether the loading of primary feeders in a specific substation area is restricted by voltage drop or the current-carrying capacity of the feeders. For each given substation or feed point kVA loading, a curve representing constant loading can be followed (from the upper left toward the lower right) as the load density increases. While following such a curve, the load density rises, leading to a decrease in the number of primary feeders required to serve that load. However, eventually, the number of primary feeders reaches the minimum necessary to accommodate the given kVA load based on the feeder's current-carrying capacity or kVA thermal capacity. Further reduction in the number of primary feeders is not permitted, resulting in a sudden change in the slope of the line representing constant feed-point loading, causing it to become horizontal.

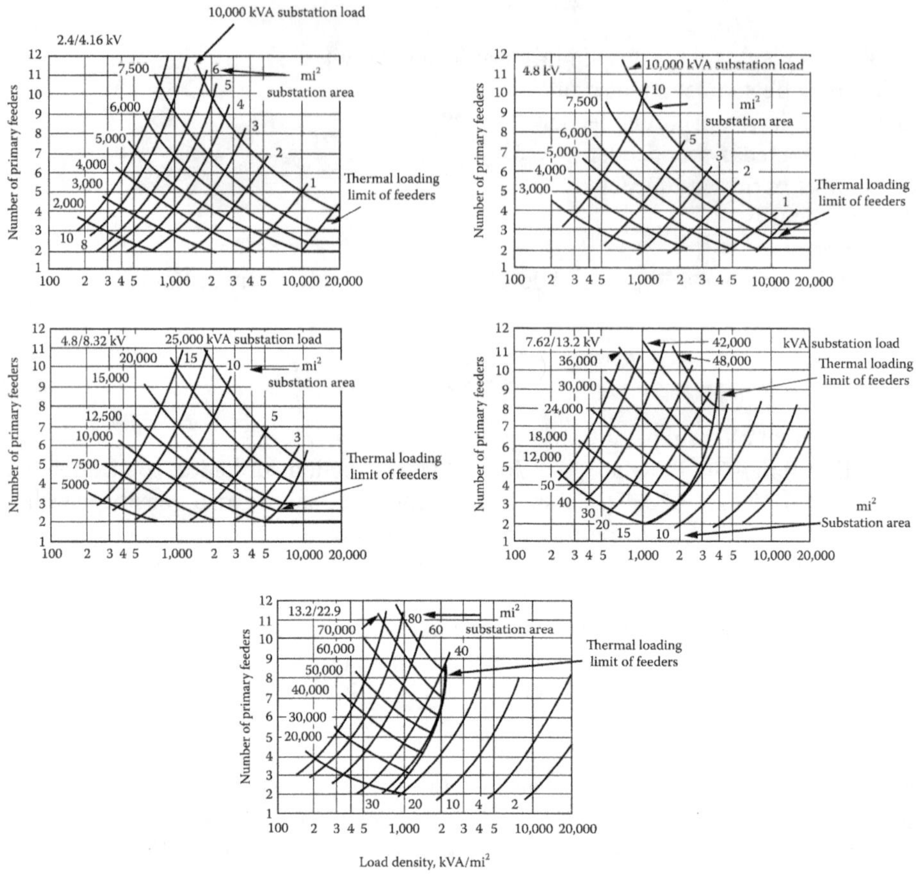

Figure 4.21 Distribution substation application curves for 3% voltage drop. (From Westinghouse *Electric Corporation, Electric Utility Engineering Reference Book-Distribution Systems*, Vol. 3, East Pittsburgh, PA, 1965.)

During the horizontal segment of the curve, feeder loading remains constant, but the percent voltage drop decreases as the load density increases. Therefore, each set of planning curves can be divided into two main regions: one region where the voltage drop remains constant and another region where the primary-feeder loading remains constant. In the region with constant primary-feeder loading, the percent voltage drop decreases as the load density increases.

Example 4.7. Referring to the preceding text, it is important to observe that the application curves for distribution substations, as presented in Figures 4.21 and 4.22, are applicable solely to the specified conductor sizes, spacing, and load power factor.

a. Use the substation application curves and the data given in Table 4.3 for eight different cases and determine (1) the substation sizes, (2) the required number of feeders, and (3) whether the feeders are TL or VDL. Tabulate the results.
b. In case thermally loaded or TL feeders are encountered, attempt to deduce if it is the #4/0 AWG copper main or the #4 AWG copper lateral that is TL. Show and explain your reasoning and calculations.

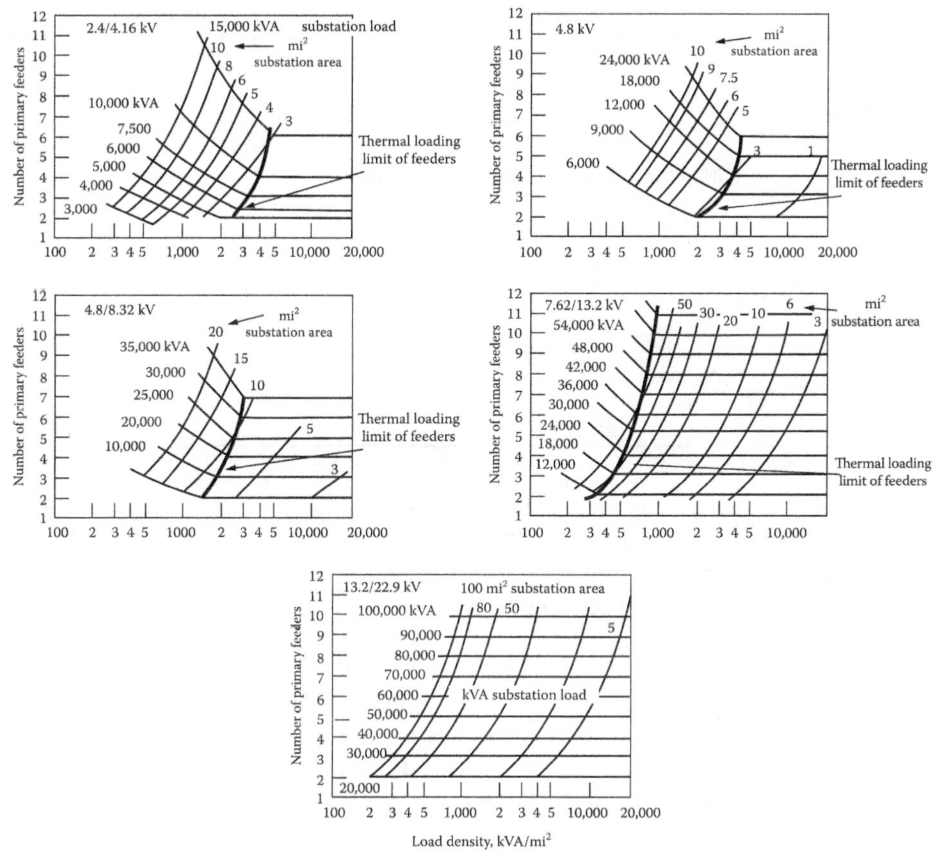

Figure 4.22 Distribution substation application curves for 6% voltage drop. (From Westinghouse Electric Corporation, *Electric Utility Engineering Reference Book-Distribution Systems*, Vol. 3, East Pittsburgh, PA, 1965.)

Table 4.3
Data for Example 4.7

Case No.	Load Density D, kVA/mi^2	Substation Area Coverage TA_n, mi^2	Maximum Total Primary Feeder, %VD	Base Feeder Voltage, kV_{L-L}
1	500	6.0	3.0	4.16
2	500	6.0	6.0	4.16
3	2,000	3.0	3.0	4.16
4	2,000	3.0	6.0	4.16
5	10,000	1.0	3.0	4.16
6	10,000	1.0	6.0	4.16
7	2,000	15.0	3.0	13.2
8	2,000	15.0	6.0	13.2

Solution. a. For case # 1, the total substation kVA load is

$$TS_n = D \times TA_n$$
$$= 500 \times 6.0 = 3000 \text{ kVA}$$

Table 4.4
Cases of Example 4.7

Case No.	Substation Size TS_n	Required No. of Feeders, *n*	VDL or TL Feeders
1	3,000	3.8 (or 4)	VDL
2	3,000	2	VDL
3	6,000	5	VDL
4	6,000	3	VDL
5	10,000	5	VDL
6	10,000	4	TL
7	30,000	5.85 (or 6)	VDL
8	30,000	5	TL

From the appropriate figure (the one with 3.0% voltage drop and 4.16 kV line-to-line voltage base) among the figures given in Figure 4.21, for 3000 kVA substation load, 500 kVA/mi^2 load density, and 6.0 mi^2 substation area coverage, the number of required feeders can be found as 3.8 or 4. Since the corresponding point in the figure is located on the left-hand side of the curve for the thermal-loading limit of feeders (the one with darker line), the feeders are VDL. The remaining cases can be answered in a similar manner as given in Table 4.4. Note that cases # 6 and 8 are of TL feeders since their corresponding points are located on the right-hand side of the thermal-loading limit curves.

b. Cases # 6 and 8 have feeders that are thermally loaded. From Table A.1, the conductor ampacities for a #4/0 copper main and a #4 copper lateral can be found as 480 and 180 A, respectively.
For case 6, the kVA load of one feeder is

$$S_n = \frac{TS_n}{n} = \frac{10,000}{4} = 2500 \text{ kVA}$$

Therefore, the load current is

$$I = \frac{S_n}{\sqrt{3} \times V_{L-L}} = \frac{2500 \text{ kVA}}{\sqrt{3} \times 4.16 \text{ kV}} = 347.4 \text{ A}$$

Since the conductor ampacity of the lateral is less than the load current, it is TL but not the main feeder.

For case # 8, the kVA load of one feeder is

$$S_n = \frac{30,000 \text{ kVA}}{5} = 6000 \text{ kVA}$$

The load current is

$$= \frac{6000 \text{ kVA}}{\sqrt{3} \times 13.2 \text{ kV}} = 262.4 \text{ A}$$

Therefore, only the lateral is TL.

4.12 INTERPRETATION OF PERCENT VOLTAGE DROP FORMULA

Equation 4.85 can be rewritten in alternative forms to illustrate the interrelationship of several parameters guiding the application of distribution substations to load areas

$$\begin{aligned}\%\mathrm{VD}_n &= \frac{((2/3)\times l_n)K(n\times D\times A_n)}{n}\\ &= \frac{((2/3)\times l_n K)TS_n}{n}\\ &= ((2/3)\times l_n\times K)S_n\end{aligned}$$

where

$\%\mathrm{VD_n}$ is the percent voltage drop in primary-feeder circuit
$2/3\times l_n$ is the effective length of primary feeder
$TS_n = n\times D\times A_n$ is the total kVA supplied from feed point
K is the %VD/(kVA · mi) of the feeder
A_n is the area served by one feeder
n is the number of primary feeders
D is the load density

To illustrate the use and interpretation of the equation, assume five different cases, as shown in Table 4.5

Case # 1 represents an increasing service area as a result of geographic extensions of a city. If the length of the primary feeder is doubled (shown in the table by ×2), holding everything else constant, the service area A_n of the feeder increases four times, which in turn increases TS_n and S_n four times, causing the $\%\mathrm{VD_n}$ n the feeder to increase eight times. Therefore, increasing the feeder length should be avoided as a remedy due to the severe penalty.

Case # 2 represents load growth due to load density growth. For example, if the load density is doubled, it causes TS_n and S_n to be doubled, which in turn increases the $\%\mathrm{VD_n}$ in the feeder to be doubled. Therefore, increasing load density also has a negative effect on the voltage drop.

Case # 3 represents the addition of new feeders. For example, if the number of the feeders is doubled, it causes S_n to be reduced by half, which in turn causes the $\%\mathrm{VD_n}$ to be reduced by half. Therefore, new feeder additions help to reduce the voltage drop.

Case # 4 represents feeder reconductoring. For example, if the conductor size is doubled, it reduces the K constant by half, which in turn reduces the $\%\mathrm{VD_n}$ by half.

Case # 5 represents the delta-to-grounded-wye conversion. It increases the line-to-line base kV by $\sqrt{3}$, which in turn decreases the K constant, causing the $\%\mathrm{VD_n}$ to decrease to one-third its previous value.

Table 4.5
Illustration of the Use and Interpretation of Equation 4.85

Case	l_n	K	Base, kV_{L-L}	N	D	An	TS_n	S_n	$\%VD_n$
1. Geographic extensions	$\times2\uparrow$	$\times1$	$\times1$	$\times1$	$\times1$	$\times4\uparrow$	$\times4\uparrow$	$\times4\uparrow$	$\times8\uparrow$
2. Load growth	$\times1$	$\times1$	$\times1$	$\times1$	$\times2\uparrow$	$\times1$	$\times2\uparrow$	$\times2\uparrow$	$\times2\uparrow$
3. Add new feeders	$\times1$	$\times1$	$\times1$	$\times2\uparrow$	$\times1$	$\times\frac{1}{2}\downarrow$	$\times1$	$\times\frac{1}{2}\downarrow$	$\times\frac{1}{2}\downarrow$
4. Feeder reconductoring	$\times1$	$\times\frac{1}{2}\downarrow$	$\times1$	$\times1$	$\times1$	$\times1$	$\times1$	$\times1$	$\times\frac{1}{2}\downarrow$
5. Δ-to-Y grounded conversion	$\times1$	$\times\frac{1}{3}\downarrow$	$\times\sqrt{3}\uparrow$	$\times1$	$\times1$	$\times1$	$\times1$	$\times1$	$\times\frac{1}{3}\downarrow$

Example 4.8. To illustrate distribution substation sizing and spacing, assume a square-shaped distribution substation service area as shown in Figure 4.25. Assume that the substation is served by four three-phase four-wire 2.4/4.16 kV grounded-wye primary feeders. The feeder mains are made of either #2 AWG copper or #1/0 ACSR conductors. The three-phase open-wire overhead lines have a geometric mean spacing of 37 in. between phase conductors. Assume a lagging-load power factor of 0.9 and a 1000 kVA/mi^2 uniformly distributed load density. Calculate the following:

a. Consider thermally loaded feeder mains and find the following:
    i. Maximum load per feeder
    ii. Substation size
    iii. Substation spacing, both ways
    iv. Total percent voltage drop from the feed point to the end of the main
b. Consider VDL feeders that have 3% voltage drop and find the following:
    i. Substation spacing, both ways
    ii. Maximum load per feeder
    iii. Substation size
    iv. Ampere loading of the main in per unit of conductor ampacity
c. Write the necessary codes to solve the problem in MATLAB.

Solution. From Tables A.1 and A.5, the conductor ampacities for #2 AWG copper and #1/0 ACSR conductors can be found as 230 A.

a. *Thermally loaded mains:*
    i. Maximum load per feeder is

$$\begin{aligned} S_n &= \sqrt{3} \times V_{L-L} \times I_{max} \\ &= \sqrt{3} \times 4.16 \times 230 = 1657.2 \text{ kVA} \end{aligned}$$

    ii. Substation size is

$$\begin{aligned} TS_n &= 4 \times S_n \\ &= 4 \times 1657.2 = 6628.8 \text{ kVA} \end{aligned}$$

    iii. Substation spacing, both ways, can be found from

$$\begin{aligned} S_n &= A_n \times D \\ &= l_4^2 \times D \end{aligned}$$

or

$$\begin{aligned} l_4 &= \left(\frac{S_n}{D}\right)^{1/2} \\ &= \left(\frac{1657.2 \text{ kVA}}{1000 \frac{\text{kVA}}{\text{mi}^2}}\right)^{1/2} \\ &= 1.287 \text{ mi} \end{aligned}$$

Therefore,

$$\begin{aligned} 2l_4 &= 2 \times 1.287 \\ &= 2.575 \text{ mi} \end{aligned}$$

iv. Total percent voltage drop in the main is

$$\begin{aligned}\%\mathrm{VD}_n &= \frac{2}{3}\times K\times D\times l_4^3\\ &= \frac{2}{3}\times 0.007\times 1000\times(1.287)^3\\ &= 9.95\%\end{aligned}$$

where K is 0.007 and found from Figure 4.26.

b. VDL *feeders*

i. Substation spacing, both ways, can be found from

$$\%\mathrm{VD}_n = \frac{2}{3}\times K\times D\times l_4^3$$

or

$$\begin{aligned}l_4 &= \left(\frac{3\times\%\mathrm{VD_n}}{2\times K\times D}\right)^{1/3}\\ &= \left(\frac{3\times 3}{2\times 0.007\times 1000}\right)^{1/3}\\ &= 0.86\ \text{mi}\end{aligned}$$

Therefore,

$$\begin{aligned}2l_4 &= 2\times 0.86\\ &= 1.72\ \text{mi}\end{aligned}$$

ii. Maximum load per feeder is

$$\begin{aligned}S_n &= D\times l_4^2\\ &= 1000\times(0.86)^2\cong 750\ \text{kVA}\end{aligned}$$

iii. Substation size is

$$\begin{aligned}TS_n &= 4\times S_n\\ &= 4\times 750 = 3000\ \text{kVA}\end{aligned}$$

iv. Ampere loading of the main is

$$\begin{aligned}I &= \frac{S_n}{\sqrt{3}\times V_{L-L}}\\ &= \frac{750\ \text{kVA}}{\sqrt{3}\times 4.16\ \text{kV}}\\ &= 104.09\ \text{A}\end{aligned}$$

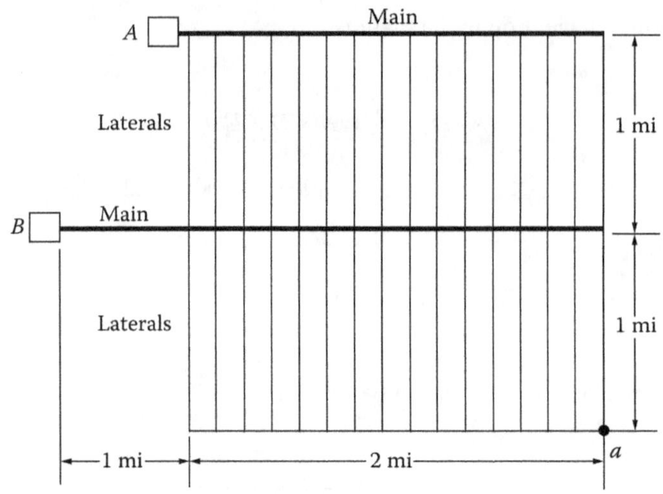

Figure 4.23 For Example 4.9.

Therefore, the ampere loading of the main in per unit of conductor ampacity is

$$I_{pu} = \frac{104.09\ \text{A}}{230\ \text{A}}$$
$$= 0.4526\ \text{pu}$$

Example 4.9. Assume a square-shaped distribution substation service area as shown in Figure 4.23. The square area is 4 mi and has numerous three-phase laterals. The designing distribution engineer has the following design data that are assumed to be satisfactory estimates.

The load is uniformly distributed, and the connected load density is 2000 kVA/mi^2. The demand factor, which is an average value for all loads, is 0.60. The diversity factor among all loads in the area is 1.20. The load power factor is 0.90 lagging, which is an average value applicable for all loads.

For some unknown reasons (perhaps, due to the excessive distance from load centers or transmission lines or other limitations, such as availability of land, its cost, and land-use ordinances and regulations), the only available substation sites are at locations A and B. If the designer selects site A as the substation location, there will be a 2 mi long feeder main and 16 three-phase 2 mi long laterals.

On the other hand, if the designer selects site B as the substation location, there will be a 3 mi long feeder main (including a 1 mi long express feeder main) and 32 three-phase 1 mi long laterals.

The designer wishes to select the better one of the given two sites by investigating the total peak-load voltage drop at the end of the most remote lateral, that is, at point a.

Assume 7.62/13.2 kV three-phase four-wire grounded-wye primary-feeder mains that are made of #2/0 copper overhead conductors. The laterals are of #4 copper conductors, and they are all three-phase, four-wire, and grounded-wye.

Using the precalculated percent voltage drop per kVA-mile curves given in Figure 4.13, determine the better substation site by calculating the percent voltage drops at point a that correspond to each substation site and select the better one.

Solution. The maximum diversified demand is

$$\text{Diversified Demand} = \frac{\sum_{i=1}^{n} \text{demand factor} \times \text{connected load}}{\text{diversity factor}}$$
$$= \frac{0.06 \times 2000\ \text{kVA/mi}^2}{1.20}$$
$$= 1000\ \text{kVA/mi}^2$$

The peak loads of the substations A and B are the same

$$TS_n = 1000\ \text{kVA/mi}^2 \times 4\ \text{mi}^2$$
$$= 4000\ \text{kVA}$$

From Figure 4.26, the K constants for #2/0 and #4 conductors are found as 0.0004 and 0.00095, respectively.

The maximum percent voltage drop for substation A occurs at point a, and it is the summation of the percent voltage drops in the main and the last lateral. Therefore,

$$\%\text{VD}_\text{a} = \frac{1}{2}K_m S_m + \frac{l}{2}K_l S_l$$
$$= \frac{2}{2} \times 0.0004 \times 4000 + \frac{2}{2} \times 0.00095 \times \frac{4000}{16}$$
$$\cong 1.84\%$$

The maximum percent voltage drop for substation B also occurs at point a. Therefore,

$$\%\text{VD}_\text{a} = 2 \times 0.0004 \times 4000 + \frac{l}{2} \times 0.00095 \times \frac{4000}{32}$$
$$\cong 3.26\%$$

Therefore, substation site A is better than substation site B from the voltage drop point of view.

Example 4.10. Assume a square-shaped distribution substation service area as shown in Figure 4.24. The four-feeder substation serves a square area of 2a × 2a mi^2.

The load density distribution is D kVA/mi^2 and is uniformly distributed. Each feeder main is three-phase four-wire grounded wye with multigrounded common neutral open-wire line.

Since dimension d is much smaller than dimension a, assume that the length of each feeder main is approximately a mi, and the area served by the last lateral, which is indicated in the figure as the cross-hatched area, is approximately $a \times d$mi^2. The power factor of all loads is cos θ lagging. The impedance of the feeder main line per phase is

$$z_m = r_m + jx_m\ \Omega/\text{mi}$$

The impedance of the lateral line per phase is

$$z_l = r_l + jx_l\ \Omega/\text{mi}$$

The V_{L-L} is the base line-to-line voltage in kVs, which is also the nominal operating voltage.

a. Assume that laterals are also three-phase four-wire grounded wye with multigrounded common neutral open-wire line. Show that the percent voltage drop at the end of the last lateral is

$$\%\text{VD} = \frac{2D \times a^3 (r_m \cos\theta + jx_m \sin\theta)}{30 \times V_{L-L}^2} + \frac{D \times a^3 \times d(r_l \cos\theta + jx_l \sin\theta)}{20 \times V_{L-L}^2} \tag{4.87}$$

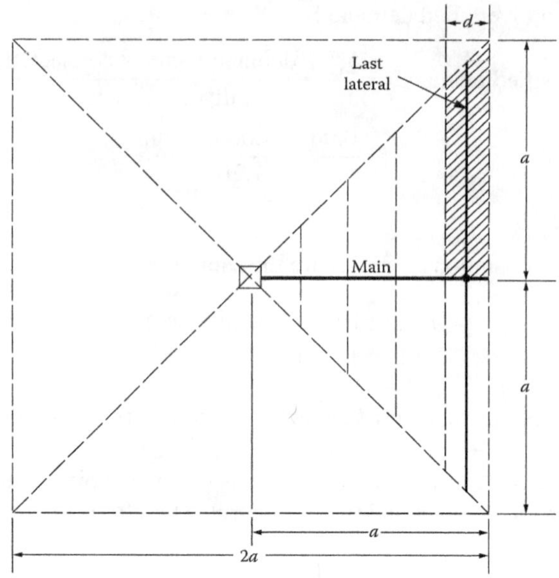

Figure 4.24 Service area for Example 4.10.

b. Assume that the laterals are single-phase two-wire with multigrounded common neutral open-wire line. Apply Morrison's approximation [105] and modify the equation given in part (a).

Solution. a. The total kVA load served by one main is

$$\begin{aligned} S_m &= D \times \frac{(2a)^2}{4} \\ &= D \times a^2 \text{ kVA} \end{aligned} \tag{4.88}$$

The current in the main of the substation is

$$I_m = \frac{D \times a^2}{\sqrt{3} \times V_{L-L}} \tag{4.89}$$

Therefore, the percent voltage drop at the end of the main is

$$\begin{aligned} \%\text{VD}_m &= \frac{D \times a^2}{\sqrt{3} \times V_{L-L}} (r_m \cos\theta + x_m \sin\theta) \frac{\sqrt{3}}{1000 \times V_{L-L}} \left(\frac{2}{3} \times a \right) 100 \\ &= \frac{2D \times a^3}{30 \times V_{L-L}} (r_m \cos\theta + x_m \sin\theta) \end{aligned} \tag{4.90}$$

The kVA load served by the last lateral is

$$S_l = D \times a \times d \text{ kVA} \tag{4.91}$$

The current in the lateral is

$$I_l = \frac{D \times a \times d}{\sqrt{3} \times V_{L-L}} \tag{4.92}$$

Thus, the percent voltage drop at the end of the lateral is

$$\begin{aligned}\%\text{VD}_l &= \frac{D\times a\times d}{\sqrt{3}\times V_{L-L}}(r_l\cos\theta + x_l\sin\theta)\frac{\sqrt{3}}{1000\times V_{L-L}}\left(\frac{1}{2}\times a\right)100 \\ &= \frac{D\times a^3\times d}{20\times V_{L-L}}(r_l\cos\theta + x_l\sin\theta)\end{aligned} \tag{4.93}$$

Therefore, the addition of Equations 4.90 and 4.93 gives Equation 4.87

b. According to Morrison [105], the percent voltage drop of a single-phase circuit is approximately four times that for a three-phase circuit, assuming the usage of the same-size conductors. Therefore,

$$\%\text{VD}_{1\phi} = 4\times(\%\text{VD}_{3\phi}) \tag{4.94}$$

Hence, the percent voltage drop in the main is the same as given in part a, but the percent voltage drop for the lateral is not the same and is

$$\begin{aligned}\%\text{VD}_{l,1\phi} &= 4\times\frac{D\times a^2\times d}{20\times V_{L-L}^2}(r_l\cos\theta + x_l\sin\theta) \\ &= \frac{D\times a^3\times d}{5\times V_{L-L}^2}(r_l\cos\theta + x_l\sin\theta)\end{aligned} \tag{4.95}$$

Thus, the total percent voltage drop will be the sum of the percent voltage drop in the three-phase main, given by Equation 4.90, and the percent voltage drop in the single-phase lateral, given by Equation 4.95. Therefore, the total voltage drop is

$$\%\text{VD} = \frac{2D\times a^3}{30\times V_{L-L}^2}(r_m\cos\theta + x_m\sin\theta) + \frac{D\times a^3\times d}{5\times V_{L-L}^2}(r_l\cos\theta + x_l\sin\theta) \tag{4.96}$$

Example 4.11. Figure 4.25 shows a pattern of service area coverage (not necessarily a good pattern) with primary-feeder mains and laterals. There are five substations shown in the figure, each with two feeder mains. For example, substation *A* has two mains like *A*, and each main has many closely spaced laterals such as $a-a$.

If the laterals are not three phase, the load in the main is assumed to be well balanced among the three phases. The load tapped off the main decreases linearly with the distance *s*, as shown in Figure 4.26.

Using the following notation and the notation given in the figures, analyze a feeder main:

D = uniformly distributed load density, kVA/mi^2
V_{L-L} = base voltage and nominal operating voltage, line-to-line kV
$A_2\ i$ = area supplied by one feeder main
TA_2 = area supplied by one substation
S_2 = kVA input at the substation to one feeder main
TS_2 = total kVA load supplied by one substation
K_2 = %VD/(kVA · mi) for conductors and load power factor being considered
z_2 = impedance of three-phase main line, Ω/(mi · phase)
VD_2 = voltage drop at end of main, for example, A_1

a. Find the differential area *dA* and the differential kVA load supplied $d(S)$ shown in Figure 4.26.

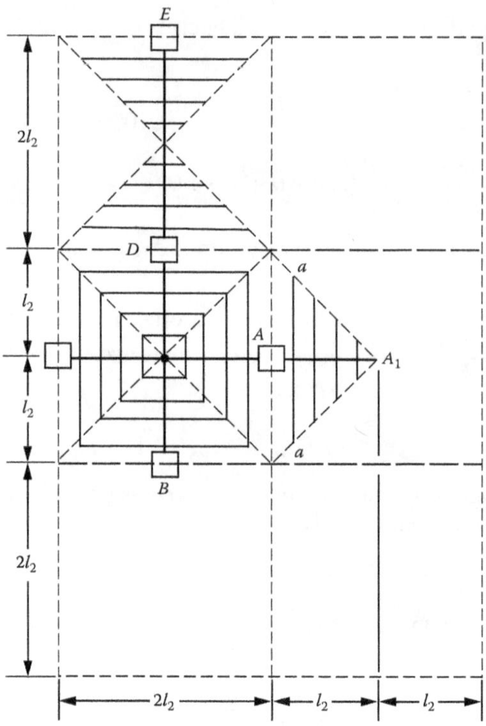

Figure 4.25 Service area for Example 4.11.

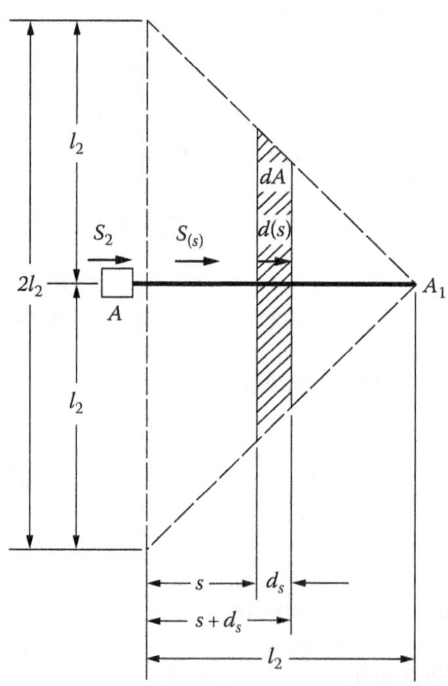

Figure 4.26 Linearly decreasing load for Example 4.11.

b. Find the kVA load flow in the main at any point s, that is, S_s. Express the S_s in terms of S_2, s, and l_2.
c. Find the differential voltage drop at point s and then show that the total load may be concentrated at $s = l_2/3$ for the purpose of computing the VD_2.
d. Suppose that this two-feeders-per-substation pattern is to be implemented with TL, that is, ampacity-loaded, feeders.

Assume that the load density is 500 kVA/mi^2, the line-to-line voltage is 12.47 kV, and the feeder mains are #4/0 AWG ACSR open-wire lines. Find the substation spacing, both ways, that is, $2l_2$, and the load on the substation transformers, that is, TS_2.

Solution. a. From Figure 4.26, the differential area is

$$dA = 2(l_2 - s)ds \text{ mi}^2 \tag{4.97}$$

Therefore, the differential kVA load supplied is

$$d(S) = 2D(l_2 - s)ds \text{ kVA} \tag{4.98}$$

b. The kVA load flow in the main at any point s is

$$\begin{aligned} S_s &= 2(l_2 - s)^2 D \\ &= 2(l_2 - s)^2 \times \frac{S_2}{2l_2^2} \text{ kVA} = \left(\frac{l_2 - s}{l_2}\right)^2 \times S_2 \text{ kVA} \end{aligned} \tag{4.99}$$

c. The differential current at any point s is

$$I_s = \frac{S_s}{\sqrt{3} \times V_{L-L}} \tag{4.100}$$

Hence, the differential voltage drop at point s is

$$\begin{aligned} d(\text{VD})_s &= I_s \times Z_2 ds \\ &= \frac{S_s}{\sqrt{3} \times V_{L-L}} \times Z_2 ds \\ &= \left(\frac{l_2 - s}{l_2}\right)^2 \left(\frac{S_2}{\sqrt{3} \times V_{L-L}}\right) Z_2 ds \\ &= \frac{S_2 \times Z_2}{\sqrt{3} \times V_{L-L} \times l_2^2} \times (l_2 - s)^2 ds \end{aligned} \tag{4.101}$$

The integration of either side of Equation 4.101 gives the voltage drop at point s:

$$\begin{aligned} (\text{VD})_s &= \int_0^s d(\text{VD})_s \\ &= \int_0^s \frac{S_2 \times Z_2}{\sqrt{3} \times V_{L-L} \times l_2^2} \times (l_2 - s)^2 ds \\ &= \frac{S_2 \times Z_2}{\sqrt{3} \times V_{L-L} \times l_2^2} \frac{l_2^3}{3} - \frac{S_2 \times Z_2}{\sqrt{3} \times V_{L-L} \times l_2^2} \frac{(l_2 - s)^3}{3} \\ &= \frac{S_2 \times Z_2}{\sqrt{3} \times V_{L-L} \times l_2^2} \times [l_2^3 - (l - s)^2] \end{aligned} \tag{4.102}$$

When $s = l_2$, Equation 4.102 becomes

$$
\begin{aligned}
(\text{VD})_2 &= \frac{S_2 \times Z_2 \times l_2^3}{\sqrt{3} \times V_{L-L} \times l_2^2} \\
&= \frac{S_2 \times Z_2 \times l_2}{3\sqrt{3} \times V_{L-L}} \\
&= \frac{S_2}{\sqrt{3} \times V_{L-L}} \times z_2 \times \frac{l_2}{3}
\end{aligned}
\tag{4.103}
$$

Therefore, the load has to be lumped at $l_2/3$.

d. From Table A.5, the conductor ampacity for #4/0 AWG ACSR conductor can be found as 340 A. Therefore

$$
\begin{aligned}
S_2 &= \sqrt{3} \times 12.47\text{kV} \times 340\ \text{A} \\
&\cong 7343.5\ \text{kVA}
\end{aligned}
$$

Since

$$S_2 = D \times l_2^2$$

then

$$
\begin{aligned}
l_2 &= \left(\frac{S_2}{D}\right) 1/2 \\
&= \left(\frac{7343.5\ \text{kVA}}{500\ \text{kVA/mi}^2}\right) \\
&= 3.83\ \text{mi}
\end{aligned}
\tag{4.104}
$$

Therefore, the substation spacing, both ways, is

$$
\begin{aligned}
2l_2 &= 2 \times 3.83 \\
&= 7.66\ \text{mi}
\end{aligned}
$$

Total load supplied by one substation is

$$
\begin{aligned}
TS_2 &= 2 \times S_2 \\
&= 2 \times 7343.5 \\
&= 14{,}687\ \text{kVA}
\end{aligned}
$$

Example 4.12. Compare the method of service area coverage given in Example 4.11 with the four-feeders-per-substation pattern of Section 4.6 (see Figure 4.12). Use the same feeder main conductors so that $K_2 = K_4$ and the same line-to-line nominal operating voltage V_{L-L}. Here, let S_4 be the kVA input to one feeder main of the four-feeder substation, and let TS_4, A_4, VD_4, K_4, etc., all pertain similarly to the four-feeder substation. Investigate the VDL feeders and determine the following:

a. Ratio of substation spacings = $2l_2/2l_4$
b. Ratio of areas covered per feeder main = A_2/A_4
c. Ratio of substation loads = TS_2/TS_4

Solution. a. Assuming the percent voltage drops and the K constants are the same in both cases,

$$\%\text{VD}_2 = \%\text{VD}_4$$

and

$$K_2 = K_4$$

where

$$\%\text{VD}_2 = (D \times l_2^2)(K_2)\left(\frac{1}{3}l_2\right)$$
$$= \frac{1}{3}K_2 \times D \times l_2^3$$

and

$$\%\text{VD}_4 = (D \times l_4^2)(K_4)\left(\frac{1}{3}l_4\right)$$
$$= \frac{1}{3}K_4 \times D \times l_4^3$$

Therefore,

$$l_2^3 = 2l_4^3$$

or the ratio of substation spacings is

$$\left(\frac{l_2}{l_4}\right)^3 = 2 \tag{4.105}$$

or, for both ways,

$$\frac{2l_2}{2l_4} \cong 2 \tag{4.106}$$

b. The ratio of areas covered per feeder main is

$$\frac{A_2}{A_4} = \frac{l_2^2}{l_4^2}$$
$$= \left(\frac{l_2}{l_4}\right)^2 \tag{4.107}$$
$$= 2^{2/3}$$
$$\cong 1.59$$

c. The ratio of substation loads is

$$\frac{TS_2}{TS_4} = \frac{2 \times D \times l_2^2}{4 \times D \times l_4^2}$$
$$= \frac{1}{2}\left(\frac{l_2}{l_4}\right)^2 \tag{4.108}$$
$$\cong 0.8$$

4.13 CAPABILITY OF FACILITIES

The ability of distribution substations to provide power to their service area load is typically determined by the capacity of the substation transformer banks. However, there are instances where the transmission facilities supplying the substation or the capabilities of the distribution feeders originating from it can impose limitations on the amount of load the substation can handle.

Both the substation transformer banks and the feeders have a normal capacity, typically measured at 100°F (40°C), and an emergency capacity, which is usually higher. These capacity values are determined by the temperature rise restrictions and the components of the transformers and feeders. Consequently, they tend to be higher in warmer regions. It is common practice to calculate the normal and emergency capacities, expressed in kVA, for both existing and proposed transformer banks using transformer capability assessment computer programs [106].

Furthermore, the component that restricts the capacity of a feeder may vary, ranging from the station breaker or associated switches to the underground or overhead conductors, current transformers, metering equipment, or the protective relay settings.

Each component should be carefully evaluated to determine its current-carrying capacity under normal and emergency conditions. In certain cases, it may be feasible to enhance this capacity at a relatively low cost by replacing the limiting component or modifying the protective scheme of the feeder [107].

As per the practices followed by certain utility companies, feeder circuit breakers and associated switches that are in good condition are typically capable of operating at 100% of their nameplate ratings for both normal summer and winter conditions, as well as summer emergency conditions. For winter emergency conditions, their capacity is extended to 110% of their nameplate ratings. However, if the equipment is not in good condition, it may be necessary to establish lower limits or replace the equipment altogether.

In general, it is advisable to multiply the ampacities of overhead conductors, switches, and single-phase feeder regulators by 0.95. This factor accounts for phase unbalance and allows for situations where the substation is limited by the circuit's capacity. It is important to note that this multiplier is not required for cables in underground ducts and risers, as the heat buildup is shared among all the cables within the duct. Additionally, the phase unbalance multiplier is not used for oil circuit breakers.

Once the normal and emergency capacities of feeders are determined in amperes, they can be converted to kVA by multiplying them by specific factors. For instance, nominal circuit voltages of 4,160, 4,800, 12,000, 17,000, and 20,780 V are multiplied by factors of 7.6, 8.73, 21.82, 30.92, and 37.8, respectively. These multiplication factors are based on an input voltage to the feeder of 126 V, with a 120 V base. However, it is worth noting that the multiplier of 0.95 to account for the effect of phase unbalance is not included in this calculation.

PROBLEMS

4.1 Verify Equation 4.17

4.2 Derive Equation 4.44

4.3 Prove that doubling feeder voltage level causes the percent voltage drop in the primary-feeder circuit to be reduced to one-fourth of its previous value.

4.4 Repeat Example 4.2, parts (a) and (b), assuming a three-phase 34.5 kV wye-grounded feeder main that has 350 kcmil 19-strand copper conductors with an equivalent spacing of 37 in between phase conductors and a lagging-load power factor of 0.9.

4.5 Repeat part (a) of Problem 4.4, assuming 300 kcmil ACSR conductors.

4.6 Repeat Problem 4.5, assuming a lagging-load power factor of 0.7.

4.7 Repeat Problem 4.6, assuming AWG #4/0 conductors.

4.8 Repeat Example 4.3, assuming ACSR conductors.

4.9 Repeat Example 4.4, assuming ACSR conductors.

4.10 Repeat Example 4.5, assuming ACSR conductors.

4.11 Repeat Example 4.6, assuming ACSR conductors.

4.12 Repeat Example 4.8, assuming a 13.2/22.9 kV voltage level.

4.13 Repeat Example 4.9 for a load density of 1000 kVA/mil.

4.14 Repeat part d of Example 4.11 for a load density of 1000 kVA/mil.

4.15 A three-phase 34.5 kV wye-grounded feeder has 500 kcmil ACSR conductors with an equivalent spacing of 60 in. between phase conductors and a lagging-load power factor of 0.8. Use 25°C and 25 Hz and find the K constant in %VD per kVA per mile.

4.16 Assume a square-shaped distribution substation service area and that it is served by four three-phase 12.47 kV wye-grounded feeders. Feeder mains are of 2/0 copper conductors are made up of three-phase open-wire overhead lines having a geometric mean spacing of 37 in. between phase conductors. The percent voltage drop of the feeder is given as 0.0005 per kVA-mile. If the uniformly distributed load has a 4 MVA per square mile load density and a lagging-load factor of 0.9, and conductor ampacity is 360 A, find the following:

(a) Maximum load per feeder
(b) Substation size
(c) Substation spacing, both ways
(d) Total percent voltage drop from the feed point to the end of the main

4.17 Repeat Problem 4.15 for a load density of 1000 kVA/mi.

4.18 Assume that a 5 mile long feeder is supplying a 2000 kVA load of increasing load density starting at a substation. If the K constant of the feeder is given as 0.00001%VD per kV · Ami, determine the following:

(a) The percent voltage drop in the main.
(b) Repeat part (a) but assume that the load is a lumped-sum load and connected at the end of the feeder.
(c) Repeat part (a) but assume that the load is distributed uniformly along the main.

4.19 Consider the two-transformer bank shown in Figure P3.1 of Problem 3.3. Connect them in open-delta primary and open-delta secondary.

(a) Draw and label the voltage-phasor diagram required for the open-delta primary and open-delta secondary on the given 0° reference line.
(b) Show the connections required for the open-delta primary and open-delta secondary. Show the dot markings.

4.20 A three-phase 12.47 kV wye-grounded feeder main has 250 kcmil with 19-strand, copper conductors with an equivalent spacing of 54 in. between phase conductors, and a lagging-load power factor of 0.85. Use 50°C and 60 Hz, and compute the K constant.

4.21 Consider the square-shaped distribution substation given in Example 4.10. The dimension of the area is 2 × 2 miles and served by a 12,470 V (line-to-line) feeder main and laterals. The load density is 1200 kVA/mi^2 and is uniformly distributed, having a lagging power factor of 0.9. A young distribution engineer is considering selection of 4/0 copper conductors with 19 strands and 1/0 copper conductors, operating 60 Hz and 50°C, for main and laterals, respectively. The geometric mean distances are 53 in. for main and 37 in. for the lateral. If the width of the service area of a lateral is 528 ft, determine the following:

(a) The percent voltage drop at the end of the last lateral, if the laterals are also three-phase four-wire wye grounded.
(b) The percent voltage drop at the end of the last lateral, if the laterals are single-phase two-wire wye grounded. Apply Morrison's approximation. (Explain what is right or wrong in the parameter selection in the problem mentioned earlier.) Any suggestions?

5 Design Considerations of Primary Systems

Imagination is more important than knowledge.

Albert Einstein

The great end of learning is nothing else but to seek for the lost mind.

Mencius, Works, 299 BC

Earn your ignorance! Learn something about everything before you know nothing about anything.

Turan Gönen

5.1 INTRODUCTION

The segment of the electrical utility system that lies between the distribution substation and the distribution transformers is referred to as the primary system. This section consists of circuits known as primary feeder or primary distribution feeders.

Figure 5.1 presents a one-line diagram illustrating a typical primary distribution feeder. A feeder comprises a "main" or main feeder, which is typically a three-phase four-wire circuit. It also includes branches or laterals, which are usually single-phase or three-phase circuits derived from the main feeder. Additionally, sublaterals may be tapped off the laterals as required. In residential and rural areas, the laterals and sublaterals are typically single-phase and consist of a single-phase conductor and a neutral conductor. The majority of distribution transformers are single-phase and are connected between the phase conductor and the neutral conductor through fuse cutouts.

A specific feeder is divided into sections using reclosing devices strategically placed to minimize service disruption and inconvenience to customers in the event of a fault. The coordination of fuses and reclosers plays a vital role in achieving this objective.

With an increasing focus on service reliability, it is anticipated that future protection schemes will become more sophisticated and complex. This includes a shift from manually operated devices to remotely controlled automatic devices, utilizing supervisory controlled or computer-controlled systems.

In metropolitan areas, congested and high-load locations are typically served by underground primary feeders. These feeders commonly employ radial three-conductor cables. The advantages of this approach include improved aesthetics and reduced frequency of issues. However, it is a more expensive option with longer repair times compared to overhead systems. In some cases, cables can be suspended from poles, which falls between the cost of open wire and fully underground installations.

The selection of a primary-feeder rating is influenced by various interconnected factors. Examples of these factors include:

1. The nature of the load connected
2. The load density of the area served
3. The growth rate of the load
4. The need for providing spare capacity for emergency operations

DOI: 10.1201/9781003129721-5

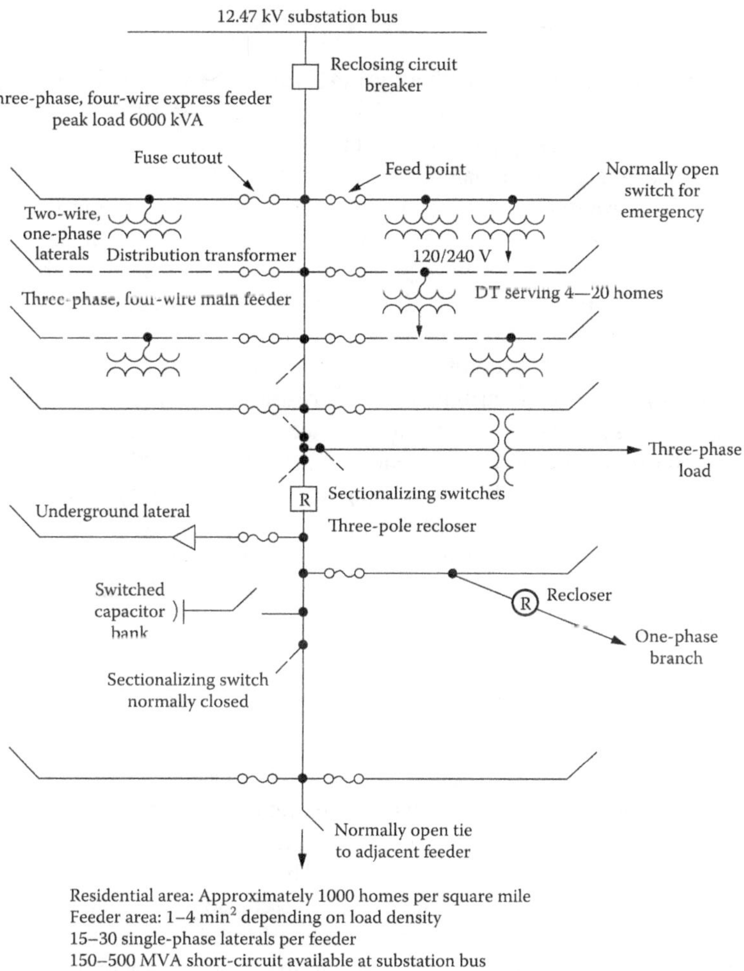

Figure 5.1 One-line diagram of typical primary distribution feeders. (From Fink, D.G. and Beaty, H.W., *Standard Handbook for Electrical Engineers*, 11th edn., McGraw-Hill, New York, 1978 [77].)

5. The type and cost of circuit construction employed
6. The design and capacity of the substation involved
7. The type of regulating equipment used
8. The quality of service required
9. The continuity of service required

The voltage conditions within distribution systems can be enhanced by incorporating shunt capacitors connected as close to the loads as feasible, thereby maximizing their benefits. Shunt capacitors serve to improve the power factor, subsequently reducing voltage drops, currents, and losses in the distribution system sections between the capacitors and the main power buses. Careful consideration should be given to selecting appropriate capacitor ratings to avoid excessive overvoltages that may occur during times of light loads due to voltage rise resulting from capacitor currents.

Alternatively, the voltage conditions within distribution systems can be improved by employing series capacitors. However, it's important to note that while series capacitors have a positive impact on voltage conditions, they do not reduce currents and the associated losses in the system.

5.2 STATE-OF-THE-ART UPDATE

The primary network is often designed with a certain level of interconnection between feeders originating from the same or different substations. This interconnection allows for the transfer of faults from a segment of a line to a healthy part of another network. During the planning phase, determining the optimal phase arrangement, unbalance, and minimizing losses are crucial aspects of primary network planning [108]. Analytical approaches are employed to determine the optimal locations of distributed generation sources in a radial network, taking into account the existing loads and switches [109, 110]. Feasibility studies have proposed upgrading traditionally radial configurations to normally closed-loop arrangements [111]. Radial topologies facilitate the efficient localization of electrical faults in medium-voltage (MV) networks [112].

However, over time, the loading conditions within a feeder may increase. Therefore, strategic placement and realistic sizing of distributed energy resources (DERs) are necessary in planning studies to assess voltage stability and potential enhancements [113, 114]. The growth of traffic in an area and the stochastic nature of vehicle charging can shift the bottleneck of a feeder, requiring consideration of these factors [115, 116]. Enhancing the reconfigurability of MV networks is crucial to effectively allocate losses in weakly meshed feeders where smart meters are deployed [117, 118]. Changes occurring in recent decades have led to net positive/negative loads along existing feeders. Consequently, an analytical reassessment of non-existent feeders should consider the aforementioned developments in tomorrow's network.

5.3 RADIAL-TYPE PRIMARY FEEDERS

The most common and cost-effective form of primary feeder is the radial-type primary feeder, as depicted in Figure 5.2. This configuration consists of a main primary feeder that branches into multiple primary laterals, which further divide into several sublaterals to supply power to distribution transformers. Typically, the main feeder and subfeeders are three-phase three- or four-wire circuits, while the laterals can be either three-phase or single-phase. The highest current magnitude is observed in the conductors leaving the substation, and it gradually decreases as laterals and sublaterals branch off from the feeder. As the current diminishes, the size of the feeder conductors is often reduced. However, any reduction in feeder size based solely on the thermal capability, i.e., current-carrying capacity, may be limited by permissible voltage regulation requirements.

The reliability of service continuity for radial primary feeders is relatively low. A fault occurring at any point along the radial primary feeder leads to a power outage for all consumers on that feeder unless the fault can be isolated from the power source using disconnecting devices such as fuses, sectionalizers, disconnect switches, or reclosers.

Figure 5.3 illustrates a modified radial-type primary feeder featuring tie and sectionalizing switches. This design enables swift restoration of service to customers by redirecting the unfaulted sections of the feeder to an adjacent primary feeder or feeders. The fault can be isolated by opening the associated disconnecting devices on both sides of the faulted section.

In Figure 5.4, a different type of radial primary feeder is displayed, known as an express feeder with backfeed. The section of the feeder between the substation low-voltage bus and the load center of the service area is referred to as the express feeder. This segment does not allow any subfeeders or laterals to be tapped off. However, a subfeeder is permitted to provide a backfeed toward the substation from the load center.

Figure 5.5 demonstrates a radial-type phase-area feeder arrangement where each phase of the three-phase feeder serves its designated service area. In both Figures 5.4 and 5.5, each dot represents a balanced three-phase load concentrated at that specific location.

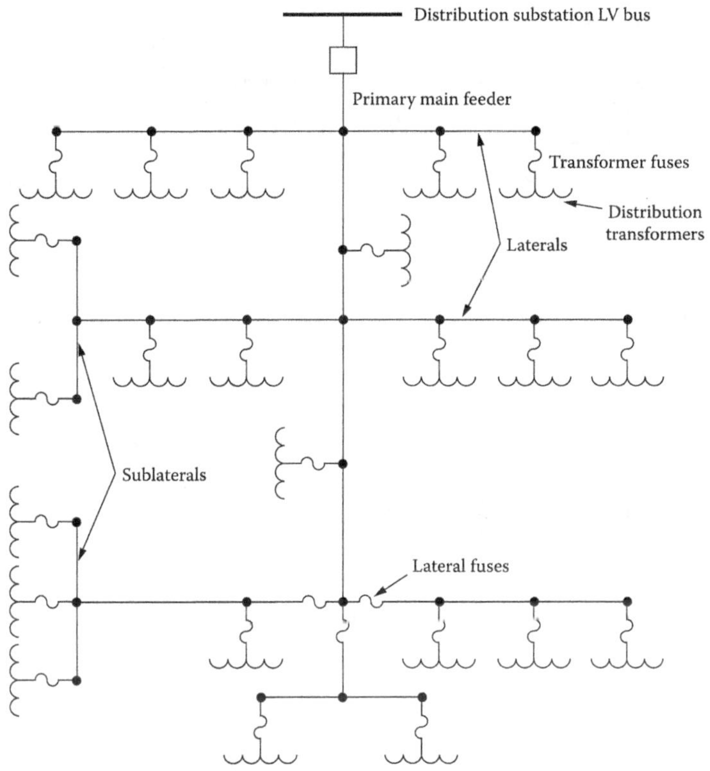

Figure 5.2 Radial-type primary feeder.

5.4 LOOP-TYPE PRIMARY FEEDER

Figure 5.6 depicts a loop-type primary feeder that forms a loop within the feeder load area and returns to the bus. In some cases, the loop tie disconnect switch may be replaced with a loop tie breaker based on the load conditions. In either scenario, the loop can operate with the tie disconnect switches or breakers either normally open (NO) or normally closed (NC).

Typically, the size of the feeder conductor remains consistent throughout the loop in a loop-type primary feeder arrangement. It is chosen to accommodate the normal load of the feeder as well as the load of the other half of the loop. This configuration ensures that two parallel paths are available from the substation to the load when the loop operates without tie breakers or disconnect switches.

In the event of a primary fault, the feeder breaker opens and remains in that state until the fault is isolated from both directions. The loop-type primary feeder arrangement is particularly advantageous for providing reliable service to loads where high service reliability is crucial. Ideally, a separate feeder breaker is installed at each end of the loop, despite the associated costs. This enables the parallel feeder paths to be connected to separate bus sections in the substation and supplied from distinct transformers. Alongside main feeder loops, lateral loops with normally open (NO) configurations are also employed, especially in underground systems.

5.5 PRIMARY NETWORK

Figure 5.7 depicts a primary network, which consists of interconnected feeders supplied by multiple substations. The radial primary feeders can be tapped off the interconnecting tie feeders, or they can be directly supplied from the substations. Each tie feeder is equipped with two circuit breakers at each end to minimize the number of customers affected in the event of a tie-feeder fault.

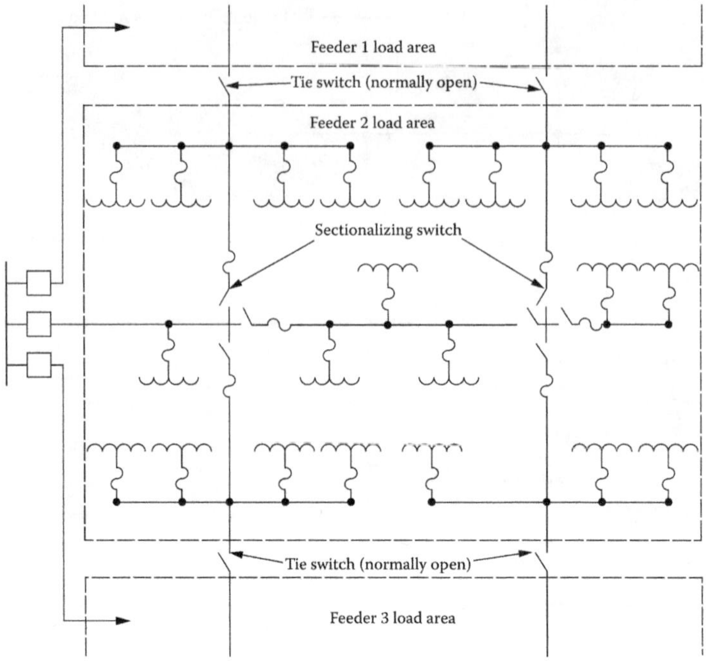

Figure 5.3 Radial-type primary feeder with tie and sectionalizing switches.

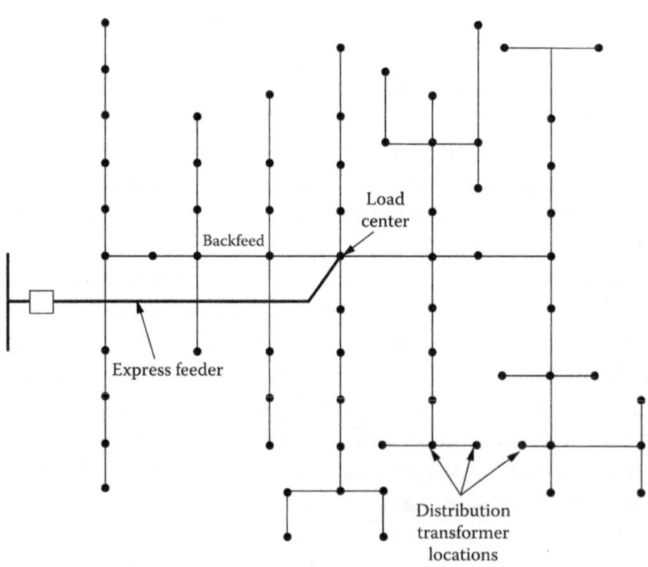

Figure 5.4 Radial-type primary feeder with express feeder and backfeed.

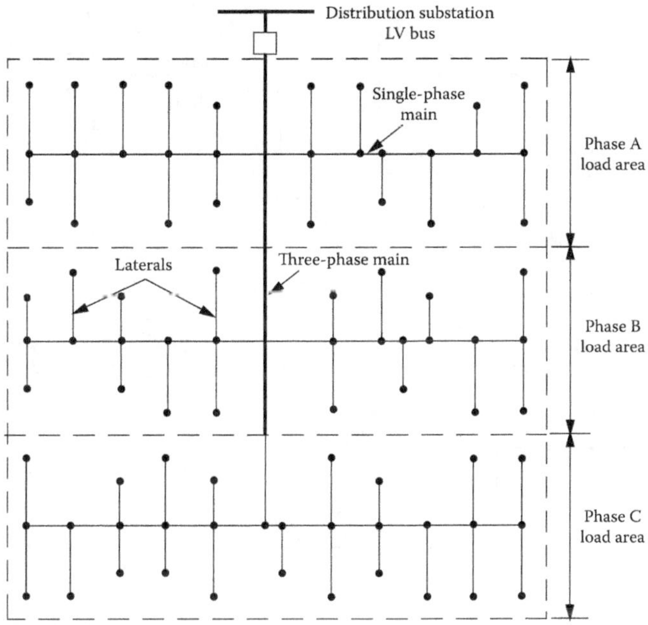

Figure 5.5 Radial-type phase-area feeder.

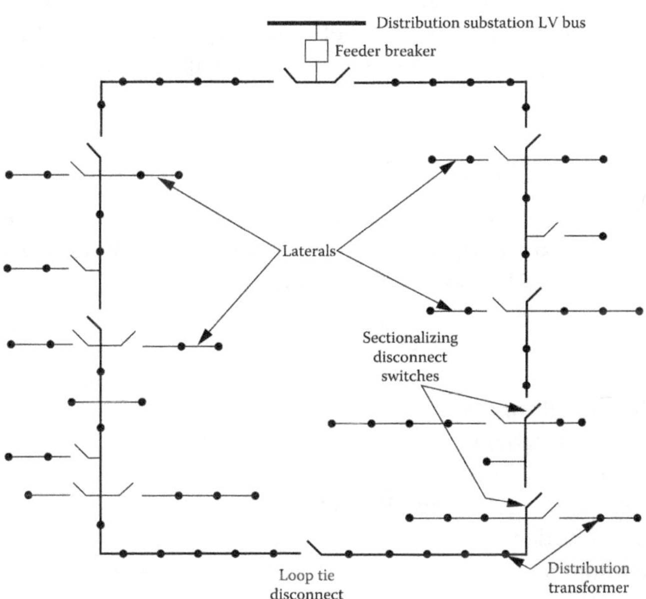

Figure 5.6 Loop-type primary feeder.

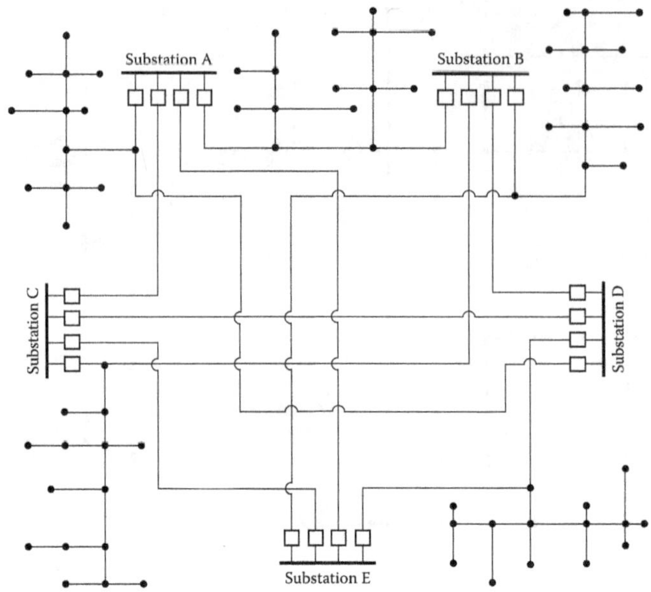

Figure 5.7 Loop-type primary feeder.

The primary-network system delivers power to the load from multiple directions. By strategically locating transformers near heavy-load centers and regulating the feeders at the substation buses, adequate voltage levels are maintained at the utilization points. Generally, primary networks exhibit lower losses compared to comparable radial systems due to load distribution and division.

The primary-network arrangement offers significantly higher reliability and service quality compared to radial and loop arrangements. However, designing and operating a primary network is more challenging than radial or loop systems.

5.6 PRIMARY-FEEDER VOLTAGE LEVELS

The voltage level of the primary feeder is a crucial factor that significantly influences the design, cost, and operation of the system. The primary-feeder voltage level impacts various aspects of the design and operation, including but not limited to [50]:

1. Primary-feeder length
2. Primary-feeder loading
3. Number of distribution substations
4. Rating of distribution substations
5. Number of subtransmission lines
6. Number of customers affected by a specific outage
7. System maintenance practices
8. The extent of tree trimming
9. Joint use of utility poles
10. Type of pole-line design and construction
11. Appearance of the pole line

The selection of primary-feeder voltage levels is influenced by several additional factors, as depicted in Figure 5.8. Typical primary voltage levels utilized in the United States are provided in Table 5.1. Three-phase four-wire systems with a multigrounded common neutral are predominantly

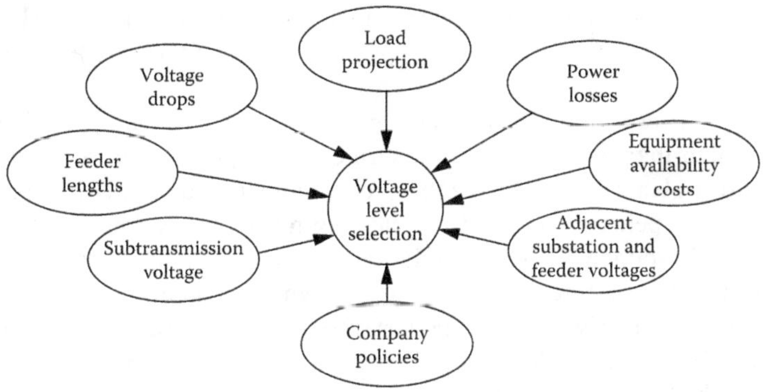

Figure 5.8 Factors affecting primary-feeder voltage-level selection decision.

employed, such as 12.47Y/7.2 kV, 24.9Y/14.4 kV, and 34.5Y/19.92 kV. The fourth wire serves as the multigrounded neutral for both the primary and secondary systems. The primary voltage levels in the 15 kV class are commonly used, with 12.47 kV being the most prevalent primary distribution voltage in North America. However, there is a current shift toward higher voltages, with the 34.5 kV class gaining rapid acceptance. On the other hand, the usage of the 5 kV class continues to decline.

Some distribution systems utilize multiple primary voltages, such as 12.47 kV and 34.5 kV. California is among the few states that employ three-phase three-wire primary systems. The four-wire

Table 5.1
Typical Primary Voltage Levels

Class, kV	3φ Voltage	
2.5	2,300	3W-A
	2,400[a]	3W-A
5	4,000	3W-A or 3W-Y
	4,160[a]	4W-Y
	4,330	3W-A
	4,400	3W-A
	4,600	3W-A
	4,800	3W-A
8.66	6,600	3W-A
	6,900	3W-A or 4W-Y
	7,200[a]	3W-A or 4W-Y
	7,500	4W-Y
	8,320	4W-Y
15	11,000	3W-A
	11,500	3W-A
	12,000	3W-A or 4W-Y
	12,470[a]	4W-Y
	13,200[a]	3W-A or 4W-Y
	13,800[a]	3W-A
	14,400	3W-A
25	22,900[a]	4W-Y
	24,940[a]	4W-Y
34.5	34,500[a]	4W-Y

[a] Most common voltage in the individual classes.

system is cost-effective, particularly for underground residential distribution (URD) systems, as each primary lateral requires only one insulated phase wire and a bare neutral wire instead of two insulated wires.

Typically, primary feeders in low-load density areas are limited in length and loading by permissible voltage drop rather than thermal constraints. In contrast, primary feeders in high-load density areas, including industrial and commercial zones, may be restricted by thermal limitations.

In general, for a given percent voltage drop, the length and loading of a feeder are directly proportional to the feeder's voltage level. This relationship is commonly referred to as the "voltage-square rule." For instance, doubling the feeder voltage, while maintaining the same percent voltage drop, allows the feeder to supply the same power over a distance four times greater. However, as explained by Lokay [50], the extended feeder length also serves more load. Consequently, the advantage gained from the higher voltage level through the voltage-square factor is

$$\text{Voltage - square factor} = \left(\frac{V_{L-N,\text{new}}}{V_{L-N,\text{old}}}\right)^2 \tag{5.1}$$

has to be allocated between the growth in load and in distance. Further, the same percent voltage drop will always result provided that the following relationship exists:

$$\text{Distance ratio} \times \text{Load ratio} = \text{Voltage - square factor} \tag{5.2}$$

where

$$\text{Distance ratio} = \frac{\text{New Distance}}{\text{Old Distance}} \tag{5.3}$$

and

$$\text{Load ratio} = \frac{\text{New feeder loading}}{\text{Old feeder loading}} \tag{5.4}$$

The relationship between the voltage-square factor rule and the feeder *distance-coverage principle* is further explained in Figure 5.9.

There is a relationship between the area served by a substation and the voltage rule. Lokay [50] defines it as the *area-coverage principle*. As illustrated in Figure 5.10, for a constant percent voltage drop and a uniformly distributed load, the feeder service area is proportional to

$$\left[\left(\frac{V_{L-N,\text{new}}}{V_{L-N,\text{old}}}\right)^2\right]^{2/3} \tag{5.5}$$

provided that both dimensions of the feeder service area change by the same proportion. For example, if the new feeder voltage level is increased to twice the previous voltage level, the new load and area that can be served with the same percent voltage drop are

$$\left[\left(\frac{V_{L-N,\text{new}}}{V_{L-N,\text{old}}}\right)^2\right]^{2/3} = (2^2)^{2/3} = 2.52 \tag{5.6}$$

times the original load and area. If the new feeder voltage level is increased to three times the previous voltage level, the new load and area that can be served with the same percent voltage drop are

$$\left[\left(\frac{V_{L-N,\text{new}}}{V_{L-N,\text{old}}}\right)^2\right]^{2/3} = (3^2)^{2/3} = 4.32 \tag{5.7}$$

times the original load and area.

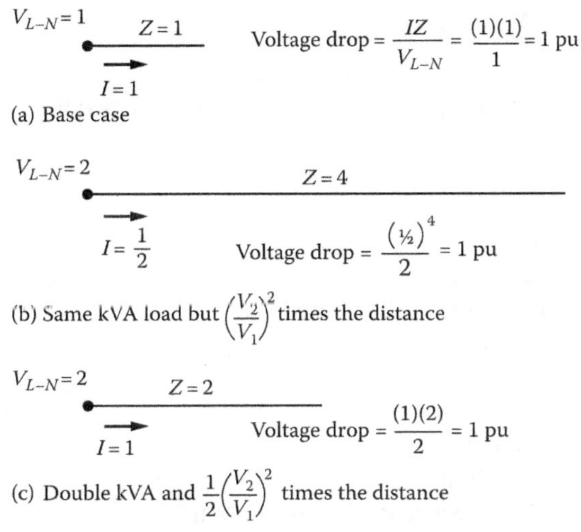

Figure 5.9 Illustration of the voltage-square rule and the feeder distance-coverage principle as a function of feeder voltage level and a single load. (From Westinghouse Electric Corporation, *Electric Utility Engineering Reference Book-Distribution Systems*, Vol. 3, East Pittsburgh, Pittsburgh, PA, 1965.)

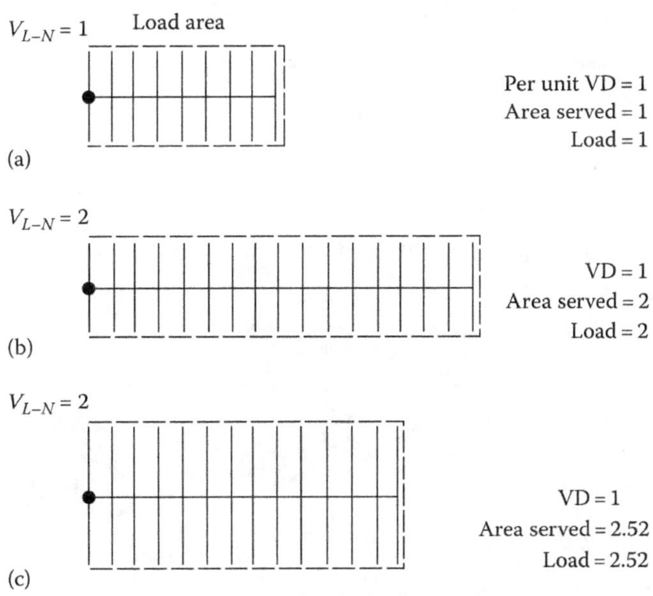

Figure 5.10 Feeder area-coverage principle as related to feeder voltage and a uniformly distributed load. (From Westinghouse Electric Corporation, *Electric Utility Engineering Reference Book-Distribution Systems*, Vol. 3, East Pittsburgh, Pittsburgh, PA, 1965.)

5.7 PRIMARY-FEEDER LOADING

The design loading of a primary feeder, measured at the substation during peak-load conditions, is referred to as primary-feeder loading [50]. Several factors influence the design loading of a feeder, including:

1. The density of the feeder load
2. The nature of the feeder load
3. The growth rate of the feeder load
4. The reserve-capacity requirements for emergency
5. The service-continuity requirements
6. The service-reliability requirements
7. The quality of service
8. The primary-feeder voltage level
9. The type and cost of construction
10. The location and capacity of the distribution substation
11. The voltage regulation requirements

There are additional factors affecting the decisions for feeder routing, the number of feeders, and feeder conductor size selection, as shown in Figures 5.11 through 5.13.

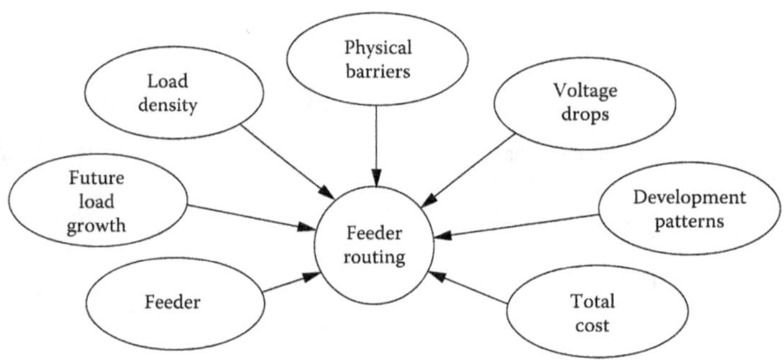

Figure 5.11 Factors affecting feeder routing decisions.

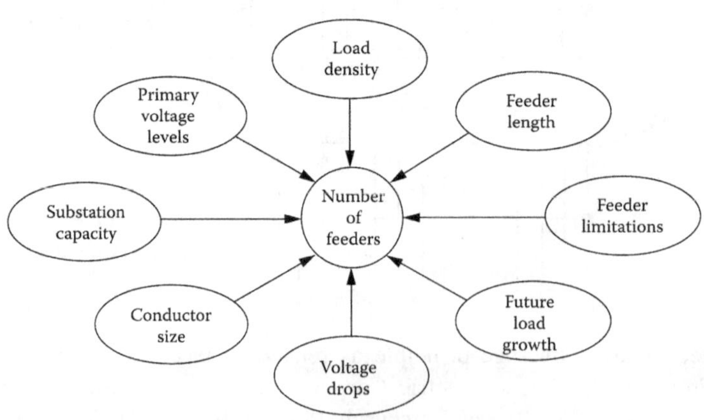

Figure 5.12 Factors affecting the number of feeders.

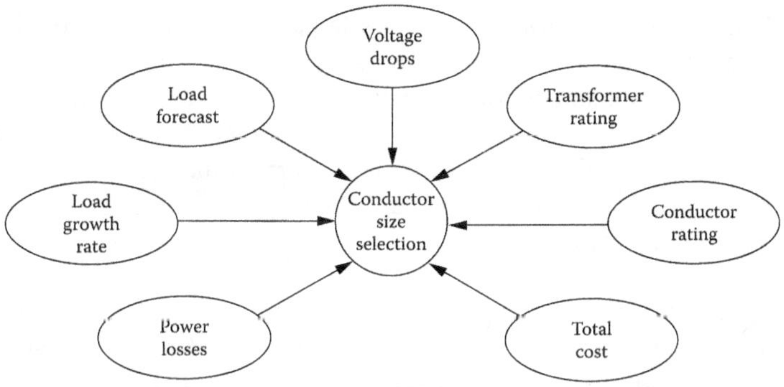

Figure 5.13 Factors affecting conductor size selection.

5.8 TIE LINES

A tie line, depicted in Figure 5.14, is a connection between two supply systems that allows for emergency service to be provided from one system to another. In addition to offering emergency support to adjacent areas or substations, tie lines also provide service to area loads along their route. Consequently, tie lines serve two primary functions:

1. To provide emergency service for an adjacent feeder for the reduction of outage time to the customers during emergency conditions.
2. To provide emergency service for adjacent substation systems, thereby eliminating the necessity of having an emergency backup supply at every substation. Tie lines should be installed when more than one substation is required to serve the area load at one primary distribution voltage.

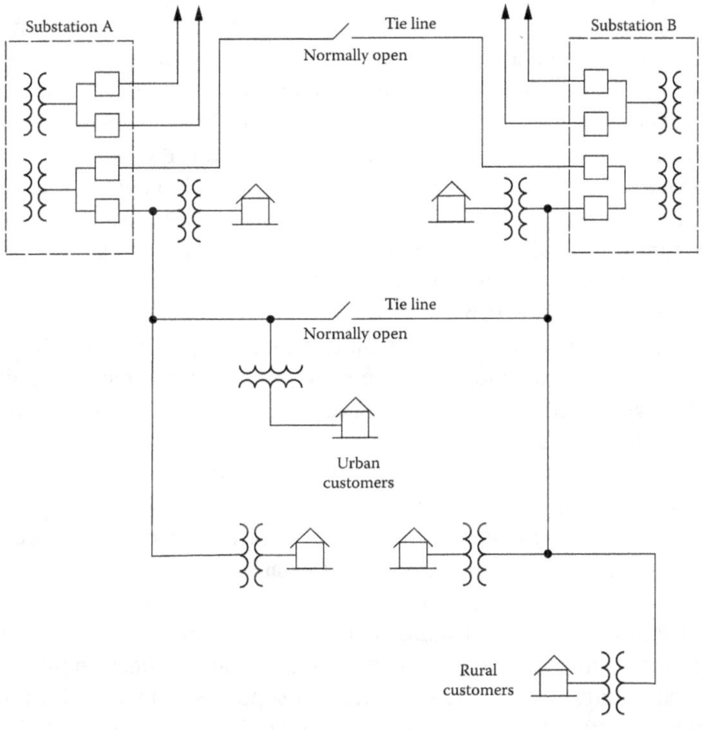

Figure 5.14 One-line diagram of typical two-substation area supply with tie lines.

The design and installation of substation primary feeders typically involve arranging the feeders in a manner where those supplied by the same transformer extend in opposite directions. This arrangement allows for the establishment of ties with circuits supplied by different transformers. For instance, in a substation with two transformers and four feeders, two feeders from one transformer may extend north and south, while the other two feeders from the second transformer may extend east and west. It is important that all tie lines are connected to circuits supplied by other transformers. This configuration facilitates the service restoration to an area affected by a transformer failure.

Disconnect switches are strategically placed at specific intervals along the main feeder tie lines to facilitate load transfer and service restoration. The location of these disconnect switches must be carefully selected to ensure maximum operational flexibility. It is crucial to consider not only the physical arrangement of the circuit but also the size and nature of loads between switches. Balancing the loads between disconnect switches is essential to minimize any adverse impact on circuit operation during load transfers. Optimal voltage conditions can only be achieved if the circuit is balanced as closely as possible throughout its entire length.

5.9 DISTRIBUTION FEEDER EXIT: RECTANGULAR-TYPE DEVELOPMENT

The purpose of this section is to present an example of a uniform area development plan aimed at minimizing circuitry changes associated with the systematic expansion of the distribution system.

In this scenario, it is assumed that underground feeder exits are extended from a distribution substation into an existing overhead system. Additionally, it is envisioned that upon reaching its ultimate development stage, the substation will serve a service area of 6 square miles with a total of 12 feeder circuits, evenly distributed with four circuits per transformer. Assuming a uniform load distribution, each of the 12 circuits would cover approximately half a square mile within the fully developed service area. This development approach is referred to as the rectangular-type development and is illustrated in Figures 5.15 through 5.18.

Typically, in order to ensure transfer capabilities and maintain service continuity during transformer outages, adjacent service areas are served from different transformer banks. As the service area develops and new feeder circuits and transformer banks are added, circuit number changes become necessary. The center transformer bank is always fully developed when the substation has eight feeder circuits. As the service area continues to develop, the remaining transformer banks reach their full capacity. There are two primary methods of development, which depend on the load density of the service area: the 1-2-4-8-12 feeder circuit method and the 1-2-4-6-8-12 feeder circuit method. It's important to note that the numbers associated with feeders and transformer banks in the following figures represent the sequence of installation as the substation develops [119].

For service areas with high-load density, it is essential to develop adjacent substations in a similar manner to ensure adequate load-transfer capability and service continuity. In such cases, a two-transformer-bank substation can handle the emergency rating of one bank, circuit ties, and reserve considerations. Since sufficient circuit ties are necessary to support the loss of a large transformer unit, the 1-2-4-8-12 feeder method is particularly suitable for high-load density areas. Figures 5.15 through 5.18 demonstrate the sequential installation of additional transformers and feeders as part of this development approach.

Method of development for low-load density areas: In low-load density areas where adjacent substations lack adequate development and circuit ties are impractical due to substantial distances between substations, the 1-2-4-6-8-12 circuit-developing substation scheme is more appropriate. These significant distances between substations impose limitations on load transfer capacity without causing extensive outage times due to circuit switching, while also ensuring that minimum voltage levels are maintained. In this method, the substation must possess all three transformer banks before utilizing larger transformers to enhance the firming capability within each individual substation.

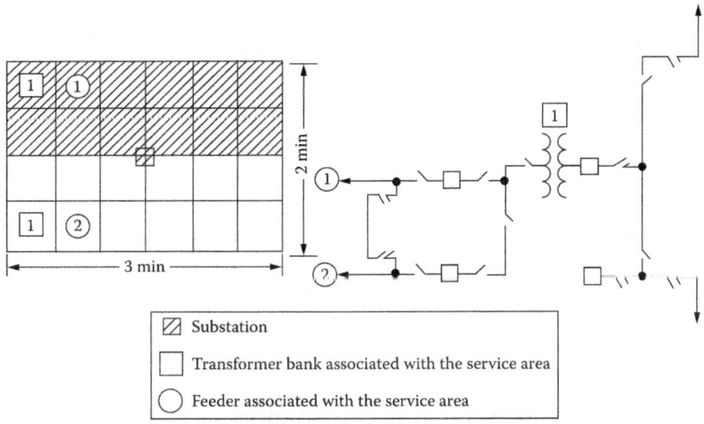

Figure 5.15 Rectangular-type development.

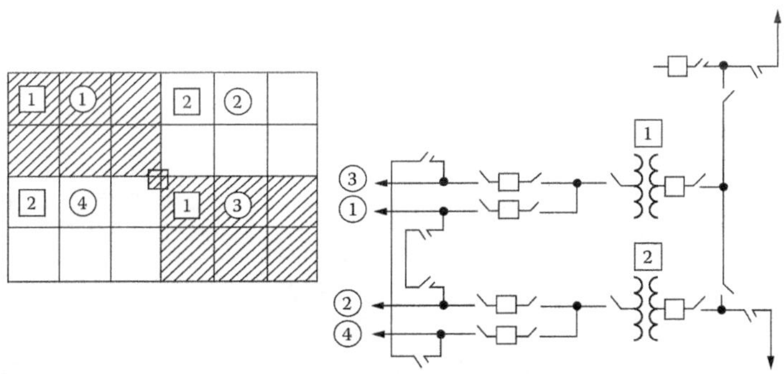

Figure 5.16 Rectangular-type development with two transformers, type 1.

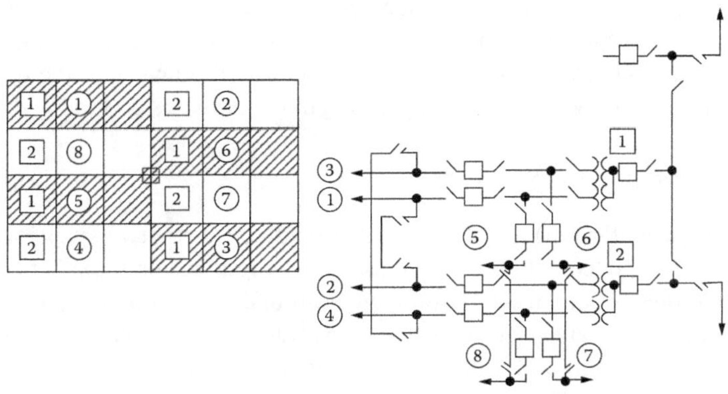

Figure 5.17 Rectangular-type development with two transformers, type 2.

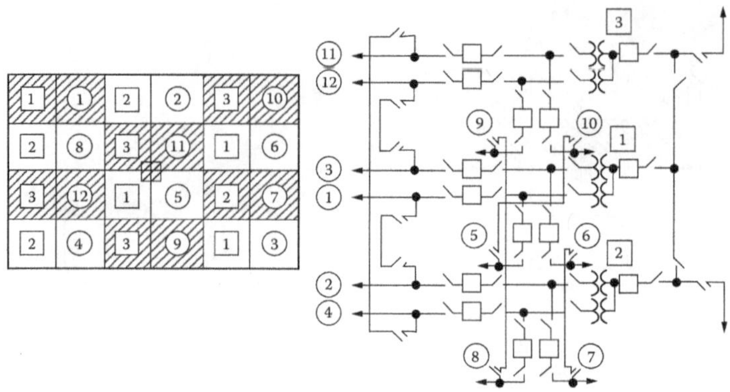

Figure 5.18 Rectangular-type development with three transformers.

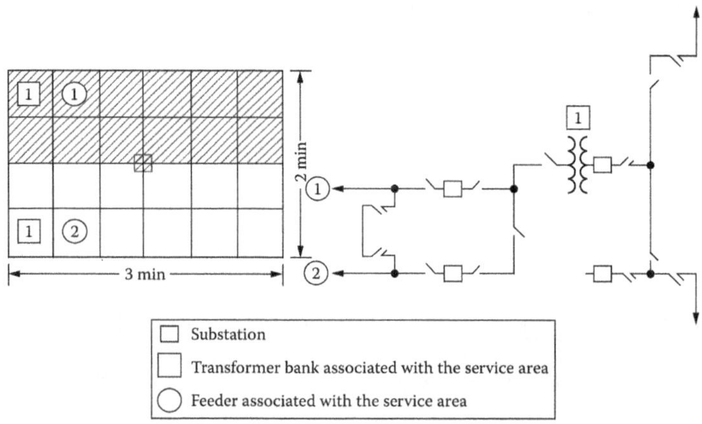

Figure 5.19 The sequence of installing additional transformers and feeders, type 1.

As depicted in Figures 5.19 through 5.23, once three transformer units, such as 12/16/20 MVA, and six feeders are installed during the development of this substation type, two alternatives for further expansion arise: (1) removing one of the banks and increasing the sizes of the remaining two banks to accommodate larger transformers, such as 24/32/40 MVA, utilizing the low-side bays of the third transformer as part of the circuitry for the development of the remaining two banks, or (2) disregarding the third transformer-bank area entirely and proceeding with the development of the two remaining sections, following a similar approach to the previous method.

5.10 RADIAL-TYPE DEVELOPMENT

In addition to the rectangular-type development that involves overhead expansion, there is another type of development associated with the growth of underground residential distribution (URD) subdivisions. This development entails the use of underground feeders that serve local loads as they extend into adjacent service areas. In these locations, the overhead feeders along quarter section lines are replaced with underground cables. As these underground lines continue to extend outward from the substation, they provide service to the area load. These underground lines traverse through the platted service area developments and typically terminate at a remote overhead feeder along a section line. This type of development is referred to as radial-type development, resembling a wagon wheel with the substation at the center and the radial spokes representing the feeders, as depicted in Figure 5.24.

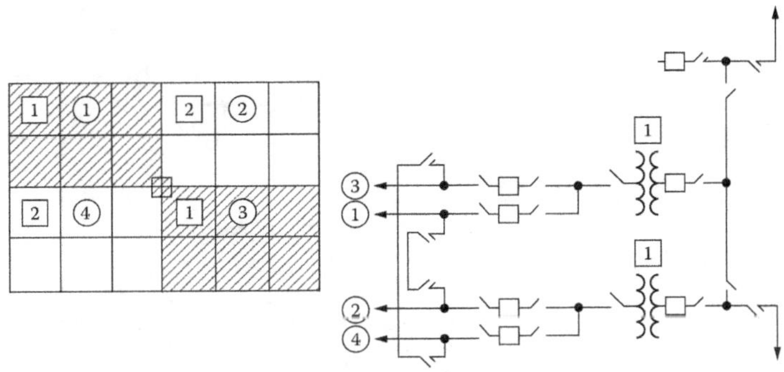

Figure 5.20 The sequence of installing additional transformers and feeders, type 2.

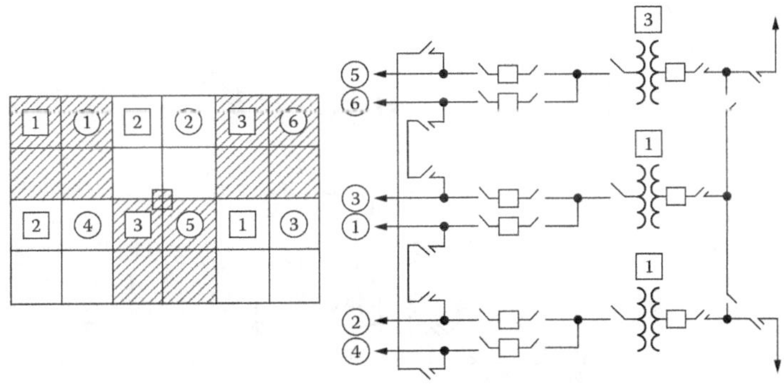

Figure 5.21 The sequence of installing additional transformers, type 3.

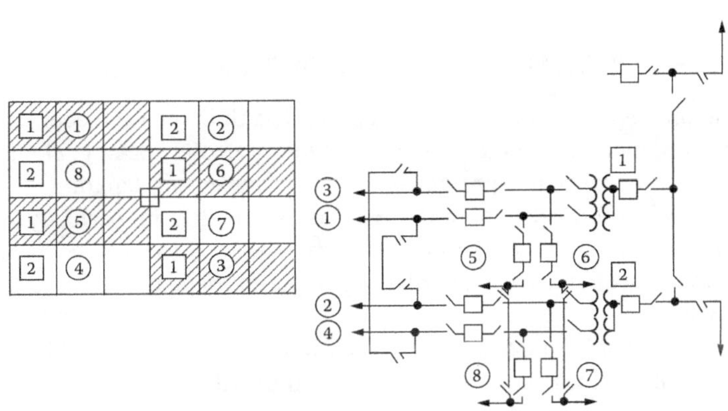

Figure 5.22 The sequence of installing additional transformers and feeders, type 4.

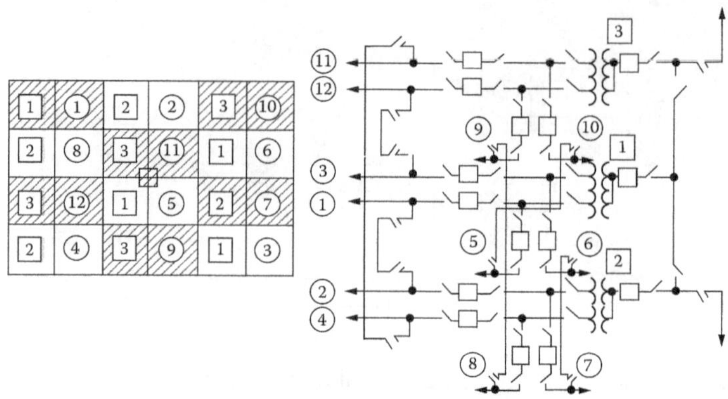

Figure 5.23 The sequence of installing additional transformers, type 5.

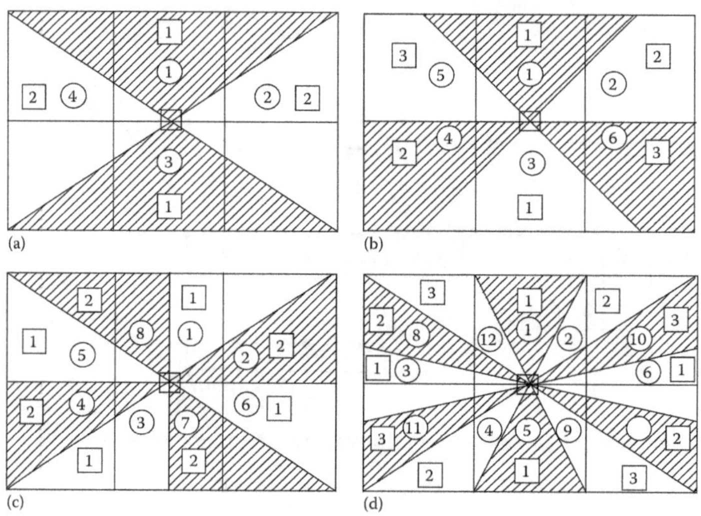

Figure 5.24 Radial-type development: (a) type 1, (b) type 2, (c) type 3, and (d) type 4.

5.11 RADIAL FEEDERS WITH UNIFORMLY DISTRIBUTED LOAD

The simplified diagram depicted in Figure 5.25 represents a three-phase feeder main that maintains consistent construction, including cable size or open-wire size and spacing, along its entire length, denoted as l. The line impedance is represented as $z = r + jx$ per unit length.

For analysis purposes, the load flow in the main is assumed to be perfectly balanced and uniformly distributed at all points along the main. In practice, a reasonably good phase balance can sometimes be achieved by strategically distributing single-phase and open-wye laterals among the three phases of the main.

It is important to note that Figure 5.25 does not show all the closely spaced loads and lateral lines that are connected to the main. Due to the uniform load distribution along the main, as illustrated in Figure 5.26, the load current in the main varies depending on the distance. To account for the numerous closely spaced small loads, a differential tapped-off load current $d\bar{I}$ is considered, which corresponds to a differential distance dx. In this context, l represents the total length of the feeder, and x represents the distance from the beginning end of the feeder to point 1. Consequently, the distance from the beginning end of the feeder to point 2 is $x + dx$. $\bar{I}s$ denotes the sending-end current at the feeder breaker, while $\bar{I}r$ represents the receiving-end current. $\bar{I}x1$ and $\bar{I}x21$ denote the currents

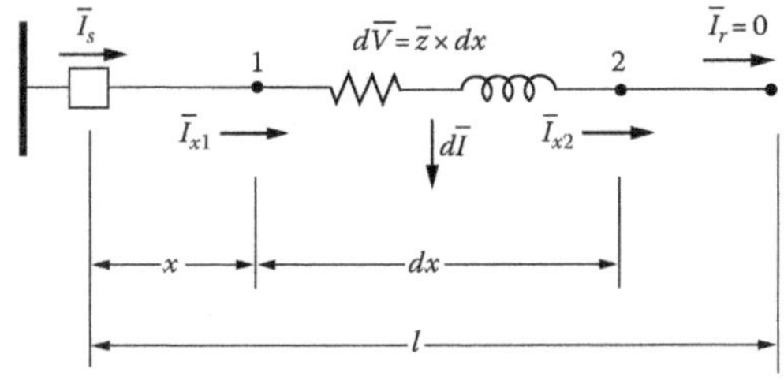

Figure 5.25 A radial feeder.

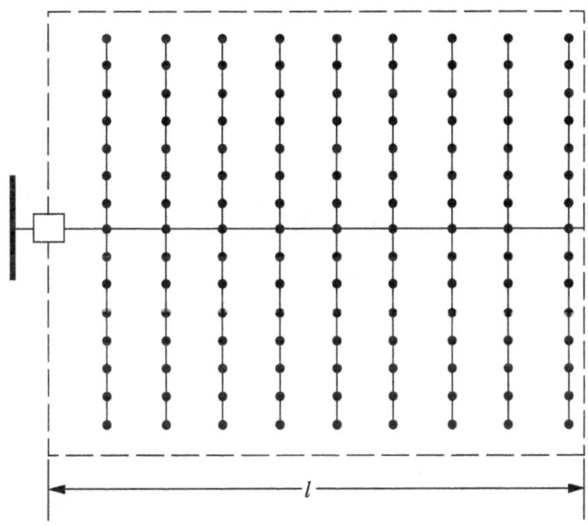

Figure 5.26 A uniformly distributed main feeder.

at points 1 and 2, respectively, along the main. Additionally, it is assumed that all loads connected to the feeder have the same power factor [120].

The equations provided below hold true for both per-unit and per-phase (line-to-neutral) dimensional variables. The circuit voltage can be either primary or secondary, thus allowing neglect of shunt capacitance currents. As the total load is uniformly distributed along the entire length of the feeder from $x = 0$ to $x = l$, the following equations are applicable.

$$\frac{d\bar{I}}{dx} = \bar{k} \tag{5.8}$$

which is a constant. Therefore $\bar{I}_x$, that is, the current in the main of some x distance away from the circuit breaker, can be found as a function of the sending-end current $\bar{I}_s$s and the distance x. This can be accomplished either by inspection or by writing a current equation containing the integration of $d\bar{I}$. Therefore, for dx distance,

$$\bar{I}_{x1} = \bar{I}_{x2} + d\bar{I} \tag{5.9}$$

or

$$\bar{I}_{x2} = \bar{I}_{x1} + -d\bar{I} \tag{5.10}$$

From Equation 5.10,

$$\begin{aligned}\bar{I}_{x1} &= \bar{I}_{x1} + d\bar{I}\frac{dx}{dx} \\ &= \bar{I}_{x1} - \frac{d\bar{I}}{dx}dx\end{aligned} \quad (5.11)$$

or

$$\bar{I}_{x2} = \bar{I}_{x1} - \bar{k}dx \quad (5.12)$$

where $\bar{k} = d\bar{I}/dx$ or, approximately,

$$\bar{I}_{x2} = \bar{I}_{x1} - kd\bar{I} \quad (5.13)$$

and

$$\bar{I}_{x1} = \bar{I}_{x2} - kd\bar{I} \quad (5.14)$$

Therefore, for the total feeder,

$$I_r = I_s - k \times l \quad (5.15)$$

and

$$I_s = I_r + k \times l \quad (5.16)$$

When $x = l$, from Equation 5.15,

$$I_r = I_s - k \times I = 0$$

hence

$$k = \frac{I_s}{l} \quad (5.17)$$

and since $x = l$,

$$I_r = I_s - k \times x \quad (5.18)$$

Therefore, substituting Equation 5.17 into Equation 5.18,

$$I_r = I_s\left(1 - \frac{x}{l}\right) \quad (5.19)$$

For a given x distance,

$$I_r = I_s$$

thus Equation 5.19 can be written as

$$I_x = I_s\left(1 - \frac{x}{l}\right) \quad (5.20)$$

which gives the current in the main at some x distance away from the circuit breaker. Note that from Equation 5.20,

$$I_s = \begin{cases} I_r = 0 & \text{at } x = l \\ I_r = I_s & \text{at } x = 0 \end{cases}$$

The differential series voltage drop $d\overline{V}$ and the differential power loss dP_{LS} due to I^2R losses can also be found as a function of the sending-end current I_s and the distance x in a similar manner. Therefore, the differential series voltage drop can be found as

$$d\overline{V} = I_x \times z dx \tag{5.21}$$

or substituting Equation 5.20 into Equation 5.21,

$$d\overline{V} = I_s \times z\left(1 - \frac{x}{l}\right) dx \tag{5.22}$$

Also, the differential power loss can be found as

$$dP_{LS} = I_x^2 \times r dx \tag{5.23}$$

or substituting Equation 5.20 into Equation 5.23,

$$dP_{LS} = \left[I_s\left(1 - \frac{x}{l}\right)\right]^2 r dx \tag{5.24}$$

The series voltage drop VD_x due to I_x current at any point x on the feeder is

$$VD_x = \int_0^x dV \tag{5.25}$$

Substituting Equation 5.22 into Equation 5.25,

$$VD_x = \int_0 I_s \times z\left(1 - \frac{x}{l}\right) dx \tag{5.26}$$

or

$$VD_x = I_s \times z \times x\left(1 - \frac{x}{2l}\right) dx \tag{5.27}$$

Therefore, the total series voltage drop $\sum VD_x$ on the main feeder when $x = l$ is

$$\sum VD_x = I_s \times z \times l\left(1 - \frac{l}{2l}\right)$$

or

$$\sum VD_x = \frac{1}{2} z \times l \times I_s \tag{5.28}$$

The total copper loss per phase in the main due to I^2R losses is

$$\sum P_{LS} = \int_0^i dP_{LS} \tag{5.29}$$

or

$$\sum P_{LS} = \frac{1}{3} I_s^2 \times r \times l \tag{5.30}$$

Therefore, from Equation 5.28, the distance x from the beginning of the main feeder at which location the total load current I_s may be concentrated, that is, lumped for the purpose of calculating the total voltage drop, is

$$x = \frac{l}{2}$$

whereas, from Equation 5.30, the distance x from the beginning of the main feeder at which location the total load current I_s may be lumped for the purpose of calculating the total power loss is

$$x = \frac{l}{3}$$

5.12 RADIAL FEEDERS WITH NONUNIFORMLY DISTRIBUTED LOAD

The single-line diagram presented in Figure 5.27 depicts a three-phase feeder main with a linear increase in the tapped-off load as the distance x progresses. It is worth noting that the load is zero when $x = 0$. The curve displayed in Figure 5.28 represents the plot of the sending-end current against the distance x along the feeder main.

From Figure 5.28, the negative slope can be written as

$$\frac{dI_x}{dx} = -k \times I_s \times x \tag{5.31}$$

Here, the constant k can be found from

$$\begin{aligned} I_s &= \int_{x=0}^{l} -dI_x \\ &= \int_{x=0}^{l} k \times I_s \times x dx \end{aligned} \tag{5.32}$$

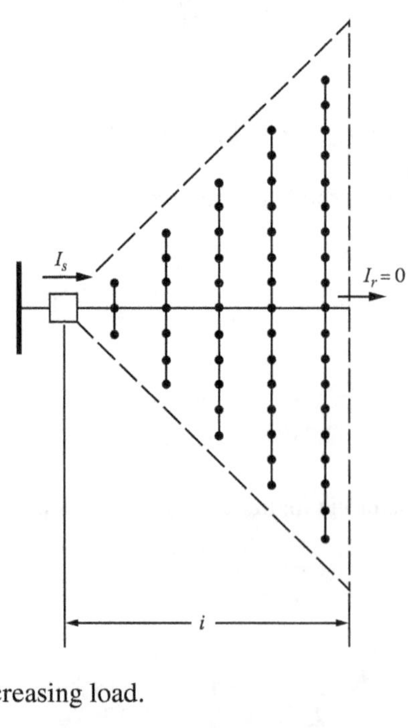

Figure 5.27 A uniformly increasing load.

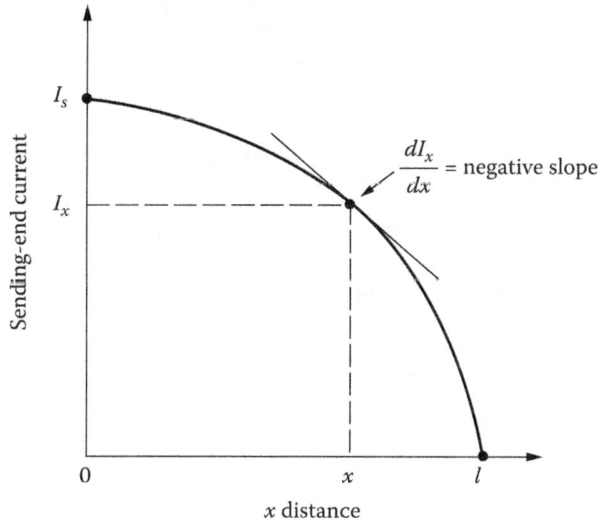

Figure 5.28 The sending-end current as a function of the distance along a feeder.

or

$$I_s = k \times I_s \times \frac{l^2}{2} \tag{5.33}$$

From Equation 5.33, the constant k is

$$k = \frac{2}{l^2} \tag{5.34}$$

Substituting Equation 5.34 into Equation 5.31,

$$\frac{dI_x}{dx} = -2I_s \times \frac{x}{l^2} \tag{5.35}$$

Therefore, the current in the main at some x distance away from the circuit breaker can be found as

$$I_x = I_s\left(1 - \frac{x^2}{l^2}\right) \tag{5.36}$$

Hence, the differential series voltage drop is

$$d\overline{V} = I_x \times z dx \tag{5.37}$$

or

$$d\overline{V} = I_s \times z\left(1 - \frac{x^2}{l^2}\right)dx \tag{5.38}$$

Also, the differential power loss can be found as

$$dP_{\text{LS}} = \text{I}_x^2 \times r dx \tag{5.39}$$

or

$$dP_{\text{LS}} = \text{I}_s^2 \times r\left(1 - \frac{x^2}{l^2}\right)^2 dx \tag{5.40}$$

The series voltage drop due to I_x current at any point x on the feeder is

$$VD_x = \int_0^x dV \tag{5.41}$$

Substituting Equation 5.38 into Equation 5.41 and integrating the result,

$$VD_x = I_s \times z \times x\left(1 - \frac{x^2}{3l^2}\right) \tag{5.42}$$

Therefore, the total series voltage drop on the main feeder when $x = 1$ is

$$\sum VD_x = \frac{2}{3} z \times l \times I_s \tag{5.43}$$

The total copper loss per phase in the main due to I^2R losses is

$$\sum P_{LS} = \int_0^1 dP_{LS} \tag{5.44}$$

or

$$\sum P_{LS} = \frac{8}{15} I_s^2 \times r \times l \tag{5.45}$$

5.13 APPLICATION OF THE A,B,C,D GENERAL CIRCUIT CONSTANTS TO RADIAL FEEDERS

Assume a single-phase or balanced three-phase transmission or distribution circuit characterized by the $\overline{A},\overline{B},\overline{C},\overline{D}$ general circuit constants, as shown in Figure 5.29. The mixed data assumed to be known, as commonly encountered in system design, are $|\overline{V}_S|$,P_r, and cos θ. Assume that all data represent either per-phase dimensional values or per unit values.

As shown in Figure 5.30, taking phasor $\overline{V}_r$ as the reference,

$$\overline{V}_r = V_r \angle 0° \tag{5.46}$$

$$\overline{V}_s = V_s \angle \delta \tag{5.47}$$

$$\overline{I}_r = I_r \angle -\theta_r \tag{5.48}$$

where
$\overline{V}_r$ = receiving-end voltage phasor
$\overline{V}_s$ = sending-end voltage phasor
$\overline{I}_r$ = receiving-end current phasor

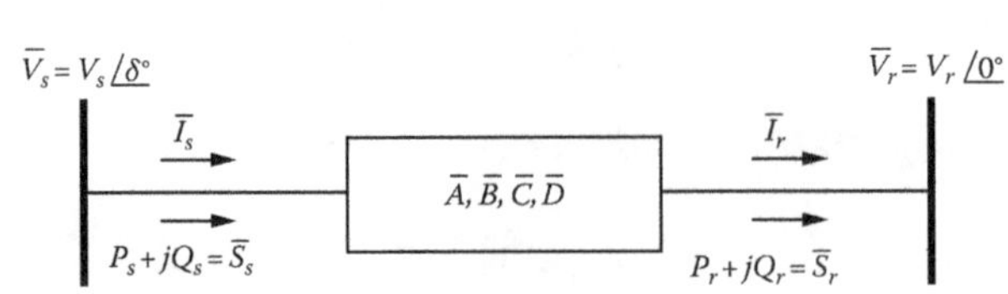

Figure 5.29 A symbolic representation of a line.

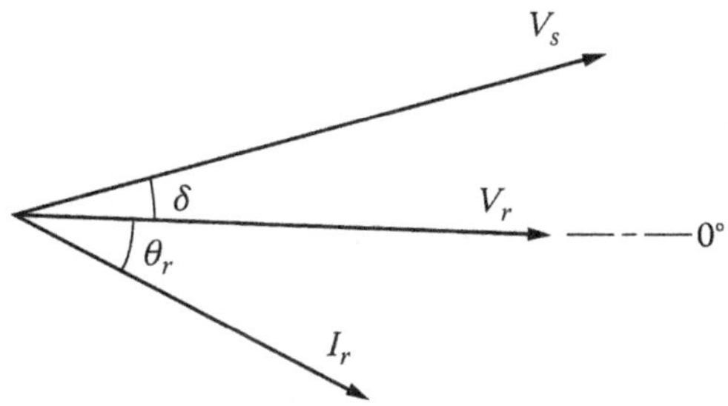

Figure 5.30 Phasor diagram.

The sending-end voltage in terms of the general circuit constants can be expressed as

$$\overline{V}_s = \overline{A} \times \overline{V}_r + \overline{B} \times \overline{I}_r \tag{5.49}$$

where

$$\overline{A} = \overline{A}_1 + j\overline{A}_2 \tag{5.50}$$

$$\overline{B} = \overline{B}_1 + j\overline{B}_2 \tag{5.51}$$

$$\overline{I}_r = I_r(\cos\theta_r - j\,\sin\theta_r) \tag{5.52}$$

$$\overline{V}_r = V_r \angle 0° = V_r \tag{5.53}$$

$$\overline{V}_r = V_s(\cos\delta - j\,\sin\delta) \tag{5.54}$$

Therefore, Equation 5.49 can be written as

$$V_s \cos\,\delta + jV_s \sin\,\delta = (A_1 + jA_2)\,V_r + (B_1 + jB_2)(I_r \cos\,\theta_r - jI_r \sin\,\theta_r)$$

from which

$$V_s \cos\,\delta = A_1 V_r + B_1 I_r \cos\,\theta_r + B_2 I_r \sin\,\theta_r \tag{5.55}$$

and

$$V_s \sin\,\delta = A_2 V_r + B_2 I_r \cos\,\theta_r - B_1 I_r \sin\,\theta_r \tag{5.56}$$

By taking squares of Equations 5.55 and 5.56, and adding them side by side,

$$V_s^2 = (A_1 V_r + B_1 I_r \cos\,\theta_r + B_2 I_r \sin\,\theta_r)^2 + (A_2 V_r + B_2 I_r \cos\,\theta_r - B_1 I_r \sin\,\theta_r)^2 \tag{5.57}$$

or

$$\begin{aligned} V_s^2 = V_r^2(A_1^2 + A_2^2) + 2V_r I_r \cos\theta_r (A_1 B_1 + A_2 B_2) + B_1^2(V_r^2 \cos^2\theta_r + I_r^2 \sin^2\theta_r) \\ + B_2^2(I_r^2 \sin^2\theta_r + I_r^2 \cos^2\theta_r) + 2V_r I_r \sin\theta_r (A_1 B_2 - B_1 A_2) \end{aligned} \tag{5.58}$$

Since

$$P_r = V_r I_r \cos\theta_r \tag{5.59}$$

$$Q_r = V_r I_r \sin\theta_r \tag{5.60}$$

and

$$Q_r = P_r \tan\theta \tag{5.61}$$

Equation 5.58 can be rewritten as

$$V_r^2(A_1^2+A_2^2)+(B_1^2+B_2^2)(1+\tan^2\theta_r)\frac{P_r^2}{V_r^2} = V_s^2 - 2P_r[(A_1B_1+A_2B_2)+(A_1B_2-B_1A_2)\tan\theta_r] \tag{5.62}$$

Let

$$\hat{K} = V_s^2 - 2P_r[(A_1B_1+A_2B_2)+(A_1B_2-B_1A_2)\tan\theta_r] \tag{5.63}$$

Then Equation 5.62 becomes

$$V_r^2(A_1^2+A_2^2)+(B_1^2+B_2^2)(1+\tan^2\theta_r)\frac{P_r^2}{V_r^2} - \hat{K} = 0 \tag{5.64}$$

or

$$V_r^2(A_1^2+A_2^2)+(B_1^2+B_2^2)(\sec^2\theta_r)\frac{P_r^2}{V_r^2} - \hat{K} = 0 \tag{5.65}$$

Therefore, from Equation 5.65, the receiving-end voltage can be found as

$$V_r = \left\{\frac{\hat{K}+[\hat{K}^2-4(A_1^2+A_2^2)(B_1^2+B_2^2)P_r^2\sec^2\theta_r]^{1/2}}{2(A_1^2+A_2^2)}\right\}^{1/2} \tag{5.66}$$

Also, from Equations 5.55 and 5.56,

$$V_s\sin\delta = A_2V_r + B_2I_r\cos\theta_r - B_1I_r\sin\theta_r$$

and

$$V_s\cos\delta = A_1V_r + B_1I_r\cos\theta_r - B_2I_r\sin\theta_r$$

where

$$I_r = \frac{P_r}{V_r\cos\theta_r} \tag{5.67}$$

Therefore,

$$V_s\sin\delta = A_2V_r + \frac{B_2P_r}{V_r} - \frac{B_1P_r}{V_r}\tan\theta_r \tag{5.68}$$

and

$$V_s\cos\delta = A_1V_r + \frac{B_1P_r}{V_r} - \frac{B_2P_r}{V_r}\tan\theta_r \tag{5.69}$$

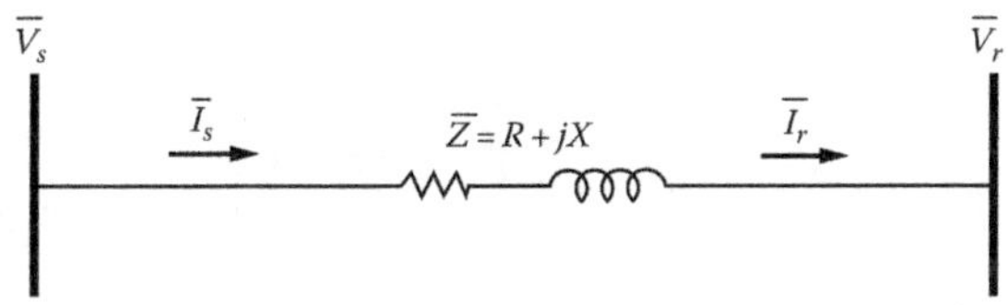

Figure 5.31 A radial feeder.

By dividing Equation 5.68 by Equation 5.69,

$$\tan\delta = \frac{A_2 V_r^2 + B_2 P_r + B_1 P_r \tan\theta_r}{A_1 V_r^2 + B_1 P_r + B_2 P_r \tan\theta_r} \tag{5.70}$$

or

$$\tan\delta = \frac{A_2 V_r^2 + P_r(B_2 - B_1 \tan\theta_r)}{A_1 V_r^2 + P_r(B_1 B_2 \tan\theta_r)} \tag{5.71}$$

Equations 5.66 and 5.71 are found for a general transmission system. They could be adapted to the simpler transmission consisting of a short primary-voltage feeder where the feeder capacitance is usually negligible, as shown in Figure 5.31.

To achieve the adaptation, Equations 5.63, 5.66, and 5.71 can be written in terms of R and X. Therefore, for the feeder shown in Figure 5.31,

$$[\overline{I}] = [\overline{Y}][\overline{V}] \tag{5.72}$$

or

$$\begin{bmatrix} \overline{I}_s \\ \overline{I}_r \end{bmatrix} = \begin{bmatrix} \overline{Y}_{11} & \overline{Y}_{12} \\ \overline{Y}_{21} & \overline{Y}_{22} \end{bmatrix} \begin{bmatrix} \overline{V}_s \\ \overline{V}_r \end{bmatrix} \tag{5.73}$$

where

$$\overline{Y}_{11} = \frac{1}{\overline{Z}} \tag{5.74}$$

$$\overline{Y}_{21} = \overline{Y}_{12} = \frac{1}{\overline{Z}} \tag{5.75}$$

$$\overline{Y}_{22} = \frac{1}{\overline{Z}} \tag{5.76}$$

Therefore,

$$A_1 = -\frac{\overline{Y}_{22}}{\overline{Y}_{21}} = 1 \tag{5.77}$$

or

$$A_1 + jA_2 = 1 \tag{5.78}$$

where

$$A_1 = 1 \tag{5.79}$$

and

$$A_2 = 0 \tag{5.80}$$

Similarly,

$$\overline{B}_1 = -\frac{1}{\overline{Y}_{21}} = \overline{Z} \tag{5.81}$$

or

$$B_1 + jB_2 = R + jX \tag{5.82}$$

where

$$B_1 = R \tag{5.83}$$

and

$$B_2 = X \tag{5.84}$$

Substituting Equations 5.79, 5.80, 5.83, and 5.84 into Equation 5.66,

$$V_r = \left\{ \frac{\hat{K} + [\hat{K}^2 - 4(R^2 + X^2)P_r^2 \sec^2 \theta_r]^{1/2}}{2} \right\}^{1/2} \tag{5.85}$$

or

$$V_r = \left(\frac{\hat{K}}{2} \left\{ 1 \pm \left[1 - \frac{4(R^2 + X^2)P_r^2}{\hat{K}^2 \cos^2 \theta_r} \right]^{1/2} \right\} \right)^{1/2} \tag{5.86}$$

or

$$V_r = \left(\frac{\hat{K}}{2} \left\{ 1 \pm \left[1 - \frac{2ZP_r}{\hat{K}^2 \cos^2 \theta_r} \right]^{1/2} \right\} \right)^{1/2} \tag{5.87}$$

where

$$\hat{K} = V_s^2 - 2 \times P_r + X \times \tan\theta_r \tag{5.88}$$

Also, from Equation 5.71,

$$\tan\delta = \frac{P_r(X - R \times \tan\theta_r)}{V_r^2 + P_r(R + X \times \tan\theta_r)} \tag{5.89}$$

Example 5.1. Assume that the radial express feeder, shown in Figure 5.31, is used on rural distribution and is connected to a lumped-sum (or concentrated) load at the receiving end. Assume that the feeder impedance is 0.10 + j0.10 pu, the sending-end voltage is 1.0 pu, P_r is 1.0 pu constant power load, and the power factor at the receiving end is 0.80 lagging. Use the given data and the exact equations for K, P_r, and tan δ given previously and determine the following:

a. Compute V_r and δ by using the exact equations and find also the corresponding values of the I_r and I_s currents.
b. Verify the numerical results found in part (a) by using those results in

$$\overline{V}_s = \overline{V}_r + (R + jX)\overline{I}_r \tag{5.90}$$

Solution. a. From Equation 5.88,

$$\begin{aligned}\hat{K}_s &= V_s^2 - 2\times P_r(R+X\times\tan\theta_r)\\ &= 1.2^2 - 2\times 1[0.10+0.1\times\tan(\cos^{-1}0.80)]\\ &= 0.65 \text{ pu}\end{aligned}$$

From Equation 5.87,

$$\begin{aligned}V_r &= \left(\frac{\hat{K}}{2}\left\{1\pm\left[1-\frac{2ZP_r}{\hat{K}\cos\theta_r}\right]^{1/2}\right\}\right)^{1/2}\\ &= \left(\frac{0.65}{2}\left\{1\pm\left[1-\frac{2\times 0.141\times 1.0}{0.65\times 0.8}\right]^{1/2}\right\}\right)^{1/2}\\ &= 0.7731 \text{ pu}\end{aligned}$$

From Equation 5.89,

$$\begin{aligned}\tan\delta &= \frac{P_r(X-R\times\tan\theta_r)}{V_r^2+P_r(R+X\times\tan\theta_r)}\\ &= \frac{1.0[0.10-0.10\times\tan(\cos^{-1}0.80)]}{0.7731^2+1.0[0.10+0.10\times\tan(\cos^{-}10.80)]}\\ &= 0.0323\end{aligned}$$

therefore

$$\begin{aligned}\bar{I}_r = \bar{I}_s &\frac{P_r}{V_r\cos\theta_r}\angle-\theta_r\\ &= \frac{1.0}{0.7731\times 0.80}\angle-36.8^\circ\\ &= 1.617\angle-36.8^\circ \text{ pu}\end{aligned}$$

b. From the given equation,

$$\begin{aligned}\overline{V}_r &= \overline{V}_s - (R+jX)\bar{I}_r\\ &= 1.0\angle 1.85^\circ - (0.10+j0.10)(1.617\angle-6.8^\circ)\\ &\cong 0.7731\angle 0^\circ \text{ pu}\end{aligned}$$

5.14 DESIGN OF RADIAL PRIMARY DISTRIBUTION SYSTEMS

Radial primary distribution systems are designed using various approaches, including the following options: (1) overhead primaries with overhead laterals, or (2) underground residential distribution (URD) systems, which involve a combination of overhead primaries and underground laterals.

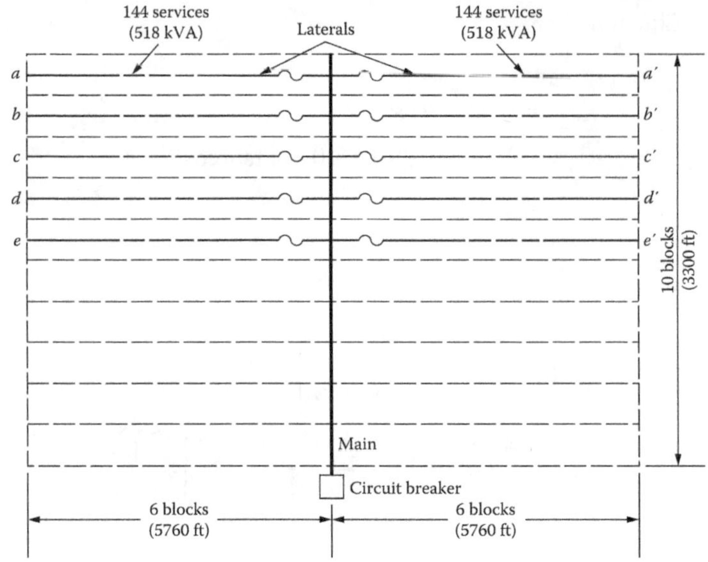

Figure 5.32 An overhead radial distribution system.

5.14.1 OVERHEAD PRIMARIES

To provide an example, Figure 5.32 illustrates an overhead distribution arrangement comprising a main feeder and 10 laterals connected to the main using sectionalizing fuses. It is important to note that the distribution substation depicted in the figure is positioned arbitrarily and could potentially serve another area, not shown in the figure, which is of equal size and situated, for instance, "below" the depicted substation site.

In this arrangement, the feeder mains consist of three phases and are relatively short, with a length of 10 blocks or less. Conversely, the laterals are all six blocks in length and are equipped with sectionalizing fuses. Typically, the laterals can be either single-phase, open-wye grounded, or three-phase configurations.

In the event of a permanent fault occurring on a lateral line, only a relatively small portion of the total area is affected by the outage. Usually, permanent faults on the overhead lines can be swiftly identified and repaired.

5.14.2 UNDERGROUND RESIDENTIAL DISTRIBUTION

Despite the higher cost, underground residential distribution (URD) systems are widely utilized due to their various advantages [121, 122]. Some of these advantages include the following:

1. The lack of outages caused by the abnormal weather conditions such as ice, sleet, snow, severe rain and storms, and lightning
2. The lack of outages caused by accidents, fires, and foreign objects
3. The lack of tree trimming and other preventative maintenance tasks
4. The aesthetic improvement

For illustrative purposes, Figure 5.33 depicts a two-way feed type underground residential distribution (URD) system, alongside a typical overhead and underground primary distribution system. The diagram assumes the existence of two substations, which are arbitrarily positioned and supplied by the same subtransmission line (not shown in the figure). This arrangement ensures that the low-voltage buses of the substations are approximately in phase. In the diagram, the total load of the

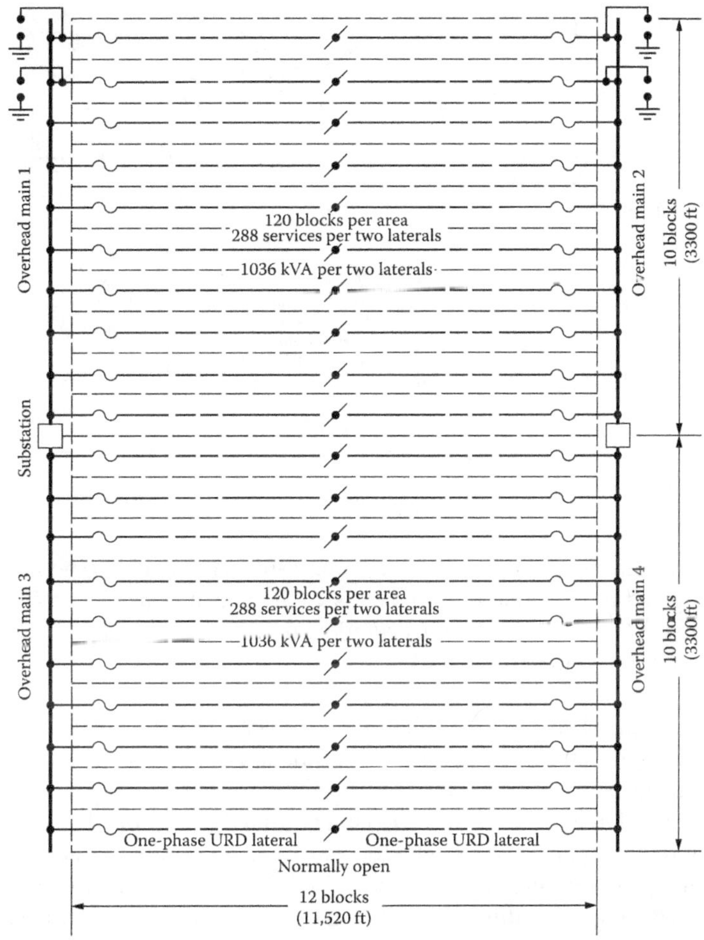

Figure 5.33 A two-way feed-type underground residential distribution system.

considered area, covering a 12-block by 10-block region, is carried by the two overhead primary feeder mains. Similarly, the other equally large area is served by the remaining two overhead feeder mains. As a result, each area consists of 120 blocks in this example.

In residential areas, the laterals in the underground residential distribution (URD) system typically consist of single-phase, directly buried cross-linked polyethylene (XLPE)-insulated cables with a concentric neutral configuration. These cables are designed for solidly grounded neutral service with a nominal insulation rating of 15 kV line-to-line. The commonly used operating voltages for single-phase line-to-neutral service are around 7200 or 7620 V [123].

The use of long cable lengths that can be directly plowed into the ground or placed in shallow trenches without the need for ducts or manholes significantly reduces installation and maintenance costs. Along the periphery of a residential development, heavy three-phase feeders are installed overhead, while the laterals leading to the pad-mounted transformers are buried approximately 40 inches deep. The secondary service lines then extend to individual dwellings at a depth of about 24 inches and enter the dwelling meter through conduit. These service conductors follow easements and are designed to avoid crossing adjacent property lines [124].

In modern URD systems, the distribution transformers commonly used are of the pad-mounted or submersible type. Pad-mounted distribution transformers are securely enclosed in robust sheet metal enclosures and mounted at ground level on a concrete slab. Submersible-type distribution

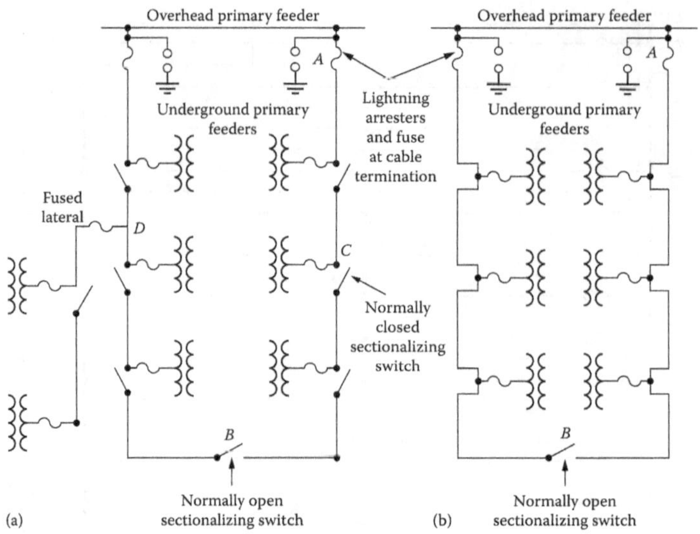

Figure 5.34 Single-line diagram of loop-type primary-feeder circuits: (a) with a disconnect switch at each transformer and (b) without a disconnect switch at each transformer. (From Westinghouse Electric Corporation, *Electric Utility Engineering Reference Book-Distribution Systems*, Vol. 3, East Pittsburgh, Pittsburgh, PA, 1965.)

transformers, on the other hand, are installed in cylindrical excavations lined with concrete bituminized fiber or corrugated sheet metal tubes. The tubular liner is sealed with a locked cover once it reaches near-grade level.

Typically, each lateral line in a directly buried underground residential distribution (URD) system is operated with normally open (NO) switches, usually located near the center as depicted in Figure 5.33. Locating and repairing faults in directly buried URD cables can be time-consuming, making it essential to have switching capabilities that allow de-energization of a specific cable run for repair or replacement while maintaining service to all or most distribution transformers.

To achieve the desired switching or sectionalizing, an arrangement suggested by Lokay [50] is commonly used, as shown in Figure 5.34. This figure presents a single-line diagram of a loop-type primary-feeder circuit for a cost-effective underground distribution system in residential areas. Figure 5.34a illustrates the setup with a disconnect switch at each transformer, while Figure 5.34b shows a similar configuration without a disconnect switch at each transformer. In Figure 5.34a, if the cable above point *C* is faulted, the switch at *C* and the switch or cutout above *C* are opened. Simultaneously, the sectionalizing switch at *B* is closed. This action isolates the faulted cable above *C* and takes the distribution transformer at *C* out of service.

Figure 5.35 demonstrates a distribution transformer equipped with an internal high-voltage fuse and stick-operated plug-in type high-voltage load-break connectors. Commonly used plug-in load-break connectors have ratings such as 8.66 kV line-to-neutral, 200 A continuous load break, and a symmetrical fault close-in rating of 10,000 A.

Figure 5.36 depicts a distribution transformer featuring an internal high-voltage fuse and stick-operated high-voltage load-break switches. These components can be utilized in conjunction with the arrangement shown in Figure 5.34a, enabling four different modes of operation, which are as follows:

1. The transformer is energized and the loop is closed
2. The transformer is energized and the loop is open to the right
3. The transformer is energized and the loop is open to the left
4. The transformer is de-energized and the loop is open

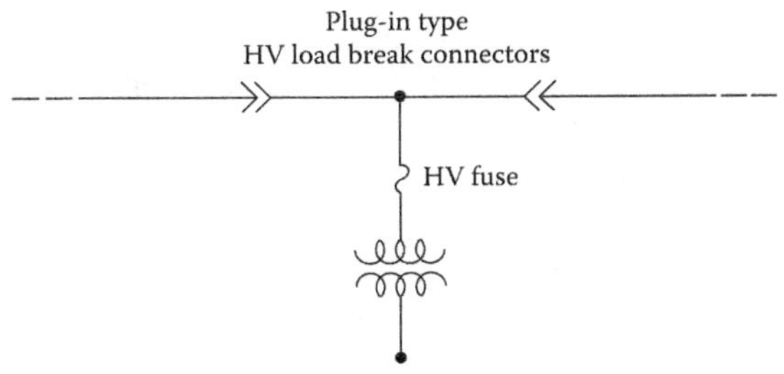

Figure 5.35 A distribution transformer with internal high-voltage fuse and load-break connectors.

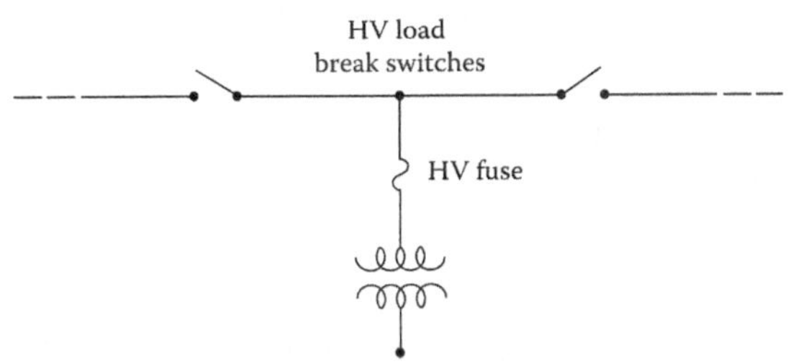

Figure 5.36 A distribution transformer with internal high-voltage fuses and load-break switches.

In Figure 5.33, it is important to consider that in the event of a fault, the open point may be situated near one of the underground feed points. Therefore, in this particular design example, it is advisable to size the single-phase underground cables with an ampacity that can accommodate the load of 12 blocks, rather than just six blocks.

Additionally, in Figure 5.33, it is worth noting the challenge of providing adequate overvoltage protection to the cables and distribution transformers by placing lightning arresters at the open cable ends. The location of the open point changes due to switching operations, whether for repair purposes or for load balancing.

Example 5.2. Consider the layout of the area and the annual peak demands shown in Figure 5.32. Note that the peak demand per lateral is found as

$$144 \text{ customers} \times 3.6 \text{ kVA/customer} \cong 518 \text{ kVA}$$

Assuming a lagging-load power factor of 0.90 at all locations within the primary circuits during the annual peak load, the following calculations are based on the specified conditions:

For the overhead copper lines, which utilize open-wire configuration, the percent voltage drops will be computed. This calculation will employ the precalculated percent voltage drop per kVA-mile curves provided in Chapter 4 of the reference material. It should be noted that a value of $D_m = 37$ inches is assumed for this calculation.

1. The load distribution among the three phases will be assumed to be perfectly balanced for the purpose of computing voltage drop in mains and three-phase laterals.
2. The voltage-drop calculations will be further idealized by assuming a uniformly distributed load along all laterals.
3. When computing the current based on the kVA load, the nominal operating voltage will be used.

In accordance with the joint report by Edison Electric Institute (EEI) and National Electrical Manufacturers Association (NEMA) [125], the voltage levels considered "favorable" at the point of utilization, inside the buildings, range from 110 to 125 V. For illustrative purposes, the lower limit is arbitrarily raised to 116 V at the meter, which is located at the end of the service-drop cable. This adjustment is made to account for any additional voltage drops that are not explicitly calculated and may occur due to the following factors:

1. Unbalanced loading in three-wire single-phase secondaries
2. Unbalanced loading in four-wire three-phase primaries
3. Load growth
4. Voltage drops in building wiring

Hence, the voltage standards to be applied in this scenario are:

$$V_{max} = 125 \text{ V} = 1.0417 \text{ pu}$$

and

$$V_{min} = 116 \text{ V} = 0.9667 \text{ pu}$$

at the meter. Therefore, the following criteria and assumptions will be considered for this problem:

1. The maximum allowable voltage drop, from the low-voltage bus of the distribution substation to the farthest meter, is 7.50%. It is expected that a maximum steady-state voltage drop of 3.5% in the secondary distribution system is achievable. Therefore, the maximum permissible primary voltage drop for this problem is limited to 4.0%.
2. The primary system consists of open-wire overhead primaries with three-phase four-wire laterals. The nominal voltage used as the base voltage is 2400/4160 V for the three-phase four-wire grounded-wye primary system. Copper conductors are employed, and the vertical clearance, denoted as D_m, is assumed to be 37 inches.
3. Only the longest primary circuit is considered, which includes a 3300 ft main and the two most distant laterals (e.g., laterals a and a' in Figure 5.32).

Ampacity-sized conductors will be used, ensuring that the conductors are not smaller than AWG #6 for mechanical strength reasons.

The following objectives will be determined based on these criteria and assumptions:

a. The percent voltage drops at the ends of the laterals and the main.
b. If the 4% maximum voltage-drop criterion is exceeded, find a reasonable combination of larger conductors for main and for lateral that will meet the voltage-drop criterion.

Solution. a. Figure 5.37 shows the “longest” primary circuit, consisting of the 3300 ft main and the most remote laterals a and a'. In Figure 5.37, the signs //// indicate that there are three phase and one neutral conductors in that portion of the one-line diagram. The current in the lateral is !5.91)

$$\begin{aligned} I_{\text{lateral}} &= \frac{S_l}{\sqrt{3} \times V_{L-L}} \\ &= \frac{518}{\sqrt{3} \times 4.16} \cong 72 \text{ A} \end{aligned} \tag{5.91}$$

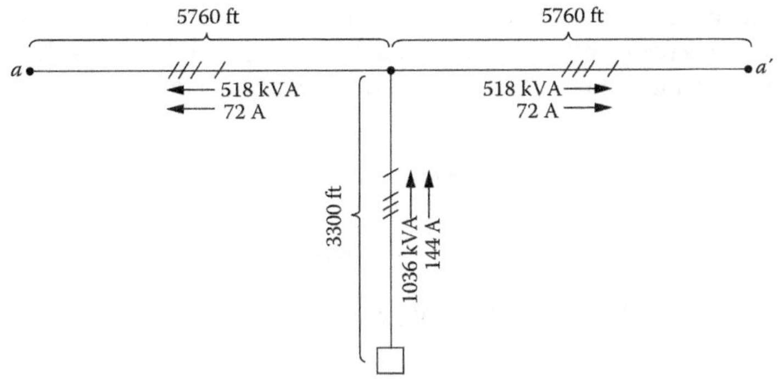

Figure 5.37 The "longest" primary circuit.

Thus, from Table A.1, AWG #6 copper conductor with 130-A ampacity is selected for the laterals. The current in the main is

$$\begin{aligned} I_{\text{main}} &= \frac{S_m}{\sqrt{3} \times V_{L-L}} \\ &= \frac{1036}{\sqrt{3} \times 4.16} \cong 144\ \text{A} \end{aligned} \tag{5.92}$$

Hence, from Table A.1, AWG #4 copper conductor with 180-A ampacity is selected for the mains. Here, note that the AWG #5 copper conductors with 150-A ampacity are not selected due to the resultant too-high total voltage drop.

From Figure 4.17, the K constants for the AWG #6 laterals and the AWG #4 mains can be found as 0.015 and 0.01, respectively. Therefore, since the load is assumed to be uniformly distributed along the lateral,

$$\begin{aligned} \%\text{VD}_{\text{lateral}} &= \frac{1}{2} \times K \times S \\ &= \frac{1}{2} \times \frac{5760\ \text{ft}}{5280\ \text{ft/min}} \times 0.015 \times 518\ \text{kVA} \\ &= 4.24 \end{aligned} \tag{5.93}$$

and since the main is considered to have a lumped-sum load of 1036 kVA at the end of its length,

$$\begin{aligned} \%\text{VD}_{\text{main}} &= l \times K \times S \\ &= \frac{3300\ \text{ft}}{5280\ \text{ft/min}} \times 0.01 \times 1036\ \text{kVA} \\ &= 6.48 \end{aligned} \tag{5.94}$$

Therefore, the total percent primary voltage drop is

$$\begin{aligned} \sum \%\text{VD} &= \%\text{VD}_{\text{main}} + \%\text{VD}_{\text{lateral}} \\ &= 6.48 + 4.24 \\ &= 10.72 \end{aligned} \tag{5.95}$$

which *exceeds* the maximum primary voltage-drop criterion of 4.00%.

Here, note that if single-phase laterals were used instead of the three-phase laterals, *according to Morrison* [126], the percent voltage drop of a single-phase circuit is approximately four times that for a three-phase circuit, assuming the use of the same-size conductors. Hence, for the laterals,

$$\begin{aligned}\sum \%\mathrm{VD}_{1\phi} &= 4(\%\mathrm{VD}_{3\phi})\\ &= 4 \times 4.24\\ &= 16.96\end{aligned} \tag{5.96}$$

Therefore, from Equation 5.95, the new total percent voltage drop would be

$$\begin{aligned}\sum \%\mathrm{VD} &= \%\mathrm{VD}_{\mathrm{main}} + \%\mathrm{VD}_{\mathrm{lateral}}\\ &= 6.48 \times 16.96\\ &= 23.44\end{aligned}$$

which would be far *exceeding* the maximum primary voltage-drop criterion of 4.00%.

b. Therefore, to meet the maximum primary voltage-drop criterion of 4.00%, from Table A.1, select 4/0 and AWG #1 copper conductors with ampacities of 480 A and 270 A for the main and laterals, respectively. Hence, from Equation 5.93,

$$\begin{aligned}\%\mathrm{VD}_{\mathrm{lateral}} &= \frac{l}{2} \times K \times S\\ &= \frac{1}{2} \times \frac{5760\ \mathrm{ft}}{5280\ \mathrm{ft/min}} \times 0.0006 \times 518\ \mathrm{kVA}\\ &= 1.695\end{aligned}$$

and from Equation 5.94,

$$\begin{aligned}\%\mathrm{VD}_{\mathrm{main}} &= l \times K \times S\\ &= \frac{3300\ \mathrm{ft}}{5280\ \mathrm{ft/min}} \times 0.003 \times 1036\ \mathrm{kVA}\\ &= 1.943\end{aligned}$$

Therefore, from Equation 5.95,

$$\begin{aligned}\sum \%\mathrm{VD} &= \%\mathrm{VD}_{\mathrm{main}} + \%\mathrm{VD}_{\mathrm{lateral}}\\ &= 1.943 + 1.695\\ &= 3.638\end{aligned}$$

which *meets* the maximum primary voltage-drop criterion of 4.00%.

Example 5.3. Repeat Example 5.2 but assume that, instead of the open-wire overhead primary system, a self-supporting aerial messenger cable with aluminum conductors is being used. This is to be considered one step toward the improvement of the aesthetics of the overhead primary system, since, in general, very few crossarms are required.

*Consider again only the "longest" primary circuit, consisting of a 3300 ft main and the two most remote laterals,** like the laterals a and a' of Figure 5.32. For the voltage-drop calculations in the self-supporting aerial messenger cable, use Table A.23 for its resistance and reactance values. For *ampacities*, use Table 5.2, which gives data for XLPE-insulated aluminum conductor, grounded neutral +3/0 aerial cables. These ampacities are based on 40°C ambient and 90°C conductor temperatures and are taken from the General Electric Company's Publication No. PD-16.

Table 5.2
Current-Carrying Capacity of XLPE Aerial Cables

	Ampacity, A	
Conductor Size	**5 kV Cable**	**15 kV Cable**
6 AWG	75	
4 AWG	99	
2 AWG	130	135
1 AWG	151	155
1/0 AWG	174	178
2/0 AWG	201	205
3/0 AWG	231	237
4/0 AWG	268	273
250 kcmil	297	302
350 kcmil	368	372
500 kcmil	459	462

Solution. a. The voltage drop, due to the uniformly distributed load, at the lateral is

$$\text{VD}_{\text{lateral}} = I(r \times \cos\theta + X_L \times \sin\theta)\frac{l}{2}\text{V} \tag{5.97}$$

where
$I = 72$ A, from Example 5.2
$r = 4.13$ Ω/min, for AWG #6 aluminum conductors from Table A.23
$x_L = 0.258$ Ω/min, for AWG #6 aluminum conductors from Table A.23
$\cos\theta = 0.90$
$\sin\theta = 0.436$

Therefore,

$$\begin{aligned}\text{VD}_{\text{lateral}} &= 72(4.13 \times 0.9 + 0.258 \times 0.436)\frac{5760 \text{ ft}}{5280 \text{ ft/min}} \times \frac{1}{2} \\ &= 150.4 \text{ V}\end{aligned}$$

or, in percent,

$$\begin{aligned}\text{VD}_{\text{lateral}} &= \frac{150.4 \text{ V}}{2400 \text{ V}} \\ &= 6.27\end{aligned}$$

The voltage drop due to the lumped-sum load at the end of main is

$$\text{VD}_{\text{main}} = I(r \times \cos\theta + X_L \times \sin\theta)l\text{V} \tag{5.98}$$

where
$I = 144$ A, from Example 5.2
$r = 1.29$ Ω/min, for AWG #1 aluminum conductors from Table A.23
$x_L = 0.211$ Ω/min, for AWG #1 aluminum conductors from Table A.23

Therefore,

$$\begin{aligned}\mathrm{VD_{main}} &= 144(1.29\times 0.9+0.211\times 0.436)\frac{3300\ \mathrm{ft}}{5280\ \mathrm{ft/min}}\\ &\cong 112.8\ \mathrm{V}\end{aligned}$$

or, in percent,

$$\begin{aligned}\mathrm{VD_{main}} &= \frac{112.8\ \mathrm{V}}{2400\ \mathrm{V}}\\ &= 4.7\end{aligned}$$

Thus, from Equation5.95, the total percent primary voltage drop is

$$\begin{aligned}\sum \%\mathrm{VD} &= \%\mathrm{VD_{main}} + \%\mathrm{VD_{lateral}}\\ &= 4.7+6.27\\ &= 10.97\end{aligned}$$

which *far exceeds* the maximum primary voltage-drop criterion of 4.00%.

b. Therefore, to meet the maximum primary voltage-drop criterion of 4.00%, from Tables 5.2 and A.23, select 4/0 and 1/0 aluminum conductors with ampacities of 268 A and 174 A for the main and laterals, respectively. Hence, from Equation 5.97,

$$\begin{aligned}\mathrm{VD_{lateral}} &= 72(1.03\times 0.9+0.207\times 0.436)\frac{5760\ \mathrm{ft}}{5280\ \mathrm{ft/min}}\times\frac{1}{2}\\ &= 39.95\ \mathrm{V}\end{aligned}$$

or, in percent,

$$\begin{aligned}\mathrm{VD_{lateral}} &= \frac{39.95\ \mathrm{V}}{2400\ \mathrm{V}}\\ &= 1.66\end{aligned}$$

From Equation 5.98,

$$\mathrm{VD_{main}} = 144(0.518\times 0.9+0.191\times 0.436)\frac{3300\ \mathrm{ft}}{5280\ \mathrm{ft/min}} = 49.45\ \mathrm{V}$$

or, in percent,

$$\begin{aligned}\mathrm{VD_{main}} &= \frac{49.95\ \mathrm{V}}{2400\ \mathrm{V}}\\ &= 2.06\end{aligned}$$

Thus, from Equation 5.95, the total percent primary voltage drop is

$$\begin{aligned}\sum \%\mathrm{VD} &= 2.06+1.66\\ &= 3.72\end{aligned}$$

which *meets* the maximum primary voltage-drop criterion of 4.00%.

Example 5.4. Repeat Example 5.2, but assume that the nominal operating voltage is used as the base voltage and is equal to 7,200/12,470 V for the three-phase four-wire grounded-wye primary system with copper conductors. Use D_m = 37 in. although D_m = 53 in. is more realistic for this voltage class. This simplification allows the use of the precalculated percent voltage drop per kVA-mile curves given in Chapter 4.

Consider serving the *total area* of 12 × 10 = 120-block area, shown in Figure 5.32, with *two feeder mains* so that the longest of the two feeders would consist of a 3300 ft main and 10 laterals, that is, the laterals a through e and the laterals a' through e'. Use ampacity-sized conductors, but not smaller than AWG #6, and determine the following:

a. Repeat part (a) of Example 5.2.
b. Repeat part (b) of Example 5.2.
c. The deliberate use of too-small D leads to small errors in what and why?

Solution. a. The assumed load on the longer feeder is

$$\frac{518\ \text{kVA}}{\text{Lateral}} \times \frac{10\ \text{laterals}}{\text{Feeder}} = 5180\ \text{kVA}$$

Therefore, the current in the main is

$$I_{\text{main}} = \frac{5180\ \text{kVA}}{\sqrt{3} \times 12.47\ \text{kV}} = 240.1\ \text{A}$$

Thus, from Table A.1, AWG #2, three-strand copper conductor, is selected for the mains. The current in the lateral is

$$I_{\text{lateral}} = \frac{518\ \text{kVA}}{\sqrt{3} \times 12.47\ \text{kV}} = 24.1\ \text{A}$$

Hence, from Table A.1, AWG #6 copper conductor is selected for the laterals.

From Figure 4.17, the K constants for the AWG #6 laterals and the AWG #2 mains can be found as 0.00175 and 0.0008, respectively. Therefore, since the load is assumed to be uniformly distributed along the lateral, from Equation 5.93,

$$\begin{aligned}\%\text{VD}_{\text{lateral}} &= \frac{1}{2} \times K \times S \\ &= \frac{1}{2} \times \frac{5760\ \text{ft}}{5280\ \text{ft/min}} \times 0.0175 \times 518\ \text{kVA} \\ &= 0.50\end{aligned}$$

and since, due to the peculiarity of this new problem, one-half of the main has to be considered as an express feeder and the other half is connected to a uniformly distributed load of 5180 kVA,

$$\begin{aligned}\%\text{VD}_{\text{main}} &= \frac{3}{4} \times K \times S \\ &= \frac{3}{4} \times \frac{3300\ \text{ft}}{5280\ \text{ft/min}} \times 0.0008 \times 5180\ \text{kVA} \\ &= 1.94\end{aligned} \tag{5.99}$$

Therefore, from Equation 5.95, the total percent primary voltage drop is

$$\sum \%\text{VD} = 1.94 + 0.50$$
$$= 2.44$$

b. It *meets* the maximum primary voltage-drop criterion of 4.00%.
c. Since the inductive reactance of the line is

$$x_L = 0.1213 \times \ln\frac{1}{D_s} + 0.1213 \times \ln D_m \ \Omega/\text{min}$$

or

$$x_L = x_a + x_d \Omega/\text{min}$$

when D_m = 37 in.,

$$x_d = 0.1213 \times \ln\frac{37\text{in}}{12\text{in.ft}}$$
$$= 1.1366\Omega/\text{min}$$

and when D_m = 53 in.,

$$x_d = 0.1213 \times \ln\frac{53 \text{ in.}}{12\text{in.ft}}$$
$$= 0.1802 \ \Omega/\text{min}$$

Hence, there is a difference of

$$\Delta x_d = 0.0436 \ \Omega/\text{min}$$

which calculates a voltage-drop value smaller than it really is.

Example 5.5. Take into account the configuration of the area and the annual peak demands depicted in Figure 5.33. The primary distribution system displayed in the figure is a combination of overhead mains and an underground residential distribution (URD) system. It is assumed that the overhead mains consist of 7,200/12,470 V three-phase four-wire grounded-wye conductors made of Aluminum Conductor Steel Reinforced (ACSR), with a vertical clearance denoted as D_m set at 53 inches. Additionally, the single-phase underground cable laterals utilize concentric neutral cross-linked polyethylene (XLPE)-insulated cables with aluminum conductors, operating at 7,200 V.

To perform voltage-drop calculations and determine the ampacity of the concentric neutral cross-linked polyethylene (XLPE)-insulated underground residential distribution (URD) cable with aluminum conductors, please refer to Table 5.3.

The provided data pertain to a commonly used 15 kV cable with a solidly grounded neutral. This cable construction comprises an Aluminum phase conductor, an extruded semiconducting conductor shield, a 175 mils thickness of cross-linked PE insulation, an extruded semiconducting sheath and insulation shield, and bare copper wires spirally applied around the exterior to function as the current-carrying grounded neutral. The data presented are specific to single-phase applications, ensuring that the number and size of concentric neutral strands are chosen to achieve 100% neutral

Table 5.3
15 kV Concentric Neutral XLPE-Insulated Al URD Cable

		Ω/1000 ft[a]		Ampacity, A[b]	
Al Conductor Size	**Cu Neutral**		**XL**	**Direct Burial**	**In Duct**
4 AWG	6-#14	0.526	0.0345	128	91
2 AWG	104 14	0.331	0.03	168	119
1 AWG	13-#14	0.262	0.029	193	137
1/0 AWG	16–114	0.208	0.0275	218	155
2/0 AWG	134 12	0.166	0.026	248	177
3/0 AWG	16-#12	0.132	0.024	284	201
4/0 AWG	20-#12	0.105	0.023	324	230
250 kcmil	25–112	0.089	0.022	360	257
300 kcmil	18–1110	0.074	0.0215	403	291
350 kcmil	204 10	0.063	0.021	440	315

Source: Data abstracted from Rome Cable Company, *URD Technical Manual*, 4th edn.
[a] For single-phase circuitry.
[b] At 90°C conductor temperature.

ampacity. However, when three such cables are installed to create a three-phase circuit, the number and/or size of copper concentric neutral strands on each cable are reduced to 33% or less of the neutral ampacity per cable [127].

An alternative insulation type, known as high-molecular-weight PE (HMWPE), is also in use. However, it is rated for a conductor temperature of only 75°C, resulting in slightly lower ampacity compared to XLPE insulation for the same conductor size. HMWPE requires a 220 mils insulation thickness instead of 175 mils. Consequently, cable reactances are slightly higher when employing HMWPE insulation. However, the impact on Δx_L (line reactance) is negligible for most practical purposes.

The accurate determination of the $r + jxL$ values for these recently developed concentric neutral cables is currently a topic of active investigation and study. In these cables, a portion of the neutral current flows through the bare concentric neutral conductors, while the remaining current returns to the earth (as represented by Carson's equivalent conductor). For more comprehensive information regarding this subject, please refer to references [123, 128]. Utilize the provided data to calculate the following:

a. Size each of the overhead mains 1 and 2, of Figure 5.33, *with enough ampacity to serve the entire 12 × 10 block area.* Size each single-phase lateral URD cable with ampacity for the load of 12 blocks.
b. Find the percent voltage drop at the ends of the most remote laterals under *normal operation*, that is, all laterals open at the center, and both mains are energized.
c. Find the percent voltage drop at the most remote lateral under *the worst possible emergency operation*, that is, one main is outaged, and all laterals are fed full length from the one energized main.
d. Is the voltage-drop criterion met for normal operation and for the worst emergency operation?

Solution. a. Since *under the emergency operation* the remaining energized main supplies the doubled number of laterals, the assumed load is

$$\frac{2 \times 518\ \text{kVA}}{\text{Lateral}} \times \frac{10\ \text{laterals}}{\text{feeder}} = 10,360\ \text{kVA}$$

Therefore, the current in the main is

$$I_{\text{main}} = \frac{10,360\ \text{kVA}}{\sqrt{3} \times 12.47\ \text{kV}} = 480.2\ \text{A}$$

Thus, from Table A.5, 300 kcmil ACSR conductors, with 500 A ampacity, are selected for the mains. Since under the emergency operation, due to doubled load, the current in the lateral is doubled,

$$I_{\text{lateral}} = \frac{2 \times 518\ \text{kVA}}{7.2\ \text{kV}} = 144\ \text{A}$$

Therefore, from Table 5.3 [129], AWG #2 XLPE Al URD cable, with 168 A ampacity, is selected for the laterals.

b. *Under normal operation*, all laterals are open at the center, and both mains are energized. Thus the voltage drop, due to uniformly distributed load, at the main is

$$\text{VD}_{\text{main}} = I[r \times \cos\theta + X_L \times \sin\theta] = \frac{l}{2}\text{V} \tag{5.100}$$

or

$$\text{VD}_{\text{main}} = I[r \times \cos\theta + (X_a + X_d) \times \sin\theta] = \frac{l}{2}\text{V} \tag{5.101}$$

where

$I = 480.2/2 = 240.1$ A

$r = 0.342$ Ω/min for 300-kcmil ACSR conductors from Table A.5

$x_a = 0.458$ Ω/min for 300-kcmil ACSR conductors from Table A.5

$x_d = 0.1802$ Ω/min for $D_m = 53$ in. from Table A.10

$\cos\theta = 0.90$ $\sin\theta = 0.436$

Therefore,

$$\begin{aligned}\text{VD}_{\text{main}} &= 240.1[0.342 \times 0.9 + (0.458 + 0.1802)0.436]\frac{3300\ \text{ft}}{5280\ \text{ft/min}} \times \frac{1}{2} \\ &\cong 44\ \text{V}\end{aligned}$$

or, in percent,

$$\begin{aligned}\%\text{VD}_{\text{main}} &= \frac{44\ \text{V}}{7200\ \text{V}} \\ &= 0.61\end{aligned}$$

The voltage drop at the lateral, due to the uniformly distributed load, from Equation 5.97 is

$$\text{VD}_{\text{lateral}} = I[r \times \cos\theta + X_L \times \sin\theta] = \frac{l}{2}\text{V}$$

where

$I = 144/2 = 72$ A

$r = 0.331$ Ω/1000 ft for AWG #2 XLPE Al URD cable from Table 5.3

$x_L = 0.0300$ ft/1000 ft for AWG #2 XLPE Al URD cable from Table 5.3

Therefore,

$$\begin{aligned}\text{VD}_{\text{lateral}} &= 72(0.331 \times 0.9 + 0.0300 \times 0.436)\frac{5760\ \text{ft}}{1000\ \text{ft}} \times \frac{1}{2} \\ &= 64.5\ \text{V}\end{aligned}$$

or, in percent,

$$\%\text{VD}_{\text{lateral}} = \frac{64.5 \text{ V}}{7200 \text{ V}} = 0.9$$

Thus, from Equation 5.95, the total percent primary voltage drop is

$$\sum \%\text{VD} = 0.61 + 0.9 = 1.51$$

c. *Under the worst possible emergency operation*, one main is outaged and all laterals are supplied full length from the remaining energized main. Thus the voltage drop in the main, due to uniformly distributed load, from Equation 5.101 is

$$\text{VD}_{\text{main}} = 480.2(0.3078 + 0.2783)\frac{3300 \text{ ft}}{5280 \text{ ft/min}} \times \frac{1}{2} = 88 \text{ V}$$

or, in percent,

$$\%\text{VD}_{\text{main}} = 1.22$$

The voltage drop at the lateral, due to uniformly distributed load, from Equation 5.97 is

$$\text{VD}_{\text{lateral}} = 144(0.331 \times 0.9 + 0.03 \times 0.435)\frac{5760 \text{ ft}}{1000 \text{ ft}} = 258 \text{ V}$$

or, in percent,

$$\%\text{VD}_{\text{lateral}} = \frac{258 \text{ V}}{7200 \text{ V}} = 3.5$$

Therefore, from Equation 5.95, the total percent primary voltage drop is

$$\sum \%\text{VD} = 1.22 + 3.5 = 4.72$$

d. The primary voltage-drop criterion is *met for normal operation but is not met for the worst emergency operation*.

5.15 PRIMARY SYSTEM COSTS

Based on 1994 prices, the construction cost of three-phase, overhead, wooden pole crossarm-type feeders with large conductors (e.g., 600 kcmil per phase) at a voltage level of approximately 12.47

kV is around \$150,000 per mile. However, it is important to note that costs can significantly vary due to factors such as labor, filing, permit costs, utility-specific design standards, as well as differences in terrain and geology. The thermal capacity of such a feeder would be approximately 15 MVA, with a recommended economic peak loading of around 10 MVA, taking into consideration losses and other associated costs. With a cost of \$150,000 per mile, this translates to approximately \$10–\$15 per kW-mile.

On the other hand, underground construction of three-phase primary distribution systems is more expensive as it requires buried ductwork and cables. The cost for underground construction typically ranges between \$30 and \$50 per kW-mile.

When it comes to lateral lines, overhead costs range between \$5 and \$15 per kW-mile. For underground lateral lines, direct-buried cables have costs between \$5 and \$15 per kW-mile, while ducted cables have costs between \$30 and \$100 per kW-mile.

It is important to note that the costs of other distribution equipment, such as regulators, capacitor banks and switches, sectionalizers, and line switches, can vary significantly depending on the specific requirements of each application. In general, the cost of a distribution system can range between \$10 and \$30 per kW-mile.

PROBLEMS

5.1 Repeat Example 5.2, assuming a 30 min annual maximum demand of 4.4 kVA per customer.

5.2 Repeat Example 5.3, assuming the nominal operating voltage to be 7,200/12,470 V.

5.3 Repeat Example 5.3, assuming a 30 min annual maximum demand of 4.4 kVA per customer for a 12.47 kV system.

5.4 Repeat Example 5.4 and find the exact solution by using $D_m = 53$ in.

5.5 Repeat Example 5.5, assuming a lagging-load power factor of 0.80 at all locations.

5.6 Assume that a radial express feeder used in rural distribution is connected to a concentrated and static load at the receiving end. Assume that the feeder impedance is $0.15 + j0.30$ pu, the sending-end voltage is 1.0 pu, and the constant power load at the receiving end is 1.0 pu with a lagging power factor of 0.85. Use the given data and the exact equations for $\overline{K}$, $\overline{V}_r$, and tan δ given in Section 5.12 and determine the following:

a. The values $\overline{V}_r$ and δ by using the exact equations

b. The corresponding values of the $\overline{I}_r$ and $\overline{I}_s$ currents

5.7 Use the results found in Problem 5.6 and Equation 5.90 and determine the receiving-end voltage $\overline{V}_r$.

5.8 Assume that a three-phase 34.5 kV radial express feeder is used in rural distribution and that the receiving-end voltages at full load and no load are 34.5 and 36.9 kV, respectively. Determine the percent voltage regulation of the feeder.

5.9 A three-phase radial express feeder has a line-to-line voltage of 22.9 kV at the receiving end, a total impedance of $5.25 + j10.95$ Ω/phase, and a load of 5 MW with a lagging power factor of 0.90. Determine the following:

a. The line-to-neutral and line-to-line voltages at the sending end

b. The load angle

5.10 Use the results of Problem 5.9 and determine the percent voltage regulation of the feeder.

5.11 Assume that a wye-connected three-phase load is made up of three impedances of $50\angle 25°$ Ω each and that the load is supplied by a three-phase four-wire primary express feeder. The balanced line-to-neutral voltages at the receiving end are

$$\overline{V}_{an} = 7630\angle 0°$$

$$\overline{V}_{bn} = 7630\angle 240°$$

$$\overline{V}_{bn} = 7630\angle 120°$$

Determine the following:

a. The phasor currents in each line
b. The line-to-line phasor voltages
c. The total active and reactive power supplied to the load

5.12 Repeat Problem 5.11, if the same three load impedances are connected in a delta connection.

5.13 Assume that the service area of a given feeder is increasing as a result of new residential developments. Determine the new load and area that can be served with the same percent voltage drop if the new feeder voltage level is increased to 34.5 kV from the previous voltage level of 12.47 kV.

5.14 Assume that the feeder in Problem 5.13 has a length of 2 min and that the new feeder uniform loading has increased to three times the old feeder loading. Determine the new maximum length of the feeder with the same percent voltage drop.

5.15 Consider a 12.47 kV three-phase four-wire grounded-wye overhead radial distribution system, similar to the one shown in Figure 5.32. The uniformly distributed area of $12 \times 10 = 120$ – block area is served by one main located in the middle of the service area. There are 10 laterals (6 blocks each) on each side of the main. The lengths of the main and the laterals are 3300 and 5760 ft, respectively. From Table A.1, arbitrarily select 4/0 copper conductor with 12 strands for the main and AWG # 6 copper conductor for the lateral. The K constants for the main and lateral are 0.0032% and 0.00175% VD per kVA-min, respectively. If the maximum diversified demand per lateral is 518.4 kVA, consider the total service area and determine the following:

a. The total load of the main feeder in kVA.
b. The amount of current in the main feeder.
c. The amount of current in the lateral.
d. The percent voltage drop at the end of the lateral.
e. The percent voltage drop at the end of the main.
f. The total voltage drop for the last lateral. Is it acceptable if the 4% maximum voltage-drop criterion is used?

5.16 After solving Problem 5.15, use the results obtained, but assume that the main is made up of 500 kcmil, 19-strand copper conductors with $D_m = 37$ in. and determine the following:

a. The percent voltage drop at the end of the main.
b. The total voltage drop to the end of last lateral. Is it acceptable and why?

5.17 After solving Problem 5.15, use the results obtained, but assume that the main is made up of 350 kcmil, 12-strand copper conductors with $D_m = 37$ in. and determine the following:

a. The percent voltage drop at the end of the main.
b. The total voltage drop to the end of last lateral. Is it acceptable and why?

5.18 After solving Problem 5.15, use the results obtained, but assume that the main is made up of 250 kcmil, 12-strand copper conductors with $D_m = 37$ in. and determine the following:

a. The percent voltage drop at the end of the main.
b. The total voltage drop to the end of last lateral. Is it acceptable and why?

5.19 Resolve Example 5.2 by using MATLAB. Use the same selected conductors and their parameters.

5.20 Resolve Example 5.3 by using MATLAB, assuming the nominal operating voltage to be 7,200/12,470 V. Use the same selected conductors and their parameters.

6 Design Considerations of Secondary Systems

Egyptian Proverb: The worst things:
To be in bed and sleep not,
To want for one who comes not,
To try to please and please not.

Francis Scott Fitzgerald, *Notebooks*, 1925

6.1 INTRODUCTION

A practical approach to power distribution systems should be based on the concept of "*gathering*" rather than "*distributing*" functions. This is because the size and locations of customer demands are determined by the customers themselves, who install various energy-consuming devices in different combinations and at their own preferred times. The distribution process starts with individual customers and loads and progresses through multiple gathering stages, where each stage encompasses groups of customers and their loads that increase in number. Ultimately, this process leads to the generating stations through a series of interconnected components such as services, secondaries, distribution transformers, primary feeders, distribution substations, subtransmission and bulk power stations, and transmission lines.

When designing a distribution system, engineers should consider both short-term and long-term factors. The system should not only address the current challenges of cost-effective construction and operation, but also anticipate future demands and ensure the most economical components and practices are in place to accommodate higher customer demands. Therefore, present design practices should be influenced by the requirements of the future system.

Distribution engineers face complex design problems that involve numerous factors, variables, and potential solutions. They require a technique that allows them to select the most cost-effective combination of distribution transformers, secondary conductors, and service drops (SDs).

The recent developments in high-speed digital computers, through the use of computer programs, have provided (1) the fast and economic consideration of many feasible alternatives and (2) the economic and engineering evaluation of these alternatives as they evolve with different strategies throughout the study period. The strategies may include, for example, cutting the secondary, changing the transformers, and possibly adding capacitors.

Naturally, each designed system should meet a specified performance criterion throughout the study period. The most optimum, that is, most economical, system design that corresponds to a load-growth projection schedule can be selected. Also, through the periodic use of the programs, distribution engineers can determine whether strategies adopted continue to be desirable or whether they require some modification as a result of some changes in economic considerations and load-growth projections.

To minimize the secondary-circuit lengths, distribution engineers locate the distribution transformers close to the load centers and try to have the secondary SDs to the individual customers as short as possible.

Since only a small percentage of the total service interruptions are due to failures in the secondary system, distribution engineers, in their system design decisions of the secondary distribution, are primarily motivated by the considerations of economy, copper losses (I^2R) in the transformer and

DOI: 10.1201/9781003129721-6

secondary circuit, permissible voltage drops, and voltage flicker of the system. Of course, there are some other engineering and economic factors affecting the selection of the distribution transformer and the secondary configurations, such as permissible transformer loading, balanced phase loads for the primary system, investment costs (ICs) of the various secondary system components, cost of labor, cost of capital, and inflation rates [130].

Distribution transformers represent a significant part of the secondary system cost. Therefore, one of the major concerns of distribution engineers is to minimize the investment in distribution transformers. In general, the present practice in the power industry is to plan the distribution transformer loading on the basis that there should not be excessive spare capacity installed, and transformers should be exchanged, or banked, as the secondary load grows.

Usually, a transformer load management (TLM) system is desirable for consistent loading practices and economical expansion plans. Distribution engineers, recognizing the impracticality of obtaining complete demand information on all customers, have attempted to combine a limited amount of demand data with the more complete, and readily available, energy consumption data available in the customer account files. A typical demand curve is scaled according to the energy consumed, and the resultant information is used to estimate the peak loading on specific pieces of equipment, such as distribution transformers, in which case it is known as TLM, feeders, and substations [131–134].

However, in general, residential, commercial, and industrial customers are categorized in customer files by rate classification only; that is, potentially useful and important subclassifications are not distinguished. Therefore, the demand data are generally collected for the purpose of generating typical curves only for each rate classification.

6.2 STATE-OF-THE-ART UPDATE

The secondary distribution system can be sensitive to significant growth with additional routes to tap on new loads. The expert system paradigm has dictated the logic/rule-based knowledge of combinations in the design, especially with potential paths of new electrical loads with respect to single-, double-, or three-phase setup [135]. The ant colony algorithm provides an automation of selected primary circuit to be extended with secondary circuit routes geographically [136]. To enhance the secondary network planning, an evolutionary algorithm (EA) is proposed with control variables to project specific characteristics of the sub-network [137]. Planning on secondary distribution network should be part of the integral solution of both MV and LV networks [138, 139]. Five different situations can first be planned differently with individual stage analyzed independently. Then, the mutistage plannig with and without DER can be determined together with the constraints on investment different loading levels of the feeders [140].

Planning on infrastructural investment can be modeled on the electrical and mechanical constraints using mixed integer nonlinear programming problem (MINPL) for possibilities of new customers connections and addressing conductor sizing in installing them of the secondary circuits. The determination of type and characteristics of the poles and supporting structures on MV and LV circuits can be complex [141]. Another extensive work is to apply Voronoi diagrams and Tabu search to sequentially search for preliminary locations of DTs in the Voronoi polygons [142]. Issues such as overloading the DTs can be mitigated by increasing the distributed generation in LV network within the household where it would reduce loss-of-life rates of the transformers and impact on voltage profiles in the residential neighborhood [143, 144]. Many reseach activities on LV network of distribution systems have emphasized on optimal dispatch and design on the photovoltaic inverters, emphasizing on the formulation on prosumers' unit commitment optimization [145, 146]. A holistic framework has been proposed to consider the optimality in system planning by incorporating the MV and LV networks with distributed generation [147]. Since networked microgrids have been extensively promoted in the electrical distribution network, the islanding mode of the microgrid with voltage control consideration has been established to incorporate the time delays [148].

6.3 SECONDARY VOLTAGE LEVELS

According to the American National Standards Institute (ANSI) Standard C84.1-1977 titled "Voltage Ratings for Electric Power Systems and Equipment" (60 Hz) (Beaty, 1978), the current standard or preferred voltage levels for electric power systems are defined [77].

For single-phase residential loads, the standard voltage level is set at 120/240 V. This voltage is supplied through three-wire single-phase services, allowing for both 120 V lighting and 240 V single-phase power connections to be made for large household appliances like ranges, clothes dryers, and water heaters. In grid- or mesh-type secondary-network systems, typically utilized in areas with high-load densities from commercial and residential customers, the voltage level is specified as 208Y/120 V. This system also employs three-wire single-phase services and supports both 120 V lighting and 208 V single-phase power connections. In the case of "spot" networks used in downtown areas with high-rise buildings and in industrial and/or commercial locations with significant load densities, the voltage level is designated as 480Y/277 V. These networks rely on four-wire three-phase services, providing 277 V for fluorescent lighting and other single-phase loads, as well as 480 V three-phase power connections.

It should be noted that there are alternative voltage levels currently in use that deviate from the ANSI standards. Examples include 120/240 V four-wire three-phase, 240 V three-wire three-phase, 480 V three-wire three-phase, 240/416 V four-wire three-phase, or 240/480 V four-wire three-phase systems. To enhance service reliability for critical loads such as hospitals, computer centers, and essential industrial loads, backup systems like emergency generators and/or batteries equipped with automatic switching devices are employed.

6.4 PRESENT DESIGN PRACTICE

The section of the electric utility system situated between the primary system and the consumer's property is known as the *secondary system*. This system encompasses various components such as step-down distribution transformers, secondary circuits (secondary mains), consumer services (SDs), and meters used to measure consumer energy consumption.

Typically, secondary distribution systems are designed with single-phase configurations for residential areas, while three-phase configurations are employed for areas with industrial or commercial customers featuring high-load densities. The secondary distribution systems can be categorized into the following types:

1. The separate-service system for each consumer with separate distribution transformer and secondary connection
2. The radial system with a common secondary main, which is supplied by one distribution transformer and feeding a group of consumers
3. The secondary-bank system with a common secondary main that is supplied by several distribution transformers, which are all fed by the same primary feeder
4. The secondary-network system with a common grid-type main that is supplied by a large number of the distribution transformers, which may be connected to various feeders for their supplies

The separate-service system is infrequently utilized and primarily serves industrial or rural service areas. In general, radial design is commonly employed for the majority of secondary systems that cater to residential, rural, and light-commercial areas. Figure 6.1 illustrates a one-line diagram depicting a radial secondary system. This design is cost-effective and straightforward to operate.

6.5 SECONDARY BANKING

In residential and light-commercial areas where services are located in close proximity to each other, a practice called "banking" is sometimes implemented for distribution transformers. This involves

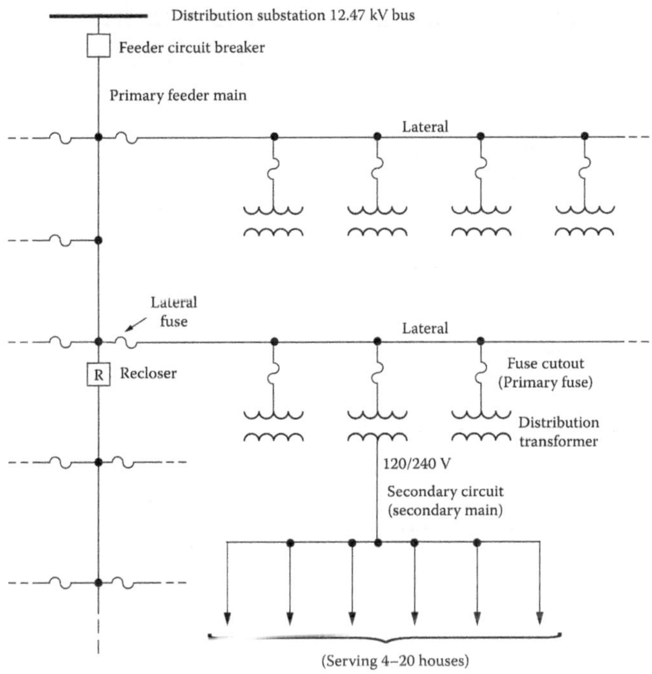

Figure 6.1 One-line diagram of a simple radial secondary system.

the parallel connection or interconnection of the secondary sides of two or more transformers that are supplied power from the same primary feeder.

However, several utilities choose to maintain the separation of each distribution transformer's secondary side from others. In essence, secondary banking represents a unique network configuration within a radial distribution system. The advantages of implementing distribution transformer banking include the following:

1. Improved voltage regulation
2. Reduced voltage dip or light flicker due to motor starting, by providing parallel supply paths for motor-starting currents
3. Improved service continuity or reliability
4. Improved flexibility in accommodating load growth, at low cost, that is, possible increase in the average loading of transformers without corresponding increase in the peak load

Banking the secondary connections of distribution transformers allows us to exploit the diversity in load patterns among a larger number of consumers. This, in turn, leads to savings in the required kVAs of the transformer, which can be significant, up to 35%, as reported by Lokay [50]. The magnitude of savings depends on the types of loads and the number of consumers.

Figure 6.2 displays two different approaches to banking the secondaries. The method depicted in Figure 6.2a is commonly employed and generally preferred because it enables the use of a lower-rated fuse on the high-voltage side of the transformer. It also prevents cascading of fuses and simplifies coordination with primary-feeder sectionalizing fuses. Additionally, this method provides the most cost-effective system.

Figure 6.3 presents two alternative methods of banking the secondaries. The method shown in Figure 6.3a is the oldest and offers the least amount of protection, while the method illustrated in Figure 6.3b offers the highest level of protection. Consequently, the methods depicted in Figures 6.2a and b, as well as 6.3a, possess certain disadvantages, which include the following:

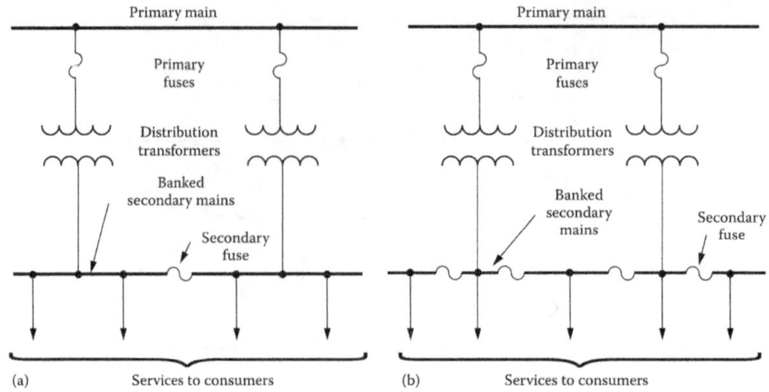

Figure 6.2 Two different methods of banking secondaries: (a) type 1 and (b) type 2.

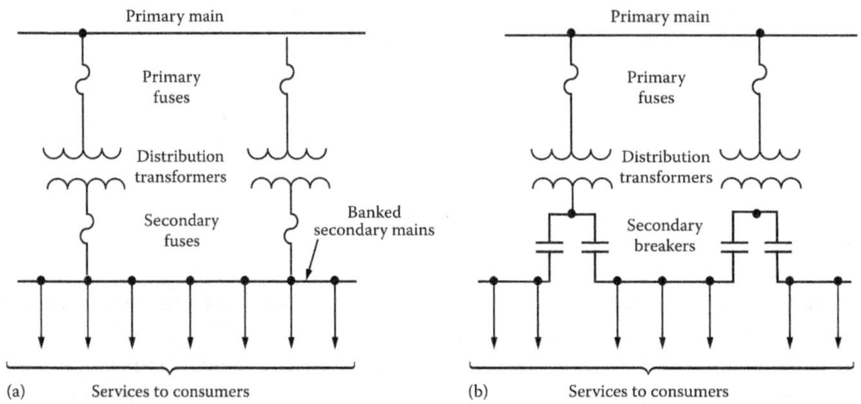

Figure 6.3 Two additional methods of banking secondaries: (a) type 3 and (b) type 4.

1. The requirement for careful policing of the secondary system of the banked transformers to detect blown fuses.
2. The difficulty in coordination of secondary fuses.
3. Furthermore, the method illustrated in Figure 6.2b has the additional disadvantage of being difficult to restore service after a number of fuses on adjacent transformers have been blown.

Owing to the aforementioned challenges, many utility companies favor the approach depicted in Figure 6.3b. This method involves the use of a specialized distribution transformer known as the completely self-protecting-bank (CSPB) transformer, which incorporates various features within its unit. These features include a built-in high-voltage protective link, secondary breakers, signal lights to indicate overload warnings, and lightning protection.

CSPB transformers are manufactured in both single-phase and three-phase configurations. They are equipped with two identical secondary breakers that operate independently in the event of excessive current flow. In case of a transformer malfunction, both the primary protective links and the secondary breakers will trip simultaneously. As a result, the service disruption will be minimized and limited only to the consumers connected to the faulty section of the secondary.

However, all methods of secondary banking possess an inherent drawback: the challenge of maintaining an accurate load balance in the face of changing load conditions. The primary concern when designing a banked secondary system is to achieve an equitable distribution of loads among the transformers. Ideally, transformers with sequentially banked secondaries should differ in size by no more than one unit.

For other types of banking, transformers can vary in size by up to two units to prevent excessive overloading in the event of a blown primary fuse in an adjacent larger transformer. Currently, secondary banking is primarily implemented with single-phase transformers, and all transformers within a bank must be supplied from the same phase of the primary feeder.

6.6 SECONDARY NETWORKS

In general, the majority of secondary systems are designed in a radial configuration, except for specific service areas such as downtown areas, business districts, military installations, and hospitals. In these areas, the focus is on reliability and uninterrupted service rather than cost considerations. Therefore, the secondary systems in such areas may be designed using grid or mesh-type network configurations.

Low-voltage secondary networks are particularly justified in areas with high-load density and can be installed underground to avoid overhead (OH) congestion. Economically, overhead low-voltage secondary networks are preferred in areas with medium-load density. However, underground secondary networks provide a very high level of service reliability. In general, when the load density justifies an underground system, it also justifies the use of a secondary network.

Figure 6.4 illustrates a simplified representation of a segment of a secondary network supplied by three primary feeders. Typically, the low-voltage (208Y/120 V) grid or mesh-type secondary network is supplied by multiple primary feeders through network-type transformers to enhance service reliability. These primary feeders are generally radial, although loop-type feeders are also utilized to a limited extent. The primary feeders are interwoven in a way that prevents adjacent transformer banks from being supplied by the same feeder. This arrangement ensures that if one primary feeder goes out of service (single contingency), the remaining feeders can supply the load without overloading or causing significant voltage drops. The primary feeder voltage levels range from 4.16 to 34.5 kV, with a growing trend toward higher primary voltages [149].

Currently, the 15 kV class is most commonly used. The design of the secondary network should allocate at least one primary feeder as spare capacity along with its transformers to achieve balanced load distribution and minimize voltage drop. Network transformers must be strategically located throughout the secondary network [150].

As mentioned earlier, smaller secondary networks are designed with a focus on single contingency, assuming the outage of one primary feeder. However, larger secondary network systems need to be designed considering double contingency, which accounts for the possibility of two feeder outages occurring simultaneously. According to Reps [50], the factors influencing the likelihood of double outages include:

1. The total number of primary feeders
2. The total mileage of the primary-feeder outages per year
3. The number of accidental feeder outages per year
4. The scheduled feeder-outage time per year
5. The time duration of a feeder outage

While in theory, primary feeders can receive supply from various sources like distribution substations, bulk power substations, or generating plants, it is generally more favorable to have the feeders sourced from the same substation. This practice is adopted to prevent discrepancies in voltage magnitude and phase angles among the feeders, which could lead to improper load division and consequently reduce the capacities of the transformers associated with them. Moreover, during periods of light load, the possibility of power flowing in a reverse direction in certain feeders connected to separate sources becomes an additional concern.

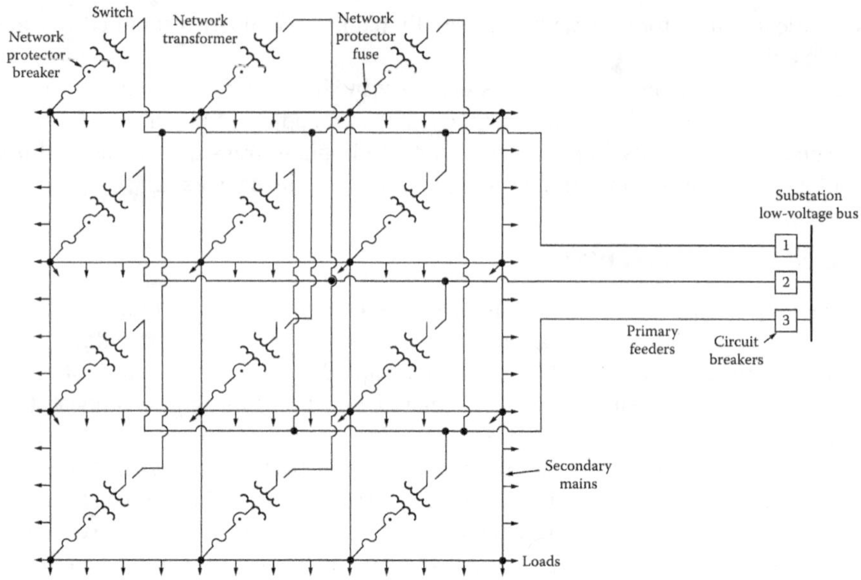

Figure 6.4 One-line diagram of the small segment of a secondary-network system.

6.6.1 SECONDARY MAINS

Seelye [51] put forth recommendations regarding the appropriate size and configuration of the secondary mains, emphasizing the need to fulfill the following requirements:

1. The proper division of the normal load among the network transformers
2. The proper division of the fault current among the network transformers
3. Good voltage regulation to all consumers
4. Burning off short circuits or grounds at any point without interrupting service

The secondary mains, whether installed underground or overhead, are typically routed alongside the streets. They are configured as three-phase, four-wire, wye-connected systems with a solidly grounded neutral conductor. In underground networks, the secondary mains usually consist of single-conductor cables, which can be either metallic or nonmetallic sheathed. While rubber-insulated cables were commonly used in the past, there has been a significant shift toward using PE (polyethylene) cables. These cables are installed within duct lines or duct banks. Manholes are strategically placed at street intersections, providing sufficient space for cable connections, limiters, and allowing workers to carry out any necessary repairs.

In overhead secondary networks, the secondary mains typically comprise open-wire circuits with weatherproof conductors. The size of the conductors is determined based on the ratings of the network transformers. For grid-type secondary mains, the minimum conductor size should be able to carry approximately 60% of the full-load current to the largest network transformer. This percentage is lower for underground secondary mains. The most commonly used cable sizes for secondary mains are 4/0 or 250 kcmil, and to a lesser extent, 350 and 500 kcmil.

The selection of main sizes is also influenced by the requirement for fault clearance through burning. In the event of a phase-to-phase or phase-to-ground short circuit, the secondary network is designed to burn itself clear without the need for sectionalizing fuses or other overload protective devices. "Burning clear" refers to the process of the faulted cable burning away the metal contact between phases or to the ground until the low voltage of the secondary network can no longer sustain the electrical arc.

To achieve rapid fault clearing, the secondary network must be capable of carrying high current values during a fault. The larger the cable size, the higher the short-circuit current required to achieve the burning clear of the faulted cable. Consequently, conductors with a size of 500 kcmil are among the largest used for secondary network mains.

The selection of conductor size also takes into account voltage drop considerations. It is important to ensure that under normal load conditions, the voltage drop along the mains does not exceed a maximum of 3%.

6.6.2 LIMITERS

Most of the time, the method permitting secondary-network conductors to burn clear, especially in 120/208 V, gives good results without loss of service. However, under some circumstances, particularly at higher voltages, for example, 480 V, this method may not clear the fault due to insufficient fault current, and, as a result, extensive cable damage, manhole fires, and service interruptions may occur.

To have fast clearing of such faults, the so-called limiters are used. The limiter is a high-capacity fuse with a restricted copper section, and it is installed in each phase conductor of the secondary main at each junction point. The limiter's fusing or time–current characteristics are designed to allow the normal network load current to pass without melting but to operate and clear a faulted section of main before the cable insulation is damaged by the heat generated in the cable by the fault current.

The fault should be cleared away by the limiters rapidly, before the network protector (NP) fuses blow. Therefore, the time–current characteristics of the selected limiters should be coordinated with the time–current characteristics of the NPs and the insulation-damage characteristics of the cable.

The distribution engineer's decision of using limiters should be based upon two considerations: (1) minimum service interruption and (2) whether the saving in damage to cables pays more than the cost of the limiters. Figure 6.5 shows the time–current characteristics of limiters used in 120/208 V systems and the insulation-damage characteristics of the underground-network cables (paper or rubber insulated).

6.6.3 NETWORK PROTECTORS

As shown in Figure 6.4, the network transformer is connected to the secondary network through a Network Protector (NP). The NP comprises an air circuit breaker with a closing and tripping mechanism, which is controlled by a network master and phasing relay, along with backup fuses.

All of these components are housed within a metal case, which can be either mounted on the transformer or separately mounted. The backup fuses serve to provide additional protection by disconnecting the network transformer from the network in case the NP fails to do so during a fault. The primary functions of an NP include the following:

1. The first objective is to automatically isolate faults that occur in the network transformer or primary feeder. When a fault occurs in a high-voltage feeder, the circuit breaker at the substation is opened, causing current to flow from the secondary network through the network transformers supplied by the faulted feeder. This reverse power flow triggers the circuit breakers of the connected NPs (network protectors) to open, effectively isolating the fault without interrupting service to consumers.
2. The second objective is to automatically close under specific conditions, namely when the voltage magnitude and phase relation of the primary feeder are correct with respect to the network voltage. For instance, the transformer voltage should be slightly higher (around 2 V) than the secondary network voltage to ensure power flows from the network transformer to the secondary

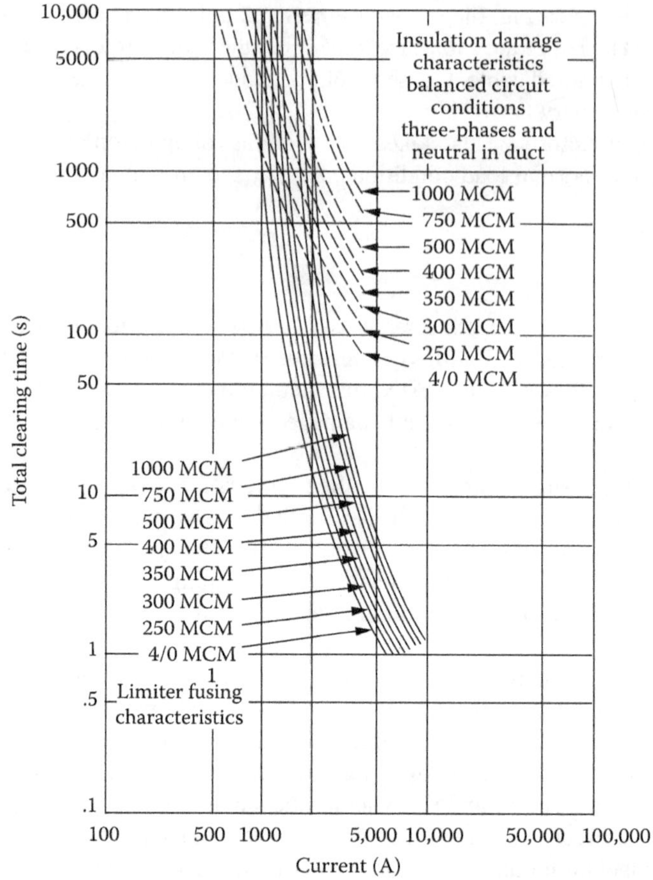

Figure 6.5 Limiter characteristics in terms of time to fuse versus current and insulation-damage characteristics of the underground-network cables. (From Westinghouse Electric Corporation, *Electric Utility Engineering Reference Book-Distribution Systems*, Vol. 3, East Pittsburgh, PA, 1965.)

network system, not the other way around. Additionally, the low-side transfer voltage should be in phase with or leading the network voltage.

3. The third objective is to ensure that the reverse power relay is sufficiently sensitive to trip the circuit breaker with currents as small as the transformer's exciting current. This sensitivity is crucial for protecting against line-to-line faults that occur in unigrounded three-wire primary feeders supplying network transformers with delta connections.
4. The fourth objective is to provide protection against reverse power flow in certain feeders connected to separate sources. For example, when a network is supplied by two different substations, power may flow from one substation to the other through the secondary network and network transformers under specific conditions. In such cases, the NPs should detect this reverse power flow and open. It's worth noting that the best protection strategy is to avoid using more than one substation as the power source.

As mentioned earlier, each network is equipped with backup fuses, with one fuse per phase. These backup fuses serve as an additional layer of protection for the network transformer in case the NP breakers fail to function.

Figure 6.6 demonstrates an ideal configuration of protective devices in the secondary network, ensuring proper *coordination*. This coordination is achieved by selecting appropriate time delays

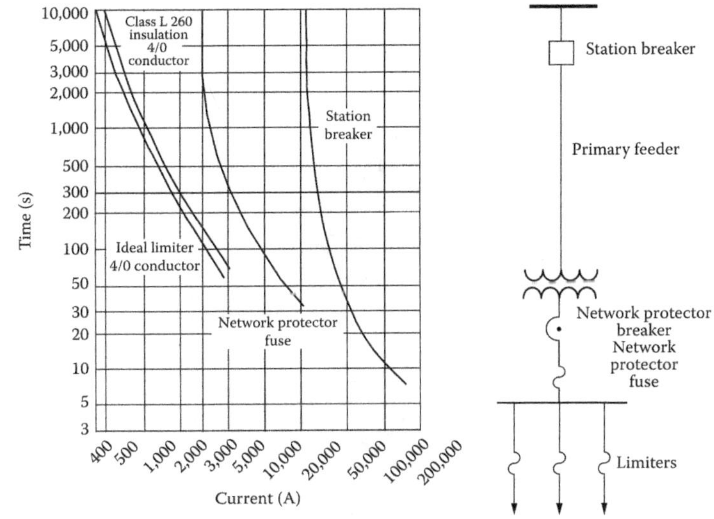

Figure 6.6 An ideal coordination of secondary-network overcurrent protection devices. (From Westinghouse *Electric Corporation, Electric Utility Engineering Reference Book-Distribution Systems*, Vol. 3, East Pittsburgh, PA, 1965.)

for the series of protective devices. Table 6.1 outlines the specific actions or operations required for each protective equipment in various fault scenarios associated with the secondary network system. For instance, when a fault occurs in a particular secondary main, only the corresponding limiters should isolate the fault. On the other hand, if there is an internal fault in the transformer, both the NP breaker and the substation breaker should trip. Three-position switches located on the high-voltage side of the network transformers are depicted in Figures 6.4 and 6.7. These switches are physically mounted at one end of the network transformer.

6.6.4 HIGH-VOLTAGE SWITCH

As depicted in Figure 6.7, position 2 is designated for normal operation, position 3 is for disconnecting the network transformer, and position 1 is for grounding the primary circuit. It's important to note that the switch is operated manually and is not intended to interrupt the current flow. The process begins by opening the primary-feeder circuit breaker at the substation before operating the switch to remove the network unit from service. Once the unit is taken out of service, the feeder circuit breaker can be closed to restore service to the remaining parts of the network.

Table 6.1
Required Operation of the Protective Apparatus

Fault Type	Limiter	NP Fuse	NP Breaker	Substation Circuit Breaker
Mains	Yes	No	No	No
Low-voltage bus	Yes	Yes	No	No
Transformer internal fault	No	No	Trips	Trips
Primary feeder	No	No	Trips	Trips

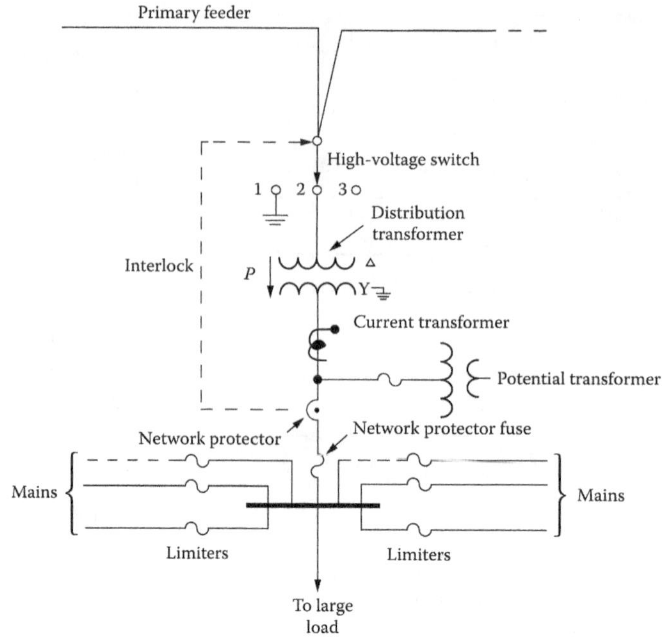

Figure 6.7 High-voltage switch.

Nevertheless, the operation of the switch is restricted by an electric interlock system, which requires the network transformer to be de-energized before the switch can be operated. The grounding position of the switch ensures the safety of workers during any maintenance or repairs on the de-energized primary feeders.

In order to simplify the disconnection process of the transformer from an energized feeder, a specialized disconnecting switch is sometimes employed. This switch is interlocked with the associated NP, as illustrated in Figure 6.7. Consequently, the switch cannot be opened unless the NP removes the load from the network transformer, ensuring safe disconnection from the network.

6.6.5 NETWORK TRANSFORMERS

In overhead (OH) secondary networks, the mounting of transformers can vary based on their sizes. Smaller transformers, such as those with capacities of 75 or 150 kVA, are typically mounted on poles. On the other hand, larger transformers, like those with a capacity of 300 kVA, are mounted on platforms. These transformers can be either single-phase or three-phase distribution transformers.

In underground secondary networks, the transformers are installed in vaults. On one side of the transformer, the NP is mounted, while the three-position high-voltage switch is mounted on the other side. This arrangement, known as a *network unit*, is commonly used in underground secondary networks.

A typical network transformer is a three-phase transformer with a low-voltage rating of 216Y/125 V and can have a maximum capacity of 1000 kVA. Standard ratings for three-phase transformers commonly used as secondary-network transformers are provided in Table 6.2. Nowadays, network transformers are predominantly constructed as three-phase units due to the advantages they offer in terms of space-saving in vaults and reduced installation costs.

Typically, network transformers are designed to be submersible and cooled using oil or askarel. However, due to environmental concerns, askarel is no longer used as an insulating medium in new installations. Depending on the specific location of the installation, network transformers can also

Table 6.2
Standard Ratings for Three-Phase Secondary-Network Transformers Transformer High Voltage

Preferred Nominal System Voltage	Rating	BIL (kV)	Taps Above	Taps Below	Standard kVA Ratings for Low-Voltage Rating of 216Y/125 V
2400/4160Y	4160[a]	60	None	None	300, 500, 750
	4160Y/2400[a,b]		None	None	
	4330		None	None	
	4330Y/2500[b]		None	None	
4800	5000	60	None	4875/4750/4625/4500	300, 500, 750
7200	7200[a]	75	None	7020/6840/6660/6480	300, 500, 750
	7500		None	7313/7126/6939/6752	
7200	11,500	95	None	11,213/10,926/10,639/10,w352	300, 500, 750, 1000
12,000	12,000[a]	95	None	11,700/11,400/11,100/10,800	300, 500, 750, 1000
	12,500		None	12,190/11,875/11,565/11,250	
7200/12,470Y	13,000Y/7500[b]	95	None	12,675/12,350/12,025/11,700	300, 500, 750, 1000
13,200	13,200[a]	95	None	12,870/12,540/12,210/11,880	300, 500, 750, 1000
7620/13,200Y	13,200Y/7620[a,b]		None	12,870/12,540/12,210/11,880	
	13,750		None	13,406/13,063/12,719/12,375	
	13,750Y/7940[b]		None	13,406/13,063/12,719/12,375	
14,440	14,400[a]	95	None	14,040/13,680/13,320/12,960	300, 500, 750, 1000
23,000	22,900[a]	150	24,100/23,500	22,300/21,700	500, 750, 1000
	24,000		25,200/24,600	23,400/22,800	

Source: Westinghouse Electric Corporation, *Electric Utility Engineering Reference Book Distribution Systems*, Vol. 3, East Pittsburgh, PA, 1965. With permission.

Note: All windings are delta connected unless otherwise indicated.

[a] Preferred ratings that should be used when establishing new networks.

[b] High-voltage and low-voltage neutrals are internally connected by a removable link.

be of the ventilated dry type, sealed dry type, or submersible design. This variation in transformer types is influenced by factors such as environmental considerations and specific requirements of the installation site.

6.6.6 TRANSFORMER APPLICATION FACTOR

According to Westinghouse [50], the application factor can be defined as "the ratio of the installed network transformer to the load." Therefore, using the same logic,

$$\text{Application Factor} = \frac{\sum S_T}{\sum S_L} \tag{6.1}$$

where

$\sum S_T$ is the total capacity of network transformers

$\sum S_L$ is the total load of secondary network

The application factor is based upon single contingency, that is, the loss of one of the primary feeders. According to Reps [50], the application factor is a function of the following:

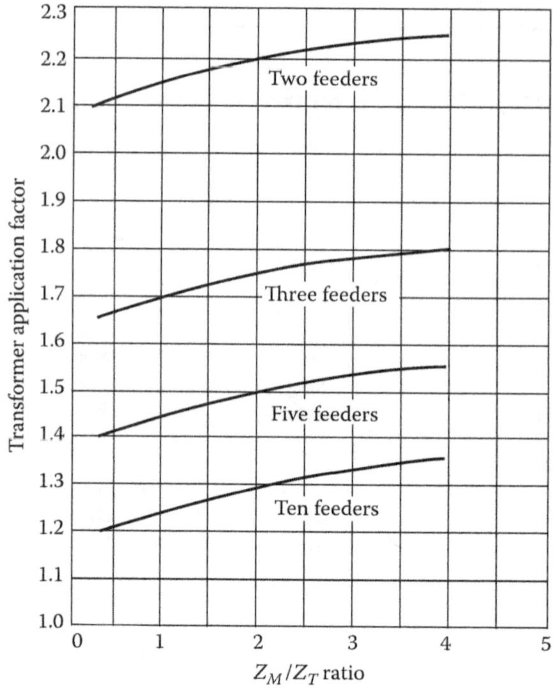

Figure 6.8 Network-transformer application factors as a function of Z_M/Z_T ratio and number of feeders used. (From Westinghouse Electric Corporation, *Electric Utility Engineering Reference Book-Distribution Systems*, Vol. 3, East Pittsburgh, PA, 1965.)

1. The number of primary feeders used
2. The ratio of Z_M/Z_T, where Z_M is the impedance of each section of secondary main and Z_T is the impedance of the secondary-network transformer
3. The extent of nonuniformity in load distribution among the network transformers under the single contingency

Figure 6.8 displays the graphs illustrating the relationship between the transformer application factor and the Z_M/Z_T ratio for various numbers of feeders. Using Figure 6.8, one can determine the necessary capacity of network transformers to meet a specific load requirement for a given Z_M/Z_T ratio and number of feeders.

6.7 SPOT NETWORKS

A spot network is a special type of network that may have two or more network units feeding a common bus from which services are tapped. The transformer capacity utilization is better in the spot networks than in the distributed networks due to equal load division among the transformers regardless of a single-contingency condition.

The impedance of the secondary main, between transformers, is zero in the spot networks. The spot networks are likely to be found in new high-rise commercial buildings. Even though spot networks with light loads can utilize 208Y/120 V as the nominal low voltage, the commonly used nominal low voltage of the spot networks is 480Y/277 V. Figure 6.9 shows a one-line diagram of the primary system for the John Hancock Center.

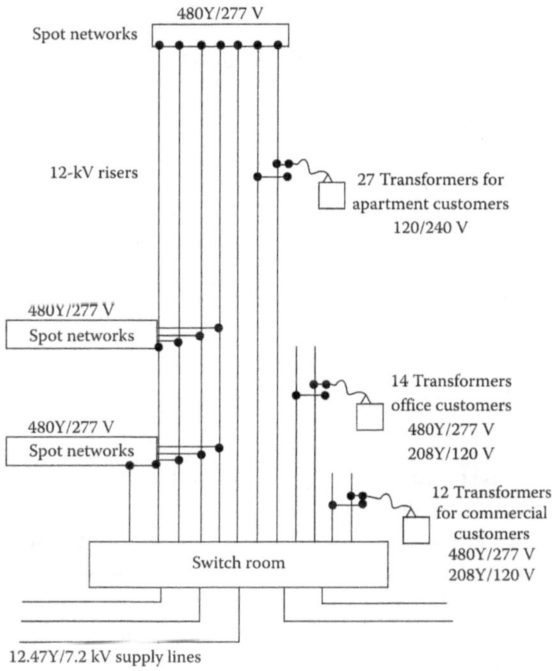

Figure 6.9 One-line diagram of the multiple primary system for the John Hancock Center. (From Westinghouse Electric Corporation, *Electric Utility Engineering Reference Book-Distribution Systems*, Vol. 3, East Pittsburgh, PA, 1965.)

6.8 ECONOMIC DESIGN OF SECONDARIES

This section introduces a method aimed at minimizing the overall annual cost (TAC) associated with owning and operating the secondary segment of a three-wire single-phase distribution system in a residential area. The method can be applied to both overhead (OH) and underground residential distribution (URD) construction scenarios. The primary objective is to achieve a design that ensures satisfactory performance in terms of voltage drop and voltage dip while also aligning reasonably well with the design that yields the minimum TAC.

6.8.1 PATTERNS AND SOME OF THE VARIABLES

Figure 6.10 depicts the layout and a specific arrangement featuring a single span of secondary line (SL) extending in each direction from the distribution transformer. The system assumes a linear configuration along an alley or rear lot lines. The lots are assumed to have a uniform width denoted as d, resulting in each SL span being twice the length of d. In cases where SLs are not utilized, a distribution transformer is present on every pole in an overhead (OH) construction, and each transformer serves four service drops (SDs).

Figure 6.10 does not include the primary line, which is naturally installed along the alley. The number of SL spans in each direction from a transformer is a significant variable. In areas with high load density, SLs may not be utilized. Conversely, in areas with low load density, it is common to encounter three or more spans of SL extending in each direction from the transformer.

In the case where Figure 6.10 represents an overhead (OH) system, the transformer, along with its arresters and fuse cutouts, is mounted on a pole. The secondary line (SL) and service drop (SD) can be constructed using either open-wire or triplex cable.

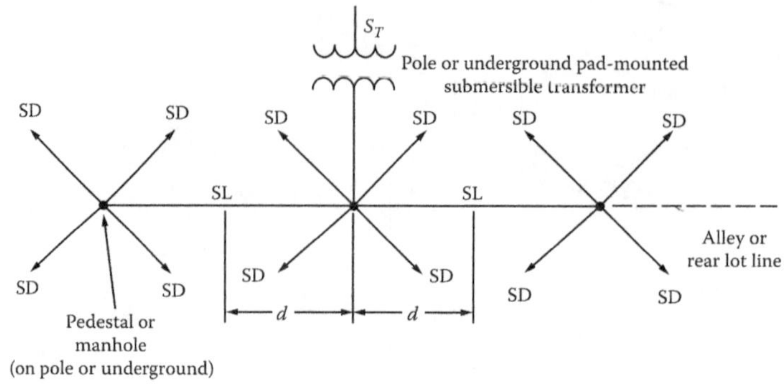

Figure 6.10 Illustration of a typical pattern.

Alternatively, if Figure 6.10 represents a typical underground residential distribution (URD) design, the transformer is grade mounted on a concrete slab and enclosed within a grounded metal housing. Another possibility is that the transformer is installed in a submersible manner, placed within a hole lined with materials such as concrete, Transite, or an equivalent substance. In URD configurations, both the SL and SDs are laid using triplexed or twin concentric neutral direct-burial cables. These cables are placed in narrow trenches and subsequently backfilled after installation.

The distribution transformers possess the parameters that are defined as follows:

S_T is the transformer capacity, continuously rated kVA.

I_{exc} is the per unit exciting current (based on ST).

$P_{T,Fe}$ is the transformer core loss at rated voltage and rated frequency, kW.

$P_{T,Cu}$ is the transformer copper loss at rated kVA load, kW.

The SL has the parameters defined in the following:

A_{SL} is the conductor area, kcmil.

ρ is the conductor resistivity, $(\Omega \cdot \text{cmil})/\text{ft}$.

= 20.5 at 65°C for aluminum cable.

The SDs have the parameters A_{SD} and ρ with meanings that correspond to those given for SLs.

6.8.2 FURTHER ASSUMPTIONS

1. All secondaries and services are single phase three wire and nominally 120/240 V.
2. Perfectly balanced loading obtains in all three-wire circuits.
3. The system is energized 100% of the time, that is, 8760 h/year.
4. The annual loss factor is estimated by using Equation 2.40a, that is,

$$F_{LS} = 0.3F_{LD} + 0.7F_{LD}^2 \tag{2.40a}$$

5. The annual peak-load kVA loading in any element of the pattern, that is, SD, section of SL, or transformer, is estimated by using the maximum diversified demand of the particular number of customers located downstream from the circuit element in question. This point is illustrated later.
6. Current flows are estimated in kVAs and nominal operating voltage, usually 240 V.
7. All loads have the same (and constant) power factor.

6.8.3 GENERAL TAC EQUATION

The TAC of owning and operating one pattern of the secondary system is a summation of *investment* (*fixed*) *costs* (ICs) and *operating* (*variable*) *costs* (OCs). The costs to be considered are contained in the following equation:

$$\text{TAC} = \sum \text{IC}_\text{T} + \sum \text{IC}_\text{SL} + \sum \text{IC}_\text{SD} + \sum \text{IC}_\text{PH} + \sum \text{OC}_\text{exe} + \sum \text{OC}_\text{T,Fe} + \sum \text{OC}_\text{T,Cu} + \sum \text{OC}_\text{SL,Cu} + \sum \text{OC}_\text{SD,Cu} \quad (6.2)$$

The summations are to be taken for the one standard pattern being considered, like Figure 6.10, but modified appropriately for the number of spans of SL being considered. It is apparent that the TAC so found may be divided by the number of customers per pattern so that the TAC can be allocated on a per customer basis.

6.8.4 ILLUSTRATING THE ASSEMBLY OF COST DATA

The subsequent cost data provided are intended for illustrative purposes and may not meet the precision required for engineering design in practical commercial applications. Some of the cost figures presented might be significantly inaccurate due to recent and substantial inflation. The purpose of these data is to serve as a representation of an overhead (OH) system employing three-conductor triplex aluminum cable for both secondary lines (SLs) and service drops (SDs). The key objective of the following procedures is to establish equations for all costs, enabling analytical methods to be employed for minimizing the total annual cost (TAC) as follows:

1. I_CT is the annual installed cost of the distribution transformer + associated protective equipment

$$= (250 + 7.26 \times S_T) \times i \ \$/\text{transformer} \quad (6.3)$$

where

$$15kVA \leq S_T \leq 100kVA$$

S_T is the transformer-rated kVA

2. IC_SL is the annual installed cost of triplex aluminum SL cable

$$= (60 + 4.50 \times A_\text{SL}) \times i \ \$/10000\text{ft} \quad (6.4)$$

where
A_{SL} is the conductor area, kcmil
i is the pu fixed charge rate on investment

Note that this cost is 1000 ft of cable, that is, 3000 ft of conductor.

3. IC_SD is the annual installed cost of triplex aluminum SD cable

$$= (60 + 4.50 \times A_\text{SL}) \times i \ \$/10000\text{ft} \quad (6.5)$$

In this example, Equations 6.4 and 6.5 are alike because the same material, that is, triplex aluminum cable, is assumed to be used for both SL and SD construction.

4. IC_PH is the annual installed cost of pole and hardware on it but excluding transformer and transformer protective equipment

$$= 160 \times i\$/\text{pole} \quad (6.6)$$

In case of URD design, the cost item IC_PH would designate the annual installed cost of a secondary pedestal or manhole.

5. OC_{exe} is the annual operating cost of transformer exciting current

$$= I_{exe} \times S_T \times IC_{cap} \times i \text{ /transformer} \tag{6.7}$$

where
IC_{cap} is the total installed cost of primary-voltage shunt capacitors = \$5.00/kVAr
I_{exc} is the an average value of transformer exciting current based on S_T kVA rating = 0.015 pu

6. $OC_{T,Fe}$ is the annual operating cost of transformer due to core (iron) losses

$$= (IC_{sys} \times i + 8760 \times EC_{off})P_{T,Fe} \text{ \$/transformer} \tag{6.8}$$

where
IC_{sys} is the average investment cost of power system upstream, that is, toward generator, from distribution transformers
= \$350/kVA
EC_{off} is the incremental cost of electric energy (off-peak)
= \$0.008/kWh
$P_{T,Fe}$ is the annual transformer core loss, kW
= 0.004 × S_T 15 kVA $\leq S_T \leq$ 100 kVA

7. $OC_{T,Cu}$ is the annual operating cost of transformer due to copper losses

$$= (IC_{sys} \times i + 8760 \times EC_{on} \times F_{LS})\left(\frac{S_{max}}{S_T}\right)^2 \times P_{T,Cu} \text{ \$/transformer} \tag{6.9}$$

where
EC_{on} is the incremental cost of electric energy (on-peak) = \$0.010/kWh
S_{max} is the annual maximum kVA demand on transformer
$P_{T,Cu}$ is the transformer copper loss, kW at rated kVA load

$$= 0.073 + 0.00905 \times S_T \text{ where } 15 \text{ kVA} \leq S_T \leq 100 \text{ kVA} \tag{6.10}$$

F_{LS} is the annual loss factor

8. $OC_{SL,Cu}$ is the annual operating cost of copper loss in a unit length of SL

$$= (IC_{sys} \times i + 8760 \times EC_{on} \times F_{LS})P_{SL,Cu} \tag{6.11}$$

where
$P_{SL,Cu}$ is the power loss in a unit of SL at time of annual peak load due to copper losses, kW
$P_{SL,Cu}$ is an I^2R loss, and it must be related to conductor area A_{SL} with $R = \rho L / A_{SL}$

One has to decide carefully whether L should represent length of conductor or length of cable. When establishing $\sum OC_{SL,Cu}$ for the particular pattern being used, one has to remember that different sections of SLs may have different values of current and, therefore, different $P_{SL,Cu}$.

9. $OC_{SD,Cu}$ is the annual operating cost of copper loss in a unit length of SD. $OC_{SD,Cu}$ is handled like $OC_{SL,Cu}$ as described in Equation 6.11. When developing $\sum OC_{SD,Cu}$, it is important to relate $P_{SD,Cu}$ properly to the total length of SDs in the entire pattern.

Table 6.3
Illustrative Load Data

No. of Customers Being Diversified	Ann. Max. Demand (kVA/Customer)
1	5
2	3.8
4	3
8	2.47
10	2.2
20	2.1
30	2
100	1.8

Source: Lawrence, R.F. et al., *AIEE Trans.,* PAS-79(pt. III), 199, June 1960, Fig. 3.

6.8.5 ILLUSTRATING THE ESTIMATION OF CIRCUIT LOADING

The previously mentioned simplifying assumptions (5) and (6) outline a method for approximating the load on each component of the pattern. It is crucial to obtain reasonable estimates for the current loads in each service drop (SD), each section of the secondary line (SL), and the transformer. These estimates are essential to ensure appropriate approximations are utilized for the copper-loss costs OCT, Cu (transformer copper-loss cost), $\sum$OCSL, Cu (total copper-loss cost for all secondary lines), and $\sum OC_{SD,\,Cu}$ (total copper-loss cost for all service drops).

In order to proceed, it is essential to obtain data on the annual maximum diversified kVA demand per customer, considering the number of customers being diversified. The illustrative data presented in Table 6.3 have been sourced from Lawrence, Reps, and Patton's paper titled *Distribution System Planning through Optimized Design, I-Distribution Transformers and Secondaries* [151, Fig. 3]. As explained in the referenced paper, the maximum diversified demand data were developed by utilizing appliance diversity curves and hourly variation factors.

One approach to utilizing the data is to plot it and then read the demand per customer for intermediate numbers of customers from the curve. Alternatively, if a digital computer is programmed to carry out the described calculations, linear interpolation could reasonably be employed to estimate the per customer demand for intermediate numbers of customers [152].

Figure 6.11 showcases a specific pattern featuring two secondary lines (SLs) extending in each direction from the transformer. By applying the aforementioned data and utilizing linear interpolation, readers can determine the flows depicted in Figure 6.11. The nominal voltage considered for this scenario is 240 V.

6.8.6 DEVELOPED TOTAL ANNUAL COST EQUATION

By expanding all the cost elements (1)-(9) as outlined in Section 6.7, considering the appropriate summations for the specific pattern being employed, and incorporating the obtained results into Equation 6.2, it is determined that

$$\text{TAC} = A + \frac{B}{S_T^2} + \frac{C}{S_T} + D \times A_{\text{SD}} + \frac{F}{A_{\text{SD}}} + G \times A_{\text{SL}} + \frac{H}{A_{\text{SL}}} \tag{6.12}$$

where the coefficients A to H are numerical constants. It is important to note that TAC has been reduced to a function of three design variables, that is,

$$\text{TAC} = f(S_T, A_{\text{SD}}, A_{\text{SL}}) \tag{6.13}$$

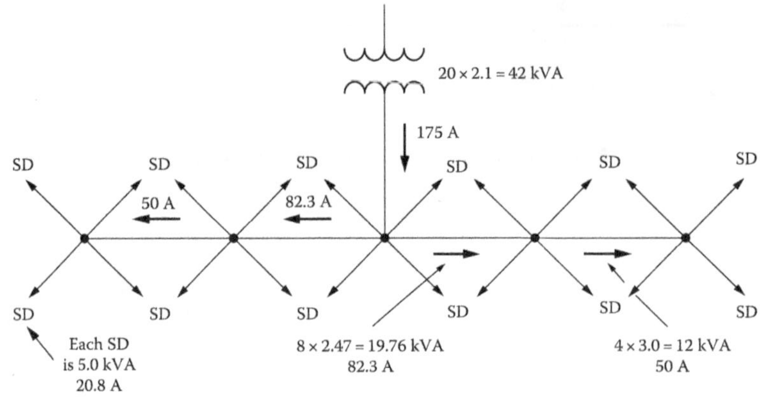

Figure 6.11 Estimated circuit loading for copper-loss determinations.

However, it is important to bear in mind that various parameters, including the fixed charge rate i, transformer core and copper losses, and the installed costs of poles and lines, are encompassed within the coefficients represented by A to H. Additionally, it should be noted that the variables S_T, ASD, and ASL are discrete variables, not continuous ones. For instance, if theoretical analysis suggests that the optimal transformer size is $S_T = 31$ kVA, the designer must make a somewhat arbitrary selection from the available standard commercial sizes, such as 25 kVA or 37.5 kVA. The same principles apply to the choice of conductor sizes for ASL and ASD.

6.8.7 MINIMIZATION OF THE TOTAL ANNUAL COSTS

We can initiate the process by utilizing Equation 6.12, computing three partial derivatives, and then equating each derivative to zero:

$$\frac{\partial(\text{TAC})}{\partial S_T} = 0 \tag{6.14}$$

$$\frac{\partial(\text{TAC})}{\partial A_{\text{SL}}} = 0 \tag{6.15}$$

$$\frac{\partial(\text{TAC})}{\partial A_{\text{SD}}} = 0 \tag{6.16}$$

The calculations involved in Equation 6.14 can be quite complex. Finding the roots of a cubic equation is a challenging task. At this stage, if we only vary the S_T variable and likewise for the A_{SL} and A_{SD} variables individually, we can obtain the minimum TAC. However, there is no guarantee that the true, overall minimum TAC will be achieved by applying the results of Equations 6.14 through 6.16 simultaneously.

Given the presence of discrete variables in this problem, continuous variable methods are no longer applicable. The results obtained from Equations 6.14 through 6.16 are henceforth used merely as indicators of the region that encompasses the minimum TAC achievable with standard commercial equipment sizes. The problem is further tackled by calculating the TAC for the standard commercial equipment sizes that are closest to the results of Equations 6.14 through 6.16. Additionally, TAC calculations are performed for one (or more) standard sizes larger and smaller than those indicated by Equations 6.14 through 6.16.

At this point, we have a reasonable number of computed TAC values, all closely aligned with the idealized continuous variable TAC. Designers can easily examine these final TAC results and select the combinations of S_T, A_{SL}, and A_{SD} that they deem most suitable for their design.

6.8.8 OTHER CONSTRAINTS

There are additional criteria that must be taken into account in the overall design of the distribution system, regardless of whether the minimum TAC is achieved or not. These additional criteria pertain to the quality of utility service. It is possible to encounter minimum TAC designs that violate one or more of the commonly used criteria, including:

1. A minimum allowable steady-state voltage at the furthest service entrance, which may be mandated by law, public utility commission orders, or company policies.
2. A minimum allowable steady-state voltage at the furthest service entrance, which may be mandated by law, public utility commission orders, or company policies.
3. A minimum allowable steady-state voltage at the furthest service entrance, which may be mandated by law, public utility commission orders, or company policies.
4. The designer should not exceed the maximum allowable loading of distribution transformers, expressed as a percentage of their continuous rating.

Example 6.1. This example focuses on the costs associated with a single-phase overhead (OH) secondary distribution system in a residential area. Figures 6.12 and 6.13 illustrate the layouts and the service arrangement that will be considered. It is assumed that the lots have equal widths, resulting in uniform load spacings. All service drops are assumed to be 70 ft in length. The calculations should be performed for a single block within the residential area.

For the OH secondary distribution system in this case, let's assume there are 12 services per transformer, meaning there are two transformers per block located at poles 2 and 5, as shown in Figure 6.12.

In the subsequent chapters, voltage-drop constraints will be addressed, which are used to establish a minimum standard for service quality. Ideally, a design that achieves satisfactory voltage-drop performance should reasonably align with the design that minimizes the TAC.

Table 6.4 provides the load data to be utilized in this example problem. For this specific problem, use the 30-min annual maximum demands for customer class 2. Use the following data and assumptions:

1. All secondaries and services are single phase three wire, nominally 120/240 V.
2. Assume perfectly balanced loading in all single-phase three-wire circuits.
3. Assume that the system is energized 100% of the time, that is, 8760 h/year.
4. Assume the annual load factor to be $F_{LD} = 0.35$.
5. Assume the annual loss factor to be

$$F_{LS} = 0.3F_{LD} + 0.7F_{LD}^2$$

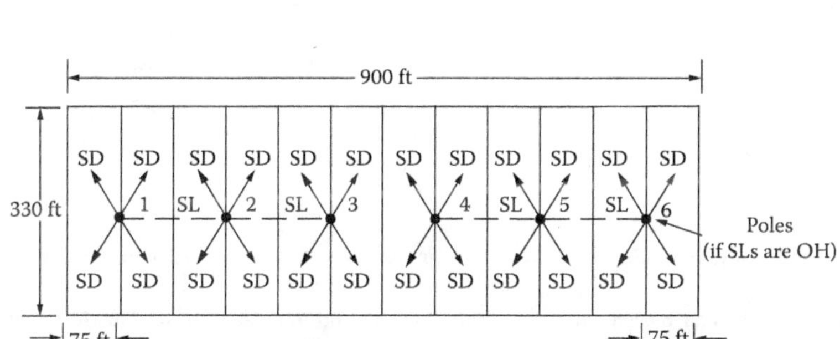

Figure 6.12 Residential area lot layout and service arrangement.

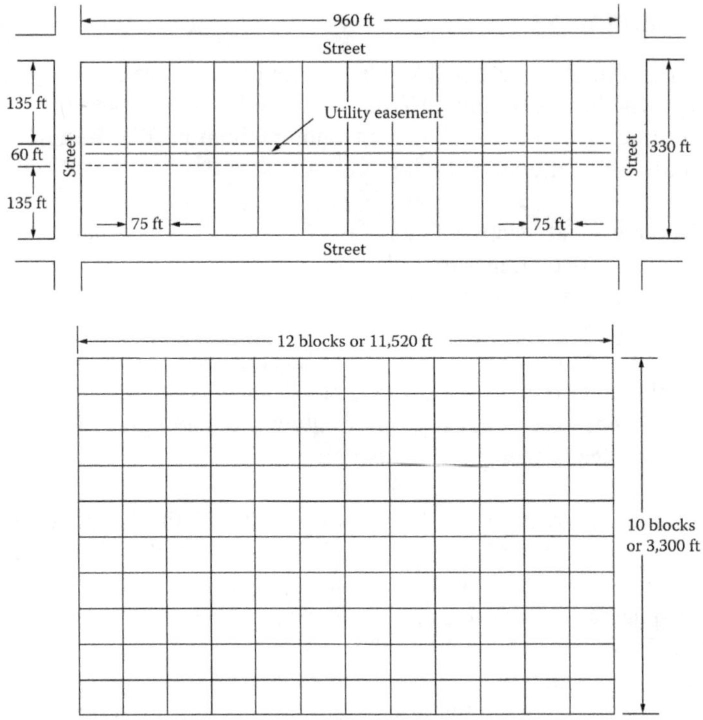

Figure 6.13 Residential area lot layout and utility easement arrangement.

6. Assume that the annual peak-load copper losses are properly evaluated ($\sum I^2R$) by applying the given class 2 loads as
   a. One consumer per SD
   b. Four consumers per section of SL
   c. Twelve consumers per transformer

Table 6.4
Load Data for Example 6.1

No. of Customers Being Diversified	30 Min Ann. Max. Demands (kVA/Customer)		
	Class 1	Class 2	Class 3
1	18	10	2.5
2	14.4	7.6	1.8
4	12	6	1.5
12	10	4.4	1.2
100	8.4	3.6	1.1

Source: Fink, D.G. and Beaty, H.W., *Standard Handbook for Electrical Engineers,* 11th edn., McGraw-Hill, New York, 1978, Figure 3.

Note: The kVA demands cited have been doubled arbitrarily in an effort to modernize the data. It is explained in the reference cited that the original maximum demand data were developed from appliance diversity curves and hourly variation factors.

Here, $P_{\text{SL,Cu}}$ is an I^2R. loss, and it must be related to conductor area A_{SL} with

$$R = \frac{\rho \times L}{1000 \times A_{\text{SL}}}$$

where

A_{SL} is the conductor area, kcmil

ρ is 20.5 (Ω·cmil)/ft at 65°C for aluminum cable

L is the length of conductor wire involved (not cable length)

(The designer must be careful to establish a correct relation between $\sum \text{OC}_{\text{SL},Cu}$, that is, the annual OC per block, and the amount of SL for which $P_{\text{SL},Cu}$ is evaluated.)

7. Assume nominal operating voltage of 240 V when computing currents.
8. Assume a 90% power factor for all loads.
9. Assume a fixed charge (capitalization) rate of 0.15.

Using the given data and assumptions, develop a numerical TAC *equation applicable to one block of these residential areas for the case of 12 services per transformer, that is, two transformers per block*. The equation should contain the variables of S_T, A_{SD}, and A_{SL}. Also determine the following:

a. The most economical SD size (A_{SD}) and the nearest larger standard AWG wire size
b. The most economical SL size (A_{SL}) and the nearest larger standard AWG wire size
c. The most economical distribution transformer size (S_T) and the nearest larger standard transformer size
d. The TAC per block for the theoretically most economical sizes of equipment
e. The TAC per block for the nearest larger standard commercial sizes of equipment
f. The TAC per block for the nearest larger transformer size and for the second larger sizes of A_{SD} and A_{SL}
g. Fixed charges per customer per month for the design using the nearest larger standard commercial sizes of equipment
h. The variable (operating) costs per customer per month for the design using the nearest larger standard commercial sizes of equipment

Solution. From Equation 6.2, the TAC is

$$\begin{aligned}\text{TAC} = &\sum \text{IC}_T + \sum \text{IC}_{\text{SL}} + \sum \text{IC}_{\text{SD}} + \sum \text{IC}_{\text{PH}} + \sum \text{OC}_{\text{exe}} + \sum \text{OC}_{T,\text{Fe}} + \sum \text{OC}_{T,Cu} \\ &+ \sum \text{OC}_{\text{SL},Cu} + \sum \text{OC}_{\text{SD},Cu}\end{aligned} \quad (6.17)$$

Since there are two transformers per block and 12 services per transformer, from Equation 6.3, the annual installed cost of the two distribution transformers and associated protective equipment is

$$\begin{aligned}\text{IC}_T &= 2(250 + 7.26 \times S_T) \times i \\ &= 2(250 + 7.26 \times S_T) \times 0.15 \\ &= 75 + 2.178\ S_T\ \$/\text{block}\end{aligned} \quad (6.18)$$

From Equation 6.4, the annual installed cost of the triplex aluminum cable used for 300 ft per transformer (since there is 150 ft SL at each side of each transformer) in the SLs is

$$\begin{aligned}\text{IC}_{\text{SL}} &= 2(60 + 4.50 \times A_{\text{SL}}) \times i \\ &= 2(60 + 4.50 \times A_{\text{SL}}) \times 0.15 \times \frac{300\ \text{ft/ transformer}}{1000\ \text{ft}} \\ &= 5.4 + 0.405\ A_{\text{SL}}\ \$/\text{block}\end{aligned} \quad (6.19)$$

From Equation 6.5, the annual installed cost of triplex aluminum 24 SDs per block (each SD is 70 ft long) is

$$\begin{aligned}\mathrm{IC_{SD}} &= 2(60 + 4.50 \times A_{\mathrm{SD}}) \times i \\ &= 2(60 + 4.50 \times A_{\mathrm{SD}}) \times 0.15 \times \frac{12 \times 70\ \mathrm{ft/SD}}{1000\ \mathrm{ft}} \\ &= 15.12 + 1.134 \times A_{\mathrm{SD}}\ \$/\mathrm{block}\end{aligned} \tag{6.20}$$

From Equation 6.6, the annual cost of pole and hardware for the six poles per block is

$$\begin{aligned}\mathrm{IC_{PH}} &= \$160 \times i \times 6\ \mathrm{poles/block} \\ &= \$160 \times 0.15 \times 6 \\ &= \$144/\mathrm{block}\end{aligned} \tag{6.21}$$

From Equation 6.7, the annual OC of transformer exciting current per block is

$$\begin{aligned}\mathrm{OC_{exe}} &= 2I_{exe} \times S_T \times \mathrm{IC_{CAP}} \times i \\ &= 2(0.015) \times \$5/\mathrm{kVAr} \times 0.15 \\ &= 0.0225\ S_T\ \$/\mathrm{block}\end{aligned} \tag{6.22}$$

From Equation 6.8, the annual OC of core (iron) losses of the two transformers per block is

$$\begin{aligned}\mathrm{OC}_{T,\mathrm{Fe}} &= 2(\mathrm{IC_{SYS}} \times i + 8760 \times \mathrm{EC_{off}})0.004 \times S_T \\ &= 2(\$350/\mathrm{kVA} \times 0.15 + 8760 \times \$0.008/\mathrm{kWh})0.004 \times S_T \\ &= 0.985\ S_T\ \$/\mathrm{block}\end{aligned} \tag{6.23}$$

From Equation 6.9, the annual OC of transformer copper losses of the two transformers per block is

$$\mathrm{OC}_{T,Cu} = (\mathrm{IC_{sys}} \times i + 8760 \times \mathrm{EC_{on}} \times F_{\mathrm{LS}})\left(\frac{\mathrm{S_{max}}}{\mathrm{S_T}}\right)^2 \times P_{T,Cu}$$

where

$$\begin{aligned}F_{\mathrm{LS}} &= 0.3F_{\mathrm{LD}} + 0.7F_{LD}^2 \\ &= 0.3(0.35) + 0.7(0.35)^2 \\ &= 0.1904\end{aligned}$$

$$\begin{aligned}S_{\max} &= 12\ \mathrm{customer/transformer} \times 4.4\ \mathrm{kVA/customer} \\ &= 52.8\ \mathrm{kVA/customer}\end{aligned}$$

Here, the figure of 4.4 kVA/customer is found from Table 6.4 for 12 class 2 customers. From Equation 6.10, the transformer copper loss in kW at rated kVA load is found as

$$P_{T,Cu} = 0.073 + 0.00905\ S_T$$

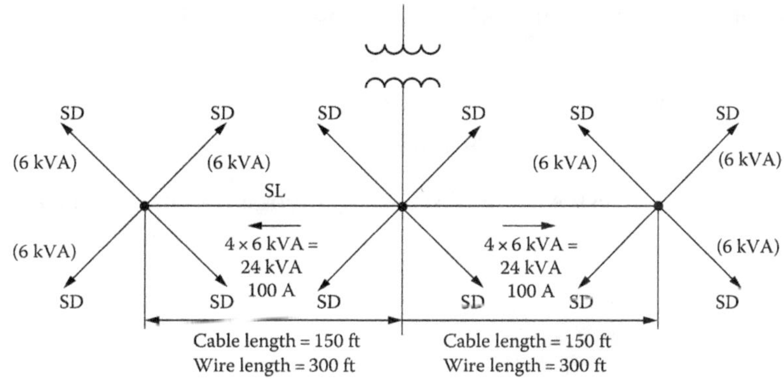

Figure 6.14 Illustration of the SLs.

Therefore,

$$\begin{aligned} OC_{T,\text{Cu}} &= 2[(\$350/\text{kVA}) \times 0.15 + 8760 \times (\$0.01/\text{kWh}) \\ &\quad \times 0.1904] \times \left(\frac{52.8\ \text{kVA/transformer}}{S_T}\right)^2 (0.073 + 0.00905 \times S_T) \\ &= \frac{28,170}{S_T^2} + \frac{3492}{S_T} \$/\text{block} \end{aligned} \tag{6.24}$$

From Equation 6.11, the annual OC of copper losses in the four SLs is

$$OC_{\text{SL,Cu}} = 2(\text{IC}_{\text{sys}} \times i + 8760 \times \text{EC}_{\text{on}} \times F_{\text{LS}})P_{\text{LS},Cu}$$

where $P_{\text{LS},Cu}$ is the copper losses in two SLs at time of annual peak load, kW/transformer (see Figure 6.14) = $I^2 \times R$

where

$$\begin{aligned} R &= \frac{\rho \times L}{1000 \times A_{\text{SL}}} \\ &= \frac{20.5(\Omega \cdot \text{cmil/ft} \times 300\text{ft wire}times2}{1000 \times A_{\text{SL}}} \\ &= \frac{12.3}{A_{\text{SL}}}(\Omega \cdot \text{kcmil})/\text{transformer} \end{aligned}$$

$$\begin{aligned} P_{LS,Cu} &= \left(\frac{24\ \text{kVA}}{240\ \text{V}}\right)^2 \times \frac{12.3}{A_{\text{SL}}} \times \frac{1}{1000} \\ &= \frac{123}{A_{\text{SL}}} \text{kW/transfomer} \end{aligned}$$

Thus,

$$\begin{aligned} OC_{\text{SL},Cu} &= 2[(\$350/\text{kVA}) \times 0.15 + 8760 \times (\$0.01/\text{kWh}) \times 0.1904] \times \frac{123}{A_{\text{SL}}} \\ &= \frac{17,018}{A_{\text{SL}}} = \$/\text{block} \end{aligned} \tag{6.25}$$

Also from Equation 6.11, the annual OC of copper losses in the 24 SDs is

$$\begin{aligned} \mathrm{OC}_{\mathrm{SD},Cu} &= (\mathrm{IC}_{\mathrm{sys}} \times i + 8760 \times \mathrm{EC}_{\mathrm{on}} \times F_{\mathrm{LS}})P_{\mathrm{SD},Cu} \\ &= (69.179)P_{\mathrm{SD},Cu} \end{aligned}$$

where $P_{\mathrm{SD},Cu}$ is the copper losses in the 24 secondary drops at the time of annual peak load, kW

$$= I^2 \times R$$

where

$$\begin{aligned} R &= \frac{\rho \times L}{1000 \times A_{\mathrm{SD}}} \\ &= \frac{20.5(\Omega \cdot \mathrm{cmil/ft})(70\ \mathrm{ft}) \times (24\mathrm{SD/block}) \times (2\mathrm{wire/SD})}{1000 \times A_{\mathrm{SD}}} \\ &= \frac{68.88}{A_{\mathrm{SD}}}(\Omega \cdot \mathrm{kcmil})/\mathrm{block} \end{aligned}$$

From Table 6.4, the 30 min annual maximum demand for one SD per one class 2 customer can be found as 10 kVA. Therefore,

$$\begin{aligned} P_{\mathrm{SD},Cu} &= \left(\frac{10\ \mathrm{kVA}}{0.240\ \mathrm{kV}}\right)^2 \times \frac{68.88}{A_{\mathrm{SD}}} \times \frac{1}{1000} \\ &= \frac{119.58}{A_{\mathrm{SD}}}\mathrm{kW/block} \end{aligned}$$

Thus,

$$\begin{aligned} \mathrm{OC}_{\mathrm{SD},Cu} &= 69.179 \times \frac{119.58}{A_{\mathrm{SD}}} \\ &\cong \frac{8273}{A_{\mathrm{SD}}}\$/\mathrm{block} \end{aligned} \tag{6.26}$$

Substituting Equations 6.17 through 6.25 into Equation 6.2, the TAC equation can be found as

$$\begin{aligned} \mathrm{TAC} = {} & (75 + 2.178 \times S_T) + (5.4 + 0.405 \times A_{\mathrm{SL}}) + (15.12 + 1.134 \times A_{\mathrm{SD}}) + (144 + 0.0225 \times S_T) \\ & + (0.98 \times S_T) + \left(\frac{28,170}{S_T^2} + \frac{3,492}{S_T}\right) + \frac{17,108}{A_{\mathrm{SL}}} + \frac{8,273}{A_{\mathrm{SD}}} \end{aligned}$$

After simplifying,

$$\mathrm{TAC} = 239.52 + 3.1805 \times S_T + \frac{3,492}{S_T} + \frac{28,170}{S_T^2} + 0.405 \times A_{\mathrm{SL}} + \frac{17,108}{A_{\mathrm{SL}}} + 1.034 \times A_{\mathrm{SD}} + \frac{8,273}{A_{\mathrm{SD}}} \tag{6.27}$$

a. By partially differentiating Equation 6.26 with respect to A_{SD} and equating the resultant to zero,

$$\frac{\partial(\mathrm{TAC})}{\partial A_{\mathrm{SD}}} = 1.134 - \frac{8273}{A_{\mathrm{SD}}^2} = 0$$

from which the most economical service-drop size can be found as

$$A_{SD} = \left(\frac{8273}{1.134}\right)^{1/2}$$
$$= 85.41 \text{ kcmil}$$

Therefore, the nearest larger standard AWG wire size can be found from the copper-conductor table (see Table A.1) as 1/0, that is, 106,500 cmil.

b. Similarly, the most economical SL size can be found from

$$\frac{\partial(\text{TAC})}{\partial A_{SL}} = 0.405 - \frac{17{,}018}{A_{SL}^2} = 0$$

as

$$A_{SL} = \left(\frac{17{,}018}{0.405}\right)^{1/2}$$
$$= 204.99 \text{ kcmil}$$

Therefore, the nearest larger AWG wire size is 4/0, that is, 211.6 kcmil.

c. The most economical distribution transformer size can be found from

$$\frac{\partial(\text{TAC})}{\partial S_T} = 3.1805 - \frac{3{,}492}{S_T^2} - \frac{56.340}{S_T^3} = 0$$

or

$$S_T \cong 39 \text{ kVA}$$

Therefore, the nearest larger standard transformer size is 50 kVA.

d. By substituting the found values of A_{SD}, A_{SL}, and S_T into Equation 6.26, the TAC per block for the theoretically most economical sizes of equipment can be found as

$$\text{TAC} = 239.52 + 3.1805 \times (39) + \frac{3{,}492}{(39^2)} + \frac{28{,}170}{(39^2)} + 0.405 \times (204.99) + \frac{17{,}108}{(204.99)}$$
$$+ 1.134 \times (85.41) + \frac{8{,}273}{(85.41)} \cong \$838/\text{block}$$

e. By substituting the found standard values of A_{SD}, A_{SL}, and S_T into Equation 6.26, the TAC per block for the nearest larger standard commercial sizes of equipment can be found as

$$\text{TAC} = 239.52 + 3.1805 \times (50) + \frac{3{,}492}{(50)} + \frac{28{,}170}{(50^2)} + 0.405 \times (211.6) + \frac{17{,}108}{(211.6)}$$
$$+ 1.134 \times (106.5) + \frac{8{,}273}{(106.5)} \cong \$844/\text{block}$$

f. The second larger sizes of A_{SD} and A_{SL} are 133.1 kcmil and 250 kcmil, respectively. Therefore,

$$\text{TAC} = 239.52 + 3.1805 \times (50) + \frac{3{,}492}{(50)} + \frac{28{,}170}{(50^2)} + 0.405 \times (250) + \frac{17{,}108}{(250)}$$
$$+ 1.134 \times (133.1) + \frac{8{,}273}{(133.1)} \cong \$862/\text{block}$$

g. The fixed charges per customer per month for the design using the nearest larger standard commercial sizes of equipment is

$$\text{TAC} = \left(\sum \text{IC}_T + \sum \text{IC}_{\text{SL}} + \sum \text{IC}_{\text{SD}} + \sum \text{IC}_{\text{PH}}\right) \times \frac{1}{24 \text{ customers/block} \times 12 \text{ month/year}}$$
$$\cong \$1.9225/\text{customers/block}$$

h. The variable (operating) cost per customer per month for the design using the nearest larger standard commercial sizes of equipment is

$$\text{TAC} = \left(\sum \text{OC}_{\text{exc}} + \sum \text{OC}_{T,Fe} + \sum \text{OC}_{T,Cu} + \sum \text{OC}_{\text{SL},Cu} + \sum \text{OC}_{\text{SD},Cu}\right) \times \frac{1}{24 \times 12}$$
$$= \left[0.0225(50) + 0.98(50) + \frac{28,170}{(50^2)} + \frac{3,492}{50} + \frac{17,108}{211.6} + \frac{8,273}{106.5}\right] \frac{1}{24 \times 12}$$
$$= \$1.0084/\text{customer/month}$$

Note that the fixed charges are larger than the OCs.

6.9 UNBALANCED LOAD AND VOLTAGES

An unbalanced single-phase three-wire circuit refers to a scenario where the neutral current is not zero. This occurs when the connected loads, such as those between line and neutral, are unequal. As a consequence, the current and voltages become asymmetrical, resulting in a non-zero current flowing through the neutral line. In such cases, the calculations can be carried out using the method of symmetrical components.

Example 6.2. This example, along with Examples 6.3 and 6.4, involves the computation of voltages in unbalanced single-phase three-wire secondary circuits, as depicted in Figure 6.15. In this scenario, both the mutual-impedance methods and the flux-linkage methods can be utilized as alternative approaches to calculate the voltage drops in the secondary line (SL) [119]. In this specific example, the focus is on determining the complex linkages resulting from the line currents in the conductors *a*, *b*, and *n*.

For this single-phase three-wire distribution, the distribution transformer being used has the following specifications: 7200/120-240 V, 25 kVA, and 60 Hz. The turn ratios n_1 and n_2 are 60 and 30, respectively. As indicated in Figure 6.15, the low-voltage winding of the distribution transformer is divided into two independently loaded halves with unequal secondary loads. As a result, the single-phase three-wire secondaries become unbalanced. The vertical spacing between the secondary wires is illustrated in Figure 6.16. In this example, it is assumed that the secondary wires are made of #4/0 seven-strand hard-drawn aluminum conductors with a line length of 400 ft. When calculating the line impedances, a resistance value of 50°C is utilized.

Furthermore, assume that (1) the load impedances $\overline{Z}_a$ and $\overline{Z}_b$ are independent of voltage, (2) the primary-side voltage is $\overline{V}_1$= 7272 V and is maintained constant, and (3) the line capacitances and transformer exciting current are negligible. Use the given-information- and develop numerical equations for the phasor expressions of the flux linkages $\overline{\lambda}_a$, $\overline{\lambda}_b$ and $\overline{\lambda}_n$ in terms of $\overline{I}_a$ and $\overline{I}_b$. In other words, find the coefficient matrix, numerically, in the equation

$$\begin{bmatrix} \overline{\lambda}_a \\ \overline{\lambda}_b \\ \overline{\lambda}_c \end{bmatrix} = \begin{bmatrix} \text{coefficient} \\ \text{matrix} \end{bmatrix} \begin{bmatrix} \overline{I}_a \\ \overline{I}_b \end{bmatrix} \tag{6.28}$$

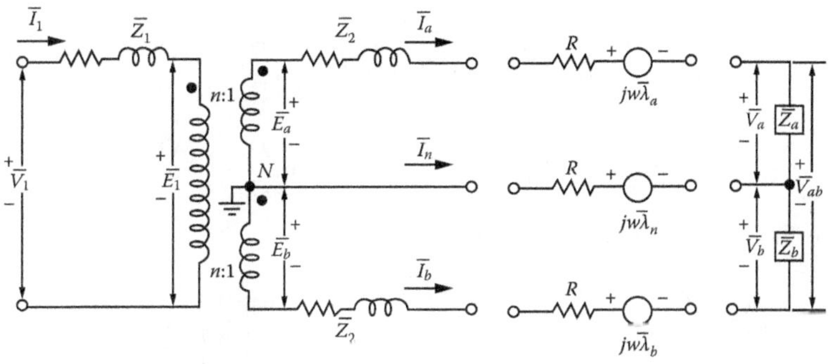

Figure 6.15 An unbalanced single-phase three-wire secondary circuit.

Solution. The phasor expressions of the complex flux linkages $\overline{\lambda}_a$, $\overline{\lambda}_b$ and $\overline{\lambda}_n$ due to the line currents in the conductors *a*, *b*, and *n* can be written as*

$$\overline{\lambda}_a = 2\times 10^{-7}\left(\overline{I}_a \times \ln\frac{1}{D_{aa}} + \overline{I}_b \times \ln\frac{1}{D_{ab}} + \overline{I}_n \times \ln\frac{1}{D_{an}}\right)\frac{\text{Wb}\cdot T}{\text{m}} \tag{6.29}$$

* The notation "ln" is used for "log to the base *e*."

$$\overline{\lambda}_b = 2\times 10^{-7}\left(\overline{I}_a \times \ln\frac{1}{D_{ab}} + \overline{I}_b \times \ln\frac{1}{D_{aa}} + \overline{I}_n \times \ln\frac{1}{D_{bn}}\right)\frac{\text{Wb}\cdot T}{\text{m}} \tag{6.30}$$

$$\overline{\lambda}_n = 2\times 10^{-7}\left(\overline{I}_a \times \ln\frac{1}{D_{na}} + \overline{I}_b \times \ln\frac{1}{D_{nb}} + \overline{I}_n \times \ln\frac{1}{D_{nn}}\right)\frac{\text{Wb}\cdot T}{\text{m}} \tag{6.31}$$

Since

$$\overline{I}_a = -\overline{I}_b - \overline{I}_n = 0 \tag{6.32}$$

the current in the neutral conductor can be written as

$$\overline{I}_n = -\overline{I}_a - \overline{I}_b \tag{6.33}$$

Thus, substituting Equation 6.32 into Equations 6.28 through 6.30,

$$\overline{\lambda}_a = 2\times 10^{-7}\left(\overline{I}_a \times \ln\frac{D_{an}}{D_{aa}} + \overline{I}_b \times \ln\frac{D_{an}}{D_{ab}}\right)\frac{\text{Wb}\cdot T}{\text{m}} \tag{6.34}$$

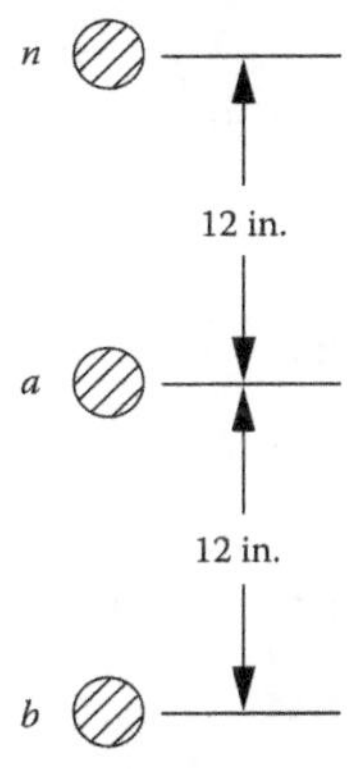

Figure 6.16 Vertical spacing between the secondary wires.

$$\overline{\lambda}_b = 2 \times 10^{-7} \left(\overline{I}_a \times \ln \frac{D_{bn}}{D_{ab}} + \overline{I}_b \times \ln \frac{D_{bn}}{D_{bb}} \right) \frac{\text{Wb} \cdot T}{\text{m}} \tag{6.35}$$

$$\overline{\lambda}_c = 2 \times 10^{-7} \left(\overline{I}_a \times \ln \frac{D_{nn}}{D_{na}} + \overline{I}_b \times \ln \frac{D_{nn}}{D_{ab}} \right) \frac{\text{Wb} \cdot T}{\text{m}} \tag{6.36}$$

Therefore, from Equations 6.33 through 6.35,

$$\begin{bmatrix} \overline{\lambda}_a \\ \overline{\lambda}_b \\ \overline{\lambda}_c \end{bmatrix} = \begin{bmatrix} 2 \times 10^{-7} \times \ln \frac{D_{an}}{D_{aa}} + 2 \times 10^{-7} \times \ln \frac{D_{an}}{D_{ab}} \\ 2 \times 10^{-7} \times \ln \frac{D_{bn}}{D_{ab}} + 2 \times 10^{-7} \times \ln \frac{D_{bn}}{D_{bb}} \\ 2 \times 10^{-7} \times \ln \frac{D_{nn}}{D_{na}} + 2 \times 10^{-7} \times \ln \frac{D_{nn}}{D_{ab}} \end{bmatrix} \begin{bmatrix} \overline{I}_a \\ \overline{I}_b \end{bmatrix} \frac{\text{Wb} \cdot \text{T}}{\text{m}} \tag{6.37}$$

Thus, from Equation 6.36, the coefficient matrix can be found numerically as

$$\begin{bmatrix} \text{coefficient} \\ \text{matrix} \end{bmatrix} = \begin{bmatrix} 2 \times 10^{-7} \times \ln \frac{1}{0.01577} & 2 \times 10^{-7} \times \ln \frac{1}{1} \\ 2 \times 10^{-7} \times \ln \frac{2}{1} & 2 \times 10^{-7} \times \ln \frac{2}{0.01577} \\ 2 \times 10^{-7} \times \ln \frac{0.01577}{1} & 2 \times 10^{-7} \times \ln \frac{0.01577}{2} \end{bmatrix}$$

$$= \begin{bmatrix} 8.2992 \times 10^{-7} & 0 \\ 1.3862 \times 10^{-7} & 9.6855 \times 10^{-7} \\ -8.2992 \times 106-7 & -9.6855 \times 10^{-7} \end{bmatrix} \frac{\text{Wb} \cdot \text{T}}{\text{m}}$$

Note that the elements in the coefficient matrix can be converted to weber-teslas per foot if they are multiplied by 0.3048 m/ft.

Example 6.3. Assume that, in Example 6.2, $\overline{I}_a$, $\overline{I}_b$ and $\overline{V}_1$ are specified but not the load impedances $\overline{Z}_a$ and $\overline{Z}_b$. Develop symbolic equations that will give solutions for the load voltages $\overline{V}_a$, $\overline{V}_b$ and $\overline{V}_{ab}$ in terms of the voltage $\overline{V}_1$, the impedances, and the flux linkages.

Solution. Since the transformation ratio of the distribution transformer is

$$\begin{aligned} n &= \frac{E_1}{E_a} = \frac{E_1}{E_b} \\ &= \frac{7200 \text{ V}}{120 \text{ V}} \\ &= 60 \end{aligned}$$

the primary-side current can be written as

$$\overline{I}_1 = \frac{\overline{I}_a - \overline{I}_b}{n} \tag{6.38}$$

Here,

$$\overline{E}_1 = \overline{V}_1 - \overline{I}_1 \overline{Z}_1 \tag{6.39}$$

Substituting Equation 6.37 into Equation 6.38,

$$\overline{E}_1 = \overline{V}_1 - \overline{Z}_1 \frac{\overline{I}_a - \overline{I}_b}{n} \tag{6.40}$$

Also,

$$\overline{E}_a = \overline{V}_b = \frac{\overline{E}_1}{n} \tag{6.41}$$

Substituting Equation 6.39 into Equation 6.40,

$$\overline{E}_a = \overline{V}_b = \frac{\overline{V}_1}{n} - \frac{\overline{Z}}{n^2}(\overline{I}_a - \overline{I}_b) \tag{6.42}$$

By writing a loop equation for the secondary side of the equivalent network of Figure 6.15,

$$-\overline{E}_a + \overline{Z}_2\overline{I}_a + R\overline{I}_a + j\omega\overline{\lambda}_a + \overline{V}_a - j\omega\overline{\lambda}_n + R(\overline{I}_a - \overline{I}_b) = 0 \tag{6.43}$$

Substituting Equation 6.41 into Equation 6.42,

$$-\frac{\overline{V}_1}{n} + \frac{\overline{Z}_1}{n^2}(\overline{I}_a - \overline{I}_b) + \overline{Z}_2\overline{I}_a + R\overline{I}_a + j\omega\overline{\lambda}_a + \overline{V}_a - j\omega\overline{\lambda}_n + R(\overline{I}_a + \overline{I}_b) = 0$$

or

$$\overline{V}_a = \frac{\overline{V}_1}{n} + \left(\frac{\overline{Z}_1}{n^2} - R\right)\overline{I}_b - \left(\frac{\overline{Z}_1}{n^2} + \overline{Z}_2 + 2R\right)\overline{I}_a - j\omega(\overline{\lambda}_a - \overline{\lambda}_n) = 0 \tag{6.44}$$

Also, by writing a second loop equation,

$$-\overline{E}_b + \overline{Z}_2\overline{I}_b + R\overline{I}_b + j\omega\overline{\lambda}_b + \overline{V}_b + R(\overline{I}_a + \overline{I}_b) - j\omega\overline{\lambda}_n = 0 \tag{6.45}$$

Substituting Equation 6.41 into Equation 6.45,

$$\overline{V}_b = \frac{\overline{V}_1}{n} - \left(\frac{\overline{Z}_1}{n^2} - R\right)\overline{I}_a - \left(\frac{\overline{Z}_1}{n^2} + \overline{Z}_2 + 2R\right)\overline{I}_b - j\omega(\overline{\lambda}_b - \overline{\lambda}_n) = 0 \tag{6.46}$$

However, from Figure 6.15,

$$\overline{V}_{ab} = \overline{V}_a + \overline{V}_b \tag{6.47}$$

Therefore, substituting Equations 6.43 and 6.45 into Equation 6.46,

$$\overline{V}_{ab} = 2\frac{\overline{V}_1}{n} - \left(\frac{2\overline{Z}_1}{n^2} + R + \overline{Z}_2\right)\overline{I}_a + \left(\frac{2\overline{Z}_1}{n^2} + R + \overline{Z}_2\right)\overline{I}_b + j\omega(\overline{\lambda}_b - \overline{\lambda}_a) \tag{6.48}$$

Example 6.4. Assuming that in Example 6.3 the given voltages are

$$\overline{V}_1 = 7272\angle 0^\circ \text{ V}$$
$$\overline{E}_a = 120\angle 0^\circ \text{ V}$$
$$\overline{E}_B = 120\angle 0^\circ \text{ V}$$

and the load impedances are

$$\overline{Z}_a = 0.80 + j0.60\Omega$$
$$\overline{Z}_b = 0.80 + j0.60\Omega$$

and

$$\overline{Z}_1 = 14.5152 + j19.90656\Omega$$
$$\overline{Z}_2 = 0.008064 + j0.0027648\Omega$$

determine the following:

a. The secondary currents $\overline{I}_a$ and$\overline{I}_b$
b. The secondary neutral current $\overline{I}_n$
c. The secondary voltages $\overline{V}_a$ and $\overline{V}_b$
d. The secondary voltage $\overline{V}_{ab}$

Solution. From Equation 6.43,

$$\overline{V}_b = \overline{I}_a\overline{Z}_b = \frac{\overline{V}_1}{n} + \left(\frac{\overline{Z}_1}{n^2} - R\right)\overline{I}_b - \left(\frac{\overline{Z}_1}{n^2} + \overline{Z}_2 + 2R\right)\overline{I}_a - j\omega(\overline{\lambda}_a - \overline{\lambda}_n) \quad (6.49)$$

or

$$\frac{\overline{V}_1}{n} = \left(\frac{\overline{Z}_1}{n^2} + \overline{Z}_2 + 2R + \overline{Z}_a\right)\overline{I}_a - \left(\frac{\overline{Z}_1}{n^2} - R\right)\overline{I}_b + j\omega(\overline{\lambda}_a - \overline{\lambda}_n) \quad (6.50)$$

Similarly, from Equation 6.45,

$$\overline{V}_b = -\overline{I}_b\overline{Z}_b = \frac{\overline{V}_1}{n} + \left(\frac{\overline{Z}_1}{n^2} - R\right)\overline{I}_a - \left(\frac{\overline{Z}_1}{n^2} + \overline{Z}_2 + 2R\right)\overline{I}_b + j\omega(\overline{\lambda}_b - \overline{\lambda}_n)$$

or

$$\frac{\overline{V}_1}{n} = \left(\frac{\overline{Z}_1}{n^2} - R\right)\overline{I}_a\left(\frac{\overline{Z}_1}{n^2} + \overline{Z}_2 + 2R + \overline{Z}_b\right)\overline{I}_b + j\omega(\overline{\lambda}_b - \overline{\lambda}_n) \quad (6.51)$$

Substituting the given values into Equation 6.48,

$$\frac{7272}{60} = \overline{I}_a\left[\frac{14.5152}{60^2} + j\frac{19.90656}{60^2} + 0.008064 + j0.027648 + 0.8 + j0.6 + \frac{2(400)(0.486)}{5280}\right] +$$
$$\overline{I}_b\left[\frac{(400)(0.486)}{5280} - \frac{14.5152}{60^2} - j\frac{19.90656}{60^2}\right] + j377(0.3048)(400)\times 10^{-7}(8.299\overline{I}_a + 8.299\overline{I}_a + 9.686\overline{I}_b)$$

or

$$121.2 = \overline{I}_a(0.8857 + j0.6846) + \overline{I}_b(0.03279 + j0.03899) \quad (6.52)$$

Also, substituting the given values into Equation 6.49,

$$\frac{7272}{60} = \overline{I}_a\left[\frac{14.5152}{60^2} + j\frac{19.90656}{60^2} - \frac{(400)(0.486)}{60^2}\right]$$
$$+\overline{I}_b\left[-0.8 + j0.6 - \frac{14.5152}{60^2} - 0.008064 - j0.027648 - \frac{2(400)(0.486)}{5280}\right] + j377(0.3048)(400)\times 10^{-7}$$
$$(1.386\overline{I}_a + 9.686\overline{I}_b + 8.299\overline{I}_a + 9.686\overline{I}_b)$$

or

$$121.2 = \overline{I}_a(-0.03279 - j0.03899) + \overline{I}_b(-0.88574 + j0.50267) \quad (6.53)$$

Therefore, from Equations 6.50 and 6.51,

$$\begin{bmatrix}121.2\\121.2\end{bmatrix}\begin{bmatrix}0.8857 + j0.6846 & 0.03279 + j0.03899\\ -0.03279 - j0.03899 & -0.88574 + j0.50267\end{bmatrix}\begin{bmatrix}\overline{I}_a\\ \overline{I}_b\end{bmatrix} \quad (6.54)$$

By solving Equation 6.52,

$$\begin{bmatrix}\overline{I}_a\\ \overline{I}_b\end{bmatrix}\begin{bmatrix}89.8347-j62.393\\ -107.387-j62.5885\end{bmatrix}\mathbf{A} \tag{6.55}$$

a. From Equation 6.53, the secondary currents are

$$\begin{aligned}\overline{I}_a &= 89.8347 - j62.393\\ &- 109.376\angle-34.78^\circ \text{ A}\end{aligned}$$

and

$$\begin{aligned}\overline{I}_b &= -107.387 - j62.5885\\ &= 124.295\angle 210.24^\circ \text{ A}\end{aligned}$$

b. Therefore, the secondary neutral current is

$$\begin{aligned}\overline{I}_n &= -\overline{I}_a - \overline{I}_b\\ &= 17.5523 + j124.98158 \text{ A}\end{aligned}$$

c. The secondary voltages are

$$\begin{aligned}\overline{V}_a &= \overline{I}_a \times \overline{Z}_a\\ &= (109.376\angle-34.78^\circ)(1\angle 36.87^\circ)\\ &= 109.376\angle 2.09^\circ \text{ V}\end{aligned}$$

and

$$\begin{aligned}\overline{V}_b &= \overline{I}_b \times \overline{Z}_b\\ &= (124.295\angle-210.24^\circ)(1\angle 36.87^\circ)\\ &= 124.295\angle-6.63^\circ \text{ V}\end{aligned}$$

d. Therefore, the secondary voltage $\overline{V}_{ab}$ is

$$\begin{aligned}\overline{V}_{ab} &= \overline{I}_a \times \overline{Z}_b\\ &= 109.376\angle-2.09^\circ + 124.295\angle-6.63^\circ)\\ &= 232.997\angle-2.55^\circ \text{ V}\end{aligned}$$

Example 6.5. Figure 6.17 illustrates an AC secondary network adapted from Ref. [50]. The loads depicted in Figure 6.17 are given in three-phase kW and kVAr, with a lagging power factor of 0.85. The nominal voltage for the network is 208 V. All distribution transformers in the network are rated at 500 kVA, three-phase, with a high voltage of 4160 V delta and a low voltage of 125/216 V wye grounded. The transformers have a leakage impedance Z_T of 0.0086 + j0.0492 per unit (pu) based on their ratings.

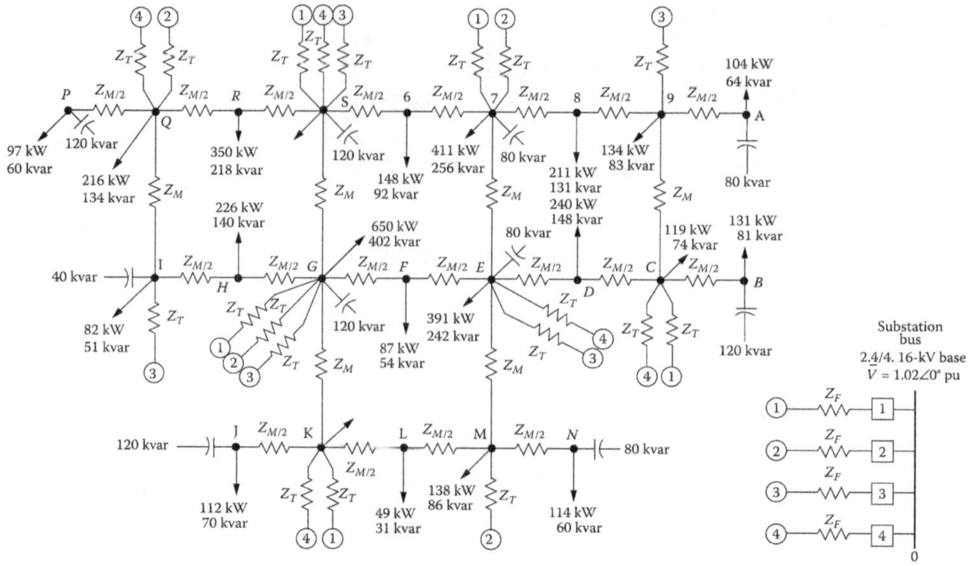

Figure 6.17 Figure 6.17 illustrates an AC secondary network. (Adapted from Westinghouse Electric Corporation, *Electric Utility Engineering Reference Book-Distribution Systems*, Vol. 3, East Pittsburgh, PA, 1965.)

For the underground mains in the network, copper cables of size 3-#4/0 per phase and 3-#3/0 neutral are used within nonmagnetic conduits. The positive-sequence impedance Z_M of a 500 ft main is $0.181 + j0.115$ pu based on a 1000 kVA base.

The primary feeder circuits in the network are 1.25 min in length. Each feeder circuit utilizes three single-conductor 500 kcmil 5 kV shielded copper PE-insulated underground cables with negligible impedance within the small network area. The positive-sequence impedance Z_F of the feeder cable is $0.01 + j0.017$ pu on a 1000 kVA base for 1.25-min long feeders. The approximate ampacities for the feeder circuits are 473 A for a single circuit per duct bank and 402 A for four equally loaded circuits per duct bank.

The following bases are used in the calculations: (1) a three-phase power base of 1000 kVA; (2) for secondary voltages, 125/216 V, 2666.7 A, and 0.04687 Ω; and (3) for primary voltages, 2400/4160 V, 138.9 A, and 17.28 Ω.

The standard network capacitor sizes used in the system are 40, 80, and 120 kVAr at 125/216 V. However, in this study, the capacitors are not switched. Typically, it is desired to avoid leading power factor operation on distribution circuits during off-peak load periods. Therefore, the total reactive power generated by the unswitched shunt capacitors should not exceed the total reactive power absorbed by the off-peak load. In this example, the total reactive load at peak load is 3150 kVAr, and it is assumed that the off-peak load is one-third of the peak load, i.e., 1050 kVAr. As a result, a total capacitor size of 960 kVAr is employed. The capacitors are distributed arbitrarily throughout the network in standard sizes, with larger capacitor banks generally located at buses with higher loads and at the ends of radial stubs from the network.

Using the given data, four separate load flow solutions have been obtained for the following operating conditions in the example secondary network:

Case 1: Normal switching: Normal loads and all shunt capacitors are off.

Case 2: Normal switching: Normal loads and all shunt capacitors are on.

Case 3: First-contingency outage: Primary feeder 1 is out. Normal loads and all shunt capacitors are on.

Table 6.5
Bus Voltage Value (pu)

Buses	Case 1	Case 2	Case 3	Case 4
A	0.951	0.967	0.954	0.915
B	0.958	0.975	0.955	0.86
C	0.976	0.986	0.966	0.873
J	0.959	0.976	0.954	0.864
K	0.974	0.984	0.962	0.875
N	0.958	0.973	0.963	0.924
P	0.96	0.977	0.966	0.926
R	0.945	0.954	0.938	0.89
S	0.964	0.972	0.951	0.898

Case 4: Second-contingency outage: Primary feeders 1 and 4 are out. Normal loads and all shunt capacitors are on. Note that this second-contingency outage is very severe, causing the largest load (at bus 5) to lose two-thirds of its transformer capacity.

To make a voltage study, Table 6.5 has been developed based on the load flow studies for the four cases. The values given in the table are per unit bus voltage values. Here, the buses selected for the study are the ones located at the ends of radials or else the ones that are badly disturbed by the second-contingency outage of case 4.

Use the given data and determine the following:

a. If the lowest "favorable" and the lowest "tolerable" voltages are defined as 114 and 111 V, respectively, what are the pu voltages, based on 125 V, that correspond to the lowest favorable voltage and the lowest tolerable voltage for nominally 120/208Y systems?
b. List the buses given in Table 6.5, for the first-contingency outage, that have (1) less than favorable voltage and (2) less than tolerable voltage.
c. List the buses given in Table 6.5, for the second-contingency outage, that have (1) less than favorable voltage and (2) less than tolerable voltage.
d. Find Z_M/Z_T, $1/2(Z_M/Z_T)$, and using Figure 6.8, find the value of the "application factor" for this example network and make an approximate judgment about the sufficiency of the design of this network.

Solution. a. The lowest favorable voltage in per unit is 114 V:

$$\frac{114\ \text{V}}{125\ \text{V}} = 0.912\ \text{pu}$$

and the lowest tolerable voltage in per unit is

$$\frac{114\ \text{V}}{125\ \text{V}} = 0.888\ \text{pu}\ \Omega$$

b. There are no buses in Table 6.5, for the first-contingency outage, that have (1) less than favorable voltage or (2) less than tolerable voltage.
c. For the second-contingency outage, the buses in Table 6.5 that have (1) less than favorable voltage are *B*, *C*, *J*, *K*, *R*, and *S* and (2) less than tolerable voltage are *B*, *C*, *J*, and *K*.

d. The given transformer impedance of 0.0086 + j0.0492 pu is based on 500 kVA. Therefore, it corresponds to

$$Z_T = 0.0172 + j0.0984 \text{ pu}\Omega$$

that is based on 1000 kVA. Therefore, the ratios are

$$\begin{aligned}\frac{Z_M}{Z_T} &= \frac{0.181 + j0.115}{0.0172 + j0.0984} \\ &= 2.147\end{aligned}$$

or

$$\frac{1}{2}\left(\frac{Z_M}{Z_T}\right) = 1.0735$$

Thus, from Figure 6.8, the corresponding average transformer application factor for four feeders can be found as 1.6. To verify this value for the given design, the actual application factor can be recalculated as

$$\begin{aligned}\text{Actual application factor} &= \frac{\text{Total installed network} - \text{transformer capacity}}{\text{Total Load}} \\ &= \frac{19\text{transformers} \times 500 \text{ kVA/transformer}}{5096 + j3158} \\ &= 1.5846\end{aligned}$$

Therefore, the design of this network is sufficient.

6.10 SECONDARY SYSTEM COSTS

As previously discussed, the secondary system comprises service transformers that convert primary voltage to utilization voltage, secondary circuits that operate at utilization voltage, and service drops (SDs) that deliver power directly to each customer. Many utilities estimate the costs of this equipment on a per customer basis. The annual costs associated with operating, maintenance, and taxes for a secondary system typically range between 1/8 and 1/30 of the capital cost.

In general, upgrading existing equipment to a higher capacity tends to be more costly than initially building to that capacity. Upgrading an existing secondary line (SL) involves removing the old conductor and installing a new one. This process usually requires new hardware, and in some cases, poles and crossarms may need to be replaced. As a result, the cost of such a conversion is often significantly higher than the cost of initially constructing the system with higher capacity. Consequently, transmission and distribution (T&D) engineers are motivated to carefully assess long-term needs and install additional capacity to accommodate future growth.

Example 6.6. It has been estimated that a 12.47 kV overhead (OH), three-phase feeder with 336 kcmil conductor costs $120,000 per mile. It has also been estimated that constructing the feeder with a 600 kcmil conductor and a 15 MVA capacity would cost approximately $150,000 per mile. However, upgrading the existing 9 MVA capacity line to a 15 MVA capacity involves removing the old conductor and installing a new one, resulting in an upgrade cost of $200,000 per mile. The following determinations need to be made:

a. The cost of building the 9 MVA capacity line in dollars per kVA-mile
b. The cost of building the 15 MVA capacity line in dollars per kVA-mile
c. The cost of the upgrade in dollars per kVA-mile

Solution. a. The cost of building the 9 MVA capacity line is

$$\text{Cost}_{9\text{ MVA line}} = \frac{\$120,000}{9,000\text{ kVA}} = \$13.33/\text{kVA} - \text{mile}$$

b. The cost of building the 15 MVA capacity is

$$\text{Cost}_{15\text{ MVA line}} = \frac{\$150,000}{15,000\text{ kVA}} = \$10/\text{kVA} - \text{mile}$$

c. The cost of the upgrade is

$$\text{Cost}_{\text{upgrade}} = \frac{\$200,000}{(15,000 - 9,000)\text{ kVA}} = \$33.33/\text{kVA} - \text{mile}$$

As it can be seen earlier, when judged against the additional capacity (15 – 9 MVA), the upgrade option is very costly, that is, over \$33/kVA-mile.

PROBLEMS

6.1 Repeat Example 6.1. Assume that there are four services per transformer, that is, one transformer on each pole so that there are six transformers per block.

6.2 Repeat Example 6.1. Assume that the annual load factor is 0.35.

6.3 Repeat Problem 6.1. Assume that the annual load factor is 0.65.

6.4 Consider Problem 6.1 and find the following:

a. The most economical service-drop size (A_{SD}) and the nearest larger commercial wire size
b. The most economical SL size (A_{SL}) and the nearest larger standard transformer size
c. The TAC per block for the nearest larger standard sizes of equipment

6.5 Repeat Example 6.4, assuming the load impedances are

$$\overline{Z}_a = 1.0 + j0.0\ \Omega$$

and

$$\overline{Z}_b = 1.5 + j0.0\ \Omega$$

6.6 Repeat Example 6.4, assuming the load impedances are

$$\overline{Z}_a = 1.0 + j0.0\ \Omega$$

and

$$\overline{Z}_b = 3.0 + j0.0\ \Omega$$

6.7 Repeat Example 6.4, assuming the load impedances are

$$\overline{Z}_a = 0.80 + j0.60\ \Omega$$

and

$$\overline{Z}_b = 1.5 + j0.0\ \Omega$$

6.8 The following table gives the total real and reactive power losses for the secondary network given in Example 6.5. Explain the circumstances that cause minimum and maximum losses. Bear in mind that the total P + *j*Q power delivered to the loads is identical in all cases.

Case No.	ΣP_L, MW	ΣQL, Mvar
1	0.16379	0.38807
2	0.1416	0.33142
3	0.19263	0.46648
4	0.36271	0.82477

6.9 The following table gives the primary-feeder circuit loading for the primary feeders given in Example 6.5.

	$P + jQ$, pu MVA			
Case No.	Feeder 1	Feeder 2	Feeder 3	Feeder 4
1	1.3575 −*j* 0.9012	1.186 −*j* 0.8131	1.3822 −*j* 0.9381	1.3341 −*j* 0.8857
2	1.3496 −*j* 0.6540	1.1854 −*j* 0.5894	1.375 −*j* 0.6936	1.3278 −*j* 0.6308
3	Out	1.5965 −*j* 0.8468	1.8427 −*j* 0.952	1.8495 −*j* 0.9354
4	Out	2.5347 −*j* 1.4587	2.924 −*j* 1.7285	Out

Determine the ampere loads of each feeder and complete the following table.

	Percent of Ampacity Rating			
Case No.	Feeder 1	Feeder 2	Feeder 3	Feeder 4
1	–	–	–	–
2	—	–	–	–
3	Out	–	–	–
4	Out	–	–	–

6.10 Assume that the following table gives the transformer loading for transformers 1, 3, and 4, using bus S data, for Example 6.5.

	Transformer Loading, kVA		
Case No.	Transformer 1	Transformer 3	Transformer 4
1	380.365	374	385.45
2	358.475	352.31	363.375
3		509.42	508.921
4		812.61	

Complete the following table. Note that bus S not only has the largest load but also loses two-thirds of its transformer capacity in the event of the second-contingency outage being considered here.

	Load-In Percent of Transformer Rating		
Case No.	Transformer 1	Transformer 3	Transformer 4
1			
2			
3			
4			

	Loading of Secondary Mains, pu MVA				
Case No.	*S*–*R*	*R*–*Q*	*S*–6	6–7	*S*–*G*
1	0.1715	0.2516	0.0699	0.1065	0.0361
2	0.1662	0.256	0.0692	0.1072	0.0364
3	0.1252	0.311	0.0816	0.0945	0.0545
4	0.0872	0.3778	0.0187	0.1901	0.143

6.11 Assume that the following table gives the loading of the secondary ins close to bus S in Example 6.5.

Determine the ampere loading of the mains close to bus S and also complete the following table.

	Loading of Secondary Mains, % of Rated Ampacity				
Case No.	*S*–*R*	*R*–*Q*	*S*–6	6–7	*S*–*G*
1					
2					
3					
4					

7 Voltage-Drop and Power-Loss Calculations

Any man may make a mistake; none but a fool will stick to it.

M.T. Cicero, 51 B.C.

Time is the wisest counselor.

Pericles, 450 B.C.

When others agree with me, I wonder what is wrong!

Author Unknown

7.1 STATE-OF-THE-ART REVIEW

Extensive research has been conducted on the correlation between voltage drop and power loss, often employing full-blown power flow calculations [153]. However, such methods require a comprehensive modeling of the network topology, which can be derived from supervisory control and data acquisition (SCADA) systems. Accurate computation of energy accounting aspects can be achieved by modeling the topology of a radial distribution feeder [154]. While data resolution for individual loads may not be available, loss calculations can be based on averaging statistics derived from case studies and determined by the percentage of system loading conditions [155]. An algorithm utilizing dynamic programming has been developed to optimize power losses in low-voltage (LV) networks based on deterministic loads [156]. Alternatively, evolutionary algorithms have been proposed for determining medium-voltage (MV) network applications in urban planning [157]. The allocation of losses has evolved over time to consider generation along the feeder, addressing nonlinearity and the impact of voltage variation in the presence of distributed generation [158].

Given the restructuring of electricity markets, the objectives of operation and planning may not align in terms of optimality. As a result, a multi-objective formulation has been devised to plan the siting and sizing of distributed generation (DG) resources within existing distribution feeders, utilizing evolutionary algorithms [159]. Various analytical methods have been simulated to compare loss estimation techniques, particularly regarding the impact of distributed generation penetration [160]. In some cases, a rough estimation of loss factors in LV networks can be obtained through a ballpark approach [161]. As load response varies in active distribution networks, planning efforts have incorporated distributed generation to ensure adequacy with estimated annualized energy losses [162, 163]. For optimal operation, renewable distributed generation has been employed to minimize energy losses over time [164]. In recent years, there has been extensive research on LV networks and voltage profiles, including strategies for peak shaving through the utilization of distributed energy storage systems and online assessment [165, 166]. Power losses can be efficiently managed through the integration of wind power [167], which can also be networked with microgrid islands for contingency events [168]. Challenges related to LV network and networked microgrids have been addressed, particularly regarding power restoration to critical loads [169, 170].

DOI: 10.1201/9781003129721-7

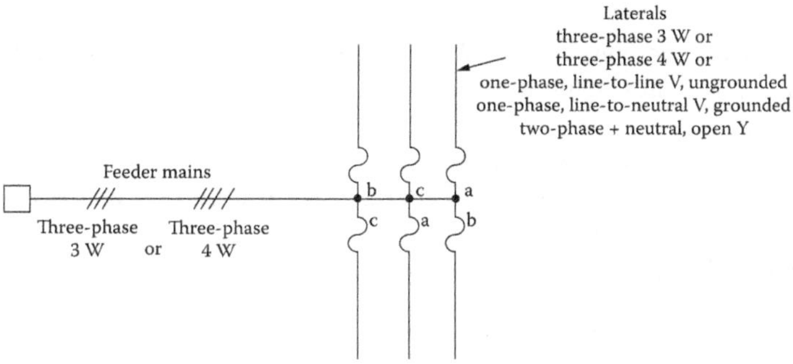

Figure 7.1 Various lateral types that exist in the United States.

7.2 THREE-PHASE BALANCED PRIMARY LINES

As emphasized in Chapter 5, utility companies strive to achieve a well-balanced distribution system to enhance system voltage regulation. Equally loading each phase plays a crucial role in this regard. Figure 7.1 illustrates a primary system featuring either a three-phase three-wire or a three-phase four-wire main. The system incorporates various types of laterals, including (1) three-phase three-wire, (2) three-phase four-wire, (3) single-phase with line-to-line voltage and ungrounded configuration, (4) single-phase with line-to-neutral voltage and grounded configuration, and (5) two-phase plus neutral with an open wye configuration.

7.3 NON-THREE-PHASE PRIMARY LINES

In most cases, primary feeders have numerous laterals that may not be exclusively three-phase, but include single-phase connections as well. This mixture of laterals leads to voltage drop and power loss, as the load current affects not only the phase conductor but also the return path.

7.3.1 SINGLE-PHASE TWO-WIRE LATERALS WITH UNGROUNDED NEUTRAL

Suppose we intend to convert an overloaded single-phase lateral into an equivalent three-phase three-wire lateral, while keeping the load constant. Given that the power input to the lateral remains unchanged,

$$S_{1\phi} = S_{3\phi} \tag{7.1}$$

Referring to Figure 7.1 [50], we use subscripts 1ϕ and 3ϕ to denote the single-phase and three-phase circuits, respectively. Equation 7.1 can be reformulated as follows:

$$(\sqrt{3} \times V_s)I_{1\phi} = 3V_s I_{3\phi} \tag{7.2}$$

where V_s is the line-to-neutral voltage. Therefore, from Equation 7.2,

$$I_{1\phi} = \sqrt{3} \times I_{3\phi} \tag{7.3}$$

which means that the current in the single-phase lateral is 1.73 times larger than the one in the equivalent three-phase lateral. The voltage drop in the three-phase lateral can be expressed as

$$VD_{3\phi} = I_{3\phi}(R\cos\theta + X\sin\theta) \tag{7.4}$$

and in the single-phase lateral as

$$VD_{1\phi} = I_{1\phi}(K_R R\cos\theta + K_X X\sin\theta) \tag{7.5}$$

where

K_R and K_X are conversion constants of R and X and are used to convert them from their three-phase values to the equivalent single-phase values

$K_R = 2.0$

$K_X = 2.0$ when underground (UG) cable is used

$K_X \cong 2.0$ when overhead line is used, with approximately a $\pm$ 10% accuracy

Therefore, Equation 7.5 can be rewritten as

$$VD_{1\phi} = I_{1\phi}(2\,R\cos\theta + 2\,X\sin\theta) \tag{7.6}$$

or substituting Equation 7.3 into Equation 7.6,

$$VD_{1\phi} = 2\sqrt{3} \times I_{3\phi}(R\cos\theta + X\sin\theta) \tag{7.7}$$

By dividing Equation 7.7 by Equation 7.4 side by side,

$$\frac{\mathrm{VD}_{1\phi}}{\mathrm{VD}_{3\phi}} = 2\sqrt{3}. \tag{7.8}$$

This indicates that the voltage drop in the single-phase ungrounded lateral is approximately 3.46 times greater than the voltage drop in the equivalent three-phase lateral. The base voltages for the single-phase and three-phase laterals are as follows:

$$V_{B(1\phi)} = \sqrt{3} \times V_{s,L-N} \tag{7.9}$$

and

$$V_{B(3\phi)} = V_{s,L-N} \tag{7.10}$$

Equation 7.8 can be expressed in per units as

$$\frac{\mathrm{VD}_{pu,1\phi}}{\mathrm{VD}_{pu,3\phi}} = 2.0. \tag{7.11}$$

This implies that the per unit voltage drop in the single-phase ungrounded lateral is twice as large as the per unit voltage drop in the equivalent three-phase lateral. For instance, if the per unit voltage drop in the single-phase lateral is 0.10, it would be 0.05 in the equivalent three-phase lateral.

The power losses caused by the load currents in the conductors of the single-phase lateral and the equivalent three-phase lateral are as follows:

$$P_{\mathrm{LS},1\phi} = 2 \times I_{1\phi}^2 R \tag{7.12}$$

and

$$P_{\mathrm{LS},3\phi} = 2 \times I_{3\phi}^2 R \tag{7.13}$$

respectively. Substituting Equation 7.3 into Equation 7.12,

$$P_{\mathrm{LS},1\phi} = 2(\sqrt{3} \times I_{3\phi})^2 R \tag{7.14}$$

and dividing the resultant Equation 7.14 by Equation 7.13 side by side,

$$\frac{P_{LS,1\phi}}{P_{LS,3\phi}} = 2.0 \tag{7.15}$$

This implies that the power loss resulting from load currents in the conductors of the single-phase lateral is twice as large as that in the equivalent three-phase lateral.

Hence, it can be inferred that when transforming a single-phase lateral into an equivalent three-phase lateral, both the per unit voltage drop and the power loss resulting from copper losses in the primary line are roughly reduced by half.

7.3.2 SINGLE-PHASE TWO-WIRE UNGROUNDED LATERALS

Currently, this system is not commonly employed due to several drawbacks. One major disadvantage is the absence of earth current in this system. It can be contrasted with a three-phase four-wire balanced lateral in the following manner: Given that the power input to the lateral remains unchanged,

$$S_{1\phi} = S_{3\phi} \tag{7.16}$$

or

$$V_s \times I_{1\phi} = 3 \times V_S \times I_{3\phi} \tag{7.17}$$

from which

$$I_{1\phi} = 3 \times V_S \times I_{3\phi} \tag{7.18}$$

The voltage drop in the three-phase lateral can be expressed as

$$VD_{3\phi} = I_{3\phi}(R \cos\theta + X \sin\theta) \tag{7.19}$$

and in the single-phase lateral as

$$VD_{1\phi} = I_{1\phi}(K_R R \cos\theta + K_X X \sin\theta) \tag{7.20}$$

where

K_R= 2.0 when a full-capacity neutral is used, that is, if the wire size used for neutral conductor is the same as the size of the phase wire

$K_R > 2.0$ when a reduced-capacity neutral is used

$K_X \cong 2.0$ when overhead line is used

Therefore, if K_R = 2.0 and K_X = 2.0, Equation 7.20 can be rewritten as

$$VD_{1\phi} = I_{1\phi}(2R \cos\theta + 2X \sin\theta) \tag{7.21}$$

or substituting Equation 7.18 into Equation 7.21,

$$VD_{1\phi} = 6 \times I_{3\phi}(R \cos\theta + X \sin\theta) \tag{7.22}$$

Dividing Equation 7.22 by Equation 7.19 side by side,

$$\frac{VD_{1\phi}}{VD_{3\phi}} = 6.0 \tag{7.23a}$$

or

$$\frac{VD_{pu,1\phi}}{VD_{pu,3\phi}} = 2\sqrt{3} = 3.46. \tag{7.23b}$$

This indicates that the voltage drop in the single-phase two-wire ungrounded lateral with a full-capacity neutral is six times greater than the voltage drop in the equivalent three-phase four-wire balanced lateral.

The power losses resulting from the load currents in the conductors of the single-phase two-wire ungrounded lateral with a full-capacity neutral and the equivalent three-phase four-wire balanced lateral are as follows:

$$P_{LS,1\phi} = I_{1\phi}^2(2R) \tag{7.24}$$

and

$$P_{LS,3\phi} = 3 \times I_{3\phi}^2 R \tag{7.25}$$

respectively. Substituting Equation 7.18 into Equation 7.24,

$$P_{LS,1\phi} = (3 \times I_{3\phi})^2(2R) \tag{7.26}$$

and dividing Equation 7.26 by Equation 7.25 side by side,

$$\frac{P_{LS,1\phi}}{P_{LS,3\phi}} = 6.0 \tag{7.27}$$

Hence, it can be concluded that the power loss resulting from the load currents in the conductors of the single-phase two-wire ungrounded lateral with a full-capacity neutral is six times greater than the power loss in the equivalent three-phase four-wire lateral.

7.3.3 SINGLE-PHASE TWO-WIRE LATERALS WITH MULTIGROUNDED COMMON NEUTRALS

Figure 7.2 illustrates a single-phase two-wire lateral configuration with a multigrounded common neutral. In this setup, the neutral wire is interconnected with the ground wire at different locations through ground electrodes. This arrangement aims to minimize the current flowing through the neutral wire. The current in the phase conductor is denoted as I_a, the return current in the neutral

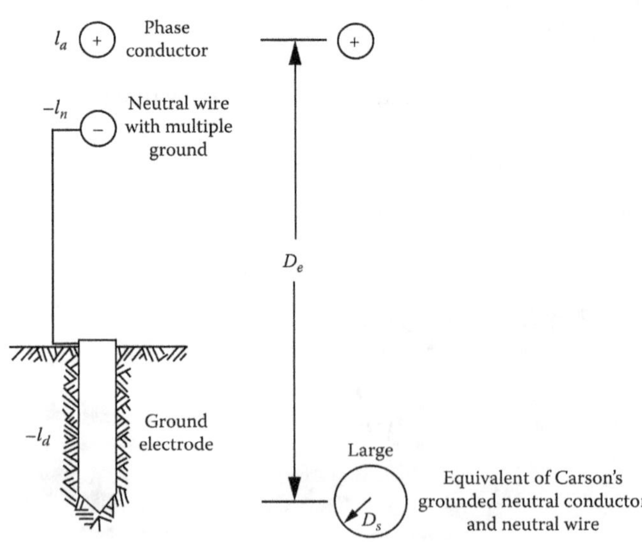

Figure 7.2 A single-phase lateral with multigrounded common neutral.

wire is represented as I_w, and the return current in Carson's equivalent ground conductor is denoted as I_d. According to Morrison [126], the return current in the neutral wire can be expressed as follows:

$$I_n = \zeta_1 I_a \quad \text{where} \zeta_1 = 0.25 - 0.33 \tag{7.28}$$

and it is almost independent of the size of the neutral conductor.

In Figure 7.2, the value of constant KR is below 2.0, while the constant K_X is approximately 2.0. This is due to the presence of significantly large D_m (mutual geometric mean distance or geometric mean radius) for Carson's equivalent ground (neutral) conductor, which creates conflicting conditions. Consequently, empirical data provided by Morrison [126] suggest that:

$$\text{VD}_{\text{pu},1\phi} = \zeta_2 \times \text{VD}_{\text{pu},3\phi} \quad \text{where} \zeta)2 = 3.8 - 4.2 \tag{7.29}$$

and

$$\text{P}_{\text{LS},1\phi} = \zeta_3 \times \text{P}_{\text{LS},3\phi} \quad \text{where} \zeta)3 = 3.5 - 3.75 \tag{7.30}$$

Therefore, assuming that the data from Morrison [126] are accurate,

$$K_R < 2.0 \text{ and } K_X < 2.0$$

the per unit voltage drops and the power losses due to load currents can be approximated as

$$\text{VD}_{\text{pu},1\phi} \cong 4.0 \times \text{VD}_{\text{pu},3\phi} \tag{7.31}$$

and

$$\text{P}_{\text{LS},1\phi} \cong 3.6 \times \text{P}_{\text{LS},3\phi} \tag{7.32}$$

for the illustrative problems.

7.3.4 TWO-PHASE PLUS NEUTRAL (OPEN-WYE) LATERALS

In Figure 7.3, an open-wye-connected lateral with two phases and a neutral is depicted. While the neutral conductor can be either unigrounded or multigrounded, the use of unigrounded neutral is generally avoided due to its disadvantages. In the case of a unigrounded neutral, all the neutral current flows through the neutral conductor itself. Theoretical expression for this situation can be stated as follows:

$$\mathbf{V} = \mathbf{ZI} \tag{7.33}$$

where

$$\overline{V}_a = \overline{Z}_a \overline{I}_a \tag{7.34}$$

$$\overline{V}_b = \overline{Z}_b \overline{I}_b \tag{7.35}$$

When equal load division is maintained between the two phases, the comparison between a two-phase plus neutral lateral and an equivalent three-phase lateral remains valid. By keeping the total kVA load constant, the following can be inferred:

$$S_{2,\phi} = S_{3,\phi} \tag{7.36}$$

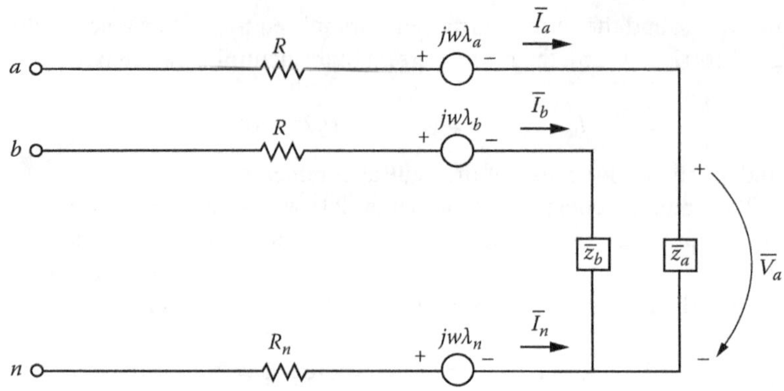

Figure 7.3 An open-wye connected lateral.

or

$$2V_5 I_{2,\phi} = 3V_5 I_{3,\phi} \tag{7.37}$$

from which

$$I_{2,\phi} = \frac{3}{2} I_{3,\phi} \tag{7.38}$$

The voltage-drop analysis can be conducted based on whether the neutral is unigrounded or multigrounded. In the case of an unigrounded neutral, assuming that the impedance of the neutral conductor (Z_n) is zero, the voltage drop in each phase can be expressed as follows:

$$\mathrm{VD}_{2\phi} = I_{2\phi}(K_R R \cos\theta + K_X X \sin\theta) \tag{7.39}$$

where
$K_R = 1.0$
$K_x = 1.0$
Therefore,

$$\mathrm{VD}_{2\phi} = I_{2\phi}(R\cos\theta + X\sin\theta) \tag{7.40}$$

or substituting Equation 7.38 into Equation 7.40,

$$\mathrm{VD}_{2\phi} = \frac{3}{2} I_{3\phi}(R\cos\theta + X\sin\theta) \tag{7.41}$$

Dividing Equation 7.41 by Equation 7.19, side by side,

$$\frac{\mathrm{VD}_{2\phi}}{\mathrm{VD}_{3\phi}} = \frac{3}{2} \tag{7.42}$$

However, when the neutral is unigrounded and the impedance of the neutral conductor (Z_n) is greater than zero,

$$\frac{\mathrm{VD}_{2\phi}}{\mathrm{VD}_{3\phi}} > \frac{3}{2} \tag{7.43}$$

Hence, in this scenario, the presence of unbalanced voltages is expected.

On the other hand, when the neutral is multigrounded and $Z_n > 0$, empirical data from Morrison [126] suggests that the per unit voltage drop in each phase is given by:

$$\mathrm{VD}_{2\phi} = 2.0 \times \mathrm{VD}_{3\phi} \tag{7.44}$$

when a full-capacity neutral is used and

$$\text{VD}_{2\phi} = 2.1 \times \text{VD}_{3\phi} \tag{7.45}$$

When a reduced-capacity neutral is used (i.e., when the neutral conductor is one or two sizes smaller than the phase conductors),

The power loss analysis also varies depending on whether the neutral is unigrounded or multigrounded. If the neutral is unigrounded, the power loss is expressed as:

$$\text{P}_{LS,2\phi} = I_{2\phi}^2 (K_R R) \tag{7.46}$$

where

$K_R = 3.0$ when a full-capacity neutral is used

$K_R > 3.0$ when a reduced-capacity neutral is used

Therefore, if $K_R = 3.0$,

$$\frac{\text{P}_{LS,2\phi}}{\text{P}_{LS,3\phi}} = \frac{3I_{2,\phi}^2 R}{3I_{3,\phi}^2 R} \tag{7.47}$$

or

$$\frac{\text{P}_{LS,2\phi}}{\text{P}_{LS,3\phi}} = 2.25 \tag{7.48}$$

On the other hand, *if the neutral is multigrounded,*

$$\frac{\text{P}_{LS,2\phi}}{\text{P}_{LS,3\phi}} < 2.25 \tag{7.49}$$

Based on the data from Morrison [126], the approximate value of this ratio is

$$\frac{\text{P}_{LS,2\phi}}{\text{P}_{LS,3\phi}} \cong 1.64. \tag{7.50}$$

This implies that the power loss resulting from load currents in the conductors of the two-phase three-wire lateral with a multigrounded neutral is approximately 1.64 times greater than that in the equivalent three-phase lateral.

Example 7.1. Consider a scenario where a uniformly distributed area is supplied by a three-phase four-wire multigrounded main, spanning a length of 6 miles and situated in the center of the service region. On each side of the main, there are six laterals. These laterals are evenly spaced, with a distance of 1 mile between each of them. The first lateral begins 1 mile away from the substation. The total three-phase load connected to the main is 6000 kVA. Each lateral is 10 miles long and consists of #6 AWG copper conductors. They serve a uniformly distributed peak load of 500 kVA at a voltage of 7.2/12.47 kV. The K constant for a #6 AWG copper conductor is 0.0016/kVA-mile. Determine the following:

a. If a lateral is a three-phase lateral with multigrounded common neutrals, what is the maximum voltage drop to the end of each lateral?
b. What is the maximum voltage drop to the end of each lateral if the lateral is a two-phase plus full-capacity multigrounded neutral (open-wye) lateral?
c. For a single-phase two-wire lateral with multigrounded common neutrals, what is the maximum voltage drop to the end of each lateral?

Solution. a. For the three-phase four-wire lateral with multigrounded common neutrals,

$$\begin{aligned}\%\mathrm{VD}_{3\phi} &= \frac{1}{2}\times K \times S \\ &= \left(\frac{10\mathrm{mi}}{2}\right)\left(0.0016\frac{\%\mathrm{VD}}{\mathrm{KVA\text{ - }mi}}\right)(500\ \mathrm{kVA}) = 4\end{aligned}$$

b. For the two-phase plus full-capacity multigrounded neutral (open-wye) lateral, according to the results of Morrison,

$$\begin{aligned}\%\mathrm{VD}_{2\phi} &= 2(\%\mathrm{VD}_{2\phi}) \\ &= 2(4\%) = 8\end{aligned}$$

c. For the single-phase two-wire lateral with multigrounded common neutrals, according to the results of Morrison,

$$\begin{aligned}\%\mathrm{VD}_{1\phi} &= 4(\%\mathrm{VD}_{3\phi}) \\ &= 4(4\%) = 16\end{aligned}$$

Example 7.2. A three-phase express feeder has an impedance of 6 + j20 ohms per phase. At the load end, the line-to-line voltage is 13.8 kV, and the total three-phase power is 1200 kW at a lagging power factor of 0.8. By using the *line-to-neutral* method, determine the following:

1. The line-to-line voltage at the sending end of the feeder (i.e., at the substation low-voltage bus)
2. The power factor at the sending end
3. The copper loss (i.e., the transmission loss) of the feeder
4. The power at the sending end in kW

Solution. Since in an express feeder, the line current is the same at the beginning or at the end of the line,

$$\begin{aligned}I_L = I_S = I_R &= \frac{P_{R(3\phi)}}{\sqrt{3}V_{R(L-L)}\cos\theta} \\ &= \frac{1{,}200\ \mathrm{kW}}{\sqrt{3}(13.8\ \mathrm{kV})0.8} = 62.75\ \mathrm{A}\end{aligned}$$

and

$$\begin{aligned}V_{R(L-N)} &= \frac{V_{R(L-L)}}{\sqrt{3}} \\ &= \frac{13{,}800\ \mathrm{V}}{\sqrt{3}} = 7{,}976.4\ \mathrm{V}\end{aligned}$$

using this as *the reference voltage*, the sending-end voltage is found from

$$\overline{V}_{S(L-N)} = \overline{V}_{R(L-N)} + \overline{I}_L\overline{Z}_L$$

where

$$\overline{V}_{R(L-N)} = 7,976.9\angle 0^\circ \text{ V}$$

$$\begin{aligned}\overline{I}_L = \overline{I}_S = \overline{I}_R &= I_L(\cos\theta_R - j\sin\theta_R)\\ &= 62.83(0.8 - j0.6) = 62.83\angle -36.87^\circ\Omega\end{aligned}$$

$$\overline{Z}_L = 6 + j20 = 20.88\angle 73.3^\circ\Omega$$

a.

$$\begin{aligned}\overline{V}_{S(L-N)} &= 7,976.9\angle 0^\circ + (62.83\angle -36.87^\circ)(20.88\angle 73.7^\circ)\\ &= 9,065.95\angle 4.93^\circ \text{ V}\end{aligned}$$

and

$$\begin{aligned}\overline{V}_{S(L-L)} &= \sqrt{3}\overline{V}_{S(L-N)}\\ &= \sqrt{3}(9,065.95)\angle 4.93^\circ + 30^\circ\\ &= 15,684.09\angle 34.93^\circ \text{ V}\end{aligned}$$

b.

$$\theta_S = |\theta_{\overline{V}_{S(L-N)}}| - |\theta_{\overline{I}_S}| = 4.9^\circ - |36.87^\circ| = 41.8^\circ \text{ and } \cos\theta_S = 0.745 \text{ lagging}$$

c.

$$\begin{aligned}P_{\text{LOSS}(3\phi)} &= 3I_L^2R = 3(62.83)^2 \times 6\\ &= 71,056.96 \text{ W} \cong 71.057 \text{ kW}\end{aligned}$$

d.

$$\begin{aligned}P_{\text{S}(3\phi)} &= P_{\text{R}(3\phi)} + P_{\text{LOSS}(3\phi)}\\ &= 1,200 + 71.057 = 1,271.057 \text{ kW}\end{aligned}$$

or

$$\begin{aligned}P_{\text{S}(3\phi)} &= \sqrt{3}V_{S(L-L)}I_S\cos\theta_S\\ &= \sqrt{3}(15,684.09)(62.83)0.745 \cong 1,270.073 \text{ kW}\end{aligned}$$

Example 7.3. Repeat Example 7.2 by using the *single-phase equivalent* method.

Solution. Here, the single-phase equivalent current is found from

$$\begin{aligned}I_{eq(1\phi)} &= \frac{P_{3\phi}}{V_{R(L-L)}\cos\theta}\\ &= \frac{1200 \text{ kW}}{(13.8 \text{ kV})(0.8)} = 108.7 \text{ A}\end{aligned}$$

where

$$I_{eq(1\phi)} = \sqrt{3} I_{3\phi}$$

or

$$\begin{aligned} I_{3\phi} = I_L &= \frac{I_{eq(1\phi)}}{\sqrt{3}} \\ &= \frac{108.7 \text{ A}}{\sqrt{3}} = 62.8 \text{ A} \end{aligned}$$

a.

$$\begin{aligned} \overline{V}_{S(L-L)} &= \overline{V}_{R(L-L)} + \overline{I}_{eq(1\phi)} \overline{Z}_L \\ &= 13{,}800\angle 0^\circ + (108.7\angle -36.9^\circ)(20.88\angle 73.3^\circ) \\ &= 15{,}684.76\angle 4.93^\circ + 30^\circ \text{ V} \end{aligned}$$

b.

$$\theta_S = |\theta_{\overline{V}_{S(L-N)}}| - |\theta_{\overline{I}_S}| = 4.9^\circ - |36.87^\circ| = 41.8^\circ \text{ and } \cos\theta_S = 0.745 \text{ lagging}$$

c.

$$\begin{aligned} P_{\text{LOSS}(3\phi)} &= I^2_{eq(1\phi)} R \\ &= 108.7^2 \times 6 = 70.89 \text{ kW} \end{aligned}$$

d.

$$\begin{aligned} P_{\text{S}(3\phi)} &= P_{\text{R}(3\phi)} + P_{\text{LOSS}(3\phi)} \\ &= 1{,}200 + 71.89 = 1{,}270.89 \text{ kW} \end{aligned}$$

7.4 FOUR-WIRE MULTIGROUNDED COMMON NEUTRAL DISTRIBUTION SYSTEM

Figure 7.4 illustrates a typical distribution system with a four-wire multigrounded common neutral. This system is widely used due to its economic and operational benefits. The secondary configurations can vary, such as (1) a 120/240 V single-phase three-wire system, (2) a 120/240 V three-phase four-wire system connected in delta, (3) a 120/240 V three-phase four-wire system connected in open delta, or (4) a 120/208 V three-phase four-wire system connected in grounded wye. In cases where both primary and secondary systems coexist, the same conductor serves as the *common* neutral for both systems. The neutral is grounded at each distribution transformer, at various locations where there are no transformers, and at water pipes or driven ground electrodes at each user's service entrance. The secondary neutral is also grounded at the distribution transformer and the service drops (SDs).

The resistances of ground electrodes typically have values of 5, 10, or 15 Ω. It is important to ensure that they do not exceed 25 Ω. In general, a metal water pipe system has a resistance value below 3 Ω. A portion of the unbalanced or zero sequence load current travels through the neutral wire, while the remaining portion flows through the ground and/or water system. Usually, the phase and neutral conductors have the same conductor size [171].

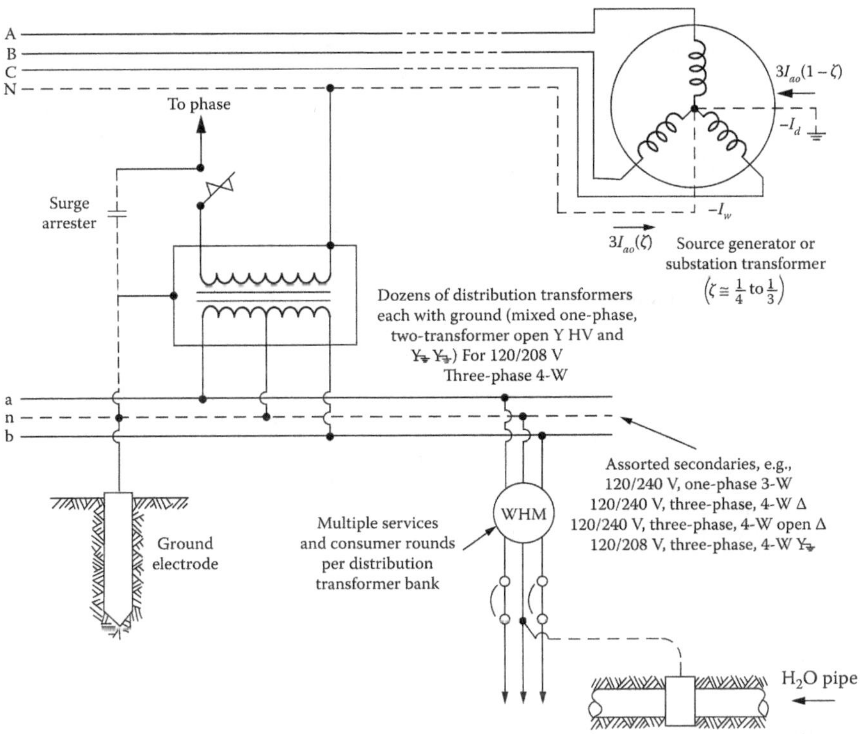

Figure 7.4 A four-wire multigrounded common neutral distribution system.

Example 7.4. Consider the circuit depicted in Figure 7.5, assuming dimensional variables for a single-phase circuit or per unit variables for a balanced three-phase circuit. The total impedance of lines and/or transformers is represented by $R + jX$. The load power factor, expressed as cos θ = $\cos(\theta_{VR} - \Omega_T)$, needs to be determined in order to find the maximum voltage drop.

Solution. The line voltage drop is

$$\text{VD} = I(R\cos\theta + X\sin\theta)$$

By taking its partial derivative with respect to the θ angle and equating the result to zero,

$$\frac{\partial(\text{VD})}{\partial\theta} = I(R\cos\theta + X\sin\theta) = 0$$

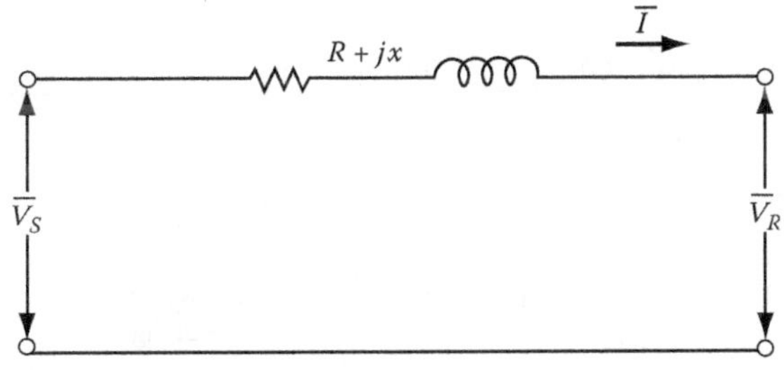

Figure 7.5 A single-phase circuit.

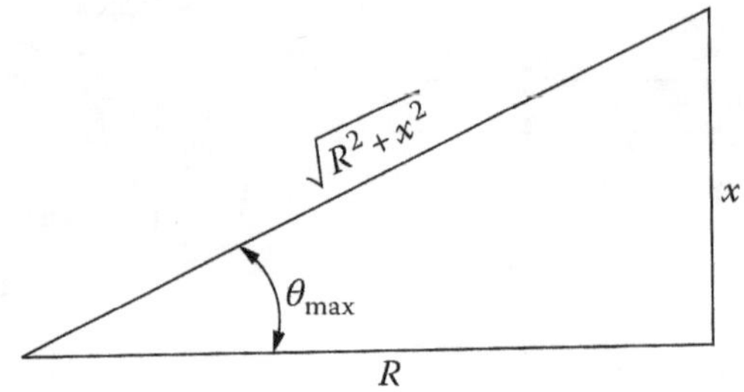

Figure 7.6 Impedance triangle.

or

$$\frac{X}{R}=\frac{\sin\theta}{\cos\theta}=\tan\theta$$

therefore

$$\theta_{max}=\tan^{-1}\frac{X}{R}$$

and from the impedance triangle shown in Figure 7.6, the load power factor for which the voltage drop is maximum is

$$\text{PF}=\cos\theta_{max}=\frac{R}{\sqrt{R^2+X^2}} \tag{7.51}$$

also

$$\cos\theta_{max}=\cos\left(\tan^{-1}\frac{X}{R}\right) \tag{7.52}$$

Example 7.5. Consider the three-phase four-wire 416-V secondary system with balanced per-phase loads at *A*, *B*, and *C* as shown in Figure 7.7. Determine the following:

a. Calculate the total voltage drop, or as it is sometimes called, *voltage regulation*, in one phase of the lateral by using the approximate method.
b. Calculate the real power per phase for each load.
c. Calculate the reactive power per phase for each load.
d. Calculate the total (*three-phase*) kVA output and load power factor of the distribution transformer.

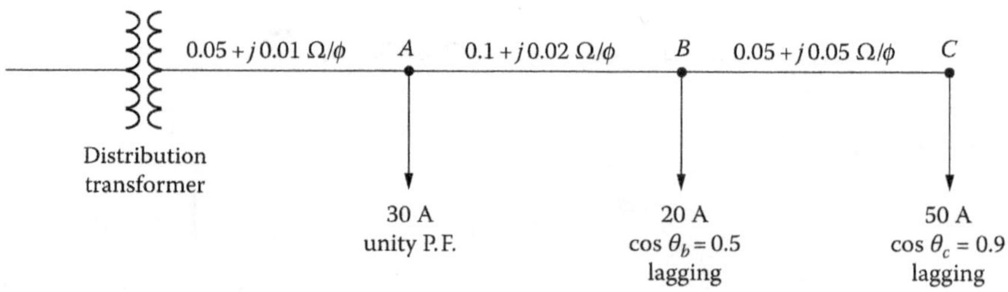

Figure 7.7 One-line diagram of a three-phase four-wire secondary system.

Solution. a. Using the approximate voltage-drop equation, that is,

$$VD = I(R\cos\theta + X\sin\theta)$$

the voltage drop for each load can be calculated as

$$VD_A = 30(0.05 \times 1.0 + 0.01 \times 0) = 1.5 \text{ V}$$
$$VD_B = 20(0.15 \times 0.5 + 0.03 \times 0.866) = 2.02\text{V}$$
$$VD_C = 50(0.20 \times 0.9 + 0.08 \times 0.436) = 10.744\text{V}$$

Therefore, the total voltage drop is

$$\begin{aligned}\sum VD &= VD_A + VD_B + VD_C \\ &= 1.5 + 2.02 + 10.744 \\ &= 14.264 \text{ V}\end{aligned}$$

or

$$\frac{14.264 \text{ V}}{240 \text{ V}} = 0.0594 \text{ puV}$$

b. The per-phase real power for each load can be calculated from

$$P = VI\cos\theta$$

or

$$P_A = 240 \times 30 \times 1.0 = 7.2 \text{ kW}$$
$$P_B = 240 \times 20 \times 0.5 = 2.4 \text{ kW}$$
$$P_C = 240 \times 50 \times 0.9 = 10.8 \text{ kW}$$

Therefore, the total per-phase real power is

$$\begin{aligned}\sum P &= P_A + P_B + P_C \\ &= 7.2 + 2.4 + 10.8 \\ &= 20.4 \text{ kW}\end{aligned}$$

c. The reactive power per phase for each load can be calculated from

$$Q = VI\sin\theta$$

or

$$Q_A = 240 \times 30 \times 0 = 0 \text{ kVAr}$$
$$Q_B = 240 \times 20 \times 0.866 = 4.156 \text{ kVAr}$$
$$Q_C = 240 \times 50 \times 0.436 = 5.232 \text{ kVAr}$$

Therefore, the total per-phase real power is

$$\begin{aligned}\sum Q &= Q_A + Q_B + Q_C \\ &= 0 + 4.156 + 5.232 \\ &= 9.389 \text{ kVAr}\end{aligned}$$

d. Therefore, the per-phase kVA output of the distribution transformer is

$$\begin{aligned}S &= \sqrt{P^2 + Q^2} \\ &= \sqrt{20.4^2 + 9.389^2} \\ &\cong 22.457 \text{ kVA/phase}\end{aligned}$$

Thus the total (or three-phase) kVA output of the distribution transformer is

$$3 \times 22.457 = 67.37 \text{ kVA}$$

Hence, the load power factor of the distribution transformer is

$$\begin{aligned}\cos\theta &= \frac{\sum P}{S} \\ &= \frac{20.4 \text{ kW}}{22.457 \text{kVA}} \\ &= 0.908 \text{ lagging}\end{aligned}$$

Example 7.6. This example is a continuation of Example 6.1. It deals with voltage drops in the secondary distribution system. In this and the following examples, a single-phase three-wire 120/240 V directly buried underground residential distribution (URD) secondary system will be analyzed, and calculations will be made for motor-starting voltage dip (VDIP) and for steady-state voltage drops at the time of annual peak load. Assume that the cable impedances given in Table 7.2 are correct for a typical URD secondary cable.

Transformer data. The data given in Table 7.1 are for modern single-phase 65°C OISC distribution transformers of the 7200–120/240 V class. The data were taken from a recent catalog of a manufacturer. All given per unit values are based on the transformer-rated kVA and voltages.

The transformers in the 2400 V-class and the ones considered are slightly different in terms of impedance. The considered 2400 V-class transformers have approximately 15% less resistance (R) and about 7% less reactance (X) compared to the 7200 V transformers. It is safe to ignore the small variation in impedance with the rated voltage, assuming that the voltage drop calculated using the given data will be sufficient regardless of the primary voltage used.

Underground Residential Distribution (URD) secondary cable data: Cable insulation and manufacturing techniques are continuously being improved, particularly for high-voltage cables. As a result, cable data can quickly become outdated. The following information and data have been extracted from recent cable catalogs.

Many 600 V-class cables commonly used for secondary lines (SLs) and services are constructed with aluminum conductors and cross-linked polyethylene (PE) insulation, which can withstand conductor temperatures of up to 90°C. Figure 7.8 illustrates a triplexed cable assembly (quadruplexed for three-phase four-wire service) consisting of three or four insulated conductors when aluminum

Table 7.1
Single-Phase 7200-120/240-V Distribution Transformer Data at 65° C

Rated (kVA, kW)	Core Loss[a] (kW)	Copper Loss[b] (kW)	*R* (pu)	*X* (pu)	Excitation Current (A)
15	0.083	0.194	0.013	0.0094	0.014
25	0.115	0.309	0.0123	0.0138	0.015
37.5	0.17	0.4	0.0107	0.0126	0.014
50	0.178	0.537	0.0107	0.0139	0.014
75	0.28	0.755	0.0101	0.0143	0.014
100	0.335	0.975	0.0098	0.0145	0.014

[a] At rated voltage and frequency.
[b] At rated voltage and kVA load.

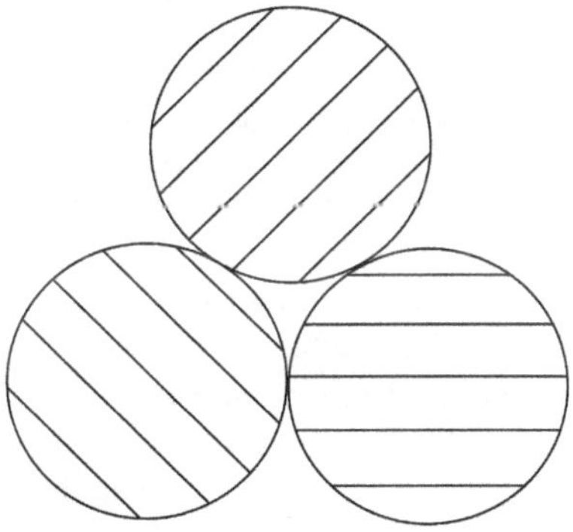

Figure 7.8 Triplexed cable assembly.

conductors are used. When copper conductors are utilized, the neutral conductor remains bare. Typically, the neutral conductor is two American Wire Gauge (AWG) sizes smaller than the phase conductors.

Figure 7.9 displays a twin concentric cable assembly that includes two insulated phase conductors made of either copper or aluminum, along with several spirally served small bare copper binding conductors that function as the current-carrying grounded neutral. The number and size of the spiral neutral wires vary to ensure that the ampacity of the neutral circuit is equivalent to two AWG wire sizes smaller than the phase conductors. Table 7.2 provides data for twin concentric aluminum/copper cross-linked polyethylene (XLPE) 600 V-class cable.

The triplex and twin concentric assemblies obviously have the same resistance for a given size of phase conductors. The triplex assembly has very slightly higher reactance than the concentric

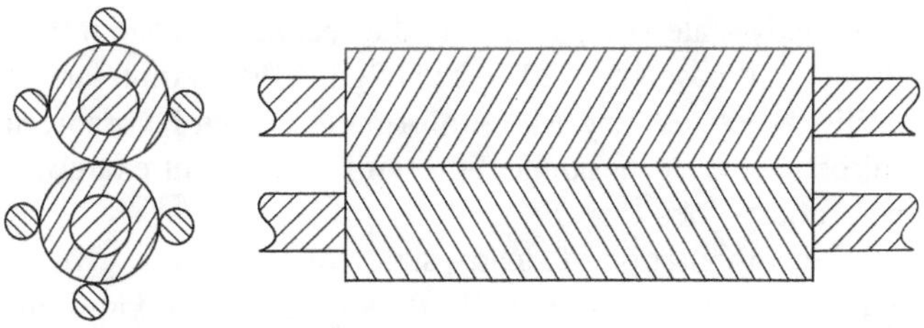

Figure 7.9 Twin concentric cable assembly.

Table 7.2
Single-Phase 7200-120/240-V Distribution Transformer Data at 65° C

	R (Ω/1000 ft) per Conductor				Direct Burial $\bar{K}$[a]	
Size	Phase Conductor 90°C	Neutral Conductor 80°C	X (Ω/1000 ft) per Phase Conductor	Ampacity (A)	90% PF	50% PF
2 AWG	0.334	0.561	0.0299	180	0.0261	0.01608
1 AWG	0.265	0.419	0.0305	205	0.021	0.01324
1/0 AWG	0.21	0.337	0.0297	230	0.0168	0.01089
2/0 AWG	0.167	0.259	0.029	265	0.0136	0.00905
3/0 AWG	0.132	0.211	0.028	300	0.0109	0.00752
4/0 AWG	0.105	0.168	0.0275	340	0.0089	0.00636
250 kcmil	0.089	0.133	0.028	370	0.0077	0.00573
350 kcmil	0.063	0.085	0.027	445	0.0057	0.00458
500 kcmil	0.044	0.066	0.026	540	0.0042	0.00371

[a] Per unit voltage drop per 10^4 A · ft (amperes per conductor times feet of cable) based on 120 V line-to-neutral or 240 V line to line. Valid for the two power factors shown and for perfectly balanced three-wire loading.

assembly. The difference in reactances is too small to be noted unless precise computations are undertaken for some special purpose. The reactances of those cables should be increased by about 25% if they are installed in iron conduit. The reactances given in the following text are valid only for balanced loading (where the neutral current is zero).

The triplex assembly has about 15% smaller ampacity than the concentric assembly, but the exact amount of reduction varies with wire size. The ampacities given are for 90°C conductor temperature, 20°C ambient earth temperature, direct burial in earth, and 10% daily load factor. When installed in buried duct, the ampacities are about 70% of those listed later. For load factors less than 100%, consult current literature or cable standards. The increased ampacities are significantly large.

Arbitrary criteria

1. Use the approximate voltage-drop equation, that is,

$$\text{VD} = I(R\cos\theta + X\sin\theta)$$

   and adapt it to per unit data when computing transformer voltage drops and adapt it to ampere and ohm data when computing SD and SL voltage drops. Obtain all voltage-drop answers in per unit based on 240 V.
2. Maximum allowable motor-starting VDIP = 3% = 0.03 pu = 3.6 V based on 120 V. This figure is arbitrary; utility practices vary.
3. Maximum allowable steady-state voltage drop in the secondary system (transformer + SL + SD) = 3.50% = 0.035 pu = 4.2 V based on 120 V. This figure also is quite arbitrary; regulatory commission rules and utility practices vary. More information about favorable and tolerable amounts of voltage drop will be discussed in connection with subsequent examples, which will involve voltage drops in the primary lines.
4. The loading data for the computation of steady-state voltage drop is given in Table 7.3.
5. As loading data for transient motor-starting VDIP, assume an air-conditioning compressor motor located most unfavorably. It has a 3 hp single-phase 240 V 80 A locked rotor current, with a 50% PF locked rotor.

Table 7.3
Load Data

Circuit Element	Load (kV)
SD	1 class 2 load (10 kVA)
SL	1 class 2 load (10 kVA) + 3 diversified class 2 loads (6.0 kVA each)
Transformer	1 class 2 load (10 kVA) + either 3 diversified class 2 loads(6.0 kVA each) or 11 diversified class 2 loads(4.4 kVA each)

Source: Lawrence, R.F. et al., *AIEE Trans.*, pt. III, PAS-79, 199, 1960.

Assumptions

1. Assume perfectly balanced loading in all three-wire single-phase circuits.
2. Assume nominal operating voltage of 240 V when computing currents from kVA loads.
3. Assume 90% lagging power factor for all loads.

Using the given data and assumptions, calculate the $\tilde{K}$ constant for any one of the secondary cable sizes, hoping to verify one of the given values in Table 7.2.

Solution. Let the secondary cable size be #2 AWG, arbitrarily. Also let the I current be 100 A and the length of the SL be 100 ft. Using the values from Table 7.2, the resistance and reactance values for 100 ft of cable can be found as

$$
\begin{aligned}
R &= 0.334\Omega/1000\text{ft} \times \frac{100\text{ ft}}{1000\text{ ft}} \\
&= 0.0334\Omega
\end{aligned}
$$

and

$$
\begin{aligned}
X &= 0.0299\Omega/1000\text{ft} \times \frac{100\text{ ft}}{1000\text{ ft}} \\
&= 0.00299\Omega
\end{aligned}
$$

Therefore, using the approximate voltage-drop equation,

$$
\begin{aligned}
\text{VD} &= I(R\cos\theta + X\sin\theta) \\
&= 100(0.0334 \times 0.9 + 0.00299 \times 0.435) \\
&= 3.136\text{ V}
\end{aligned}
$$

or, in per unit volts,

$$
\frac{3.136\text{ V}}{120\text{ V}} = 0.0261\text{ puV}
$$

which is very close to the value given in Table 7.2 for the $\tilde{K}$ constant, that is, 0.02613 pu V/(10^4 A ·ft) of cable.

Example 7.7. Use the information and data given in Examples 6.1 and 7.3. Assume a URD system. Therefore, the SLs shown in Figure 6.12 are made of UG secondary cables. Assume 12 services per distribution transformer and two transformers per block that are at the locations of poles 2 and 5, as shown in Figure 6.12. Service pedestals are at the locations of poles 1, 3, 4, and 6. Assume that the selected equipment sizes (for S_T, A_{SL}, A_{SD}) are of the nearest standard size that are larger than the theoretically most economical sizes and determine the following:

a. Find the steady-state voltage drop in per units at the most remote consumer's meter for the annual maximum system loads given in Table 7.3.
b. Find the VDIP in per units for motor starting at the most unfavorable location.
c. If the voltage-drop and/or VDIP criteria are not met, select larger equipment and find a design that will meet these arbitrary criteria. Do not, however, immediately select the largest sizes of ST, ASL, and ASD equipment and call that a worthwhile design. In addition, contemplate the data and results and attempt to be wise in selecting ASL or ASD (or both) for enlarging to meet the voltage criteria.

Solution. a. Due to the diversity factors involved, the load values given in Table 7.3 are different for SDs, SLs, and transformers. For example, the load on the transformer is selected as

$$\begin{aligned}\text{transformer load} &= 10 + 11 \times 4.4\\ &= 58.4\ \text{kVA}\end{aligned}$$

Therefore, selecting a 50 kVA transformer,

$$\begin{aligned}I &= \frac{58.4\ \text{kVA}/240\ \text{V}}{S_T/240\ \text{V}}\\ &= \frac{58.4\ \text{kVA}}{50\ \text{kVA}}\\ &= 1.168\ \text{pu A}\end{aligned}$$

Thus, the per unit voltage drop in the transformer is

$$\begin{aligned}\text{VD}_T &= I(R\cos\theta + X\sin\theta)\\ &= (1.168\ \text{pu A})(0.0107 \times 0.9 + 0.0139 \times 0.435)\\ &= 0.0183\ \text{pu V}\end{aligned}$$

As shown in Figure 7.10, the load on each SL (that portion of the wiring between the transformer and the service pedestal) is similarly calculated as

$$\begin{aligned}\text{SL load} &= 10 + 3 \times 6\\ &= 28\ \text{kVA}\end{aligned}$$

or 116.7 A. If the SL is selected to be #4/0 AWG with the $\tilde{K}$ constant of 0.0088 from Table 7.2, the per unit voltage drop in each SL is

$$\begin{aligned}VD_{SL} &= \tilde{K}\left(\frac{I \times l}{10^4}\right)\\ &= 0.0088\left(\frac{116.7 \times 150\ \text{ft}}{10^4}\right)\\ &= 0.01554\ \text{pu V}\end{aligned}$$

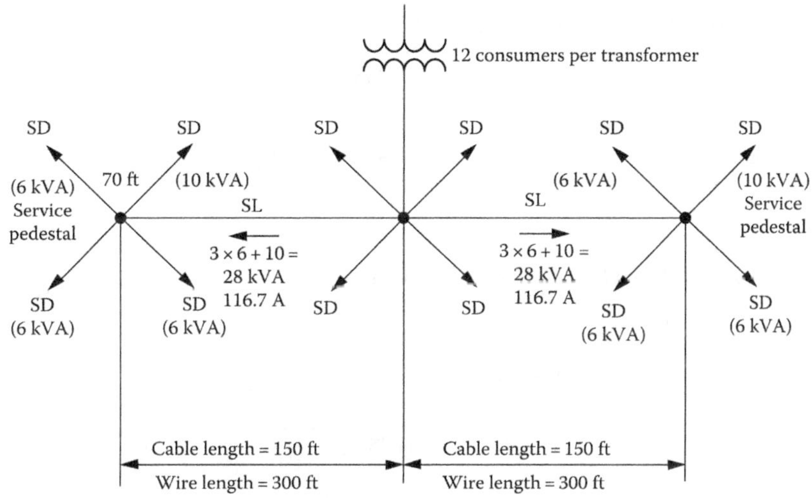

Figure 7.10 Circulation of the secondary-line currents, i.e., a zoom out from a distribution transformer (DT).

The load on each SD is given to be 10 kVA or 41.6 A from Table 7.3. If each SD of 70 ft length is selected to be #1/0 AWG with the $\tilde{K}$ constant of 0.01683 from Table 7.2, the per unit voltage drop in each SD is

$$\begin{aligned} VD_{SD} &= \tilde{K}\left(\frac{I \times l}{10^4}\right) \\ &= 0.01683\left(\frac{41.6\text{ A} \times 70\text{ ft}}{10^4}\right) \\ &= 0.0049 \text{ pu V} \end{aligned}$$

Therefore, the total steady-state voltage drop in per units at the most remote consumer's meter is

$$\begin{aligned} \sum VD &= VD_T + VD_{SL} + VD_{SD} \\ &= 0.0183 + 0.01554 + 0.0049 \\ &= 0.0388 \text{ pu V} \end{aligned}$$

which exceeds the given criterion of 0.035 pu V.

b. To find the VDIP in per units for motor starting at the most unfavorable location, the given starting current of 80 A can be converted to a kVA load of 19.2 kVA (80 A × 240 V). Therefore, the per unit VDIP in the 50 kVA transformer is

$$\begin{aligned} VDIP_T &= (R\cos\theta + X\sin\theta)\left(\frac{19.2\text{ kVA}}{50\text{ kVA}}\right) \\ &= (0.0107 \times 0.5 + 0.0139 \times 0.866)\left(\frac{19.2\text{ kVA}}{50\text{ kVA}}\right) \\ &= 0.00668 \text{ pu V} \end{aligned}$$

The per unit VDIP in the SL of #4/0 AWG cable is

$$\begin{aligned} VDIP_{\text{SL}} &= \tilde{K}\left(\frac{80\ \text{A} \times 150\ \text{ft}}{10^4}\right) \\ &= 0.00636\left(\frac{80 \times 150}{10^4}\right) \\ &= 0.00763\ \text{pu V} \end{aligned}$$

The per unit VDIP in the SD of #1/0 AWG cable is

$$\begin{aligned} VDIP_{\text{SD}} &= \tilde{K}\left(\frac{80\ \text{A} \times 70\ \text{ft}}{10^4}\right) \\ &= 0.01089\left(\frac{80 \times 70}{10^4}\right) \\ &= 0.0061 \text{pu V} \end{aligned}$$

Therefore, the total VDIP in per units due to motor starting at the most unfavorable location is

$$\begin{aligned} \sum \text{VDIP} &= \text{VDIP}_T + \text{VDIP}_{\text{SL}} + \text{VDIP}_{\text{SD}} \\ &= 0.00668 + 0.00763 + 0.0061 \\ &= 0.024\ \text{pu V} \end{aligned}$$

which meets the given criterion of 0.03 pu V.

c. Since in part (a) the voltage-drop criterion has not been met, select the SL cable size to be one size larger than the previous #4/0 AWG size, that is, 250 kcmil, keeping the size of the transformer the same. Therefore, the new per unit voltage drop in the SL becomes

$$\begin{aligned} VD_{\text{SL}} &= 0.00769\left(\frac{116.7\ \text{A} \times 150\ \text{ft}}{10^4}\right) \\ &= 0.01347\ \text{puV} \end{aligned}$$

Also, selecting one-size-larger cable, that is, #2/0 AWG, for the SD, the new per unit voltage drop in the SD becomes

$$\begin{aligned} VD_{\text{SD}} &= 0.0136\left(\frac{41.6\ \text{A} \times 70\ \text{ft}}{10^4}\right) \\ &= 0.00396\ \text{puV} \end{aligned}$$

Therefore, the new total steady-state voltage drop in per units at the most remote consumer's meter is

$$\begin{aligned} \text{VD} &= \text{VD}_T + \text{VD}_{\text{SL}} + \text{VD}_{\text{SD}} \\ &= 0.0183 + 0.01347 + 0.00396 \\ &= 0.03573 \text{pu V} \end{aligned}$$

which is still larger than the criterion. Thus, select 350 kcmil cable size for the SLs and #2/0 AWG cable size for the SDs to meet the criteria.

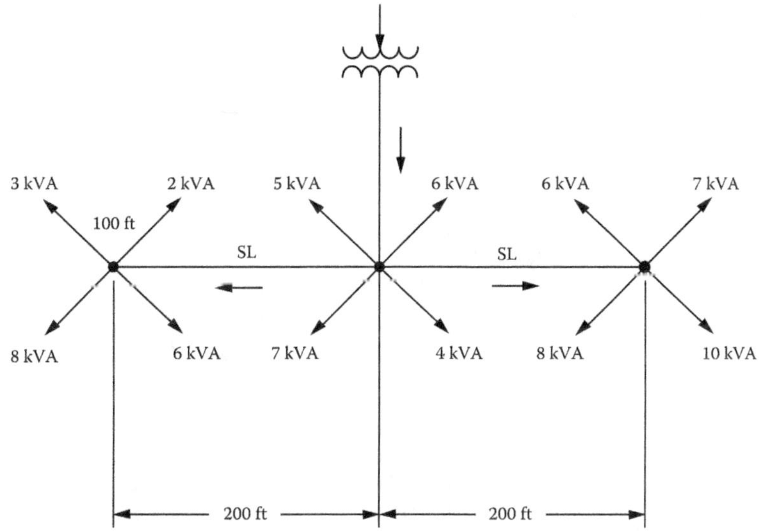

Figure 7.11 A residential secondary distribution system.

Example 7.8. Figure 7.11 shows a residential secondary distribution system. Assume that the distribution transformer capacity is 75 kVA (use Table 7.1), all secondaries and services are single-phase three wire, nominally 120/240 V, and all SLs are of#2/0 A1/Cu XLPE cable, and SDs are of #1/0 Al/Cu XLPE cable (use Table 7.2). All SDs are 100 ft long, and all SLs are 200 ft long. Assume an average lagging-load power factor of 0.9 and 100% load diversity factors and determine the following:

a. Find the total load on the transformer in kVAs and in per units.
b. Find the total steady-state voltage drop in per units at the most remote and severe customer's meter for the given annual maximum system loads.

Solution. a. Assuming a diversity factor of 100%, the total load on the transformer is

$$\begin{aligned} S_T &= (3+2+8+6)+(5+6+7+4)+(6+7+8+10) \\ &= 19+22+31 \\ &= 72 \text{ kVA} \end{aligned}$$

or, in per units,

$$\begin{aligned} I &= \frac{S_r}{S_B} \\ &= \frac{72 \text{ kVA}}{75 \text{ kVA}} \\ &= 0.96 \text{ puA} \end{aligned}$$

b. To find the total voltage drop in per units at the most remote and severe customer's meter, calculate the per unit voltage drops in the transformer, the service line, and the SD of the most remote and severe customer. Therefore,

$$\begin{aligned}VD_T &= I(R\cos\theta + X\sin\theta)\\&= 0.96(0.0101\times 0.90 + 0.0143\times 0.4359)\\&= 0.0147 \text{ pu V}\end{aligned}$$

$$\begin{aligned}VD_{SL} &= \tilde{K}\left(\frac{I\times l}{10^4}\right)\\&= 0.0136\left(\frac{129.17\text{ A}\times 200\text{ ft}}{10^4}\right)\\&= 0.03513 \text{ pu V}\end{aligned}$$

$$\begin{aligned}VD_{SD} &= \tilde{K}\left(\frac{I\times l}{10^4}\right)\\&= 0.01683\left(\frac{41.67\text{ A}\times 100\text{ ft}}{10^4}\right)\\&= 0.0070 \text{ pu V}\end{aligned}$$

Therefore, the total voltage drop is

$$\begin{aligned}\sum \text{VD} &= \text{VD}_T + \text{VD}_{\text{SL}} + \text{VD}_{\text{SD}}\\&= 0.0147 + 0.03513 + 0.0070\\&= 0.0568 \text{ pu V}\end{aligned}$$

Example 7.9. Figure 7.12 shows a three-phase four-wire grounded-wye distribution system with multigrounded neutral, supplied by an express feeder and mains. In the figure, d and s are the width and length of a primary lateral, where s is much larger than d. Main lengths are equal to cb =ce= s/2. The number of the primary laterals can be found as s/d. The square-shaped service area (s^2) has a uniformly distributed load density, and all loads are presumed to have the same lagging power factor. Each primary lateral, such as ba, serves an area of length s and width d. Assume that

D = uniformly distributed load density, kVA/(unit length)^2
V_{L-L} = nominal operating voltage, which is also the base voltage, line-to-line kV
$r_m + jx_m$ = impedance of three-phase express and mains, Ω/(phase · unit length)
$r_l + jx_l$ = impedance of a three-phase lateral line, Ω/(phase · unit length)

Use the given information and data and determine the following:

1. Assume that the laterals are in three phase and find the per unit voltage-drop expressions for
   a. The express feeder f_c, that is, $\text{VD}_{pu,fc}$
   b. The main cb, that is, $\text{VD}_{pu,cb}$
   c. The primary lateral ba, that is, $\text{VD}_{pu,ba}$

   Note that the equations to be developed should contain the constants D, s, d, impedances, θ, load power-factor angle, V_{L-L}, etc., but not current variable l.
2. Change all the laterals from the three-phase four-wire system to an open-wye system so that investment costs will be reduced, but three-phase secondary service can still be rendered where needed. Assume that the phasing connections of the many laterals are well balanced on the mains. Use Morrison's approximations and modify the equations derived in part (1).

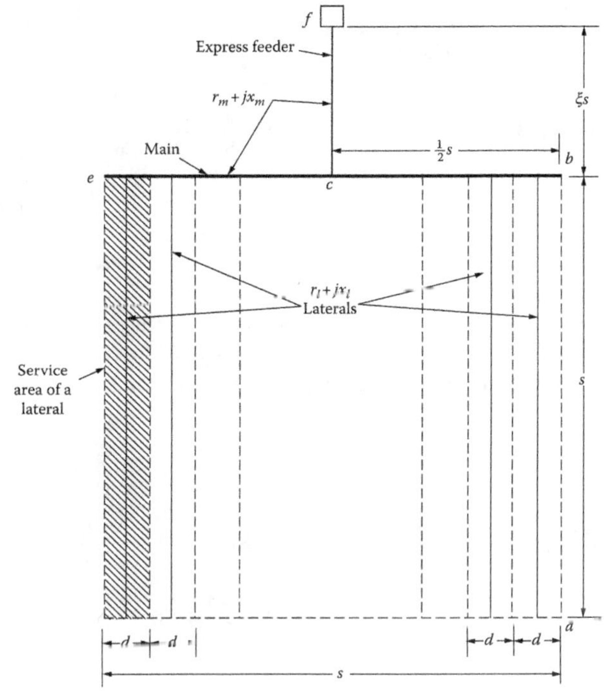

Figure 7.12 The distribution system of Example 7.9.

Solution. 1.

$$\text{Total kVA load served} = D \times s^2 \text{ kVA} \tag{7.53}$$

$$\text{Current point } f = \frac{D \times S^2}{\sqrt{3} \times V_{L-L}} \tag{7.54}$$

$$\text{Voltage drop} = I \times z \times I_{eff} \tag{7.55}$$

Therefore,

a.

$$\begin{aligned} VD_{\text{pu},fc} &= \frac{D \times S^2}{\sqrt{3} \times V_{L-L}}(r_m \cos\theta + X_m \sin\theta)\frac{\sqrt{3}}{1000 \times V_{L-L}}(\zeta \times s) \\ &= \frac{\zeta \times D \times S^3}{1000 \times V_{L-L}^2}(r_m \cos\theta + X_m \sin\theta) \end{aligned} \tag{7.56}$$

b.

$$\begin{aligned} VD_{\text{pu},cb} &= \frac{\frac{1}{2}D \times S^2}{\sqrt{3} \times V_{L-L}}(r_m \cos\theta + X_m \sin\theta)\frac{\sqrt{3}}{1000 \times V_{L-L}}\left(\frac{1}{4} \times S\right) \\ &= \frac{\times D \times S^3}{8000 \times V_{L-L}^2}(r_m \cos\theta + X_m \sin\theta) \end{aligned} \tag{7.57}$$

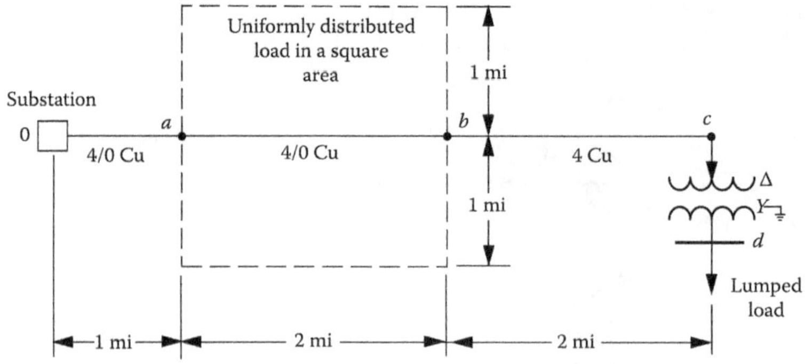

Figure 7.13 A square-shaped service area and a lumped-sum load.

c.

$$\begin{aligned} VD_{\text{pu},ba} &= \frac{D(d \times S)}{\sqrt{3} \times V_{L-L}}(r_l \cos\theta + X_l \sin\theta)\frac{\sqrt{3}}{1000 \times V_{L-L}}\left(\frac{1}{2} \times S\right) \\ &= \frac{D \times d \times S^2}{2000 \times V_{L-L}^2}(r_l \cos\theta + X_l \sin\theta) \end{aligned} \tag{7.58}$$

There would not be any change in the equations given in part (1).

Example 7.10. Figure 7.13 shows a square-shaped service area (A = 4 mi^2) with a uniformly distributed load density of D kVA/mi^2 and 2 mi of #4/0 AWG copper overhead main from *a* to *b*. There are many closely spaced primary laterals that are not shown in the square-shaped service area of the figure. In this voltage-drop study, use the precalculated voltage-drop curves of Figure 4.17 when applicable. Use the nominal primary voltage of 7,620/13,200 V for a three-phase four-wire wye-grounded system. Assume that at peak loading, the load density is 1000 kVA/mi2 and the lumped load is 2000 kVA, and that at off-peak loading, the load density is 333 kVA/mi^2 and the lumped load is still 2000 kVA. The lumped load is of a small industrial plant working three shifts a day. The substation bus voltages are 1.025 pu V of 7620 base volts at peak load and 1.000 pu V during off-peak load.

The transformer located between buses *c* and *d* has a three-phase rating of 2000 kVA and a delta-rated high voltage of 13,200 V and grounded-wye-rated low voltage of 277/480 V. It has 0 + j0.05 per unit impedance based on the transformer ratings. It is tapped up to raise the low voltage 5.0% relative to the high voltage, that is, the equivalent turns ratio in use is (7620/277) × 0.95. Use the given information and data for peak loading and determine the following:

a. TThe percent voltage drop from the substation to point *a*, from *a* to *b*, from *b* to *c*, and from *c* to *d* on the main
b. The per unit voltages at the points *a*, *b*, *c*, and *d* on the main
c. The line-to-neutral voltages at the points *a*, *b*, *c*, and *d*

Solution. a. The load connected in the square-shaped service area is

$$\begin{aligned} S_n &= D \times A_n \\ &= 1000 \times 4 \\ &= 4000 \text{ kVA} \end{aligned}$$

Thus the total kVA load on the main is

$$S_m = 4000 + 2000$$
$$= 6000 \text{ kVA}$$

From Figure 4.17, for #4/0 copper, the Kconstant is found to be 0.0003. Therefore, the percent voltage drop from the substation to point a is

$$\begin{aligned}\text{VD}_{\text{oa}} &= K \times S_m \times 1\\ &= 0.0003 \times 6000 \times 1\\ &= 148\%\text{V or } 0.018 \text{ pu V}\end{aligned}$$

The percent voltage drop from point a to point b is

$$\begin{aligned}\text{VD}_{\text{ab}} &= K \times S_n \times 1 + K \times S_{lump} \times 1\\ &= 0.0003 \times 4000 \times 1 + 0.0003 \times 2000 \times 2\\ &= 2.4\%\text{V or } 0.024 \text{ pu V}\end{aligned}$$

The percent voltage drop from point b to point c is

$$\begin{aligned}\text{VD}_{\text{ab}} &= K \times S_{lump} \times 1\\ &= 0.0009 \times 2000 \times 2\\ &= 3.6\%\text{V or } 0.036 \text{ pu V}\end{aligned}$$

To find the percent voltage drop from point c to bus d,

$$\begin{aligned}I &= \frac{2000\text{kVA}}{\sqrt{3} \times V_{L-L} \text{ at point C}}\\ &= \frac{2000\text{kVA}}{\sqrt{3} \times (0.947 \times 13.2 \text{ kV}}\\ &= 92.373\text{A}\end{aligned}$$

$$\begin{aligned}I_B &= \frac{2000\text{kVA}}{\sqrt{3} \times 13.2 \text{ kV}}\\ &= 92.373 \text{ A}\end{aligned}$$

$$\begin{aligned}I_{\text{pu}} &= \frac{I}{I_B}\\ &= 1.056 \text{ pu A}\end{aligned}$$

Note that usually in a simple problem like this, the reduced voltage at point c is ignored. In that case, for example, the per unit current would be 1.0 pu A rather than 1.056 pu A. Since

$$Z_{T,pu} = 0 + j0.05 \text{ pu } \Omega$$

and

$$\cos\theta = 0.9 \quad \text{or} \quad \theta = 25.84^\circ \text{ lagging}$$

therefore

$$I_{pu} = 1.056\angle 25.84^\circ \text{ pu A}$$

Thus, to find the percent voltage drop at bus d, first it can be found in per unit as

$$VD_{cd} = \frac{I(R\cos\theta + X\sin\theta)}{V_B} \text{pu V}$$

but since the low voltage has been tapped up 5%,

$$VD_{cd} = \frac{I(R\cos\theta + X\sin\theta)}{V_B} - 0.05\text{pu V}$$

Therefore,

$$\begin{aligned} VD_{cd} &= \frac{1.056(0\times 0.9 + 0.05\times 0.4359)}{1.0} - 0.05\text{pu V} \\ &= -0.0267 \text{ pu V} \end{aligned}$$

or

$$\%\text{VD}_{cd} = -2.67\% \text{ V}$$

Here, the negative sign of the voltage drop indicates that it is in fact a voltage rise rather than a voltage drop.

b. The per unit voltages at the points *a*, *b*, *c*, and *d* on the main are

$$\begin{aligned} V_a &= V_0 - V_{0a} \\ &= 1.025 - 0.018 \\ &= 1.007 \text{ pu V or } 100.7\% \text{ V} \end{aligned}$$

$$\begin{aligned} V_b &= V_a - V_{ab} \\ &= 1.007 - 0.024 \\ &= 0.983 \text{ pu V or } 98.3\% \text{ V} \end{aligned}$$

$$\begin{aligned} V_c &= V_b - V_{bc} \\ &= 0.983 - 0.036 \\ &= 0.947 \text{ pu V or } 94.7\% \text{ V} \end{aligned}$$

$$\begin{aligned} V_d &= V_c - V_{cd} \\ &= 0.947 - (-0.0267) \\ &= 0.9737 \text{ pu V or } 97.37\% \text{ V} \end{aligned}$$

c. The line-to-neutral voltages are

$$V_a = 7620 \times 1.007$$
$$= 7673.3 \text{ V}$$

$$V_b = 7620 \times 0.983$$
$$= 7490.5 \text{ V}$$

$$V_a = 7620 \times 0.947$$
$$= 7216.1 \text{ V}$$

$$V_a = 277 \times 0.9737$$
$$= 269.7 \text{ V}$$

Example 7.11. Use the relevant information and data given in Example 7.10 for off-peak loading and repeat Example 7.10, and find the V_d voltage at bus d in line-to-neutral volts. Also write the necessary codes to solve the problem in MATLAB.

Solution. a. At off-peak loading, the load connected in the square-shaped service area is

$$S_n = D \times A_n$$
$$= 333 \times 4$$
$$= 1332 \text{ kVA}$$

Thus the total kVA load on the main is

$$S_m = 1332 + 2000$$
$$= 3332 \text{ kVA}$$

Therefore, the percent voltage drop from the substation to point a is

$$\%\text{VD}_{\text{oa}} = K \times S_m \times 1$$
$$= 0.0003 \times 3332 \times 1$$
$$= 1.0\% \text{ V or } 0.01 \text{ pu V}$$

The percent voltage drop from point a to point b is

$$\%\text{VD}_{ab} = K \times S_n \times 1/2 + K \times S_{lump} \times 1$$
$$= 0.0003 \times 1332 \times 1 + 0.0003 \times 2000 \times 2$$
$$= 1.6\% \text{ V or } 0.016 \text{ pu V}$$

The percent voltage drop from point b to point c is

$$\%\text{VD}_{bc} = K \times S_{\text{lump}} \times I$$
$$= 0.0009 \times 2000 \times 2$$
$$= 3.6\% \text{ V or } 0.036 \text{ pu V}$$

To find the percent voltage drop from point *c* to bus *d*, the percent voltage drop at bus *d* can be found as before

$$\%\mathrm{VD}_{cd}0.0267 \text{ pu V or } -2.67\%\mathrm{V}$$

b. The per unit voltages at points *a*, *b*, *c*, and *d* on the main are

$$\begin{aligned} V_a &= V_0 - V_{0a} \\ &= 1.0 - 0.01 \\ &= 0.99 \text{ pu V or } 99\% \text{ V} \end{aligned}$$

$$\begin{aligned} V_b &= V_a - V_{ab} \\ &= 0.99 - 0.016 \\ &= 0.974 \text{ pu V or } 97.4\% \text{ V} \end{aligned}$$

$$\begin{aligned} V_c &= V_b - V_{bc} \\ &= 0.974 - 0.036 \\ &= 0.938 \text{ pu V or } 93.8\% \text{ V} \end{aligned}$$

$$\begin{aligned} V_d &= V_c - V_{cd} \\ &= 0.938 - (-0.0267) \\ &= 0.9647 \text{ pu V or } 96.47\% \text{ V} \end{aligned}$$

c. The line-to-neutral voltages are

$$\begin{aligned} V_a &= 7620 \times 0.99 \\ &= 7543.8 \text{ V} \end{aligned}$$

$$\begin{aligned} V_b &= 7620 \times 0.974 \\ &= 7421.9 \text{ V} \end{aligned}$$

$$\begin{aligned} V_a &= 7620 \times 0.938 \\ &= 7147.6 \text{ V} \end{aligned}$$

$$\begin{aligned} V_a &= 277 \times 0.9647 \\ &= 267.2 \text{ V} \end{aligned}$$

Note that the voltages at bus d during peak and off-peak loading are nearly the same.

Here is the MATLAB script:

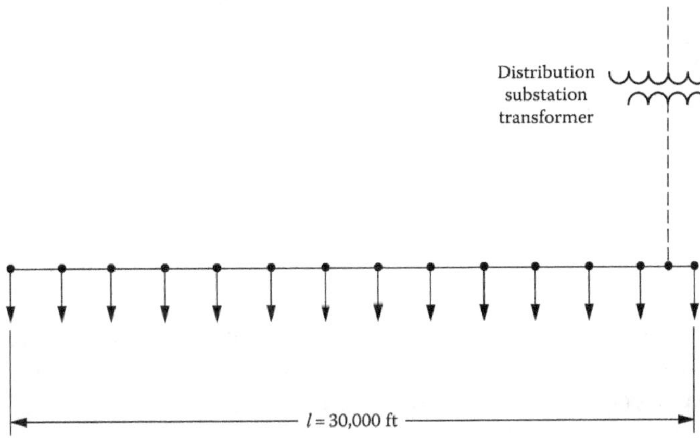

Figure 7.14 The distribution system of Example 7.12.

Example 7.12. Figure 7.14 shows that a large number of small loads are closely spaced along the length l. If the loads are single phase, they are assumed to be well balanced among the three phases.

A three-phase four-wire wye-grounded 7.62/13.2 kV primary line is to be built along the length l and fed through a distribution substation transformer from a high-voltage transmission line. Assume that the uniform (or linear) distribution of the connected load along the length l is

$$\frac{S_{\text{connected load}}}{l} = 0.45 \text{ kVA/ft}$$

The 30-min annual demand factor of all loads is 0.60, the diversity factor (F_D) among all loads is 1.08, and the annual loss factor (F_{LS}) is 0.20. Assuming a lagging power factor of 0.9 for all loads and a 37 in. geometric mean spacing of phase conductors, use Figure 4.17 for voltage-drop calculations for copper conductors. Use the relevant tables in Appendix A for additional data about copper and ACSR conductors and determine the following:

Locate the distribution substation where you think it would be the most economical, considering only the 13.2 kV system, and then find

a. The minimum ampacity-sized copper and ACSR phase conductors
b. The percent voltage drop at the location having the lowest voltage at the time of the annual peak load, using the ampacity-sized copper conductor found in part (a)

Solution. To achieve the minimum voltage drop, the substation should be located at the middle of the line 1, and therefore

a. From Equation 2.13, the diversified maximum demand of the group of the load is

$$\begin{aligned} D_g &= \frac{\sum_{i-1}^{n}}{\text{TCD}_i \times \text{DF}_i} \\ &= \frac{0.45 \text{ kVA/ft} \times 0.60}{1.08} \\ &= 0.250 \text{ kVA/ft} \end{aligned}$$

Thus the peak load of each main on the substation transformer is

$$\begin{aligned} S_{\text{PK}} &= 0.250 \text{ kVA/ft} \times 15{,}000 \text{ ft} \\ &= 3750 \text{ kVA} \end{aligned}$$

or

$$3750\ \text{kVA} = \sqrt{3} \times 13.2\ \text{kV} \times I$$

hence

$$I = \frac{3750\ \text{kVA}}{\sqrt{3} \times 13.2\ \text{kV}}$$
$$= 164.2\ \text{A}$$

in each main out of the substation. Therefore, from the tables of Appendix A, it can be recommended that either #4 AWG copper conductor or #2 AWG ACSR conductor be used.

b. Assuming that #4 AWG copper conductor is used, the percent voltage drop at the time of the annual peak load is

$$\%\text{VD} = [K\%\text{VD}/(\text{kVA} \times \text{mi})] \times [\text{S}_{\text{PK}}kVA] \times \frac{I\ \text{ft}}{2}\frac{1}{5280\ \text{ft/mi}}$$
$$= 0.0009 \times 3750 \times \frac{15{,}000}{2 \times 5820}$$
$$= 5.3\%V$$

Example 7.13. Now suppose that the line in Example 7.12 is arbitrarily constructed with #4/0 AWG ACSR phase conductor and that the substation remains where you put it in part (a). Assume 50°C conductor temperature and find the total annual I^2R energy loss (TAEL_{Cu}), in kWh, in the entire line length.

Solution. The total I^2R loss in the entire line length is

$$\sum I^2R = 3I^2\left(r \times \frac{I}{2}\right)$$
$$= 3(164.2)^2(0.592\Omega/\text{mi})\frac{30{,}000\ \text{ft}}{2 \times 5280\ \text{ft/mi}}$$
$$= 136{,}033.729\ \text{W}$$

Therefore, the total I^2R energy loss is

$$\text{TAEL}_{\text{Cu}} = \left[\left(\sum I^2R\right)F_{LS}\right](8760\ \text{h/year})$$
$$= \frac{136{,}033.729}{10^3} \times 0.20 \times 8760$$
$$= 238{,}331.09\ \text{kWh}$$

Example 7.14. Figure 7.15 shows a single-line diagram of a simple three-phase four-wire wye-grounded primary feeder. The nominal operating voltage and the base voltage are given as 7,200/12,470 V. Assume that all loads are balanced three phase and all have 90% power factor, lagging. The given values of the constant K in Table 7.4 are based on 7,200/12,470 V. There is a total of a 3000 kVA uniformly distributed load over a 4 mi line between b and c. Use the given data and determine the following:

a. Find the total percent voltage drop at points a, b, c, and d.
b. If the substation bus voltages are regulated to 7,300/12,650 V, what are the line-to-neutral and line-to-line voltages at point a?

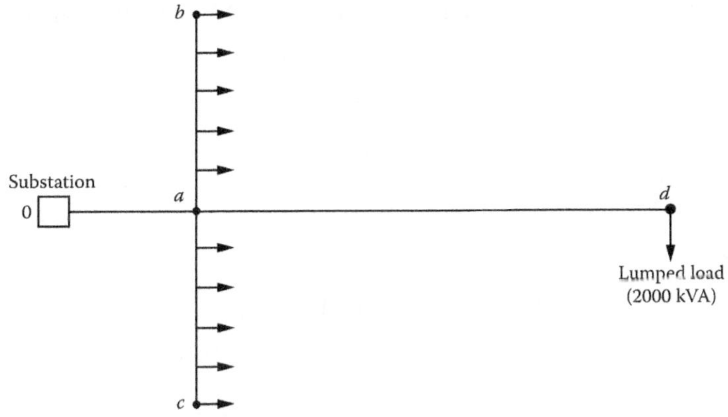

Figure 7.15 The distribution system of Example 7.14.

Table 7.4
K Constants

Run	Conductor Type	Distance (mi)	K, % VD/(kVA · mi)
Sub. to a	#4/0 ACSR	1	0.0005
a to b	# 1 ACSR	2	0.001
a to c	#1 ACSR	2	0.001
a to d	#1 ACSR	2	0.001

Solution. a. The total load flowing through the line between points 0 and *a* is

$$\sum S = 2000\ \text{kVA} + 3000\ \text{kVA} = 5000\ \text{kVA}$$

therefore the percent voltage drop at point a is

$$\begin{aligned}\%VD_a &= K\left(\sum S\right) l \\ &= 0.0005 \times 5000 \times 1.0 \\ &= 2.5\%\ \text{V}\end{aligned}$$

Similarly, the load flowing through the line between points *a* and *b* is

$$S = 1500\ \text{kVA}$$

Therefore,

$$\begin{aligned}\%VD_b &= K \times S \times l/2 + \%\text{VD}_a \\ &= 0.0010 \times 1500 \times 1 + 2.5\% \\ &= 4\%\ \text{V}\end{aligned}$$

$$\%\text{VD}_\text{c} = \%\text{VD}_\text{b} = 4\%\ \text{V}$$

$$\begin{aligned}\%VD_d &= K \times S_\text{lump} \times l + \%\text{VD}_a \\ &= 0.0010 \times 2000 \times 2 + 2.5\% \\ &= 6.5\%\ \text{V}\end{aligned}$$

b. If the substation bus voltages are regulated to 7,300/12,650 V at point a, the line-to-neutral voltage is

$$
\begin{aligned}
V_{a,L-N} &= 7300 - VD_{a,L-N} \\
&= 7300 - 7300 \times 0.025 \\
&= 7117.5 \text{ V}
\end{aligned}
$$

and the line-to-line voltage is

$$
\begin{aligned}
V_{a,L-L} &= 12,650 - VD_{a,L-L} \\
&= 12,650 - 12,650 \times 0.025 \\
&= 12,333.8 \text{ V}
\end{aligned}
$$

7.5 PERCENT POWER (OR COPPER) LOSS

The percent power (or conductor) loss of a circuit can be expressed as

$$
\begin{aligned}
\%I^2R &= \frac{P_{LS}}{P_r} \times 100 \\
&= \frac{I^2R}{P_r} \times 100
\end{aligned}
\tag{7.59}
$$

where P_{LS} is the power loss of a circuit, kW = I^2R
P_r is the power delivered by the circuit, kW

The conductor I^2R losses at a load factor of 0.6 can readily be found from Table 7.5 for various voltage levels.

At times, in AC circuits, the ratio of percent power, or conductor, loss to percent voltage regulation can be used, and it is given by the following approximate expression:

$$
\frac{\%I^2R}{\%VD} = \frac{\cos\phi}{\cos\theta \times \cos(\phi - \theta)} \tag{7.60}
$$

where
$\%I^2R$ is the percent power loss of a circuit
%VD is the percent voltage drop of the circuit
ϕ is the impedance angle = $\tan^{-1}(X/R)$
θ is the power-factor angle

7.6 METHOD TO ANALYZE DISTRIBUTION COSTS

To make any meaningful feeder-size selection, the distribution engineer should make a cost study associated with feeders in addition to the voltage-drop and power-loss considerations. The cost analysis for each feeder size should include (1) investment cost of the installed feeder, (2) cost of energy lost due to I^2R losses in the feeder conductors, and (3) cost of demand lost, that is, the cost of useful system capacity lost (including generation, transmission, and distribution systems), in order to maintain adequate system capacity to supply the I^2R losses in the distribution feeder conductors [172]. Therefore, the total annual feeder cost of a given size feeder can be expressed as

$$
\text{TAC} = \text{AIC} + \text{AEC} + \text{ADC\$/mi} \tag{7.61}
$$

Table 7.5

Conductor I^2R Losses, kWh/(mi year), at 7.2/12.5 kV and a Load Factor of 0.6

Annual Peak	Single Phase				"V" Phase				Three Phase					
		6	4	2		6	4	2	6	4	2	1	1/0	2/0
Load (kW)	8 Copper	Copper 4 ACSR	Copper 2 ACSR	Copper 1/0 ACSR	8 Copper	Copper 4 ACSR	Copper 2 ACSR	Copper 1/0 ACSR	Copper 4 ACSR	Copper 2 ACSR	Copper 1/0 ACSR	Copper 2/0 ACSR	Copper 3/0 ACSR	Copper 4/0 ACSR
20	124	82	55	37	62	41	27	19	25	16	10			
40	495	329	218	149	248	164	109	75	99	63	39	31		
60	1,110	740	491	335	557	370	246	168	224	141	88	70	56	
80	1,980	1,320	873	596	990	658	437	298	398	250	157	125	99	78
100	3,100	2,060	1,360	932	1,550	1,030	682	466	621	391	245	195	154	122
120	4,460	2,960	1,960	1,340	2,230	1,480	982	671	895	563	353	280	222	176
140	6,070	4,030	2,670	1,830	3,030	2,010	1,340	913	1,220	766	481	382	302	240
160	7,920	5,260	3,490	2,390	3,960	2,630	1,750	1,190	1,590	1,000	628	498	395	313
180	10,000	6,660	4,420	3,020	5,010	3,330	2,210	1,510	2,010	1,270	795	631	500	396
200	12,400	8,220	5,460	3,730	6,190	4,110	2,730	1,860	2,490	1,560	982	779	617	489
225	15,700	10,400	6,910	4,720	7,830	5,200	3,450	2,360	3,150	1,980	1240	986	780	619
250	19,300	12,800	8,530	5,820	9,670	6,420	4,260	2,910	3,880	2,440	1530	1220	964	764
275	23,400	15,500	10,300	7,050	11,700	7,770	5,160	3,520	4,700	2,960	1860	1470	1170	925
300		18,500	12,300	8,390	13,900	9,250	6,140	4,190	5,590	3,520	2210	1750	1390	1100
325		21,700	14,400	9,840	16,300	10,900	7,210	4,920	6,560	4,130	2590	2060	1630	1280
350			16,700	11,400	18,900	12,600	8,360	5,710	7,610	4,790	3010	2380	1890	1500
375			19,200	13,100	21,800	14,400	9,590	6,550	8,740	5,500	3450	2740	2170	1720
400			21,800	14,900	24,800	16,400	10,900	7,450	9,940	6,260	3930	3120	2470	1960
450				18,900		20,800	13,800	9,430	12,600	7,920	4970	3940	3120	2480
500				23,300		25,700	17,100	11,600	15,500	9,780	6140	4870	3850	3060
550							20,600	14,100	18,800	11,800	7420	5890	4660	3700
600							24,600	16,800	22,400	14,100	8840	7010	5550	4400

Source: Rural Electrification Administration, *U.S. Department of Agriculture: Economic Design of Primary Lines for Rural Distribution Systems*, REA Bulletin, 60, 1960.

Note: This table is calculated for a PF of 90%. To adjust for a different PF, multiply these values by the factor of $k = (90)^2/(PF)^2$. For 7.62/13.2 kV, multiply these values by 0.893; for 14.4/24.9 kV, multiply by 0.25.

where
TAC is the total annual equivalent cost of feeder, \$/mi
AIC is the annual equivalent of investment cost of installed feeder, \$/mi
AEC is the annual equivalent of energy cost due to I^2R losses in feeder conductors, \$/mi
ADC is the annual equivalent of demand cost incurred to maintain adequate system capacity to supply $I2R$ losses in feeder conductors, \$/mi

7.6.1 ANNUAL EQUIVALENT OF INVESTMENT COSTS

The annual equivalent of investment cost of a given size feeder can be expressed as

$$\mathrm{AIC} = \mathrm{IC_F} \times i_F \$/\mathrm{mi} \tag{7.62}$$

where
AIC is the annual equivalent of investment cost of a given size feeder, \$/mi
IC_F is the cost of installed feeder, \$/mi
i_F is the annual fixed charge rate applicable to feeder

The general utility practice is to include cost of capital, depreciation, taxes, insurance, and operation and maintenance (O&M) expenses in the *annual fixed charge rate* or so-called *carrying charge rate.* It is given as a decimal [173].

7.6.2 ANNUAL EQUIVALENT OF ENERGY COST

The annual equivalent of energy cost due to I^2R losses in feeder conductors can be expressed as

$$\mathrm{AEC} = 3I^2R \times \mathrm{EC} \times \mathrm{F_{LL}} \times F_{\mathrm{LSA}} \times 8760 \ \$/\mathrm{mi} \tag{7.63}$$

where
AEC is the annual equivalent of energy cost due to I^2R losses in feeder conductors, \$/mi
EC is the cost of energy, \$/kWh
$\mathrm{F_{LL}}$ is the load-location factor
$\mathrm{F_{LS}}$ is the loss factor
$\mathrm{F_{LSA}}$ is the loss-allowance factor
The load-location factor of a feeder with uniformly distributed load can be defined as

$$F_{LL} = \frac{S}{l} \tag{7.64}$$

where
$\mathrm{F_{LL}}$ is the load-location factor in decimal
s is the distance of point on feeder where total feeder load can be assumed to be concentrated for the purpose of calculating
l^2R losses l is the total feeder length, mi

The loss factor can be defined as the ratio of the average annual power loss to the peak annual power loss and can be found approximately for urban areas from

$$F_{\mathrm{LS}} = 0.3F_{\mathrm{LD}} + 0.7F_{\mathrm{LD}}^2 \tag{7.65}$$

and for rural areas [174],

$$F_{\mathrm{LS}} = 0.16F_{\mathrm{LD}} + 0.84F_{\mathrm{LD}}^2$$

The loss-allowance factor is a factor used to allocate additional losses that occur in the overall power system during the transmission of power from the generating plant to the distribution substation.

7.6.3 ANNUAL EQUIVALENT OF DEMAND COST

The annual equivalent of demand cost incurred to maintain adequate system capacity to supply the I^2R losses in the feeder conductors can be expressed as

$$\text{ADC} = 3I^2R \times F_{\text{LL}} \times F_{\text{PR}} \times F_{\text{R}} \times F_{\text{LSA}}[(C_G \times i_G) + (C_T \times i_T) + (C_S \times i_S)] \ \$/\text{mi} \tag{7.66}$$

where

ADC is the annual equivalent of demand cost incurred to maintain adequate system capacity to supply 1^2R losses in feeder conductors, \$/mi

F_{LL} is the load-location factor

F_{PR} is the peak-responsibility factor

F_R is the reserve factor

F_{LSA} is the loss-allowance factor

C_G is the cost of (peaking) generation system \$/kVA

C_T is the cost of transmission system, \$/kVA

C_S is the cost of distribution substation, \$/kVA

i_G is the annual fixed charge rate applicable to generation system

i_T is the annual fixed charge rate applicable to transmission system

i_S is the annual fixed charge rate applicable to distribution substation

The reserve factor is the ratio of total generation capability to the total load and losses to be supplied. The peak-responsibility factor is a per unit value of the peak feeder losses that are coincident with the system peak demand.

7.6.4 LEVELIZED ANNUAL COST

Typically, the expenses associated with energy, demand, and operations and maintenance (O&M) fluctuate from year to year within a specific timeframe, as depicted in Figure 7.16a. Consequently, it becomes essential to "levelize" these costs over the anticipated economic lifespan of the feeder, as illustrated in Figure 7.16b.

Assume that the costs occur discretely at the end of each year, as shown in Figure 7.16a. The *levelized annual cost** of equal amounts can be calculated as

$$A = \left[F_1\left(\frac{P}{F}\right)_1^i + F_2\left(\frac{P}{F}\right)_2^i + F_3\left(\frac{P}{F}\right)_3^i + \cdots + F_n\left(\frac{P}{F}\right)_n^i\right]\left(\frac{A}{P}\right)_n^i \tag{7.67}$$

or

* Also called the *annual equivalent* or *annual worth.*

$$A = \left[\sum_{j=1}^{n} F_i\left(\frac{P}{F}\right)_j^i\right]\left(\frac{A}{P}\right)_n^i \tag{7.68}$$

where

A is the levelized annual cost, \$/year

F_i is the unequal (or actual or unlevelized) annual cost, \$/year

n is the economic life, year

i is the interest rate

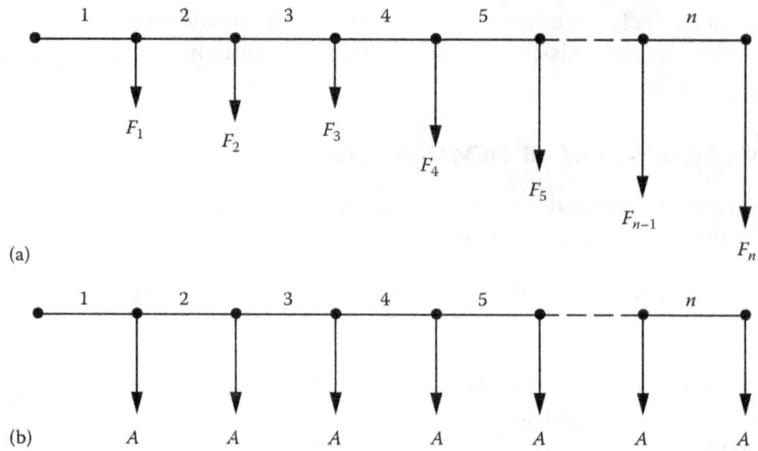

Figure 7.16 Illustration of the levelized annual cost concept: (a) unlevelized annual cost flow diagram and (b) levelized cost flow diagram.

The present worth or present equivalent of a future sum factor, also referred to as the *single-payment discount factor*, denoted as $(P/F)_n^i$, represents the value of a future sum considering an interest rate of i and an economic life of n years.

Similarly, the uniform series worth of a present sum factor, known as the *capital-recovery factor*, denoted as $(A/P)_n^i$, represents the value of a present sum considering a uniform series of payments over an economic life of n years.

These factors can be obtained from compounded-interest tables or by using the following equations, respectively.

$$\left(\frac{P}{F}\right)_n^i = \frac{1}{(1+i)} \tag{7.69}$$

and

$$\left(\frac{A}{P}\right)_n^i = \frac{i(1+i)^n}{(1+i)^n - 1} \tag{7.70}$$

Example 7.15. Assume that the following data have been gathered for the system of the NL&NP Company.

Feeder length = 1 mi
Cost of energy = 20 mills/kWh (or \$0.02/kWh)
Cost of generation system = \$200/kW
Cost of transmission system = \$65/kW
Cost of distribution substation = \$20/kW
Annual fixed charge rate for generation = 0.21
Annual fixed charge rate for transmission = 0.18
Annual fixed charge rate for substation = 0.18
Annual fixed charge rate for feeders = 0.25
Interest rate = 12%
Load factor = 0.4
Loss-allowance factor = 1.03
Reserve factor = 1.15
Peak-responsibility factor = 0.82

Table 7.6
Typical ACSR Conductors Used in Rural Areas

Conductor Ground Wire Size	Conductor Ground Wire Size	Conductor Ground Wire Weight (lb)	Conductor Ground Wire Weight (lb)	Cost $/lb	Installation and Hardware Installed Cost ($)	Total Feeder Cost ($)
			At 12.5 kV			
#4	#4	356	356	0.6	6,945.60	7,800
1/0	#2	769	566	0.6	7,176.20	8,900
3/0	1/0	1223	769	0.6	7,737.20	10,400
4/0	1/0	1542	769	0.6	8,563	11,800
266.8 kcmil	1/0	1802	769	0.6	9,985	13,690
477 kcmil	1/0	3642	769	0.6	10,967	17,660
			At 24.9 kV			
#4	#4	356	356	0.6	7,605.60	8,460
1/0	#2	769	566	0.6	7,856.20	9,580
3/0	1/0	1223	769	0.6	8,217.20	10,880
4/0	1/0	1542	769	0.6	8,293	11,530
266.8 kcmil	1/0	1802	769	0.6	9,615	13,320
477 kcmil	1/0	3462	769	0.6	11,547	18,240

Table 7.6 gives cost data for typical ACSR conductors used in rural areas at 12.5 and 24.9 kV. Table 7.7 gives cost data for typical ACSR conductors used in urban areas at 12.5 and 34.5 kV. Using the given data, develop nomographs that can be readily used to calculate the total annual equivalent cost of the feeder in dollars per mile.

Solution. Using the given and additional data and appropriate equations from Section 7.5, the following nomographs have been developed. Figures 7.17 and 7.18 give nomographs to calculate the total annual equivalent cost of ACSR feeders of various sizes for rural and urban areas, respectively, in thousands of dollars per mile.

Table 7.7
Typical ACSR Conductors Used in Urban Areas

Conductor Ground Wire Size	Conductor Ground Wire Size	Conductor Ground Wire Weight (lb)	Conductor Ground Wire Weight (lb)	Cost $/lb	Installation and Hardware Installed Cost ($)	Total Feeder Cost ($)
			At 12.5 kV			
#4	#4	356	356	0.6	21,145.60	22,000
1/0	#4	769	356	0.6	22,402.20	24,000
3/0	#4	1223	356	0.6	24,585	27,000
477 kcmil	1/0	3462	769	0.6	28,307	35,000
			At 34.5 kV			
#4	#4	356	356	0.6	21,375.60	22,230
1/0	#4	769	356	0.6	22,632.20	24,230
3/0	#4	1223	356	0.6	24,815	27,230
477 kcmil	1/0	3462	769	0.6	28,537	35,230

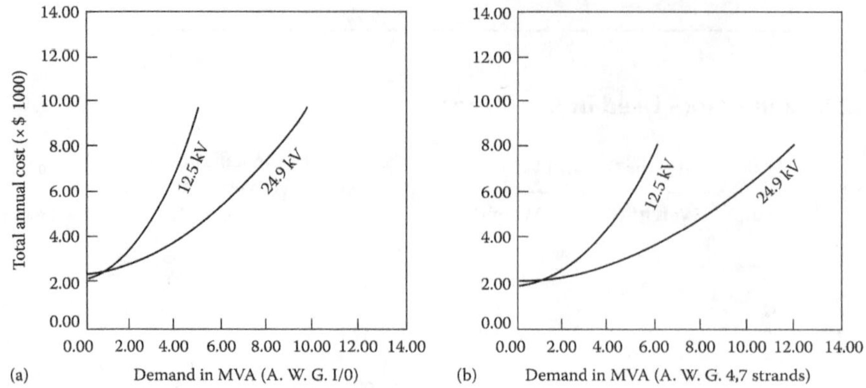

Figure 7.17 Total annual equivalent cost of ACSR feeders for *rural* areas in thousands of dollars per mile: (a) 477 cmil, 26 strands, (b) 266.8 cmil, 6 strands, AWG 4/0, and AWG 3/0.

Example 7.16. The NL&NP power and light company is required to serve a newly developed residential area. There are two possible routes for the construction of the necessary power line. Route *A* is 18 miles long and goes around a lake. It has been estimated that the required overhead power line will cost \$8000 per mile to build and \$350 per mile per year to maintain. Its salvage value will be \$1500 per mile at the end of its useful life of 20 years.

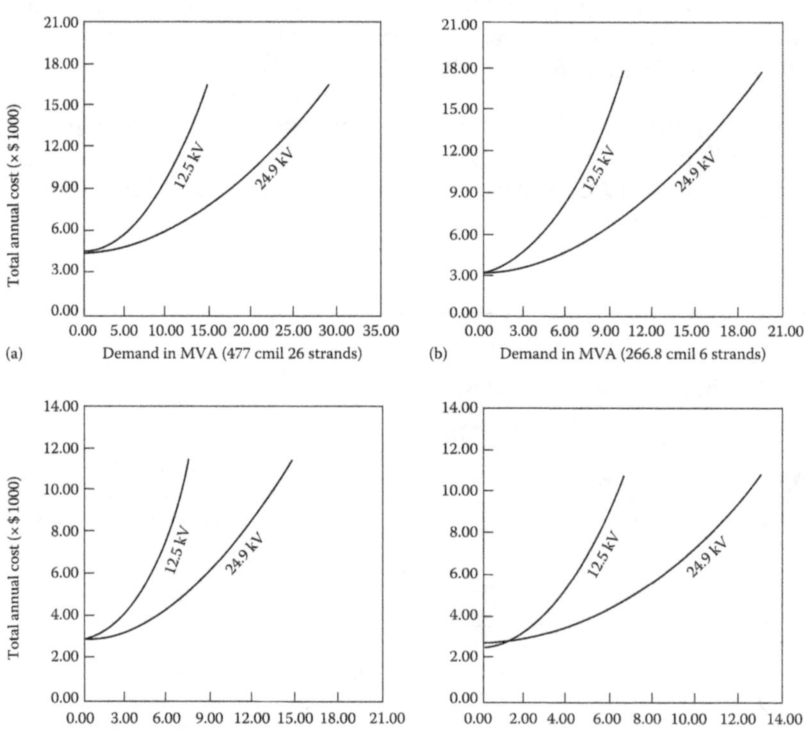

Figure 7.18 Total annual equivalent cost of ACSR feeders for *urban* areas in thousands of dollars per mile: (a) 477 cmil, 26 strands, (b) AWG 3/0, (c) AWG 1/0, and (d) AWG 4, 7 strands.

On the other hand, route B is 6 miles long and is an underwater line that goes across the lake. It has been estimated that the required underwater line using submarine power cables will cost \$21,000 per mile to build and \$1,200 per mile per year to maintain. Its salvage value will be \$6000 per mile at the end of 20 years. Assume that the fixed charge rate is 10% and that the annual ad valorem (property) taxes are 3% of the first cots of each power line. Use any engineering economy interest tables and determine the economically preferable alternative.

Solution. *Route A*: The first cost of the overhead power line is

$$P = (\$8,000/\text{mile})(18\text{ miles}) = \$144,000$$

and its estimated salvage value is

$$F = (\$1,500/\text{mile})(18\text{ miles}) = \$27,000$$

The annual equivalent cost of capital invested in the line is

$$\begin{aligned} A_1 &= \$144,000\left(\frac{A}{P}\right)_{20}^{10\%} - \$27,000\left(\frac{A}{F}\right)_{20}^{10\%} \\ &= \$144,000(0.11746) - \$27,000(0.01746) = \$16,443 \end{aligned}$$

The annual equivalent cost of the tax and maintenance is

$$A_2 = (\$144,000)(3\%) + (\$350/\text{mile})(18\text{ miles}) = \$10,620$$

Route B: The first cost of the submarine power line is

$$P = (\$21,000/\text{mile})(6\text{ miles}) = \$126,000$$

and its estimated salvage value is

$$F = (\$6,000/\text{mile})(6\text{ miles}) = \$36,000$$

Its annual equivalent cost of capital invested is

$$\begin{aligned} A_1 &= \$126,000\left(\frac{A}{P}\right)_{20}^{10\%} - \$36,000\left(\frac{A}{F}\right)_{20}^{10\%} \\ &= \$14,171 \end{aligned}$$

The annual equivalent cost of the tax and maintenance is

$$A_2 = (\$126,000)(3\%) + (\$1,200/\text{mile})(6\text{ miles}) = \$10,980$$

The total annual equivalent cost of the submarine power line is

$$\begin{aligned} A &= A_1 + A_2 \\ &= \$14,171 + \$10,980 = \$25,151 \end{aligned}$$

Hence, the *economically preferable alternative* is route B. Of course, if the present worth of the costs are calculated, the conclusion would still be the same. For example, the present worth of costs for A and B are

$$\text{PW}_A = \$27{,}063 \left(\frac{P}{A}\right)_{20}^{10\%} = \$230{,}414 = \$14{,}171$$

and

$$\text{PW}_B = \$25{,}151 \left(\frac{P}{A}\right)_{20}^{10\%} = \$214{,}136$$

Example 7.17. Use the data given in Example 6.6 and assume that the fixed charge rate is 0.15, and zero salvage values are expected at the end of useful lives of 30 years for each alternative. But the salvage value for 9 MVA capacity line is \$2000 at the end of the 10th year. Use a study period of 30 years and determine the following:

a. The annual equivalent cost of 9 MVA capacity line.
b. The annual equivalent cost of 15 MVA capacity line.
c. The annual equivalent cost of the upgrade option if the upgrade will take place at the end of 10 years. Use an average value of \$5000 at the end of 20 years for the new 15 MVA upgrade line.

Solution. a. The annual equivalent cost of 9 MVA capacity line is

$$A_1 = \$120{,}000 \left(\frac{A}{P}\right)_{30}^{15\%} = \$120{,}000(0.15230) = \$18{,}276 \text{ per mile per year}$$

b.

$$A_2 = \$150{,}000 \left(\frac{A}{P}\right)_{30}^{15\%} = \$150{,}000(0.15230) = \$22{,}845$$

c. The annual equivalent cost of 15 MVA capacity line is

$$\begin{aligned} A_2 &= \left[\$120{,}000 - \$2{,}000 \left(\frac{P}{F}\right)_{10}^{15\%} + \$200{,}000 \left(\frac{P}{F}\right)_{10}^{15\%} - \$5{,}000 \left(\frac{P}{F}\right)_{10}^{15\%}\right] \left(\frac{A}{P}\right)_{30}^{15\%} \\ &= \$120{,}000 - \$2{,}000(0.2472) + \$200{,}000(0.2472) - \$5{,}000(0.0151)](0.15230) \\ &= \$25{,}718.92 \end{aligned}$$

As it can be seen, the upgrade option is still the bad option. Furthermore, if one considers the 9 MVA vs. 15 MVA capacities, building the 15 MVA capacity line from the start is still the best option.

7.7 ECONOMIC ANALYSIS OF EQUIPMENT LOSSES

In the present day, the rising costs of plant, equipment, energy, and capital have made it increasingly important to assess the losses of electric equipment, such as power or distribution transformers, before making any final decisions regarding the purchase of new equipment or the replacement or retirement of existing ones [175]. For instance, it is now common to find that a transformer with lower losses but a higher initial price may be more cost-effective over the lifespan of the transformer compared to one with higher losses but a lower initial price [176].

However, in replacement or retirement decisions, the cost savings in operations and maintenance (O&M) expenses in a comprehensive life cycle analysis or life cycle cost study must exceed the total purchase price of the more efficient replacement transformer. This consideration is based on the concept of *sunk cost* in engineering economics. Regardless of the age of the existing unit, the carrying charges of the existing equipment do not influence the retirement decision. In other words, the fixed or carrying charges associated with existing equipment must be amortized (written off) regardless of whether the unit is retired or not.

It is worth noting that certain governmental agencies and organizations specifically require the use of phrases like *life cycle analysis* or *life cycle cost study* to ensure that bid evaluations or purchase decisions take into account all relevant factors, including future operating costs, rather than solely focusing on the initial cost, in order to determine the most economically viable alternative.

The transformer cost study should include the following factors:

1. Annual cost of copper losses
2. Annual cost of core losses
3. Annual cost of exciting current
4. Annual cost of regulation
5. Annual cost of fixed charges on the first cost of the installed equipment

The annual expenses for the equipment may vary from year to year throughout its economic lifespan. Consequently, it might be necessary to equalize these costs, as explained in Section 7.5. For more information regarding the cost analysis of distribution transformers, please refer to Section 6.7. As for the economic evaluation of power transformers for replacement purposes, the subsequent simplified approach, as outlined by Smith [177], can be adequate:

1. Determine the power ratings for the transformers as well as the peak and average system loads.
2. Obtain the load and no-load losses for the transformers under rated conditions.
3. Determine the original cost of the old transformer and the purchase price of the new one.
4. Obtain the carrying charge rate, system capital cost rate, and energy cost rate for your particular utility.
5. Calculate the transformer carrying charge and the cost of losses for each transformer. The cost of losses is equal to the system carrying charge plus the energy charge.
6. Compare the total cost per year for each transformer. The total cost is equal to the sum of the transformer carrying charge and the cost of losses.
7. Compare the total cost per year of the old and new transformers. If the total cost per year of the new transformer is less, replacement of the old transformer can be economically justified.

PROBLEMS

7.1 Consider Figure 7.19 and repeat Example 7.5.

7.2 Repeat Example 7.7, using a transformer with 75 kVA capacity.

7.3 Repeat Example 7.7, assuming four services per transformer. Here, omit the UG SL. Assume that there are six transformers per block, that is, one transformer at each pole location.

7.4 Repeat Problem 7.3, using a 75 kVA transformer.

7.5 Repeat Example 7.8, using a 100 kVA transformer and #3/0 AWG and #2 AWG cables for the SLs and SDs, respectively.

7.6 Repeat Example 7.10. Use the nominal primary voltage of 19,920/34,500 V and assume that the remaining data are the same.

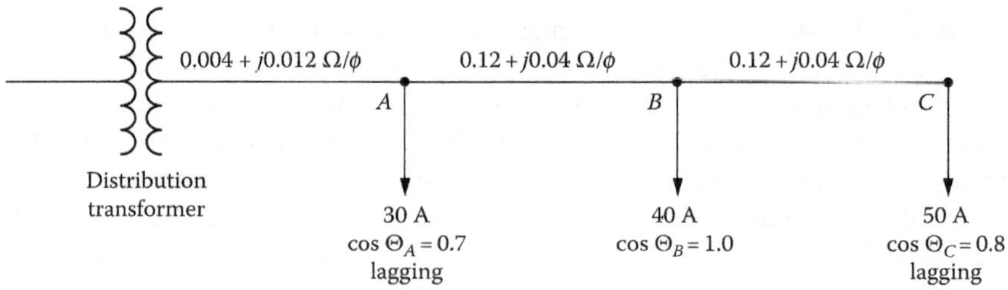

Figure 7.19 One-line diagram for Problem 7.1.

Figure 7.20 Illustration for Example 7.7.

7.7 Assume that a three-conductor DC overhead line with equal conductor sizes (see Figure 7.20) is considered to be employed to transmit three-phase three-conductor AC energy at 0.92 power factor.

If voltages to ground and transmission line efficiencies are the same for both direct and alternating currents, and the load is balanced, determine the change in the power transmitted in percent.

7.8 Assume that a single-phase feeder circuit has a total impedance of 1 + j3Ω for lines and/or transformers. The receiving-end voltage and load current are $2400\angle 0°$ V and $50 \angle -30°$ A, respectively. Determine the following:

a. The power factor of the load.
b. The load power factor for which the voltage drop is maximum, using Equation 7.51.
c. Repeat part (b), using Equation 7.52.

7.9 An unbalanced three-phase wye-connected and grounded load is connected to a balanced three-phase four-wire source. The load impedance Z_a, Z_b, and Z_c are given as $70 \angle 30°$, $85 \angle -40°$, and $50 \angle 35°$ Ω/phase, respectively, and the phase a line voltage has an effective value of 13.8 kV: Use the line-to-neutral voltage of phase a as the reference and determine the following:

a. The line and neutral currents
b. The total power delivered to the loads

7.10 Consider Figure 7.19 and assume that the impedances of the three line segments from left to right are 0.1 + j0.3, 0.1 + j0.1, and 0.08 + j0.12 Ω/phase, respectively. Also assume that this three-phase three-wire 480-V secondary system supplies balanced loads at *A*, *B*, and *C*. The loads at A, B, and C are represented by 50 A with a lagging power factor of 0.85, 30 A with a lagging power factor of 0.90, and 50 A with a lagging power factor of 0.95, respectively. Determine the following:

a. The total voltage drop in one phase of the lateral using the approximate method
b. The real power per phase for each load

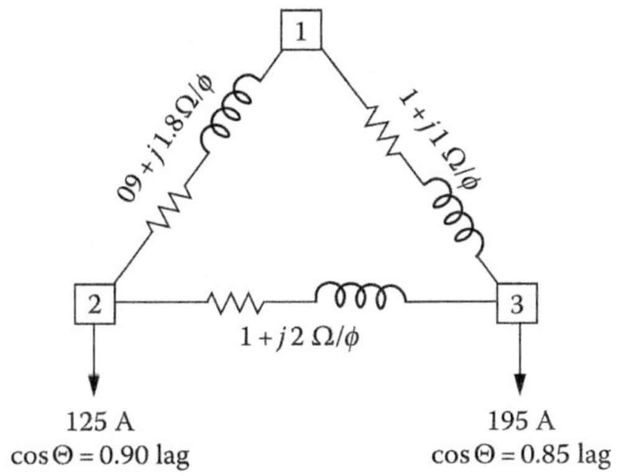

Figure 7.21 Distribution system for Problem 7.11.

c. The reactive power per phase for each load
d. The kVA output and load power factor of the distribution transformer

7.11 *Assume* that bulk power substation 1 supplies substations 2 and 3, as shown in Figure P7.11, through three-phase lines. Substations 2 and 3 are connected to each other over a tie line, as shown. Assume that the line-to-line voltage is 69 kV and determine the following:

a. The voltage difference between substations 2 and 3 when tie line 23 is open-circuited
b. The line currents when all three lines are connected as shown in the figure
c. The total power loss in part (a)
d. The total power loss in part (b).

7.12 Repeat Example 7.6, assuming 50% lagging power factor for all loads.

7.13 Resolve Example 7.4 by using MATLAB and assuming four services per transformer. Here, omit the UG SL. Assume that there are six transformers per block, that is, one transformer at each pole location.

Part IV

Regulating Voltage with/without Capacitance

8 Application of Capacitors to Distribution Systems

Who neglects learning in his youth, loses the past and is dead for the future.

Euripides, 438 BC

Where is there dignity unless there is honesty?

Cicero

8.1 STATE-OF-THE-ART REVIEW

The addition of capacitance can rectify the power factor of a circuit and enhance the voltage profile over long distances. In the distribution management system, a centralized volt/VAr control algorithm has been implemented to regulate switched capacitance, thereby improving voltage along the feeder [178]. The real-time implementation of the compensation method has proven to be effective [179]. Typically, this setup involves a parallel connection to the feeder, but the possibility of a series connection to the feeder circuit has also been analyzed [180]. In certain applications, shunt active filters are utilized to detect, analyze, and mitigate the occurrence known as "whack-a-mole" [181]. Fuzzy logic applications have been explored in the literature to compensate for radial feeders [182]. The placement of larger capacitors has been proposed using ant direction hybrid differential evolution to minimize investment costs [183]. Extensive studies have been conducted on the harmonic components of voltage and current in industrial plants [184]. The scheduling of load tap changer (LTC) switching features can be achieved through evolutionary-based algorithms [185]. Estimating voltage profiles along a feeder can be accomplished with a limited number of metered observations [186]. Network reconfiguration between feeders can be accomplished by utilizing normally open (NO) switches, and the placement of capacitors can help reduce power losses with an improved reconfiguration method, thereby impacting the voltage profile under specific circumstances [187]. A compensator is implemented for electric vehicles to enhance power quality [188]. A partitioning method has been studied to divide a feeder into subsystems, enabling distributed control over capacitive elements to improve performance in voltage profile management [189]. Several linearized approaches to power flow formulation have focused on optimization to ensure optimality, and mixed-integer quadratic programming (MIQP) can facilitate solution convergence and coordinate reactive power control [190, 191]. To address potential overvoltage along the feeder, preventive strategies can be established in the low-voltage distribution network for reactive power management [192]. Energy savings can be maximized by employing optimal strategies that cater to the reactive power requirements and identify critical locations for power injection [193–195]. Approaches utilizing genetic algorithms on a zonal basis have been applied to optimization problems in reactive power supply [196, 197].

8.2 BASIC DEFINITIONS

Capacitor element: A fundamental component of a capacitor comprises electrodes that are separated by a dielectric material, forming an inseparable unit.

Capacitor unit: A grouping of one or multiple capacitor elements enclosed within a single container, with terminals extending outward.

DOI: 10.1201/9781003129721-8

Capacitor segment: A set of capacitor units integrated with a protection and control system, designed for single-phase operation.
Capacitor module: A collection of capacitor segments organized in a three-phase configuration.
Capacitor bank: A complete arrangement of interconnected capacitor modules forming a cohesive unit.

8.3 POWER CAPACITORS

At first glance, a capacitor may appear to be a simple and uncomplicated device, consisting of two metal plates separated by a dielectric insulating material. It has no moving parts and operates through the influence of electric stress [198].

However, in reality, a power capacitor is a highly technical and intricate device that involves very thin dielectric materials, high electric stresses, and sophisticated processing techniques. Figure 8.1 provides a cross-sectional view of a power factor correction capacitor, while Figure 8.2 illustrates a typical application of a capacitor in a switched pole-top rack.

In the past, most power capacitors were constructed using two sheets of pure aluminum foil separated by multiple layers of chemically impregnated kraft paper. Over the past three decades, significant advancements have been made in power capacitors, driven by improvements in dielectric materials, enhanced utilization techniques, and advancements in processing methods. Capacitor sizes have increased from the range of 15–25 kVAr to 200–300 kVAr (capacitor banks are typically available in sizes ranging from 300 to 1800 kVAr).

Modern power capacitors are much more efficient compared to those from 30 years ago, offering electric utilities a more cost-effective solution in terms of kVAr. Capacitors are receiving increased attention today, partly due to the addition of a new dimension in analysis: changeout economics. In certain cases, even the replacement of older capacitors can be justified based on evaluations of lower losses associated with modern capacitor designs.

Capacitor technology has evolved to incorporate extremely low-loss designs, employing the all-film concept. This allows utilities to make economic evaluations when choosing between different existing capacitor technologies [199].

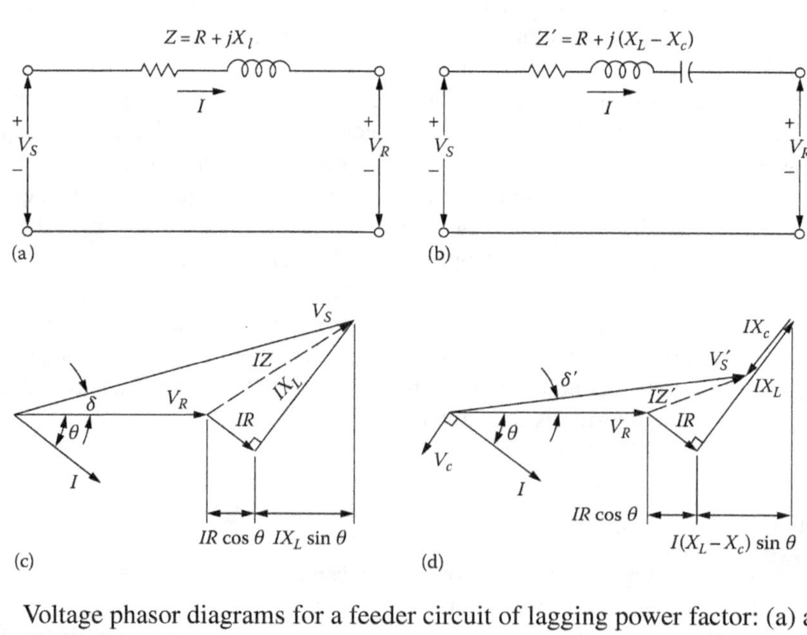

Figure 8.1 Voltage phasor diagrams for a feeder circuit of lagging power factor: (a) and (c) without and (b) and (d) with series capacitors.

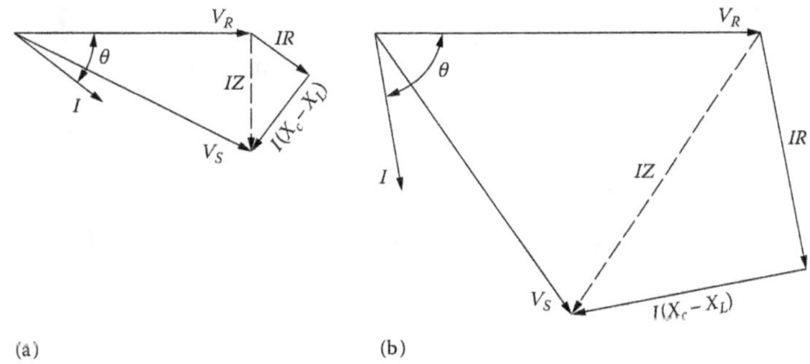

Figure 8.2 Overcompensation of the receiving-end voltage: (a) at normal load and (b) at the start of a large motor.

8.4 EFFECTS OF SERIES AND SHUNT CAPACITORS

As previously discussed, capacitors, whether installed in series or shunt configuration, individually or as a bank, serve the primary purpose of regulating voltage and reactive power flow at their installation points. Shunt capacitors achieve this by adjusting the power factor of the load, while series capacitors accomplish it by directly counteracting the inductive reactance of the circuit to which they are connected.

8.4.1 SERIES CAPACITORS

Series capacitors are capacitors that are connected in series with lines and have found limited application in distribution circuits. This is primarily due to their specialized nature and the narrow range of their usage. Additionally, the unique challenges associated with each application require extensive engineering investigations. As a result, utilities generally hesitate to install series capacitors, especially in smaller sizes.

Figure 8.1 demonstrates the role of a series capacitor in compensating for inductive reactance. Essentially, it introduces a negative (capacitive) reactance in series with the positive (inductive) reactance of the circuit, thereby compensating for a portion or the entire inductive reactance. Consequently, the primary purpose of a series capacitor is to minimize or eliminate the voltage drop caused by the inductive reactance in the circuit.

In certain situations, a series capacitor can be considered a voltage regulator that produces a voltage boost proportional to the magnitude and power factor of the current flowing through it. Consequently, as the load increases, a series capacitor enables an automatic and immediate voltage increase.

Furthermore, a series capacitor provides a greater overall voltage rise compared to a shunt capacitor, particularly at lower power factors where larger voltage drops occur. However, it does not improve the system power factor to the same extent as a shunt capacitor and has minimal impact on the source current.

Consider the feeder circuit and its voltage phasor diagram as shown in Figure 8.1a and c. The voltage drop through the feeder can be expressed approximately as

$$\text{VD} = IR\cos\theta + IX_L\sin\theta \tag{8.1}$$

where

R is the resistance of the feeder circuit
X_L is the inductive reactance of the feeder circuit
cos θ is the receiving-end power factor
sin θ is the sine of the receiving-end power factor angle

As can be observed from the phasor diagram, the magnitude of the second term in Equation 8.1 is much larger than the first. The difference gets to be much larger when the power factor is smaller and the ratio of R/X_L is small.

However, when a series capacitor is applied, as shown in Figure 8.1b and d, the resultant lower voltage drop can be calculated as

$$\text{VD} = IR\cos\theta + I(X_L - X_c)\sin\theta \tag{8.2}$$

where X_c is the capacitive reactance of the series capacitor.

8.4.1.1 Overcompensation

Typically, when selecting the size of a series capacitor for a distribution feeder application, the goal is to ensure that the resulting capacitive reactance is smaller than the inductive reactance of the feeder circuit. This helps to minimize voltage drop. However, in specific situations where the resistance of the feeder circuit is greater than its inductive reactance, the opposite approach may be preferred, resulting in a larger resultant voltage drop.

$$\text{VD} = IR\cos\theta + I(X_L - X_c)\sin\theta \tag{8.3}$$

The resulting condition that occurs when the capacitive reactance of a series capacitor exceeds the inductive reactance of the feeder circuit is referred to as *overcompensation*. In Figure 8.2a, a voltage phasor diagram illustrates the situation of overcompensation at normal load. However, when the level of overcompensation is determined solely based on normal load, the resulting overcompensation of the receiving-end voltage may not be desirable. This is particularly true when a large motor starts and generates a significant lagging current, leading to an exceptionally large voltage rise, as shown in Figure 8.2b. Such voltage fluctuations can be detrimental to lighting equipment, shortening their lifespan, and causing light flicker that leads to consumer complaints.

8.4.1.2 Leading Power Factor

using series capacitors, the load current should have a lagging power factor. For instance, Figure 8.3a illustrates a voltage phasor diagram without series capacitors in the line, where the load has a leading power factor. In contrast, Figure 8.3b depicts the resultant voltage phasor diagram with the same leading power factor, but this time with series capacitors incorporated in the line. It is evident from the diagram that the receiving-end voltage is diminished due to the presence of series capacitors. When the power factor ($\cos\theta$) equals 1.0, the value of the sinusoidal component ($\sin\theta$) is less than or equal to zero, resulting in a reduced voltage at the receiving end. Therefore,

$$I(X_L - X_c)\sin\theta \cong 0$$

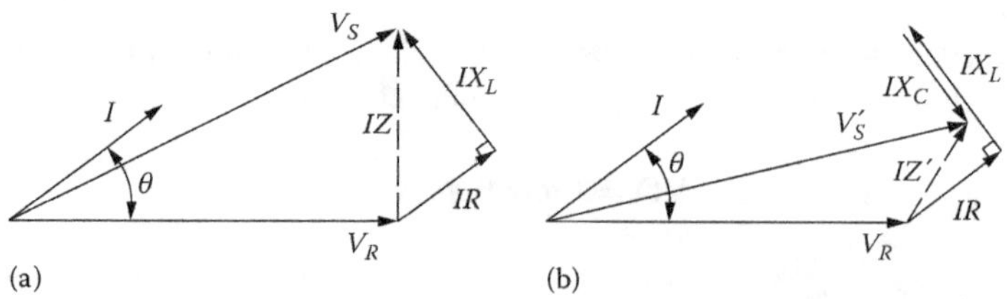

Figure 8.3 Voltage phasor diagram with leading power factor: (a) without series capacitors and (b) with series capacitors.

hence, equation above becomes

$$\mathrm{VD} \cong IR \tag{8.4}$$

Hence, in such scenarios, series capacitors offer minimal benefits.

Due to the reasons mentioned above, as well as other factors such as ferroresonance in transformers, sub-synchronous resonance during motor starting, shunting of motors during normal operation, and challenges in protecting capacitors from system fault current, the application of series capacitors in distribution systems is limited.

However, they find usage in subtransmission systems for adjusting the load distribution between parallel lines. For instance, when a new subtransmission line with higher thermal capacity runs parallel to an existing line, it can be challenging, if not impossible, to load the new line without overloading the old line. In such cases, series capacitors can be employed to offset some of the line reactance and enhance the thermal capability. They are also utilized in subtransmission systems to reduce voltage regulation.

8.4.2 SHUNT CAPACITORS

Shunt capacitors, that is,*capacitors connected in parallel with lines*, are used extensively in distribution systems. Shunt capacitors supply the type of reactive power or current to counteract the out-of-phase component of current required by an inductive load. In a sense, shunt capacitors modify the characteristic of an inductive load by drawing a leading current that counteracts some or all of the lagging component of the inductive load current at the point of installation. Therefore, a shunt capacitor has the same effect as an overexcited synchronous condenser, generator, or motor.

As shown in Figure 8.4, by the application of shunt capacitor to a feeder, the magnitude of the source current can be reduced, the power factor can be improved, and consequently the voltage drop between the sending end and the load is also reduced. However, shunt capacitors do not affect current or power factor beyond their point of application. Figure 8.4a and c shows the single-line diagram of a line and its voltage phasor diagram before the addition of the shunt capacitor, and Figure 8.6b and d shows them after the addition.

Voltage drop in feeders, or in short transmission lines, with lagging power factor can be approximated as

$$\mathrm{VD} = I_R R + I_X X_L \tag{8.5}$$

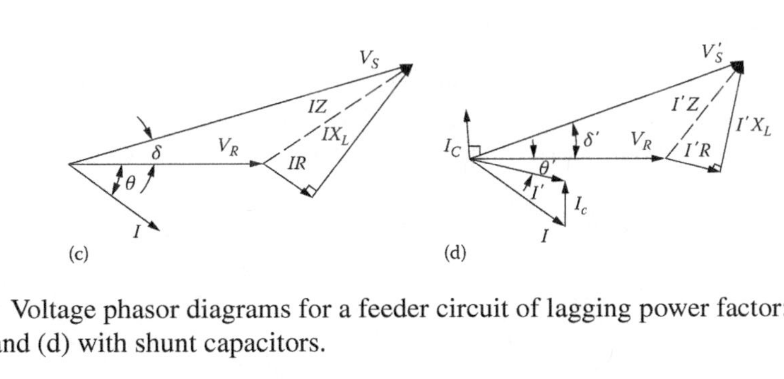

Figure 8.4 Voltage phasor diagrams for a feeder circuit of lagging power factor: (a) and (c) without and (b) and (d) with shunt capacitors.

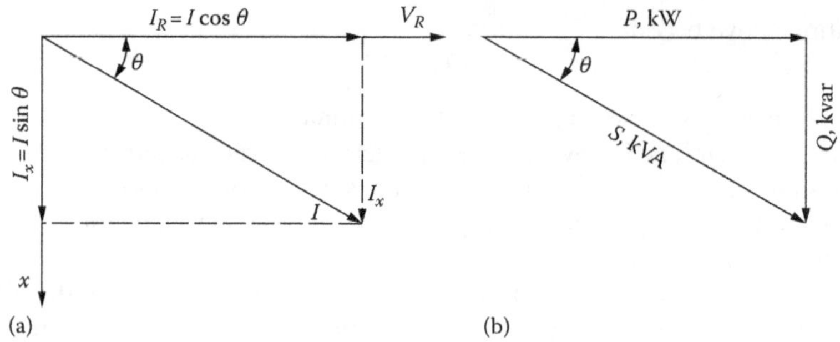

Figure 8.5 (a) Phasor diagram and (b) power triangle for a typical distribution load.

where
R is the total resistance of the feeder circuit, Ω
X_L is the total inductive reactance of the feeder circuit, Ω
I_R is the real power (or in-phase) component of the current, A
I_X is the reactive (or out-of-phase) component of the current lagging the voltage by 90°, A

Example 8.1. Consider the right-angle triangle shown in Figure 8.5b. Determine the power factor of the load on a 460 V three-phase system, if the ammeter reads 100 A and the wattmeter reads 70 kW.

Solution.

$$\begin{aligned} S &= \frac{\sqrt{3}(V)(I)}{1000} \\ &= \frac{\sqrt{3}(460\text{V})(100\text{A})}{1000} \\ &\cong 79.67 \text{ kVA} \end{aligned}$$

$$\begin{aligned} P = \cos\theta &= \frac{P}{S} \\ &= \frac{70 \text{ kW}}{79.67 \text{ kVA}} \\ &\cong 0.88 \text{ or } 88\% \end{aligned}$$

When a capacitor is installed at the receiving end of the line, as shown in Figure 8.4b, the resultant voltage drop can be calculated approximately as

$$\text{VD} = I_R R_R + I_X X_L - I_C X_L \tag{8.6}$$

where I_c is the reactive (or out-of-phase) component of current leading the voltage by 90°, A.

The difference between the voltage drops calculated by using Equations 8.5 and 8.6 is the voltage rise due to the installation of the capacitor and can be expressed as

$$\text{VR} = I_C X_L \tag{8.7}$$

8.5 POWER FACTOR CORRECTION

8.5.1 IMPACT AND CAPACITOR SOLUTIONS

In a typical utility system, the reactive load operates at an 80% power factor during the summer months. As a result, the current lags behind the voltage in typical distribution loads, as depicted in Figure 8.5a. The cosine of the angle between the current and sending voltage is known as the *power factor* of the circuit. By multiplying the in-phase and out-of-phase components of the current I with the receiving-end voltage V_R, a triangular relationship known as the *power triangle* can be established, as shown in Figure 8.5b. Figure 8.5b displays the triangular relationship between kW, kVA, and kVAr.

It is important to note that the addition of capacitors allows for a reduction or complete elimination of the reactive power component Q of the load's apparent power S. Figures 8.6a and 8.7 illustrate how the reactive power component Q increases with each 10% change in power factor. Figure 8.6a also demonstrates how the leading reactive power of the capacitor Q_c cancels out a portion of the lagging reactive power Q_{old}.

Furthermore, as shown in Figure 8.6, even at an 80% power factor, the size of the reactive power (kVAr) component is significant, resulting in a 25% increase in the total apparent power (kVAs) of the line. To offset the 75 kVAr lagging component, 75 kVAr of capacitors are required.

As previously mentioned, generating reactive power at a power plant and supplying it to a load located at a considerable distance is not economically feasible. However, capacitors (or overexcited synchronous motors) located at the load centers can easily provide the required reactive power.

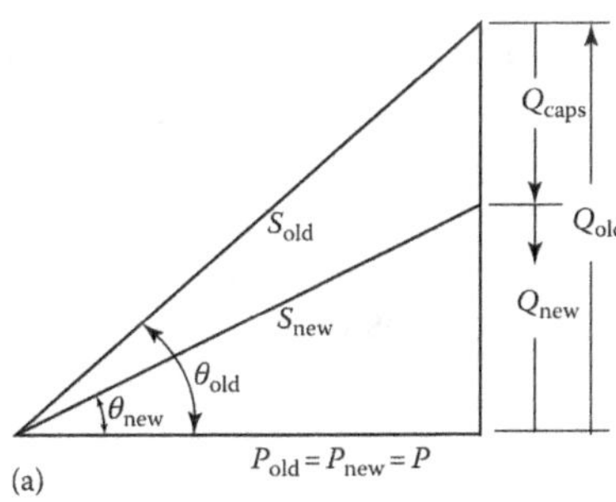

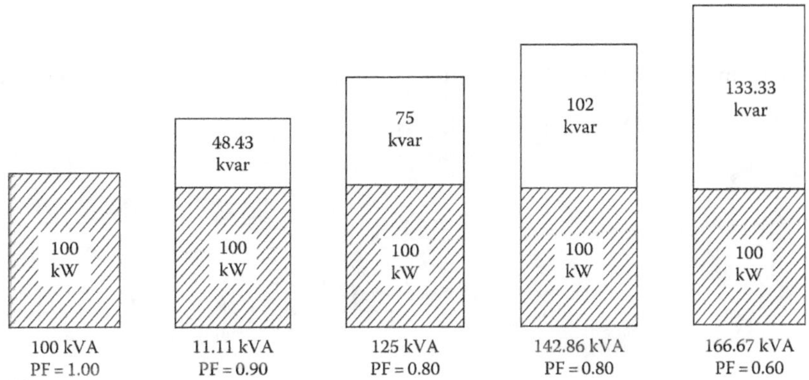

Figure 8.6 Illustration of (a) the use of a power triangle for power factor correction by employing capacitive reactive power and (b) the required increase in the apparent and reactive powers as a function of the load power factor, holding the real power of the load constant.

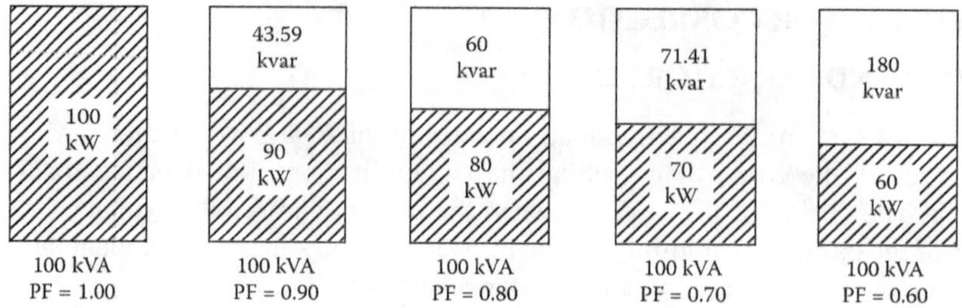

Figure 8.7 Illustration of the change in the real and reactive powers as a function of the load power factor, holding the apparent power of the load constant.

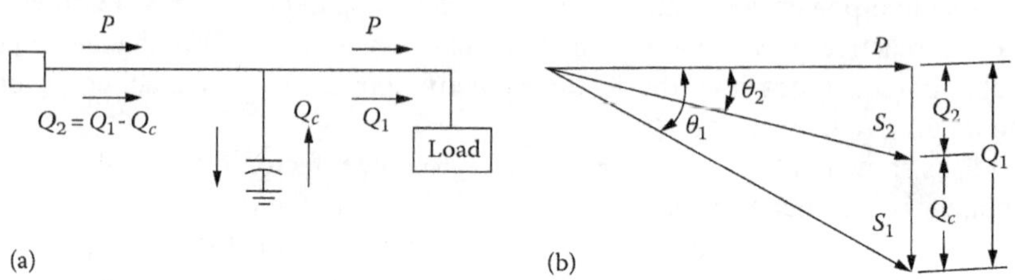

Figure 8.8 Illustration of power factor correction.

Figure 8.8 illustrates power factor correction for a given system. As depicted in the figure, capacitors draw leading reactive power from the source, thereby supplying lagging reactive power to the load. Assuming a load is supplied with real power P, lagging reactive power Q_1, and apparent power S_1 at a lagging power factor of

$$\cos\,\theta_1 = \frac{P}{S_1}$$

or

$$\cos\,\theta_1 = \frac{P}{(P^2 + Q_1^2)^{1/2}} \tag{8.8}$$

When a shunt capacitor of Q_c kVA is installed at the load, the power factor can be improved from $\cos\,\theta_1$ to $\cos\,\theta_2$,where

$$\cos\,\theta_2 = \frac{P}{S_2}$$

$$= \frac{P}{(P^2 + Q_2^2)^{1/2}}$$

or

$$\cos\,\theta_2 = \frac{P}{[P^2 + (Q_1 - Q_c)^2]^{1/2}} \tag{8.9}$$

8.5.2 CONCEPT OF LEADING AND LAGGING POWER FACTORS

The terminology of "lagging" and "leading" power factors can be confusing and lacks meaning without knowledge of the direction of real and reactive power flows. In general, for a given load,

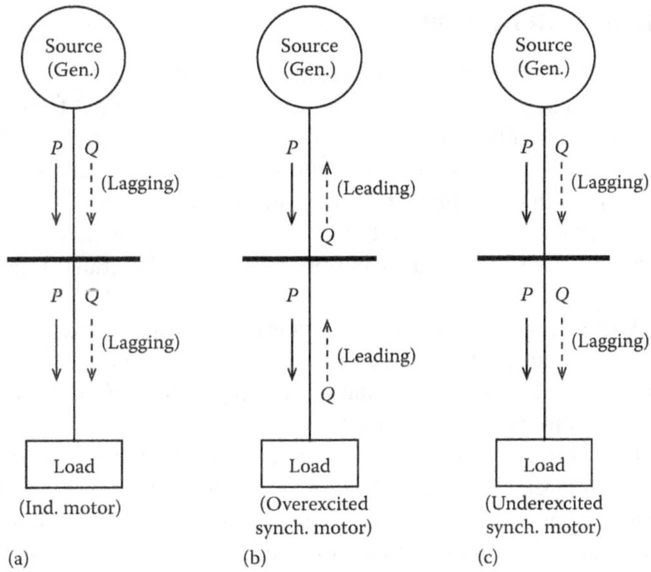

Figure 8.9 Examples of some of the sources of leading and lagging reactive power at the load.

the power factor is considered *lagging* when the load consumes reactive power, while it is deemed *leading* when the load supplies reactive power.

For instance, an induction motor has a lagging power factor as it draws reactive power from the source to meet its magnetizing requirements. Conversely, a capacitor (or an overexcited synchronous motor) supplies reactive power and therefore has a leading power factor, as depicted in Figure 8.9 and indicated in Table 8.1.

In contrast, an underexcited synchronous motor consumes both real and reactive power from the source, as indicated. To avoid confusion regarding the terms "lagging" and "leading," the use of varmeters instead of power factor meters is recommended. A varmeter has a zero center point with scales on both sides, one labeled "in" and the other labeled "out." This helps eliminate ambiguity in the understanding of power factor.

Table 8.1
Power Factor of Load and Source

	At Load			At Generator		
Load Type	***P***	***Q***	**Power Factor**[a]	***P***	***Q***	**Power Factor**[b]
Induction motor	In	Out	Lagging			
Induction generator				Out	Out	Lagging
Synchronous motor (Underexcited)	In	In	Lagging	Out	Out	Lagging
Synchronous motor (Overexcited)	In	Out	Leading	Out	In	Leading

[a] Power factor measured at the load.
[b] Power factor measured at the generator.

8.5.3 ECONOMIC POWER FACTOR

As depicted in Figure 8.8b, the introduction of reactive power (Q) leads to a decrease in both apparent power, reducing it from S_1 to S_2 kVA, and reactive power, reducing it from Q_1 to Q_2 kVAr. This reduction in reactive power results in a decrease in the total current, thereby reducing power losses.

Hence, power factor correction yields economic benefits by generating savings in capital expenditures and fuel costs. This is achieved through the release of kVA capacity and reduction of power losses in all the components situated between the capacitor installation point and the power plant source. These components include distribution lines, substation transformers, and transmission lines.

The economic power factor refers to the point where the economic advantages of incorporating shunt capacitors are equivalent to the cost of the capacitors. Previously, this economic power factor was approximately 95%. However, with the current high costs of plant operation and fuel, the economic power factor has moved closer to unity.

Nevertheless, as the corrected power factor approaches unity, the effectiveness of capacitors in improving the power factor, reducing the kVAs transmitted through the lines, enhancing load capacity, and minimizing line copper losses by reducing line current sharply diminishes. Consequently, achieving power factor correction to unity becomes more expensive in relation to the incremental cost of installed capacitors.

8.5.4 USE OF A POWER FACTOR CORRECTION TABLE

A power factor correction table, presented in Table 8.2, provides a simplified approach for calculating the required capacitor size to improve the power factor of a specific load from its original value to a desired value. The table offers a multiplier that aids in determining the kVAr requirement. The calculations are based on the following formula:

$$Q = P(\tan\theta_{\text{orig}} - \tan\theta_{\text{new}}) \tag{8.10}$$

or

$$Q = P\left(\sqrt{\frac{1}{\text{PF}_{\text{orig}}^2} - 1} - \sqrt{\frac{1}{\text{PF}_{\text{new}}^2} - 1}\right) \tag{8.11}$$

where
Q is the required compensation in kVAr
P is the real power kW
PF_{orig} is the original power factor
PFnew is the desired power factor

8.5.5 ALTERNATING CYCLES OF A MAGNETIC FIELD

In order to comprehend the improvement of power factor in a device, it is necessary to understand the underlying electrical processes involved. Let's consider the scenario of an induction motor being supplied with real power P and reactive power Q. The real power P is dissipated as losses, while the reactive power Q is not lost but rather utilized to store energy within the motor's magnetic field.

Due to the alternating nature of the current, the magnetic field experiences cycles of establishment and decay. During the establishment phase of the field, reactive current flows from the power source to the motor. Conversely, during the decay phase of the field, the reactive current flows out of the motor back to the power source. In such applications, a device is required to temporarily store the reactive power when the motor's magnetic field decays.

The most suitable device for this purpose is a *capacitor*, which stores energy in an electric field. By connecting a capacitor *in parallel with the supply line of the load*, a cyclic flow of reactive power occurs between the motor and the capacitor. In this setup, the supply lines only carry the current necessary to provide real power to the motor. It is important to note that this configuration is applicable only under conditions of unity power factor. For power factors other than unity, the supply lines would carry a certain amount of reactive power.

8.5.6 POWER FACTOR OF A GROUP OF LOADS

In general, the power factor of an individual load can be determined. However, there are instances where it becomes necessary to determine the power factor of a group of diverse loads. This can be achieved by applying the known power relationship.

Example 8.2. Let's consider a substation that supplies three different types of loads: incandescent lights, induction motors, and synchronous motors, as illustrated in Figure 8.10. To find the power factor of the substation, we need to analyze the total reactive and real powers of the connected loads. Using the provided data in Figure 8.10, we will determine the following:

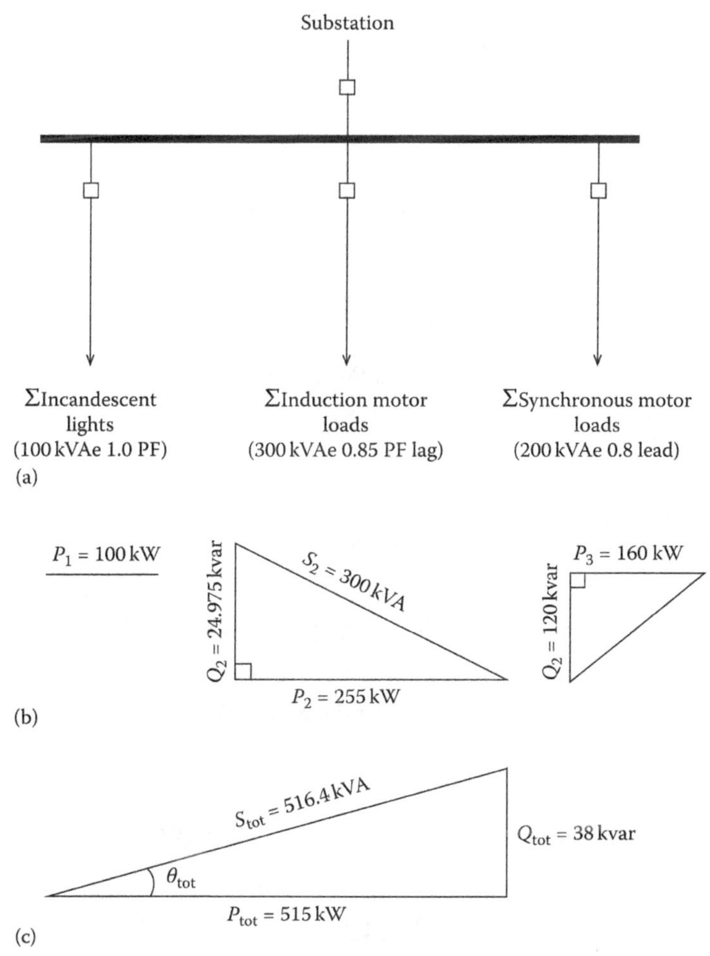

Figure 8.10 For Example 8.2: (a) connection diagram, (b) phasor diagrams of individual loads, and (c) phasor diagram of combined loads.

a. The apparent, real, and kVArs of each load
b. The total apparent, real, and reactive powers of the power that should be supplied by the substation
c. The total power factor of the substation

Solution. a.

1. *For a 100 kVA lighting load*
   Since incandescent lights are basically a unity power factor load, it is assumed that all the current is kW current. Hence,

$$S_1 = P_1$$

$$100\ \text{kVA} \cong 100\ \text{kW}$$

2. *For 400 hp of connected induction motor loads* Assume that for the motor loads,

$$\text{kVA load} = 0.75 \times (\text{Connected motor horse power})$$

with an opening power factor of 85% lagging:

$$S_2 = \left(0.75 \frac{\text{kW}}{\text{hp}}\right)(400\ \text{hp}) = 300\ \text{kVA}$$

$$P_3 = (0.75 \times 400) \times 0.85 = 255\ \text{kW}$$

$$\begin{aligned} Q_2 &= \sqrt{(300)^2 - (255)^2} \\ &= \sqrt{90,000 - 65,025} \\ &= 24,975 \\ &\cong 158\ \text{kVAr} \end{aligned}$$

3. *200 hp motor with a 0.8 leading power factor*
   At full load, assume kVA = motor-hp rating = 200 kVA:

$$\begin{aligned} P_3 &= (200\ \text{kVA}) \cos\theta \\ &= 200 \times 0.8 \\ &= 160\ \text{kW} \end{aligned}$$

$$\begin{aligned} Q_2 &= \sqrt{(200\ \text{kVA})^2 - (160\ \text{kVA})^2} \\ &= \sqrt{40,000 - 25,600} \\ &= \sqrt{14,400} \\ &= 120\ \text{kVAr} \end{aligned}$$

b. At the substation, the total real power is

$$\begin{aligned} P_{total} &= P_{lights} + P_{ind.mot.} + P_{sync.mot.} \\ &= 100 + 255 + 160 \\ &= 515 \text{ kW} \end{aligned}$$

The total reactive power is

$$\begin{aligned} Q_{total} &= Q_{lights} + Q_{ind.mot.} \\ &= 0 + 158 \\ &= 158 \text{ kVAr} \end{aligned}$$

Therefore, when an overexcited synchronous motor operates without any mechanical load connected to its shaft, it is capable of providing leading reactive power. Consequently, the net amount of lagging reactive power that needs to be supplied by the substation is determined by the difference between the reactive power supplied by the synchronous motor and the reactive power demanded by the induction motor loads.

Induction motor load required = 158 kVAr
Synchronous motor supplied = 120 kVAr
Substation must supply = 38 kVAr

c. The kVA of the substation is

$$S_{total} = \sqrt{P_{tot}^2 + Q_{tot}^2} \quad (8.12)$$

or

$$\begin{aligned} S_{total} &= \sqrt{515^2 + 38^2} \\ &= \sqrt{266,669} \\ &= 516.4 \text{ kVA} \end{aligned}$$

The power factor of the substation is

$$\begin{aligned} \text{PF} = \text{power factor} &= \frac{P}{S} \\ &= \frac{515 \text{ kW}}{516.4 \text{ kVA}} \\ &= 0.997 \text{ lagging} \end{aligned}$$

8.5.7 PRACTICAL METHODS USED BY THE POWER INDUSTRY FOR POWER FACTOR IMPROVEMENT CALCULATIONS

In the power industry, formulas commonly employ units such as kW, kVA, or kVAr instead of the symbols P, S, Q, which are more academically correct. However, there are certain advantages to using these industry-specific units as they eliminate the need for constantly determining which symbol represents P, S, or Q.

Based on the right-triangle relationship, various straightforward and valuable mathematical expressions can be derived:

$$PF = \cos\theta = \frac{kW}{kVA} \tag{8.13}$$

$$\tan\theta = \frac{kVAr}{kW} \tag{8.14}$$

$$\sin\theta = \frac{kVAr}{kVA} \tag{8.15}$$

Since the kW component normally stays the same (the kVA and kVAr components change with power factor), it is convenient to use Equation 8.11 involving the kW component. The relationship can be re-expressed as

$$kVAr = kW \times \tan\theta \tag{8.16}$$

For instance, if it is necessary to determine the capacitor rating to improve the load's power factor, one would use the following relationships:

$$\text{kVAr at original PF} = kW \times \tan\,\theta_1 \tag{8.17}$$

$$\text{kVAr at improved PF} = kW \times \tan\,\theta_2 \tag{8.18}$$

Thus, the capacitor rating required to improve the power factor can be expressed as

$$cKVAr^* = kW \times (\tan\,\theta_1 - \tan\,\theta_2) \tag{8.19}$$

or

$$\Delta\tan = \tan\,\theta_1 - \tan\,\theta_2 \tag{8.20}$$

then

$$cKVAr^* = kW \times \Delta\tan\,\theta \tag{8.21}$$

Table 8.2 provides a "kW multiplier" that can be utilized to determine the capacitor size according to the expression mentioned earlier. Additionally, it is worth noting that the prefix "c" in cKVAr is used specifically to indicate the capacitor's reactive power (kVAr), distinguishing it from the load's reactive power.

To find irrespective currents of kVA, kW, and kVAr, use the following relationships:

$$kVA = \sqrt{(kW)^2 + (kVAr)^2} \tag{8.22}$$

$$kW = \sqrt{(KVA)^2 + (kVAr)^2} \tag{8.23}$$

$$kVAr = \sqrt{(kVA)^2 + (kW)^2} \tag{8.24}$$

Example 8.3. Assume that a load withdraws 80 kW and 60 kVAr at a 0.8 power factor. It is required that its power factor is to be improved from 80% to 90% by using capacitors. Determine the amount of the reactive power to be provided by using capacitors.

Table 8.2

Determination of kW Multiplies to Calculate kVAr Requirement for Power Factor Correction

Reactive Factor	Original Power Factor (%)	Correcting Factor																				
		Desired Power Factor (%)																				
		80	81	82	83	84	85	86	87	88	89	90	91	92	93	94	95	96	97	98	99	100
0.8	60	0.584	0.61	0.636	0.662	0.688	0.714	0.741	0.767	0.794	0.822	0.85	0.878	0.905	0.939	0.971	1.01	1.04	1.08	1.311	1.192	1.334
0.791	61	0.549	0.575	0.601	0.627	0.653	0.679	0.706	0.732	0.759	0.787	0.815	0.843	0.87	0.904	0.936	0.97	1.01	1.05	1.096	1.157	1.299
0.785	62	0.515	0.541	0.567	0.593	0.619	0.645	0.672	0.698	0.725	0.753	0.781	0.809	0.836	0.87	0.902	0.94	0.97	1.01	1.062	1.123	1.265
0.776	63	0.483	0.509	0.535	0.561	0.587	0.613	0.64	0.666	0.693	0.721	0.749	0.777	0.804	0.838	0.87	0.9	0.94	0.98	1.03	1.091	1.233
0.768	64	0.45	0.476	0.502	0.528	0.554	0.58	0.607	0.633	0.66	0.688	0.716	0.744	0.771	0.805	0.837	0.87	0.91	0.95	0.997	1.058	1.2
0.759	65	0.419	0.445	0.471	0.479	0.523	0.549	0.576	0.602	0.629	0.657	0.685	0.713	0.74	0.774	0.806	0.84	0.88	0.92	0.966	1.027	1.169
0.751	66	0.388	0.414	0.44	0.466	0.492	0.518	0.545	0.571	0.598	0.626	0.654	0.682	0.709	0.743	0.775	0.81	0.85	0.89	0.935	0.996	1.138
0.744	67	0.358	0.384	0.41	0.436	0.462	0.488	0.515	0.541	0.568	0.596	0.624	0.652	0.679	0.713	0.745	0.78	0.82	0.86	0.905	0.966	1.108
0.733	68	0.329	0.355	0.381	0.407	0.433	0.459	0.486	0.512	0.539	0.567	0.595	0.623	0.65	0.684	0.716	0.75	0.79	0.83	0.876	0.937	1.079
0.725	69	0.299	0.325	0.351	0.377	0.403	0.429	0.456	0.482	0.509	0.537	0.565	0.593	0.62	0.654	0.686	0.72	0.76	0.8	0.84	0.907	1.049
0.714	70	0.27	0.296	0.322	0.348	0.374	0.4	0.427	0.453	0.48	0.508	0.536	0.564	0.591	0.625	0.657	0.69	0.73	0.77	0.811	0.878	1.02
0.704	71	0.242	0.268	0.294	0.32	0.346	0.372	0.399	0.425	0.452	0.48	0.508	0.536	0.563	0.597	0.629	0.66	0.7	0.74	0.783	0.85	0.992
0.694	72	0.213	0.239	0.265	0.291	0.317	0.343	0.37	0.396	0.423	0.451	0.479	0.507	0.534	0.568	0.6	0.63	0.67	0.71	0.754	0.821	0.963
0.682	73	0.186	0.212	0.238	0.264	0.29	0.316	0.343	0.369	0.396	0.424	0.452	0.48	0.507	0.541	0.573	0.61	0.65	0.69	0.727	0.794	0.936
0.673	74	0.159	0.185	0.211	0.237	0.263	0.289	0.316	0.342	0.369	0.397	0.425	0.453	0.48	0.514	0.546	0.58	0.62	0.66	0.7	0.767	0.909
0.661	75	0.132	0.158	0.184	0.21	0.236	0.262	0.289	0.315	0.342	0.37	0.398	0.426	0.453	0.487	0.519	0.55	0.59	0.63	0.673	0.74	0.882
0.65	76	0.105	0.131	0.157	0.183	0.209	0.235	0.262	0.288	0.315	0.343	0.371	0.399	0.426	0.46	0.492	0.53	0.56	0.6	0.652	0.713	0.855
0.637	77	0.079	0.105	0.131	0.157	0.183	0.209	0.236	0.262	0.289	0.317	0.345	0.373	0.4	0.434	0.466	0.5	0.54	0.58	0.62	0.687	0.829
0.626	78	0.053	0.079	0.105	0.131	0.157	0.183	0.21	0.236	0.263	0.291	0.319	0.347	0.374	0.408	0.44	0.47	0.51	0.55	0.594	0.661	0.803
0.613	79	0.026	0.052	0.078	0.104	0.13	0.156	0.183	0.209	0.236	0.264	0.292	0.32	0.347	0.381	0.413	0.45	0.49	0.53	0.567	0.634	0.776
0.6	80	0	0.026	0.052	0.078	0.104	0.13	0.157	0.183	0.21	0.238	0.266	0.294	0.321	0.355	0.387	0.42	0.46	0.5	0.541	0.608	0.75
0.588	81		0	0.026	0.052	0.078	0.104	0.131	0.157	0.184	0.212	0.24	0.268	0.295	0.329	0.361	0.4	0.43	0.47	0.515	0.528	0.724
0.572	82			0	0.026	0.052	0.078	0.105	0.131	0.158	0.186	0.214	0.242	0.269	0.303	0.335	0.37	0.41	0.45	0.489	0.556	0.698
0.559	83				0	0.026	0.052	0.079	0.105	0.132	0.16	0.188	0.216	0.243	0.277	0.309	0.34	0.38	0.42	0.463	0.53	0.672
0.543	84					0	0.026	0.053	0.079	0.106	0.134	0.162	0.19	0.217	0.251	0.283	0.32	0.36	0.4	0.437	0.504	0.646
0.529	85						0	0.027	0.053	0.08	0.108	0.136	0.164	0.191	0.225	0.257	0.29	0.33	0.37	0.417	0.478	0.62
0.51	86							0	0.026	0.053	0.081	0.109	0.137	0.167	0.198	0.23	0.27	0.3	0.34	0.39	0.451	0.593
0.497	87								0	0.027	0.055	0.083	0.111	0.141	0.172	0.204	0.24	0.23	0.32	0.364	0.425	0.567
0.475	88									0	0.028	0.056	0.083	0.113	0.144	0.176	0.21	0.25	0.29	0.336	0.397	0.54
0.455	89										0	0.028	0.055	0.086	0.117	0.149	0.18	0.22	0.26	0.309	0.37	0.512
0.443	90											0	0.028	0.058	0.089	0.121	0.16	0.19	0.23	0.281	0.342	0.484
0.427	91												0	0.03	0.061	0.093	0.13	0.17	0.21	0.253	0.314	0.456
0.392	92													0	0.031	0.063	0.1	0.14	0.18	0.223	0.284	0.426
0.386	93														0	0.032	0.07	0.1	0.15	0.192	0.253	0.395
0.341	94															0	0.04	0.07	0.11	0.16	0.221	0.363
0.327	95																0	0.04	0.08	0.125	0.186	0.328
0.28	96																	0	0.04	0.089	0.15	0.292
0.242	97																		0	0.048	0.109	0.251
0.199	98																			0	0.061	0.203
0.137	99																				0	0.142

Solution. *Without capacitors at PF = 0.8*

$$\text{kW} = 80$$

$$\text{kVAr} = 60$$

Thus, the kVA requirement of the load is

$$\text{kVA} = (80^2 + 60^2)^{1/2} = 100\ \text{kVA}$$

With capacitors at PF = 0.9

$$\text{kW} = 80$$

$$\text{kVA} \cong 88.9$$

$$\text{Line kVAr} = (88.9^2 - 80^2)^{1/2}(7903 - 6400)^{1/2} = 38.7$$

Hence, the line supplies 38.7 kVAr and the load needs 60 kVAr, and the capacitor supplies the difference, or

$$\text{cKVAr} = 60 - 38.7 = 21.3\ \text{kVAr}$$

as it is illustrated in Figure 8.11.

Example 8.4. Determine the capacitor rating in Example 8.3 by using Table 8.2.

Solution. From Table 8.2, the "kW multiplier" or $\Delta tan\ \theta$ is read as 0.266. Therefore,

$$\begin{aligned}\text{cKVAr} &= \text{kW} \times \Delta\tan\theta \\ &= (80\ \text{kW}) \times (0.266) \\ &= 21.3\ \text{kVAr}\end{aligned}$$

which is the same value determined by the calculation in Example 8.3.

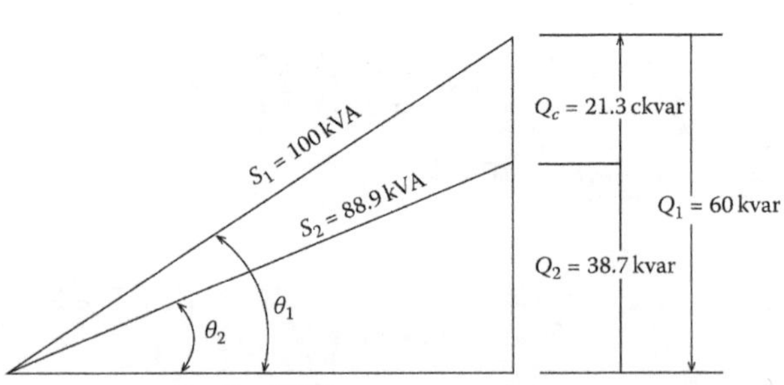

Figure 8.11 Illustration of power factor correction using a shunt capacitor in Example 8.3.

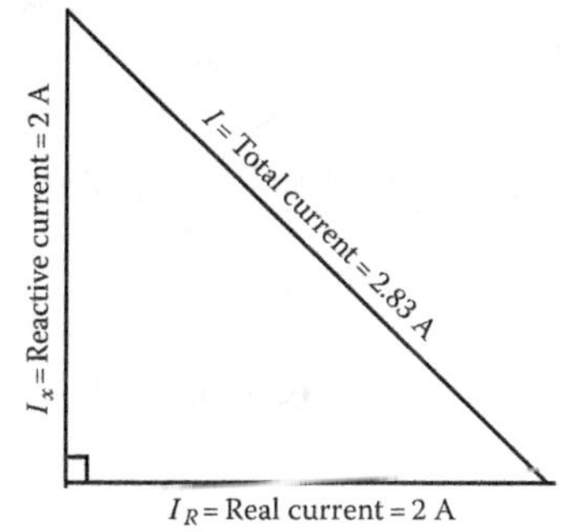

Figure 8.12 Component current diagram.

Example 8.5. Assume that a certain load withdraws a kW current of 2 A and kVAr current of 2 A. Determine the amount of total current that it withdraws.

Solution. The answer is not the following!

$$(2\text{A}) + (2\text{A}) = 4\text{A}$$

The correct answer can be found from the following right-triangle relationships:

$$(\text{kVAr current})^2 + (\text{kW current})^2 = (\text{total current})^2$$

$$(2\text{A})^2 + (2\text{A})^2 = (\text{total current})^2$$

or

$$4 + 4 = (\text{total current})^2$$

Hence,

$$\text{Total current} = 8^{1/2} = 2.83 \text{ A}$$

Thus,

$$2 + 2 \neq 4!$$

Figure 8.12 shows the component current diagram.

Example 8.6. Assume that a 460 V cable circuit is rated at 240 A but is carrying a load of 320 A at 0.65 power factor. Determine the kVAr of capacitor that is needed to reduce the current to 240 A.

Solution.

$$\begin{aligned}\text{kVA} &= \frac{\sqrt{3} \times (460\text{V}) \times (320\text{A})}{1000} \\ &= 254.96 \text{ kVA}\end{aligned}$$

$$\begin{aligned}\text{kW} &= (265.95 \text{ kVA})65 \\ &= 165.72 \text{ kW}\end{aligned}$$

The kVA corresponding to 240 A is

$$\begin{aligned}\text{kVA} &= \frac{\sqrt{3}\times(460\text{V})\times(240\text{A})}{1000}\\ &= 191.2\ \text{kVA}\end{aligned}$$

Thus, the operating power factor corresponding to the new load is

$$\begin{aligned}\text{PF}_2 = \cos\theta_2 &= \frac{P}{S_2}\\ &= \frac{165.72\ \text{kW}}{191.2\ \text{kVA}}\\ &= 0.8667\end{aligned}$$

The capacitor kVAr required is

$$\begin{aligned}\text{cKVAr} &= (165.72\ \text{kW})\tan(\cos^{-1}0.8667)\\ &= (165.72\ \text{kW})\tan(29.92^\circ)\\ &\cong (165.72)(0.5755)\\ &\cong 95.38\end{aligned}$$

8.5.8 REAL POWER LIMITED EQUIPMENT

Certain equipment, such as turbogenerators (turbine generators) and engine generator sets, have limitations on both the real power (P) of the prime mover and the kVA limit of the generator. Typically, the real power limit corresponds to the generator's S rating, and the generator set is rated at that P value under unity power factor operation.

For power factor operations below unity, other real power (P) values are determined based on the power factor and real power (S) rating of the generator. This ensures that the P and S ratings of the load remain within the generator's S rating. Improving the power factor can increase both the P and S capacities.

Example 8.7. Let's consider a 1000 kW turbine unit (turbogenerator set) with a turbine capability of 1250 kW. Currently, it is operating at a rated load of 1250 kVA with a power factor of 0.85. We need to add an additional load of 150 kW at the same power factor of 0.85. To prevent overloading the turbine and the generator, we need to determine the value of capacitors required.

Solution. *Original load*

$$\begin{aligned}P &= 1000\ \text{kW}\\ Q &= \sqrt{(\text{kVA})^2-(\text{kW})^2}\\ &= \sqrt{(1250)^2-(1000)^2}\\ &= 750\ \text{kVAr}\end{aligned}$$

Additional load

$$P = \text{kW} = 150\ \text{kW}$$

$$\begin{aligned} S &= \text{kVA} \\ &= \frac{150\ \text{kW}}{0.85} \\ &= 200\ \text{kVA} \end{aligned}$$

$$\begin{aligned} Q &= \sqrt{(200)^2 - (1000)^2} \\ &= 132.29\ \text{kVAr} \end{aligned}$$

Total load

$$\begin{aligned} P_{\text{tot}} &= \text{kW} = 1000 + 150 \\ &= 1150\ \text{kW} \end{aligned}$$

$$\begin{aligned} Q_{\text{tot}} &= 750 + 132.39 \\ &= 882.29\ \text{kVAr} \end{aligned}$$

The minimum operating power factor for a load of 1150 kW and not exceeding the kVA rating of the generator is

$$\begin{aligned} \text{PF} = \cos\theta &= \frac{1150\ \text{kW}}{1250\ \text{kVA}} \\ &= 0.92 \end{aligned}$$

The maximum load kVAr for this situation is

$$\begin{aligned} Q &= (1150\ \text{kW})\ \tan^{-1}\theta \\ &= 1150 \times \tan^{-1}(23.073918^\circ) \\ &\cong 489.9\ \text{kVAr} \end{aligned}$$

where 0.426 is the tangent corresponding to the maximum power factor of 0.935.

Thus, the capacitors must provide the difference between the total load kVAr and the permissible generator kVAr, or

$$\begin{aligned} \text{cKVAr} &= 882.29 - 489.9 \\ &= 392.39\ \text{kVAr} \end{aligned}$$

Example 8.8. Assume that a 700 k VA load has a 65% power factor. It is desired to improve the power factor to 92%. Using Table 8.2, determine the following:

a. The correction factor required.
b. The capacitor size required.
c. What would be the resulting power factor if the next higher standard capacitor size is used?

Solution. a. From Table 8.2, the correction factor required can be found as 0.74.
b. The 700 kVA load at 65% power factor is

$$\begin{aligned} P_L &= S_L \times \cos\theta \\ &= 700 \times 0.65 \\ &= 455 \text{ kW} \end{aligned} \tag{8.25}$$

The capacitor size necessary to improve the power factor from 65% to 92% can be found as

$$\begin{aligned} \text{Capacitor size} &= P_L(\text{correlation factor}) \\ &= 455(0.74) \\ &= 336.7 \text{ kVAr} \end{aligned} \tag{8.26}$$

c. Assume that the next higher standard capacitor size (or rating) is selected to be 360 kVAr. Therefore, the resulting new correction factor can be found from

$$\begin{aligned} \text{New correction factor} &= \frac{\text{Standard capacitor rating}}{P_L} \\ &= \frac{360 \text{ kVAr}}{455 \text{ kW}} \\ &= 0.7912 \end{aligned} \tag{8.27}$$

From the table by linear interpolation, the resulting corrected percent power factor, with an original power factor of 65% and a correction factor of 0.7912, can be found as

$$\begin{aligned} \text{New corrected \% power factor} &= 93 + \frac{172}{320} \\ &\cong 93.5 \end{aligned}$$

8.5.9 COMPUTERIZED METHOD TO DETERMINE THE ECONOMIC POWER FACTOR

As proposed by Hopkinson [200], a load flow digital computer program can be utilized to determine the kVA, kV, and kVAr at the annual peak level for the entire system, spanning from generation through the distribution substation buses, while varying the power factor.

To begin, shunt capacitors are installed at each substation bus to correct the power factor to an initial value, such as 90%. Subsequently, a load flow analysis is conducted to determine the total system kVA and kW losses (from generator to load) at this level, and the capacitor kVAr are recorded. Later, additional capacitors are added to each substation bus to raise the power factor by 1%, followed by another load flow analysis. This iterative process is repeated until the power factor reaches unity.

As a final step, the benefits and costs are computed at each power factor level. The economic power factor is determined as the value at which the benefits and costs are equal. Once the economic power factor is established, the required size of additional capacitors can be calculated as follows:

$$\Delta Q_c = P_{PK}(\tan\,\phi - \tan\,\theta) \tag{8.28}$$

where
ΔQ_c is the required capacitor size, kVAr
P_{PK} is the system demand at annual peak, kW
$\tan\phi$ is the tangent of original power factor angle
$\tan\theta$ is the tangent of economic power factor angle

An illustration of this method is given in Example 8.12.

8.6 APPLICATION OF CAPACITORS

Generally, capacitors can be utilized at various voltage levels. As depicted in Figure 8.15, individual capacitor units can be connected in parallel to achieve the desired kVAr capacity, while series connection allows for reaching the required kV voltage. It is typically economical to operate them at or near their rated voltage.

Based on cumulative industry data, approximately 60% of capacitors are installed on feeders, 30% on substation buses, and the remaining 10% on the transmission system [200].

The application of capacitors to secondary systems is infrequent due to their limited economic benefits. Zimmerman [201] developed a nomograph, shown in Figure 8.16, that helps determine the economic feasibility, if any, of secondary capacitors, considering only the savings in distribution transformer costs.

Example 8.9. Let's consider a wye-connected induction motor operating at 4160 V, 60 Hz, with a power rating of 500 hp. The motor has a full-load efficiency (η) of 88% and operates at a lagging power factor of 0.75. It is connected to a feeder. If the goal is to improve the power factor of the load to 0.9 lagging by connecting three capacitors at the load, we need to determine the following:

a. The rating of the capacitor bank, in kVAr
b. The capacitance of each unit if the capacitors are connected in delta, in microfarads
c. The capacitance of each unit if the capacitors are connected in wye, in microfarads

Solution. a. The input power of the induction motor can be found as

$$
\begin{aligned}
P &= \frac{(\text{HP})(0.7457\ \text{HP/hp})}{\eta} \\
&= \frac{(500\ \text{hp})(0.7457\ \text{kW/hp})}{0.88} \\
&= 423.69\ \text{kW}
\end{aligned}
$$

The reactive power of the motor at the uncorrected power factor is

$$
\begin{aligned}
Q_1 &= P \tan\theta_1 \\
&= 423.69 \tan(\cos^{-1} 0.75) \\
&= 423.69 \times 0.8819 \\
&= 373.7\ \text{kVAr}
\end{aligned}
$$

The reactive power of the motor at the corrected power factor is

$$
\begin{aligned}
Q_2 &= P \tan\theta_2 \\
&= 423.69 \tan(\cos^{-1} 0.90) \\
&= 423.69 \times 0.4843 \\
&= 205.2\ \text{kVAr}
\end{aligned}
$$

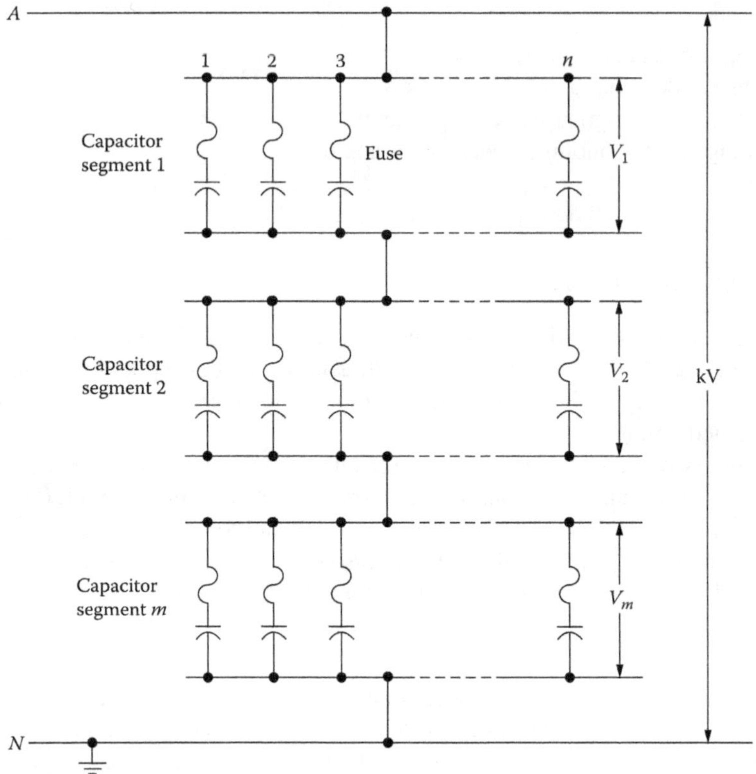

Figure 8.13 Connection of capacitor units for one phase of a three-phase wye-connected bank.

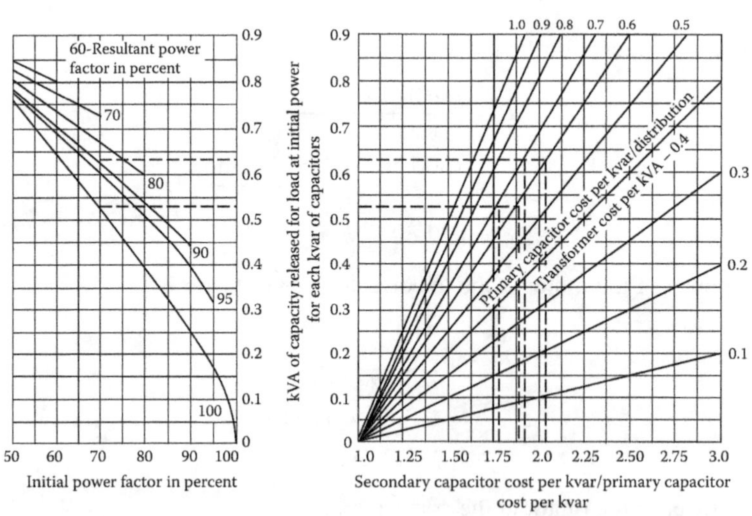

Figure 8.14 Secondary capacitor economics considering only savings in distribution transformer cost. (From Zimmerman, R.A., *AIEE Trans.*, 72, 694, Copyright 1953 IEEE. Used with permission.)

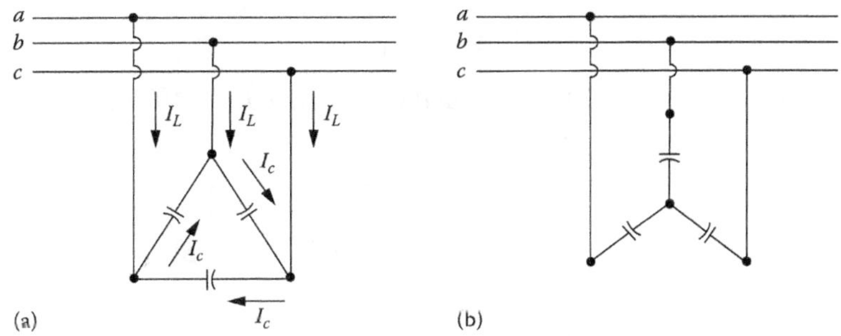

Figure 8.15 Capacitor connected (a) in delta and (b) in wye.

Therefore, the reactive power provided by the capacitor bank is

$$
\begin{aligned}
Q_c &= Q_1 - Q_2 \\
&= 373.7 - 205.2 \\
&= 168.5 \text{ kVAr}
\end{aligned}
$$

Hence, assuming the losses in the capacitors are negligible, the rating of the capacitor bank is 168.5 kVAr.

b. If the capacitors are connected in delta as shown in Figure 8.15a, the line current is

$$
\begin{aligned}
I_L &= \frac{Q_c}{\sqrt{3} \times V_{L-L}} \\
&= \frac{168.5}{\sqrt{3} \times 4.16} \\
&= 23.39 \text{ A}
\end{aligned}
$$

and therefore,

$$
\begin{aligned}
I_c &= \frac{I_L}{\sqrt{3}} \\
&= \frac{23.39}{\sqrt{3}} \\
&= 13.5 \text{ A}
\end{aligned}
$$

Thus, the reactance of each capacitor is

$$
\begin{aligned}
X_c &= \frac{V_{L-L}}{I_c} \\
&= \frac{4160}{13.5} \\
&= 308.11\ \Omega
\end{aligned}
$$

and hence, the capacitance of each unit,* if the capacitors are connected in delta, is

$$
C = \frac{10^6}{\omega X_c} \mu\text{F}
$$

or

*→ Note that $C = 1/(\omega X_c)F$. If the equation is divided by 10^6 side by side, then $C = 10^6/(\omega X_c)(10^{-6}F)$ or $C = 10^6/(\omega X_c)\mu F$:

$$C/10^6 = (1/\omega X_c)/10^6 \text{ F}$$

$$\begin{aligned} C &= \frac{10^6}{\omega X_c} \\ &= \frac{10^6}{2\pi \times 60 \times 308.11} \\ &= 8.61 \mu\text{F} \end{aligned}$$

c. If the capacitors are connected in wye as shown in Figure 8.15b,

$$I_c = I_L = 23.39 \text{ A}$$

and therefore,

$$\begin{aligned} X_c &= \frac{V_{L-L}}{I_c} \\ &= \frac{4160}{\sqrt{3} \times 23.39} \\ &= 102.70 \ \Omega \end{aligned}$$

Thus, the capacitance of each unit, if the capacitors are connected in wye, is

$$\begin{aligned} C &= \frac{10^6}{\omega X_c} \\ &= \frac{10^6}{2\pi \times 60 \times 102.70} \\ &= 25.82 \mu\text{F} \end{aligned}$$

Example 8.10. Assume that a 2.4 kV single-phase circuit feeds a load of 360 kW (measured by a wattmeter) at a lagging load factor and the load current is 200 A. If it is desired to improve the power factor, determine the following:

a. The uncorrected power factor and reactive load.
b. The new corrected power factor after installing a shunt capacitor unit with a rating of 300 kVAr.
c. Also write the necessary codes to solve the problem in MATLAB.

Solution. a. Before the power factor correction,

$$\begin{aligned} S_1 &= V \times I \\ &= 2.4 \times 200 \\ &= 480 \text{ kVA} \end{aligned}$$

therefore, the uncorrected power factor can be found as

$$\begin{aligned}\cos\theta_1 &= \frac{P}{S_1}\\ &= \frac{360\ \text{kW}}{480\ \text{kVA}}\\ &= 0.75\end{aligned}$$

and the reactive load is

$$\begin{aligned}Q_1 &= S_1 \times \sin(\cos^{-1}\theta_1)\\ &= 480 \times 0.661\\ &= 317.5\ \text{kVAr}\end{aligned}$$

b. After the installation of the 300 kVAr capacitors,

$$\begin{aligned}Q_2 &= Q_1 \quad Q_c\\ &= 317.5 - 300\\ &= 17.5\ \text{kVAr}\end{aligned}$$

and therefore, the new power factor can be found from Equation 8.9 as

$$\begin{aligned}\cos\theta_2 &= \frac{P}{[P^2 + (Q_1 - Q_c)^2]^{1/2}}\\ &= \frac{360}{(360^2 + 17.5^2)^{1/2}}\\ &= 0.9989 \text{ or } 99.89\%\end{aligned}$$

Example 8.11. Let's assume that the Riverside Substation, owned by the NL&NP Company, is equipped with a bank of three 2000 kVA transformers. This transformer bank caters to a peak load of 7800 kVA, operating at a lagging power factor of 0.89. Each of the transformers has a thermal capacity that exceeds 120% of its nameplate rating. Furthermore, as part of the plan to enhance voltage regulation, it has been decided to install 1000 kVAr of shunt capacitors on the feeder.

Determine the following:

a. Whether or not to install additional capacitors on the feeder to decrease the load to the thermal capability of the transformer
b. The rating of the additional capacitors

Solution. a. Before the installation of the 1000 kVAr capacitors,

$$\begin{aligned}P &= S_1 \times \cos\theta\\ &= 7800 \times 0.89\\ &= 6942\ \text{kW}\end{aligned}$$

and

$$\begin{aligned} Q_1 &= S_1 \times \sin\theta \\ &= 7800 \times 0.456 \\ &= 3556.8 \text{ kVAr} \end{aligned}$$

Therefore, after the installation of the 1000 kVAr capacitors,

$$\begin{aligned} Q_2 &= Q_1 - Q_c \\ &= 3556.8 - 1000 \\ &= 2556.8 \text{ kVAr} \end{aligned}$$

and using Equation 8.9,

$$\begin{aligned} \cos\theta_2 &= \frac{P}{[P^2 + (Q_1 - Q_c)^2]^{1/2}} \\ &= \frac{6942}{(6942^2 + 2556.8^2)^{1/2}} \\ &= 0.938 \text{ or } 93.8\% \end{aligned}$$

and the corrected apparent power is

$$\begin{aligned} S_2 &= \frac{P}{\cos\theta_2} \\ &= \frac{6942}{0.938} \\ &= 7397.9 \text{ kVA} \end{aligned}$$

On the other hand, the transformer capability is

$$\begin{aligned} S_T &= 6000 \times 1.20 \\ &= 7200 \text{ kVA} \end{aligned}$$

Therefore, the capacitors installed to improve the voltage regulation are not adequate; additional capacitor installation is required.

b. The new or corrected power factor required can be found as

$$\begin{aligned} \text{PF}_{2,\text{new}} = \cos\theta_{2,\text{new}} &= \frac{P}{S_T} \\ &= \frac{6942}{7200} \\ &= 0.9642 \text{ or } 96.42\% \end{aligned}$$

and thus, the new required reactive power can be found as

$$\begin{aligned} Q_{2,\text{new}} &= P \times \tan\theta_{2,\text{new}} \\ &= P \times \tan(\cos^{-1}\text{PF}_{2,\text{new}}) \\ &= 6942 \times 0.2752 \\ &= 1910 \text{ kVAr} \end{aligned}$$

Therefore, the rating of the additional capacitors required is

$$\begin{aligned} Q_{c,\text{add}} &= Q_2 - Q_{2,\text{new}} \\ &= 2556.8 - 1910 \\ &= 646.7 \text{ kVAr} \end{aligned}$$

Example 8.12. In a power system with a capacity of 10,000 kVA and operating at a power factor of 0.7, let's determine the investment required to correct the power factor using synchronous capacitors (synchronous condensers) at a cost of $10 per kVA. We need to calculate the investment required to correct the power factor to a desired value.

a. 0.85 lagging power factor
b. Unity power factor

Solution. *At original cost*

$$\theta_{\text{old}} = \cos^{-1}\text{PF} = \cos^{-1}0.7 = 45.57°$$

$$P_{\text{old}} = S\cos\theta_{\text{old}} = (10{,}000\text{kVA})\ 0.7 = 7{,}000 \text{ kW}$$

$$Q_{\text{old}} = S\cos\theta_{\text{old}} = (10{,}000\text{kVA})\sin(45.57°) = 7{,}141.43 \text{ kVAr}$$

a. *For PF = 0.85 lagging*

$$P_{\text{new}} = P_{\text{old}} = 7000 \text{ kW (as before)}$$

$$S_{\text{new}} = \frac{P_{\text{new}}}{\cos\theta_{\text{new}}} = \frac{7000 \text{ kW}}{0.85} = 8235.29 \text{ kVA}$$

$$Q_{\text{new}} = S_{\text{new}}\sin(\cos^{-1}\text{PF}) = (8235.29 \text{ kVA}\sin(\cos^{-1}0.85) = 4338.21 \text{ kVAr}$$

$$Q_c = Q_{\text{renuired}} = Q_{\text{old}} - Q_{\text{new}} = 7141.43 - 4338.21$$

$$= 2803.22 \text{ kVAr correction needed}$$

Hence, the *theoretical* cost of the synchronous capacitor is

$$\text{Cost}_{\text{capacitor}} = (2{,}803.22 \text{ kVA})\left(\frac{\$10}{\text{kVA}}\right) = \$28{,}032.20$$

Note that it is customary to give the cost of capacitors in dollars per kVA rather than in dollars per kVAr.

b. *For PF = 1.0*

$$Q_c = Q_{\text{required}} = Q_{\text{old}} - Q_{\text{new}} = 7141.43 - 0.0 = 7141.43 \text{ kVAr}$$

Thus, the *theoretical* cost of the synchronous capacitor is

$$\text{Cost}_{\text{capacitor}} = (7,141.43 \text{ kVA})\left(\frac{\$10}{\text{kVA}}\right) = \$71,414.30$$

Note that P_{new} = 7000 kW is the same as before.

Example 8.13. Let's consider a power system with a capacity of 15,000 kVA operating at a lagging power factor of 0.65. The cost of synchronous capacitors to correct the power factor is \$12.5 per kVA. We need to determine the costs involved in correcting the power factor and create a table that shows the required leading reactive power to increase the power factor to a desired value.

a. 0.85 lagging power factor
b. 0.95 lagging power
c. Unity power factor

Solution. *At original power factor or 0.65*

$$P = S\cos\theta = (15,000 \text{ kVA})0.65 = 9750 \text{ kW at a power factor angle of } 49.46^\circ$$

$$Q = S\sin\theta = (15,000 \text{ kVA})\sin(\cos^{-1}0.65) = 11,399 \text{ kVAr}$$

The following table shows the amount of reactive power that is required to improve the power factor from one level to the next at 0.05 increments.

PF	P (kW)	Q (kVAr)	Q to Correct from Next Lower PF (kVAr)	Cumulative Q Required for Correction (kVAr)
0.65	9,750	11,399	—	—
0.7	10,500	10,712	687	687
0.75	11,250	9,922	790	1,477
0.8	12,000	9,000	922	2,399
0.85	12,750	7,902	1098	3,497
0.9	13,500	6,538	1364	4,861
0.95	14,250	4,684	1854	6,715
1	15,000	0	4684	11,399

a. *For PF = 0.85 lagging*
$P = S\cos\theta = (15,000\text{kVA}) \times 0.65 = 9,750\text{kW}$. It will be the same at a power factor of 0.85.

$$S = \frac{P}{\cos\theta} = \frac{9,750 \text{ kW}}{0.85} = 11,470 \text{ kVA}$$

and

$$Q = \frac{S}{\sin\theta} = (11,470 \text{ kVA})\sin(\cos^{-1}0.85) = 6,042 \text{ kVAr}$$

The amount of additional reactive power correction required is

$$\text{Additional var correction} = 11,399 - 6,042 = 5,357 \text{ kVAr}$$

The cost of this correction is

$$\text{Cost of correction} = (5,357 \text{ kVA})\left(\frac{\$12.5}{\text{KVA}}\right) = \$66,962.50$$

b. *For PF = 0.95 lagging:*

$$S = \frac{9,750 \text{ kW}}{0.95} = 10,263 \text{ kVA}$$

and

$$Q = (10,263 \text{ kVA})\sin(\cos^{-1}0.95) = 3,204 \text{ kVAr}$$

The amount of additional reactive power correction required is

$$\text{Additional var correction} = 11,399 - 3,204 = 8,195 \text{ kVAr}$$

The cost of this correction is

$$\text{Cost of correction} = (8,195 \text{ kVA})\left(\frac{\$12.5}{\text{KVA}}\right) = \$102,438$$

c. *For unity PF*
The amount of additional reactive power correction required is

$$\text{Additional var correction} = 11,399 \text{ kVAr}$$

The cost of this correction is

$$\text{Cost of correction} = (11,399 \text{ kVA})\left(\frac{\$12.5}{\text{KVA}}\right) = \$142,487.50$$

8.6.1 CAPACITOR INSTALLATION TYPES

Typically, capacitors installed on feeders are in the form of pole-top banks, equipped with appropriate group fusing to ensure safe operation. The use of fuses imposes limitations on the size of the capacitor bank that can be employed. As a result, the maximum sizes commonly utilized are approximately 1800 kVAr at 15 kV and 3600 kVAr at higher voltage levels. Utilities typically avoid installing more than four capacitor banks, all of equal sizes, on each feeder.

Figure 8.16 demonstrates the impact of a fixed capacitor on the voltage profiles of a feeder with a uniformly distributed load under heavy load and light load conditions. When only fixed capacitors are installed, as depicted in Figure 8.16c, the utility faces an excessive leading power factor and

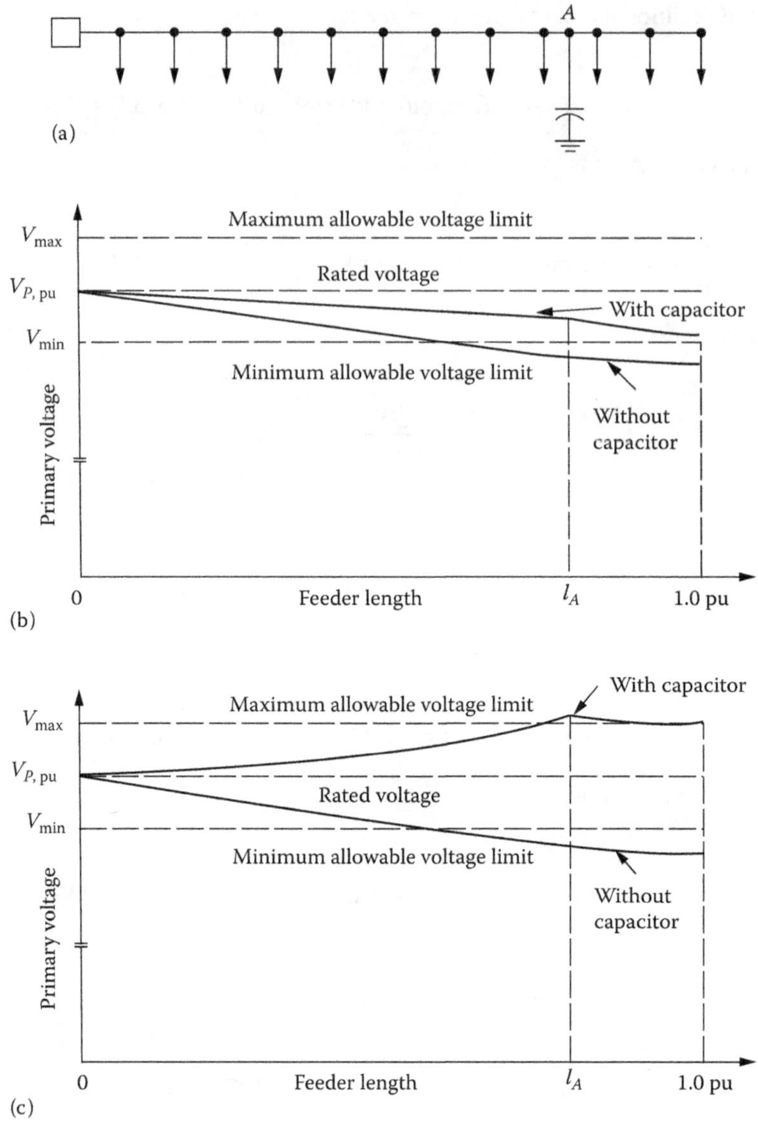

Figure 8.16 The effects of a fixed capacitor on the voltage profile of (a) feeder with uniformly distributed load (b) at heavy load and (c) at light load.

voltage rise at that feeder. To address this issue, some capacitors are installed as switched capacitor banks, as shown in Figure 8.17, allowing them to be switched off during light-load conditions.

As illustrated, the fixed capacitors are sized for light load and remain connected permanently. The switched capacitors, on the other hand, can be switched as a block or in several consecutive steps as the reactive load increases from the light-load level to the peak load, with their sizes adjusted accordingly [202].

In practice, however, the number of steps or blocks selected is usually fewer than those depicted in the figure due to the additional expenses associated with installing the necessary switchgear and control equipment.

Choosing the appropriate type of capacitor installation requires conducting a system survey. By performing load flow program runs or manual load studies on feeders or distribution substations,

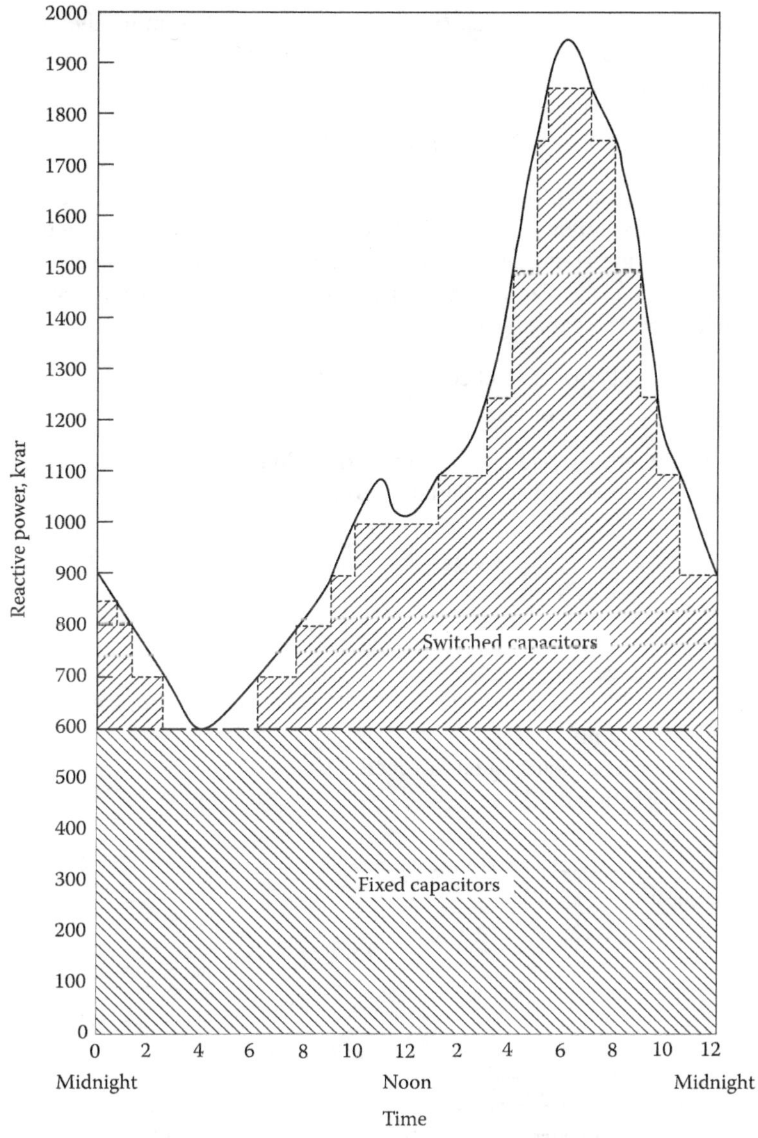

Figure 8.17 Sizing of the fixed and switched capacitors to meet the daily reactive power demands.

the lagging reactive loads (i.e., power demands) of the system can be determined and plotted on a curve, known as the reactive load-duration curve, as shown in Figure 8.17. This curve represents the cumulative sum of the reactive loads (e.g., fluorescent lights, household appliances, and motors) of consumers and the reactive power requirements of the system (e.g., transformers and regulators). Through visual inspection of this curve, the size of the fixed capacitors needed to meet the minimum reactive load can be determined. For instance, based on Figure 8.17, it can be inferred that the required size of fixed capacitors is 600 kVAr.

The remaining reactive power requirements of the loads are fulfilled either by the generator or, preferably, by switched capacitors. However, relying on the generator to meet the entire reactive power demand can be costly and may introduce stability issues in the system. Therefore, capacitors are employed. Capacitor sizes are carefully chosen to align with the varying characteristics of the remaining load throughout different time periods.

A commonly employed guideline used by many utilities to determine the sizing of switched capacitors is as follows: Keep adding switched capacitors until

$$\frac{\text{kVAr from switched + fixed capacitors}}{\text{kVAr of peak reactive feeder load}} \geq 0.70 \tag{8.29}$$

In terms of voltage regulation, the reactive power (kVAr) required to raise the voltage at the end of the feeder to the maximum permissible level during minimum load conditions (at 25% of the peak load) determines the sizing of the fixed capacitors to be used. Conversely, when multiple capacitor banks are installed, the size of each capacitor bank at every location should maintain the same proportion, which can be expressed as follows:

$$\frac{\text{kVAr of load center}}{\text{kVAr of total feeder}} = \frac{\text{kVA of load center}}{\text{kVA of total feeder}} \tag{8.30}$$

However, the resultant voltage rise must not exceed the light-load voltage drop. The approximate value of the percent voltage rise can be calculated from

$$\%\text{VR} = \frac{Q_c, 3\phi \times x \times l}{10 \times V_{L-L}^2} \tag{8.31}$$

where
%VR is the percent voltage rise
$Q_{c,3\phi}$ is the three-phase reactive power due to fixed capacitors applied, kVAr
x is the line reactance, Ω/min
l is the length of feeder from sending end of feeder to fixed capacitor location, min
V_{L-L} is the line-to-line voltage, kV

The percent voltage rise can also be found from

$$\%\text{VR} = \frac{I_c \times x \times l}{10 \times V_{L-L}^2} \tag{8.32}$$

where

$$I_c = \frac{Q_{c,3\phi}}{\sqrt{3} \times V_{L-L}}$$

$$= \text{current drawn by fixed} - \text{capacitor bank}$$

If the fixed capacitors are applied to the end of the feeder and if the percent voltage rise is already determined, the maximum value of the fixed capacitors can be determined from

$$\text{Max}Q_{c,3\phi} = \frac{10(\%\text{VR})V_{L-L}^2}{x \times l}\text{kVAr} \tag{8.33}$$

Equations 8.31 and 8.32 can also be used to calculate the percent voltage rise due to the switched capacitors. Therefore, once the percent voltage rises due to both fixed and switched capacitors, the total percent voltage rise can be calculated as

$$\sum \%\text{VR} = \%\text{VR}_{\text{NSW}} + \%\text{VR}_{\text{SW}} \tag{8.34}$$

where
$\sum \%\text{VR}$ is the total percent voltage rise
$\%\text{VR}_{\text{NSW}}$ is the percent voltage rise due to fixed (or nonswitched) capacitors

$\%VR_{SW}$ is the percent voltage rise due to switched capacitors

Certain utilities follow the following rule of thumb: "The total combined capacity of fixed and switched capacitors for a feeder should be sufficient to raise the voltage at the receiving end of the feeder to its maximum level when the feeder load is at 50% of its peak."

After determining the required kVAr of capacitors for the system, the next step is to determine their appropriate locations. For feeders with uniformly distributed loads, a general guideline is to position the fixed capacitors approximately two-thirds of the distance from the substation to the end of the feeder.

In the case of uniformly decreasing loads, fixed capacitors are typically placed around the midpoint of the feeder. On the other hand, the placement of switched capacitors primarily depends on voltage regulation requirements, and it typically corresponds to the last one-third of the feeder, furthest from the power source.

8.6.2 TYPES OF CONTROLS FOR SWITCHED SHUNT CAPACITORS

The switching operation of capacitors can be performed either through manual control or automatic control utilizing various intelligent control techniques. Manual control, which can be implemented locally or remotely, is often used at distribution substations. Automatic control methods involve different types of control intelligence, such as time-switch, voltage, current, voltage-time, voltage-current, and temperature controls.

Among these, the most widely used control types are time-switch control, voltage control, and voltage-current control. Time-switch control is the most cost-effective option. In some cases, combinations of these control methods are employed to more closely align with the reactive load-duration curve, as demonstrated in Figure 8.20.

8.6.3 TYPES OF THREE-PHASE CAPACITOR-BANK CONNECTIONS

There are three possible connection options for a three-phase capacitor bank on a distribution feeder: (1) delta connection, (2) grounded wye connection, or (3) ungrounded wye connection. The selection of the connection type depends on the following factors:

1. System type, that is, whether it is a grounded or an ungrounded system
2. Fusing requirements
3. Capacitor-bank location
4. Telephone interference considerations

A resonance condition can arise in delta and ungrounded-wye (floating neutral) capacitor banks when a one- or two-line open-type fault occurs on the source side of the capacitor bank. This fault can be problematic as the maintained voltage on the open phase can backfeed any transformers positioned on the load side of the open conductor through the series capacitor. This situation can potentially lead to damage to single-phase distribution transformers in four-wire systems. As a result, it is not advisable to use ungrounded-wye capacitor banks in the following conditions:

1. On feeders with light load where the minimum load per phase beyond the capacitor bank does not exceed 150% of the per phase rating of the capacitor bank
2. On feeders with single-phase breaker operation at the sending end
3. On fixed capacitor banks
4. On feeder sections beyond a sectionalizing-fuse or single-phase recloser
5. On feeders with emergency load transfers

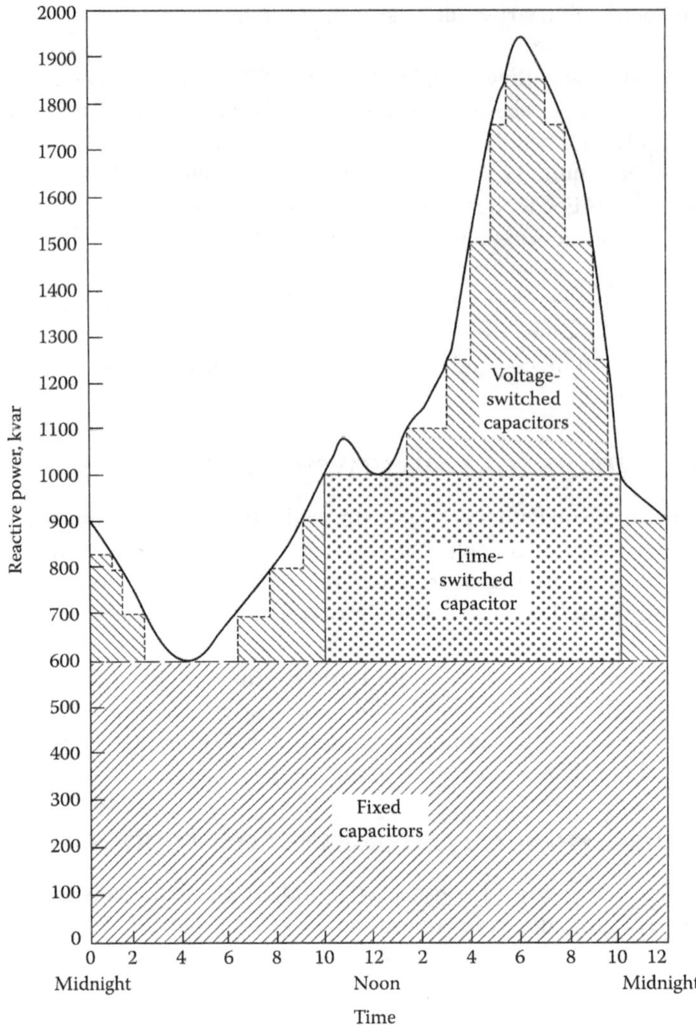

Figure 8.18 Meeting the reactive power requirements with fixed, voltage-controlled, and time-controlled capacitors.

However, the *ungrounded-wye capacitor banks are recommended if one or more of the following conditions exist*:

1. Excessive harmonic currents in the substation neutral can be precluded.
2. Telephone interferences can be minimized.
3. Capacitor-bank installation can be made with two single-phase switches rather than with three single-pole switches.

Grounded-wye capacitor banks are typically utilized exclusively in four-wire three-phase primary systems. However, if a grounded-wye capacitor bank is employed in a three-phase three-wire ungrounded-wye or delta system, it can introduce a ground current source that has the potential to disrupt sensitive ground relays.

8.7 ECONOMIC JUSTIFICATION FOR CAPACITORS

Loads in electric utility systems consist of two components: active power, measured in kW, and reactive power, measured in kVAr. Active power needs to be generated by power plants, while reactive power can be supplied by power plants or capacitors. It is widely recognized that shunt power capacitors are the most cost-effective solution to fulfill the reactive power requirements of inductive loads and transmission lines operating at a lagging power factor.

When reactive power is solely provided by power plants, every component in the system, including generators, transformers, transmission and distribution lines, switchgear, and protective equipment, must be enlarged accordingly. Capacitors alleviate this issue by reducing the demand for reactive power all the way back to the generators. This reduction in reactive power demand leads to decreased line currents from capacitor locations back to the generation equipment. Consequently, losses and loadings are reduced in distribution lines, substation transformers, and transmission lines.

Depending on the uncorrected power factor of the system, capacitor installations can increase the capability of generators and substations for additional load by at least 30%, and from a voltage regulation standpoint, they can increase individual circuit capacity by approximately 30% to 100%.

Furthermore, the reduction in current flow through transformers, distribution equipment, and lines alleviates the load on these apparatuses, which are limited in kVA, thereby delaying the need for new facility installations. Generally, the economic advantages of capacitor banks necessitate their installation on the primary distribution system rather than the secondary.

A well-established rule of thumb is that the optimal amount of capacitor kVArs to utilize is the amount where the economic benefits obtained from adding the last kVAr equal the installed cost of the capacitors.

The methods employed by utilities to determine the economic benefits derived from capacitor installations may vary, but calculating the total installed cost of a kVAr of capacitors is straightforward and uncomplicated.

In general, the economic benefits obtained from the installation of capacitors can be summarized as follows:

1. Released generation capacity
2. Released transmission capacity
3. Released distribution substation capacity
4. Additional advantages in distribution system
    a. Reduced energy (copper) losses
    b. Reduced voltage drop and consequently improved voltage regulation
    c. Released capacity of feeder and associated apparatus
    d. Postponement or elimination of capital expenditure due to system improvements and/or expansions
    e. Revenue increase due to voltage improvements

8.7.1 BENEFITS DUE TO RELEASED GENERATION CAPACITY

The released generation capacity due to the installation of capacitors can be calculated approximately from

$$\Delta S_G = \begin{cases} \left[\left(1 - \frac{Q_c^2 \times \cos^2\theta}{S_G^2}\right)^{1/2} + \frac{Q_c \times \sin\theta}{S_G} - 1\right] S_G & \text{when } Q_c > 0.10 S_G \\ Q_c \times \sin\theta & \text{when } Q_c \leq 0.10 S_G \end{cases} \quad (8.35)$$

where

ΔS_G is the released generation capacity beyond maximum generation capacity at original power factor, kVA

S_G is the generation capacity, kVA

Q_c is the reactive power due to corrective capacitors applied, kVAr

$\cos\theta$ is the original (or uncorrected or old) power factor before application of capacitors

Therefore, the annual benefits due to the released generation capacity can be expressed as

$$\Delta\$_G = \Delta S_G \times C_G \times i_G \tag{8.36}$$

where

$\Delta\$_G$ is the annual benefits due to released generation capacity, \$/year

ΔS_G is the released generation capacity beyond maximum generation capacity at original power factor, kVA

C_G is the cost of (peaking) generation, \$/kW

i_G is the annual fixed charge rate* applicable to generation

8.7.2 BENEFITS DUE TO RELEASED TRANSMISSION CAPACITY

The released transmission capacity due to the installation of capacitors can be calculated approximately as

$$\Delta S_T = \begin{cases} \left[\left(1 - \frac{Q_c^2 \times \cos^2\theta}{S_T^2}\right)^{1/2} + \frac{Q_c \times \sin\theta}{S_T} - 1\right] S_T & \text{when } Q_c > 0.10 S_T \\ Q_c \times \sin\theta & \text{when } Q_c \leq 0.10 S_T \end{cases} \tag{8.37}$$

where

ΔS_T is the released transmission capacity† beyond maximum transmission capacity at original power factor, kVA

S_T is the transmission capacity, kVA

* → Also called *carrying charge rate*. It is defined as that portion of the annual revenue requirements that results from a plant investment. Total carrying charges include (1) return (on equity and debt), (2) book depreciation, (3) taxes (including amount paid currently and amounts deferred to future years), (4) insurance, and (5) operations and maintenance. It is expressed as a decimal.

† → Note that the symbol S_T now stands for transmission capacity rather than transformer capacity.

Thus, the annual benefits due to the released transmission capacity can be found as

$$\Delta\$_T = \Delta S_T \times C_T \times i_T \tag{8.38}$$

where

$\Delta\$_T$ is the annual benefits due to released transmission capacity, \$/year

ΔS_T is the released transmission capacity beyond maximum transmission capacity at original power factor, kVA

C_T is the cost of transmission line and associated apparatus, \$/kVA

i_T is the annual fixed charge rate applicable to transmission

8.7.3 BENEFITS DUE TO RELEASED DISTRIBUTION SUBSTATION CAPACITY

The released distribution substation capacity due to the installation of capacitors can be found approximately from

$$\Delta S_S = \begin{cases} \left[\left(1 - \frac{Q_c^2 \times \cos^2\theta}{S_S^2}\right)^{1/2} + \frac{Q_c \times \sin\theta}{S_S} - 1\right] S_S & \text{when } Q_c > 0.10 S_S \\ Q_c \times \sin\theta & \text{when } Q_c \leq 0.10 S_S \end{cases} \tag{8.39}$$

where

ΔS_S is the released distribution substation capacity beyond maximum substation capacity at original power factor, kVA

S_S is the distribution substation capacity, kVA

Hence, the annual benefits due to the released substation capacity can be calculated as

$$\Delta\$_S = \Delta S_S \times C_S \times i_S \tag{8.40}$$

where

$\Delta\$_S$ is the annual benefits due to the released substation capacity, \$/year

ΔS_S is the released substation capacity, kVA

C_S is the cost of substation and associated apparatus, \$/kVA

i_S is the annual fixed charge rate applicable to substation

8.7.4 BENEFITS DUE TO REDUCED ENERGY LOSSES

The annual energy losses are reduced as a result of decreasing copper losses due to the installation of capacitors. The conserved energy can be expressed as

$$\Delta\text{ACE} = \frac{Q_{c,3\phi} R(2S_{L,3\phi}\sin\theta - Q_{c,3\phi})8760}{1000 \times V_{L-L}^2} \tag{8.41}$$

where

ΔACE is the annual conserved energy, kWh/year

$Q_{c,3\phi}$ is the three-phase reactive power due to corrective capacitors applied, kVAr

R is the total line resistance to load center, Ω

$Q_{L,3\phi}$ is the original, that is, uncorrected, three-phase load, kVA

sin θ is the sine of original (uncorrected) power factor angle

V_{L-L} is the line-to-line voltage, kV

Therefore, the annual benefits due to the conserved energy can be calculated as

$$\Delta\$_{\text{ACE}} = \Delta\text{ACE} \times \text{EC} \tag{8.42}$$

where

ΔACE is the annual benefits due to conserved energy, \$/year

EC is the cost of energy, \$/kWh

8.7.5 BENEFITS TO REDUCED VOLTAGE DROPS

The following advantages can be obtained by the installation of capacitors into a circuit:

1. The effective line current is reduced, and consequently, both IR and IX_L voltage drops are decreased, which results in improved voltage regulation.

2. The power factor improvement further decreases the effect of reactive line voltage drop.

The percent voltage drop that occurs in a given circuit can be expressed as

$$\%\text{VD} = \frac{S_{L,3\phi}(r\cos\theta + x\sin\theta)l}{10\times V_{L-L}^2} \tag{8.43}$$

where
$\%VD$ is the percent voltage drop
$S_{L,3\phi}$ is the three-phase load, kVA
r is the line resistance, 0/min
x is the line reactance, 0/min
l is the length of conductors, min
V_{L-L} is the line-to-line voltage, kV

The calculation of voltage drop, as determined by Equation 8.44, forms the foundation for capacitor application. Following the installation of capacitors, the system experiences a voltage increase attributable to the enhanced power factor and reduced effective line current. Consequently, the voltage drops caused by IR and IX_L are minimized. The approximate percentage increase in voltage along the line can be estimated as:

$$\%\text{VR} = \frac{Q_{c,3\phi}\times x\times l}{10\times V_{L-L}^2} \tag{8.44}$$

Moreover, the introduction of capacitors leads to an additional voltage-rise effect across each transformer from the power generation source to the capacitor installation point. This phenomenon is unrelated to the load and power factor of the line and can be expressed as follows:

$$\%\text{VR}_T = \left(\frac{Q_{c,3\phi}}{S_{T,3\phi}}\right)x_T \tag{8.45}$$

where
$\%VR_T$ is the percent voltage rise through the transformer
$S_{T,3\phi}$ is the total three-phase transformer rating, kVA
x_T is the percent transformer reactance (approximately equal to the transformer's nameplate impedance).

8.7.6 BENEFITS DUE TO RELEASED FEEDER CAPACITY

Typically, the capacity of a feeder is constrained by the permissible voltage drop rather than thermal limitations, as discussed in Chapter 4. Therefore, by installing capacitors, the voltage drop is reduced, resulting in an increase in feeder capacity.

Excluding the impact of released regulators or substation capacity, the additional capacity of the feeder can be calculated as follows:

$$\Delta S_F = \frac{(Q_{c,3\phi})x}{x\sin\theta + r\cos\theta}\text{kVA} \tag{8.46}$$

Consequently, the yearly advantages resulting from the increased feeder capacity can be computed as follows:

$$\Delta\$_F = \Delta S_F \times C_F \times i_F \tag{8.47}$$

Table 8.3
Additional kWh Energy Increase after Capacitor Addition

V_{av}, after / V_{av}, before	ΔkWh Increase, %
1	0
1.05	8
1.1	16
1.15	25
1.2	34
1.25	43
1.3	52

where
$\Delta\$_F$ is the annual benefits due to released feeder capacity, \$/year
ΔS_F is the released feeder capacity, kVA
C_F is the cost of installed feeder, \$/kVA
i_F is the annual fixed charge rate applicable to the feeder

8.7.7 FINANCIAL BENEFITS DUE TO THE VOLTAGE IMPROVEMENTS

The utility experiences an increase in revenues due to the rise in kWh energy consumption resulting from the installation of corrective capacitor banks, particularly on residential feeders.

The magnitude of the energy consumption increase depends on the specific equipment used. For instance, energy consumption for lighting follows a square relationship with voltage. To illustrate this, Table 8.3 provides the percentage of additional kWh energy increase based on the ratio of the average voltage after capacitor installation to the average voltage before capacitor installation, taking into account a typical load diversity.

Therefore, the calculation of the revenue increase resulting from the augmented kWh energy consumption can be expressed as

$$\Delta\$_{\text{BEC}} = \Delta\text{BEC} \times \text{BEC} \times \text{EC} \tag{8.48}$$

where
$\Delta\$_{\text{BEC}}$ is the additional annual revenue due to increased kWh energy consumption, \$/year
ΔBEC is the additional kWh energy consumption increase
BEC is the original (or base) annual kWh energy consumption, kWh/year

8.7.8 TOTAL FINANCIAL BENEFITS DUE TO THE CAPACITOR INSTALLATIONS

Therefore, the total benefits due to the installation of capacitor banks can be summarized as

$$\begin{aligned}\sum\Delta\$ &= (\text{Demand reduction}) + (\text{Energy reduction}) + (\text{Revenue increase})\\ &= (\Delta\$_G + \Delta\$_T + \Delta\$_S + \Delta\$_F) + \Delta\$_{\text{ACE}} + \Delta\$\text{BEC}\end{aligned} \tag{8.49}$$

The total benefits obtained from Equation 8.50 should be compared against the annual equivalent of the total cost of the installed capacitor banks. The total cost of the installed capacitor banks can be found from

$$\Delta\text{EIC}_c = \Delta Q_c \times \text{IC}_c \times i_c \tag{8.50}$$

Table 8.4
For Example 8.19

Comment	At 90% PF	At 98% PF
Total loss reduction due to capacitors applied to substation buses, kW	495,165	491,738
Additional loss reduction due to capacitors applied to feeders, kW	85,771	75,342
Total demand reduction due to capacitors applied to substation buses and feeders, kVA	22,506,007	21,172,616
Total required capacitor additions at buses and feeders, kVAr	9,810,141	4,213,297

* Based on Ref. [3].

where

$\Delta \mathrm{EIC}_c$ is the annual equivalent of the total cost of installed capacitor banks, \$/year

ΔQ_c is the required amount of capacitor-bank additions, kVAr

IC_c is the cost of installed capacitor banks, \$/kVAr

i_c is the annual fixed charge rate applicable to capacitors

In conclusion, capacitors serve as an effective means of cost reduction in the utility industry. As plant costs and fuel expenses continue to rise, electric utilities benefit from deferring or eliminating new plant investments and reducing energy requirements.

By minimizing operating expenses, capacitors enable utilities to serve new loads and customers with minimal system investments. Presently, utilities in the United States have installed approximately 1 kVAr of power capacitors for every 2 kW of installed generation capacity, taking advantage of the associated economic benefits [203].

Example 8.19. Let's consider a large power pool operating at a 90% power factor. The objective is to improve the power factor to 98%. Several load flow runs have been conducted, and the results are summarized in Table 8.4.

Assuming an average fixed charge rate of 0.20, an average demand cost of \$250/kW, an energy cost of \$0.045/kWh, a system loss factor of 0.17, and an average capacitor cost of \$4.75/kVAr. The responsibility factors for capacitors installed on substation buses and feeders are 1.0 and 0.9, respectively. The following determinations are required:

a. The resulting additional savings in kW losses at the 98% power factor when all capacitors are applied to substation buses.
b. The resulting additional savings in kW losses at the 98% power factor when some capacitors are applied to feeders.
c. The total additional savings in kW losses.
d. The additional savings in the system kVA capacity.
e. The additional capacitors required, kVArs.
f. The total annual savings in demand reduction due to additional capacitors applied to substation buses and feeders, \$/year.
g. The annual savings due to the additional released transmission capacity, \$/year.
h. The total annual savings due to the energy loss reduction, \$/year.
i. The total annual cost of the additional capacitors, \$/year.
j. The total net annual savings, \$/year.
k. Is the 98% power factor the economic power factor?

Solution. a. From Table 8.4, the resulting additional savings in kW losses due to the power factor improvement at the substation buses is

$$\Delta P_{LS} = 495,165 - 491,738 = 3427 \text{ kW}$$

b. From Table 8.4 for feeders,

$$\begin{aligned}\Delta P_{LS} &= 85,771 - 75,342 \\ &= 10,429 \text{ kW}\end{aligned}$$

c. Therefore, the total additional kW savings is

$$\begin{aligned}\Delta P_{LS} &= 3,427 + 10,429 \\ &= 13,856 \text{ kW}\end{aligned}$$

As can be observed, the additional kW savings due to capacitors applied to the feeders is more than three times that of capacitors applied to the substation buses. This is due to the fact that power losses are larger at the lower voltages.

d. From Table 8.4, the additional savings in the system kVA capacity is

$$\begin{aligned}\Delta S_{sys} &= 22,506,007 - 21,172,616 \\ &= 1,333,391 \text{ kVA}\end{aligned}$$

e. From 8.4, the additional capacitors required are

$$\begin{aligned}\Delta Q_c &= 9,810,441 - 4,213,297 \\ &= 5,596,844 \text{ kVAr}\end{aligned}$$

f. The annual savings in demand reduction due to capacitors applied to distribution substation buses is approximately

$$(3427 \text{ kW})(1.0)(\$250/\text{kW})(0.20/\text{year}) = \$171,350/\text{year}$$

and due to capacitors applied to feeders is

$$(10,429 \text{ kW})(0.9)(\$250/\text{kW})(0.20/\text{year}) = \$469,305/\text{year}$$

Therefore, the total annual savings in demand reduction is

$$\$171,350 + \$469,305 = \$640,655/\text{year}$$

g. The annual savings due to the additional released transmission capacity is

$$(1,333,391 \text{ kVA})(\$27/\text{kVA})(0.20/\text{year}) = \$7,200,311/\text{year}$$

h. The total annual savings due to the energy loss reduction is

$$(\$13,856 \text{ kW})(8760\text{h/year})(0.17)(\$0.045/\text{kWh}) = \$928,546/\text{year}$$

i. The total annual cost of the additional capacitors is

$$(5,596,844 \text{ kVAr})(\$4.75/\text{kVAr})(0.20/\text{year}) = \$5,317,002/\text{year}$$

j. The total annual savings is summation of the savings in demand, capacity, and energy:

$$\$640,655 + \$7,200,311 + \$928,546 = \$8,769,512/\text{year}$$

Therefore, the total net annual savings is

$$\$8,769,512 - \$5,317,002 = \$3,452,510/\text{year}$$

k. No, since the total net annual savings is not zero.

8.8 PRACTICAL PROCEDURE TO DETERMINE THE BEST CAPACITOR LOCATION

Typically, the optimal placement of capacitors is determined by balancing power loss reduction and voltage regulation improvement. Conducting a comprehensive study of the feeder's voltage profile is essential to identify the most effective capacitor locations and ensure that the voltage remains within recommended limits.

In urban areas, a maximum voltage increase of around 2 V is generally permitted when a switched capacitor bank is activated. For circuits serving rural areas, a maximum voltage increase of approximately 3 V is typically allowed. The overall iterative process involved in determining the best capacitor placement can be summarized in the following steps:

1. Collect the following circuit and load information:
   a. Any two of the following for each load: kVA, kVAr, kW, and load power factor
   b. Desired corrected power of circuit
   c. Feeder-circuit voltage
   d. A feeder-circuit map that shows locations of loads and presently existing capacitor banks
2. Determine the kW load of the feeder and the power factor.
3. From Table 8.2, determine the kVAr per kW of load (i.e., the correction factor) necessary to correct the feeder-circuit power factor from the original to the desired power factor. To determine the kVAr of capacitors required, multiply this correction factor by the total kW of the feeder circuit.
4. Determine the individual kVA and power factor for each load or group of loads.
5. To determine the kVAr on the line, multiply individual load or groups of loads by their respective reactive factors that can be found from Table 8.2.
6. Develop a nomograph to determine the line loss in W/1000 ft due to the inductive loads tabulated in steps 4 and 5. Multiply these line losses by their respective line lengths in thousands of feet. Repeat this process for all loads and line sections and add them to find the total inductive line loss.
7. In the case of having presently existing capacitors on the feeder, perform the same calculations as in step 6, but this time subtract the capacitive line loss from the total inductive line loss. Use the capacitor kVAr determined in step 3 and the nomograph developed for step 6 and find the line loss in each line section due to capacitors.
8. To find the distance to capacitor location, divide the total inductive line loss by capacitive line loss per thousand feet. If this quotient is greater than the line section length
   a. Divide the remaining inductive line loss by the capacitive line loss in the next line section to find the location
   b. If this quotient is still greater than the line section length, repeat step 8a
9. Prepare a voltage profile by hand calculations or by using a computer program for voltage profile and load analysis to determine the circuit voltages. If the profile shows that the voltages are inside the recommended limits, then the capacitors are installed at the location of minimum loss. If not, then use engineering judgment to locate them for the most effective voltage control application [204].

8.9 MATHEMATICAL PROCEDURE TO DETERMINE THE OPTIMUM CAPACITOR ALLOCATION

Numerous papers have extensively investigated the optimal deployment of shunt capacitors on distribution feeders to minimize losses. Some notable studies include works by Neagle and Samson [205], Schmidt [206], Maxwell [50, 207], Cook [208], Schmill [209], Chang [210–212], Bae [213],

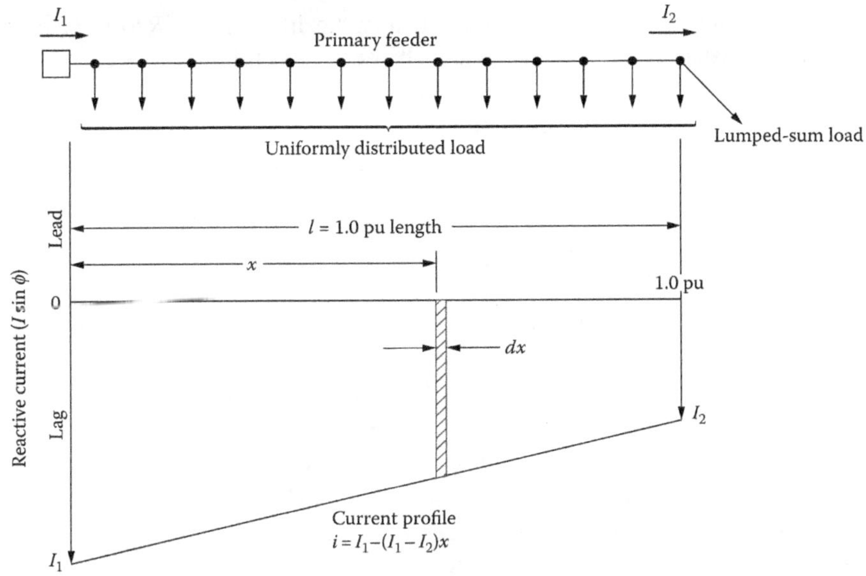

Figure 8.19 Primary feeder with lumped-sum (or concentrated) and uniformly distributed loads and reactive current profile before adding the capacitor.

Gönen and Djavashi [214,215], and Grainger et al. [216–219]. Figure 8.21 presents a realistic representation of a feeder comprising line segments with a combination of concentrated (or lumped-sum) and uniformly distributed loads, as suggested by Chang [211]. Each line segment represents a section of the feeder between sectionalizing devices, voltage regulators, and other significant points. For simplicity, the load or line current and the resulting I^2R loss can be assumed to have two components: (1) the component due to the in-phase or active current and (2) the component due to the out-of-phase or reactive current. As the losses stemming from the in-phase or active component of line current are minimally affected by the application of shunt capacitors, they are not taken into consideration. This can be verified as follows.

Let's consider that the I^2R losses are attributed to a lagging line current I passing through the resistance of the circuit, denoted as R. Consequently, we can demonstrate that

$$I^2R = (I\cos\phi)^2R + (I\sin\phi)^2R \tag{8.51}$$

After adding a shunt capacitor with current I_c, the resultants are a new line current I_1 and a new power loss I_1^2R. Hence,

$$I_1^2R = (I\cos\phi)^2R + (I\sin\phi - I_c)^2R \tag{8.52}$$

Therefore, the loss reduction as a result of the capacitor addition can be found as

$$\Delta P_{LS} = I^2R - I_1^2R \tag{8.53}$$

or by substituting Equations 8.56 and 8.57 into Equation 8.58,

$$\Delta P_{LS} = 2(I\sin\phi)I_cR - I_c^2R \tag{8.54}$$

Hence, only the out-of-phase or reactive component of the line current, represented as $I \sin \theta$, should be considered for reducing I^2R losses through the addition of capacitors.

Let's assume that the length of a feeder segment is 1.0 per unit (pu) length, as depicted in Figure 8.21. The current profile of the line current at any point along the feeder depends on the distance of that point from the beginning of the feeder. Consequently, the differential I^2R loss of a dx differential segment located at a distance x can be mathematically defined as follows:

$$dP_{\text{LS}} = 3[I_1 - (I_1 - I_2)x]^2 R\,dx \tag{8.55}$$

Therefore, the total I^2R loss of the feeder can be found as

$$\begin{aligned} P_{\text{LS}} &= \int_{x=0}^{1.0} dP_{\text{LS}} \\ &= 3\int_{x=0}^{1.0} [I_1 - (I_1 - I_2)x]^2 R dx \\ &= (I_1^2 + I_1 I_2 + I_2^2)R \end{aligned} \tag{8.56}$$

where
P_{LS} is the total I^2R loss of the feeder before adding the capacitor
I_1 is the reactive current at the beginning of the feeder segment
I_2 is the reactive current at the end of the feeder segment
R is the total resistance of the feeder segment
x is the per unit distance from the beginning of the feeder segment

8.9.1 LOSS REDUCTION DUE TO CAPACITOR ALLOCATION

8.9.1.1 Case 1: One Capacitor Bank

The insertion of one capacitor bank on the primary feeder causes a break in the continuity of the reactive load profile, modifies the reactive current profile, and consequently reduces the loss, as shown in Figure 8.20.

Therefore, the loss equation after adding one capacitor bank can be found as before:

$$P'_{\text{LS}} = 3\int_{x=0}^{x_1} [I_1 - (I_1 - I_2)x]^2 R\,dx + 3\int_{x=x_1}^{1.0} [I_1 - (I_1 - I_2)x]^2 R\,dx \tag{8.57}$$

or

$$P'_{\text{LS}} = (I_1^2 + I_1 I_2 + I_2^2)R + 3x_1[(x_1 - 2)I_1 I_c - x_1 I_2 I_c + I_c^2]R \tag{8.58}$$

Thus, the per unit power loss reduction as a result of adding one capacitor bank can be found from

$$\Delta P_{\text{LS}} = \frac{P_{\text{LS}} - P_{\text{LS}'}}{P_{\text{LS}}} \tag{8.59}$$

or substituting Equations 8.57 and 8.58 into Equation 8.60,

$$\Delta P_{\text{LS}} = \frac{-3x_1[(x_1 - 2)I_1 I_c - x_1 I_2 I_c + I_c^2]R}{(I_1^2 + I_1 I_2 + I_2^2)R} \tag{8.60}$$

or rearranging Equation 8.61 by dividing its numerator and denominator by I_1^2 so that

$$\Delta P_{\text{LS}} = \frac{3x_1}{1 + (I_2/I_1) + (I_2/I_1)^2}\left[(2 - x_1)\left(\frac{I_c}{I_1}\right) + x_1\left(\frac{I_2}{I_1}\right)\left(\frac{I_c}{I_1}\right) - \left(\frac{I_c}{I_1}\right)^2\right] \tag{8.61}$$

If c is defined as the ratio of the capacitive kilovolt-amperes (ckVAs) of the capacitor bank to the total reactive load, that is,

$$c = \frac{\text{ckVA of capacitor installed}}{\text{Total reactive load}} \tag{8.62}$$

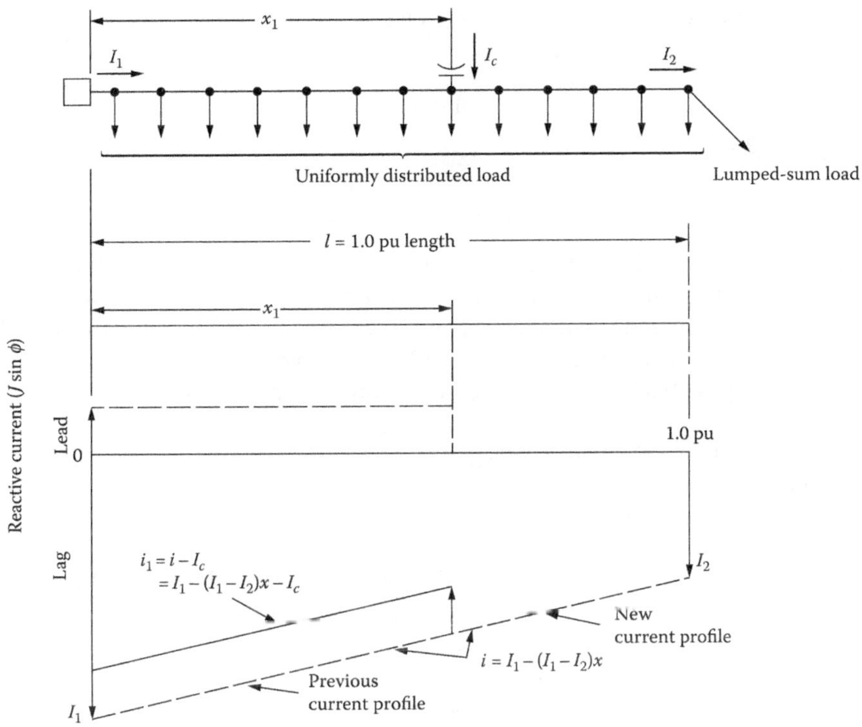

Figure 8.20 Loss reduction with one capacitor bank.

then

$$c = \frac{I_c}{I_1} \tag{8.63}$$

and if λ is defined as the ratio of the reactive current at the end of the line segment to the reactive current at the beginning of the line segment, that is,

$$\lambda = \frac{\text{Reactive current at the end of line segment}}{\text{Reactive current at the beginning of line segment}} \tag{8.64}$$

then

$$\lambda = \frac{I_2}{I_1} \tag{8.65}$$

Therefore, substituting Equations 8.64 and 8.66 into Equation 8.62, the per unit power loss reduction can be found as

$$\Delta P_{\text{LS}} = \frac{3x_1}{1+\lambda+\lambda^2}[(2-x_1)c + x_1\lambda c - c^2] \tag{8.66}$$

or

$$\Delta P_{\text{LS}} = \frac{3cx_1}{1+\lambda+\lambda^2}[(2-x_1) + x_1\lambda - c] \tag{8.67}$$

where x_1 is the per unit distance of the capacitor-bank location from the beginning of the feeder segment (between 0 and 1.0 pu).

If α is defined as the reciprocal of $1+\lambda+\lambda^2$, that is,

$$\alpha = \frac{1}{1+\lambda+\lambda^2} \tag{8.68}$$

then Equation 8.68 can also be expressed as

$$\Delta P_{LS} = 3\alpha c x_1[(2-x_1)+\lambda x_1 - c] \tag{8.69}$$

Figures 8.21 to 8.25 provide insights into the reduction in losses achievable by relocating a single capacitor bank of a specific size at different positions along the feeder. These figures consider various capacitor compensation ratios and representative load patterns, including uniformly distributed loads ($\lambda = 0$), concentrated or lumped-sum loads ($\lambda = 1$), or a combination of both ($0 < \lambda < 1$). To utilize these nomographs effectively for a specific scenario, the following factors need to be known:

1. Original losses due to the reactive current
2. Capacitor compensation ratio
3. The location of the capacitor bank

For instance, let's consider a scenario where the load on the line segment is uniformly distributed, and the desired compensation ratio is 0.5. Using Figure 8.21, it can be determined that the maximum loss reduction can be achieved by locating the capacitor bank at a position 0.75 per unit (pu) length away from the source. In this case, the associated loss reduction amounts to 0.85 pu or 85%. However, if the capacitor bank is positioned elsewhere on the feeder, the loss reduction will be less than 85%.

In essence, there is only one optimal location on the feeder, specific to each capacitor bank size, that yields the maximum loss reduction. Table 8.5 provides information on the optimum location

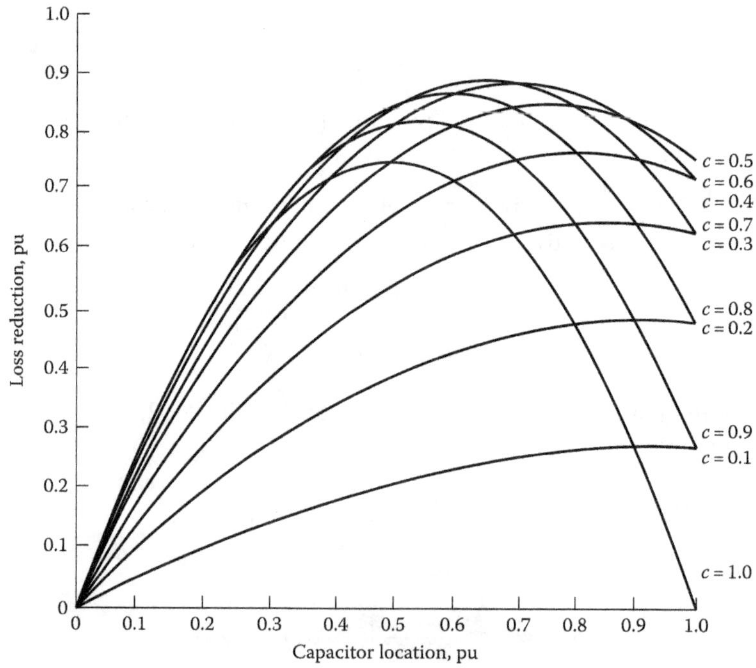

Figure 8.21 Loss reduction as a function of the capacitor-bank location and capacitor compensation ratio for a line segment with uniformly distributed loads ($\lambda = 0$).

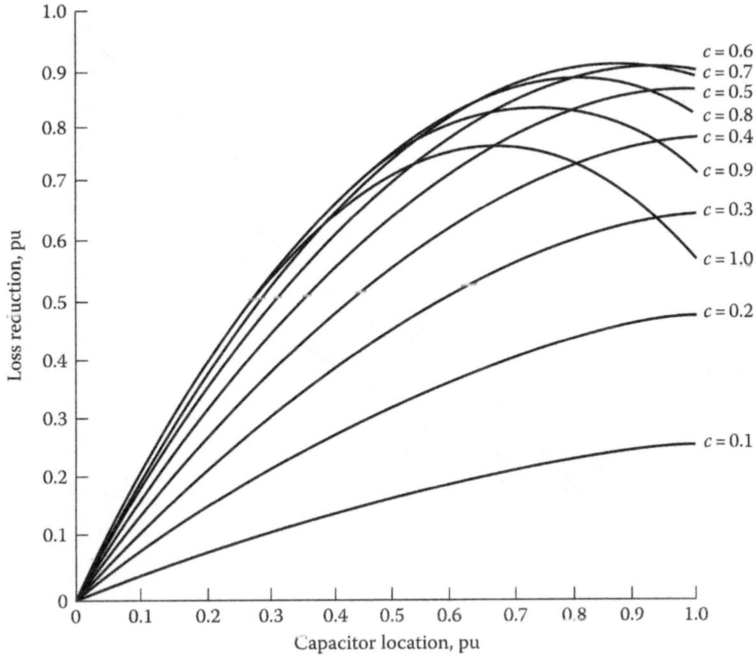

Figure 8.22 Loss reduction as a function of the capacitor-bank location and capacitor compensation ratio for a line segment with a combination of concentrated and uniformly distributed loads ($\lambda = 1/4$).

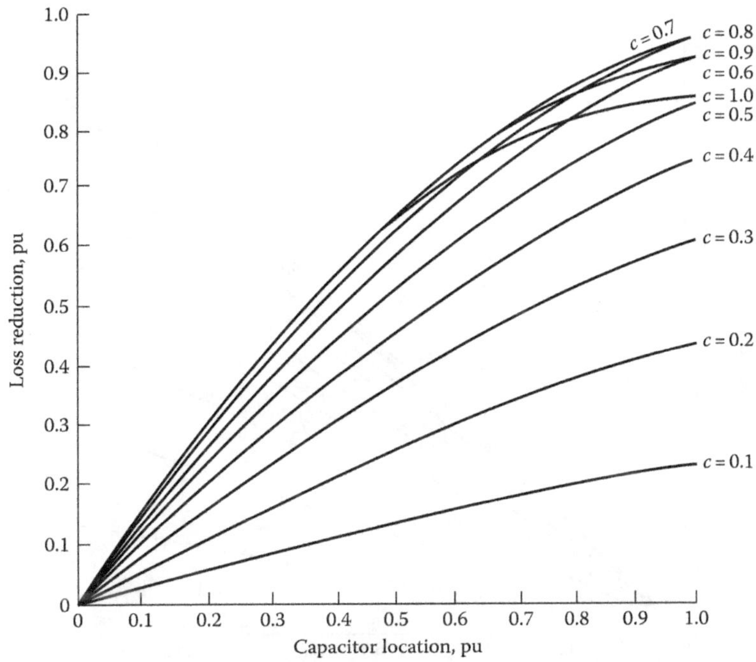

Figure 8.23 Loss reduction as a function of the capacitor-bank location and capacitor compensation ratio for a line segment with a combination of concentrated and uniformly distributed loads ($\lambda = 1/2$)

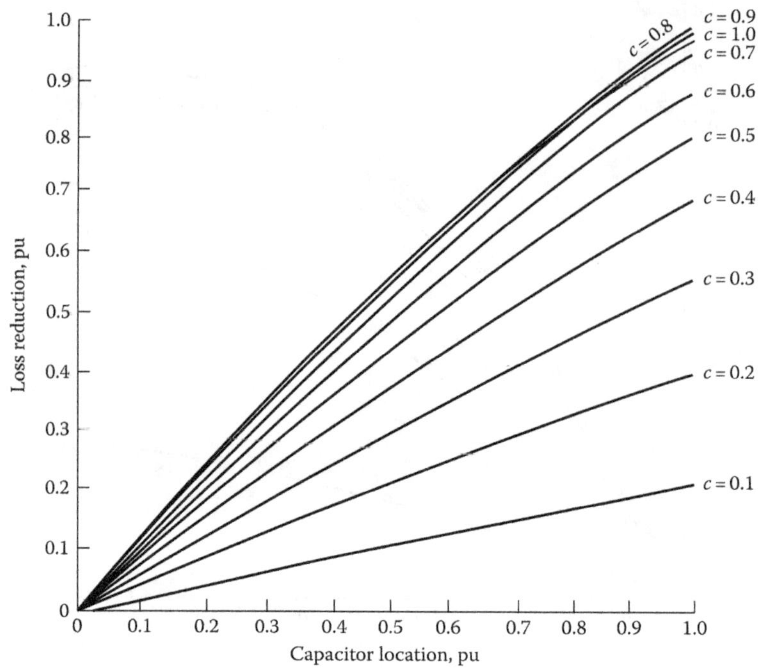

Figure 8.24 Loss reduction as a function of the capacitor-bank location and capacitor compensation ratio for a line segment with a combination of concentrated and uniformly distributed loads ($\lambda = 3/4$).

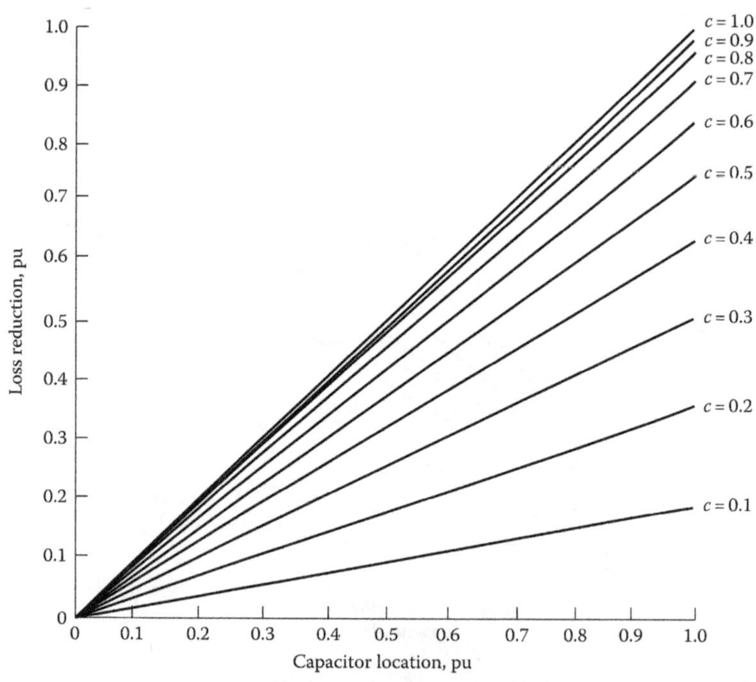

Figure 8.25 Loss reduction as a function of the capacitor-bank location and capacitor compensation ratio for a line segment with concentrated loads ($\lambda = 1$).

Table 8.5
Optimum Location and Optimum Loss Reduction

Capacitor-Bank Rating, pu	Optimum Location, pu	Optimum Loss Reduction, %
0	1	0
0.1	0.95	27
0.2	0.9	49
0.3	0.85	65
0.4	0.8	77
0.5	0.75	84
0.6	0.7	88
0.7	0.65	89
0.8	0.6	86
0.9	0.55	82
1	0.5	75

and the corresponding percentage loss reduction for capacitor banks of different sizes on a feeder with uniformly distributed load ($\lambda = 0$). The table highlights that positioning the single capacitor bank at a distance of two-thirds of the feeder length from the source achieves the maximum loss reduction.

Figure 8.26 depicts the loss reduction achievable by a capacitor bank of any given size when placed at the optimal location on a feeder with various load combinations, as calculated using Equation 8.70. Furthermore, Figure 8.27 illustrates the loss reduction achieved by an optimal-sized capacitor bank positioned on a feeder with different load combinations.

8.9.1.2 Case 2: Two Capacitor Banks

Let's consider a scenario where two capacitor banks of equal size are added to the feeder, as depicted in Figure 8.28. Similar to the previous approach, we can follow the same procedure, resulting in the following updated loss equation:

$$\Delta P_{\text{LS}'} = 3\int_{x=0}^{x_1}[I_1-(I_1-I_2)x-I_c]^2 R\,dx + 3\int_{x=x_1}^{x_2}[I_1-(I_1 I_2)x-I_c]^2 R\,dx + 3\int_{x=x_2}^{1.0}[I_1-(I_1-I_2)x]^2 R\,dx \tag{8.70}$$

Therefore, substituting Equations 8.57 and 8.71 into Equation 8.60, the new per unit loss reduction equation can be found as

$$\Delta P_{\text{LS}} = 3\alpha c x_1[(2-x_1)+\lambda x_1-3c]+3\alpha c x_2[(2-x_2)+\lambda x_2-c] \tag{8.71}$$

or

$$\Delta P_{\text{LS}} = 3\alpha c\{x_1[(2-x_1)+\lambda x_1-3c]+x_2[(2-x_2)+\lambda x_2-c]\} \tag{8.72}$$

8.9.1.3 Case 3: Three Capacitor Banks

Assume that three capacitor banks of equal size are inserted on the feeder, as shown in Figure 8.31. The relevant per unit loss reduction equation can be found as

$$\Delta P_{\text{LS}} = 3\alpha c\{x_1[(2-x_1)+\lambda x_1-5c]+x_2[(2-x_2)+\lambda x_2-3c]+x_3[(2-x_3)+\lambda x_3-c]\} \tag{8.73}$$

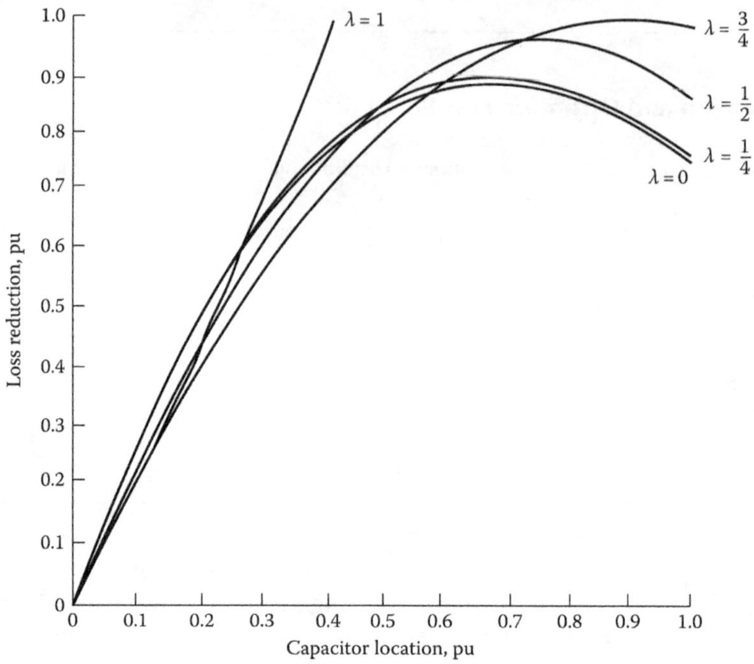

Figure 8.26 Loss reduction due to a capacitor bank located at the optimum location on a line section with various combinations of concentrated and uniformly distributed loads.

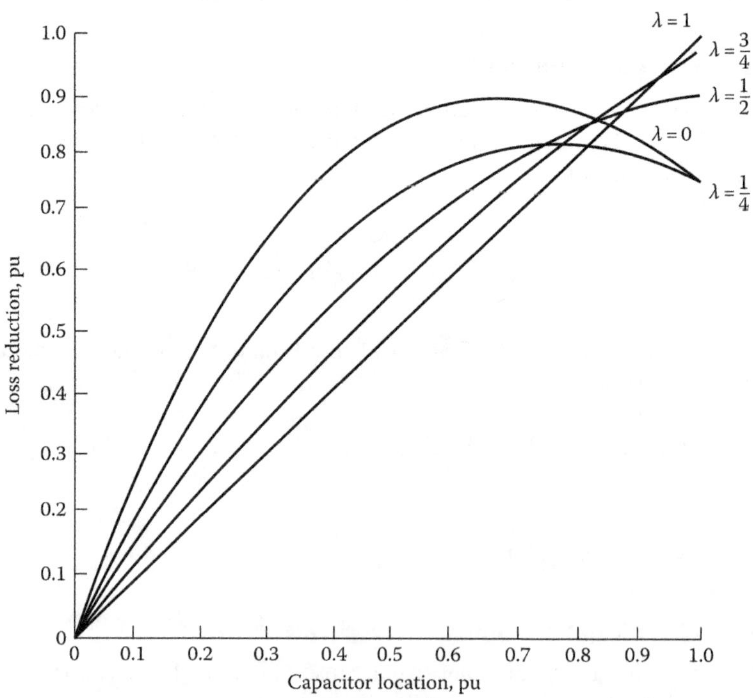

Figure 8.27 Loss reduction due to an optimum-sized capacitor bank located on a line segment with various combinations of concentrated and uniformly distributed loads.

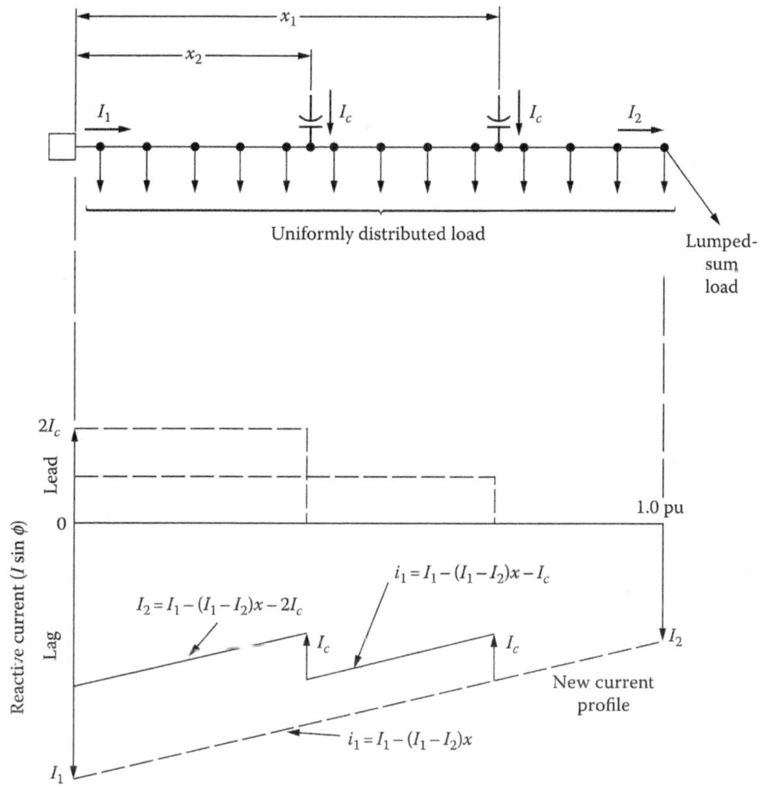

Figure 8.28 Loss reduction with two capacitor banks.

8.9.1.4 Case 4: Four Capacitor Banks

Assume that four capacitor banks of equal size are inserted on the feeder, as shown in Figure 8.30. The relevant per unit loss reduction equation can be found as

$$\Delta P_{LS} = 3\alpha c\{x_1[(2-x_1)+\lambda x_1 - 7c] + x_2[(2-x_2)+\lambda x_2 - 5c] + x_3[(2-x_3)+\lambda x_3 - 3c] + x_4[(2-x_4+\lambda x_4 - c]\} \quad (8.74)$$

8.9.1.5 Case 5: n Capacitor Banks

As the aforementioned results indicate, the per unit loss reduction equations follow a definite pattern as the number of capacitor banks increases. Therefore, the general equation for per unit loss reduction, for an n capacitor-bank feeder, can be expressed as

$$\Delta P_{LS} = 3\alpha c \sum_{i=1}^{n} x_i[(2-x_i)+\lambda x_i - (2i-1)c] \quad (8.75)$$

where

c is the capacitor compensation ratio at each location (determined from Equation 8.63)
x_i is the per unit distance of the ith capacitor-bank location from the source
n is the total number of capacitor banks

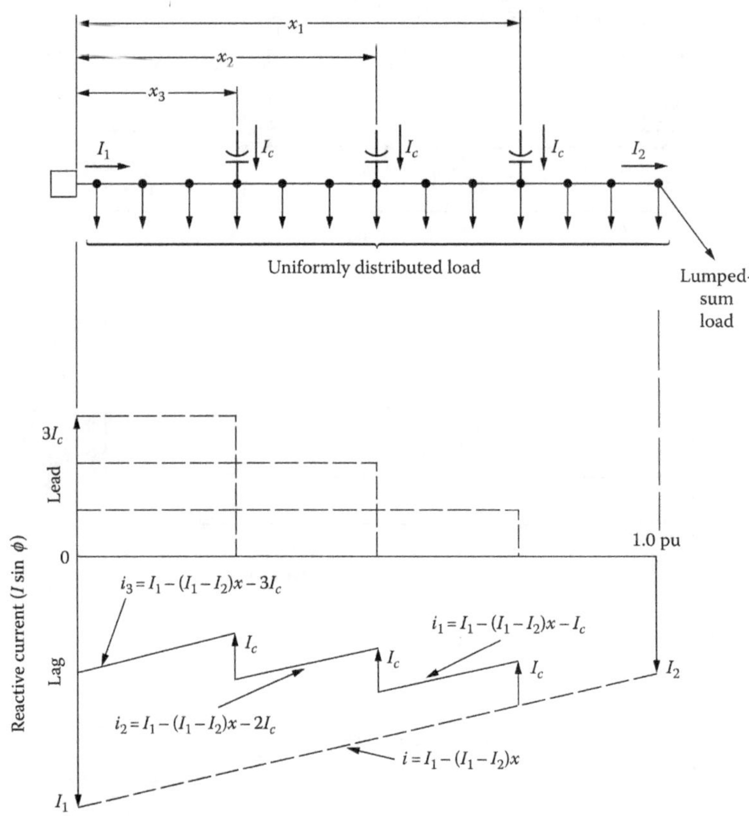

Figure 8.29 Loss reduction with three capacitor banks.

8.9.2 OPTIMUM LOCATION OF A CAPACITOR BANK

The optimum location for the ith capacitor bank can be found by taking the first-order partial derivative of Equation 8.76 with respect to x_i and setting the resulting expression equal to zero. Therefore,

$$x_{i,\text{opt}} = \frac{1}{1-\lambda} - \frac{(2i-1)c}{2(1-\lambda} \tag{8.76}$$

where $x_{i,\text{opt}}$ is the optimum location for the ith capacitor bank in per unit length.

By substituting Equation 8.81 into Equation 8.80, the optimum loss reduction can be found as

$$P_{\text{LS,opt}} = 3\alpha c \sum_{i=1}^{n} \left[\frac{1}{1-\lambda} - \frac{(2i-1)c}{2(1-\lambda}) + \frac{i^2c^2}{1-\lambda} - \frac{c^2}{4(1-\lambda)} - \frac{ic^2}{1-\lambda} \right] \tag{8.77}$$

Equation 8.78 is an infinite series of algebraic form that can be simplified by using the following relations

$$\sum_{i=1}^{n} (2i-1) = n^2 \tag{8.78}$$

$$\sum_{i=1}^{n} i = \frac{n(n+1)}{2} \tag{8.79}$$

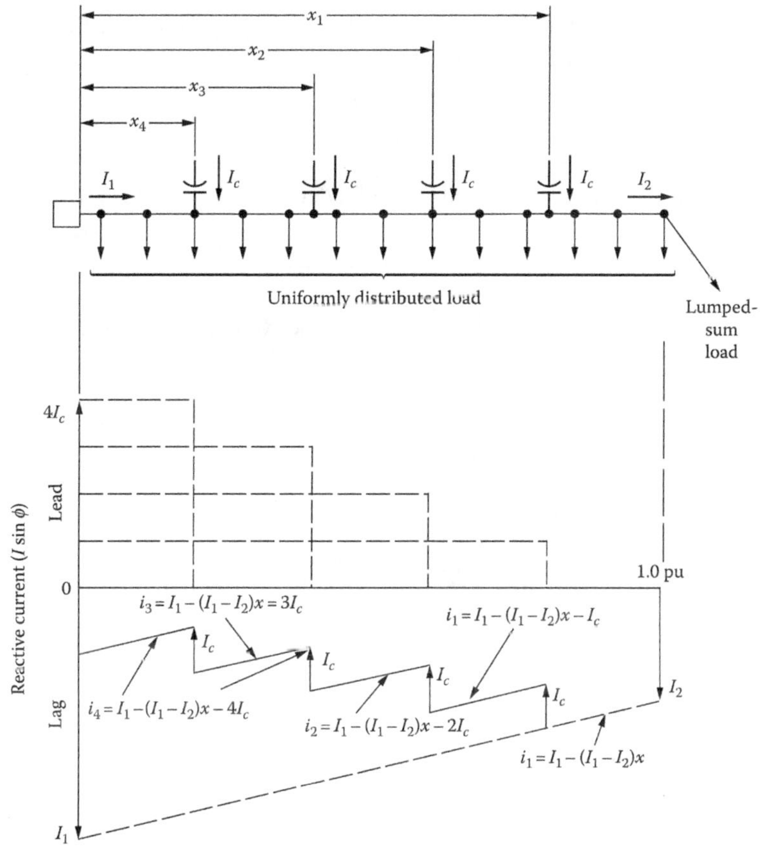

Figure 8.30 Loss reduction with four capacitor banks.

$$\sum_{i=1}^{n} i^2 = \frac{n(n+1)(2n+1)}{6} \tag{8.80}$$

$$\sum_{i=1}^{n} \frac{1}{1-\lambda} = \frac{n}{1-\lambda} \tag{8.81}$$

Therefore,

$$P_{\text{LS,opt}} = 3\alpha c \sum_{i=1}^{n} \left[\frac{n}{1-\lambda} - \frac{n^2 c}{(1-\lambda)} + \frac{nc^2(n+1)(2n+1)}{6} - \frac{nc^2}{4(1-\lambda} - \frac{nc^2(n+1)}{2(1-\lambda)} \right] \tag{8.82}$$

$$P_{\text{LS,opt}} = \frac{3\alpha c}{1-\lambda} \left[n - cn^2 + \frac{c^2 n(4n^2-1)}{12} \right] \tag{8.83}$$

The capacitor compensation ratio at each location can be found by differentiating Equation 8.88 with respect to c and setting it equal to zero as

$$c = \frac{2}{2n+1} \tag{8.84}$$

Equation 8.86 can be called the 2/(2n + 1) *rule*. For example, for n = 1, the capacitor rating is two-thirds of the total reactive load that is located at

$$x_1 = \frac{2}{3(1-\lambda)} \tag{8.85}$$

of the distance from the source to the end of the feeder, and the peak loss reduction is

$$\Delta P_{LS,opt} = \frac{2}{3(1-\lambda)} \tag{8.86}$$

For a feeder with a uniformly distributed load, the reactive current at the end of the line is zero (i.e., $I_2 = 0$); therefore,

$$\lambda = 0 \text{ and } \alpha = 1$$

Thus, for the optimum loss reduction of

$$\Delta P_{LS,opt} = \frac{8}{9}\text{pu} \tag{8.87}$$

the optimum value of x_1 is

$$x_1 = \frac{2}{3}\text{pu} \tag{8.88}$$

and the optimum value of c is

$$c = \frac{2}{3}\text{pu} \tag{8.89}$$

Figure 8.31 presents a comparison of maximum loss reduction for capacitor banks under different total reactive compensation levels. The capacitor banks are optimally located on a line segment with uniformly distributed load ($\lambda = 0$), based on Equation 8.84. The provided curves represent scenarios with one, two, three, and an infinite number of capacitor banks.

For instance, referring to the curve representing one capacitor bank, it can be observed that a capacitor bank rated at two-thirds of the total reactive load and positioned at two-thirds of the distance along the feeder from the source achieves a loss reduction of 89%. In the case of two capacitor banks, with a total reactive compensation of four-fifths, positioned at four-fifths of the distance along the feeder, the maximum loss reduction reaches 96%. Figure 8.32 displays similar curves for a combination of concentrated and uniformly distributed loads ($\lambda = 1/4$).

8.9.3 ENERGY LOSS REDUCTION DUE TO CAPACITORS

The per unit energy loss reduction in a three-phase line segment with a combination of concentrated and uniformly distributed loads due to the allocation of fixed shunt capacitors is

$$\Delta\text{EL} = 3\alpha c \sum_{i=1}^{n} x_i[(2-x_i)F'_{LD} + x_i\lambda F'_{LD} - (2i-1)c]T \tag{8.90}$$

where

F'_{LD} is the reactive load factor = Q/S

T is the total time period during which fixed-shunt-capacitor banks are connected

ΔEL is the energy loss reduction, pu

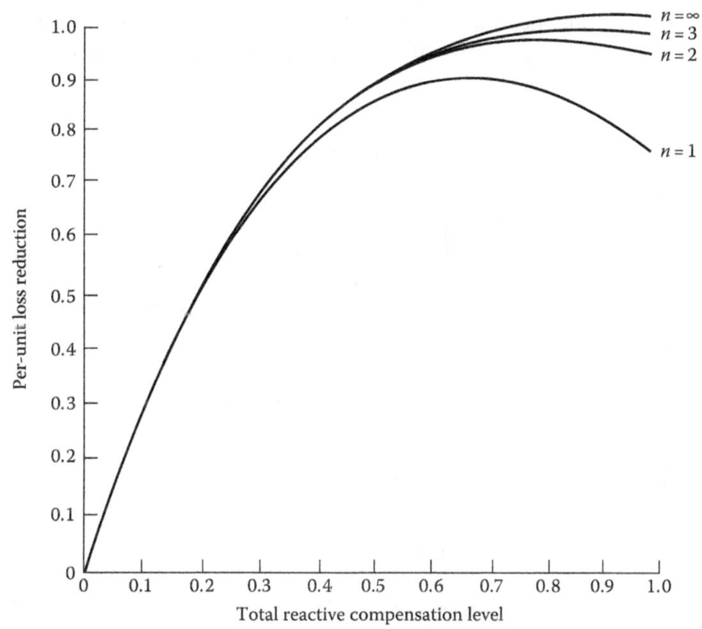

Figure 8.31 Comparison of loss reduction obtainable from $n = 1, 2, 3$, and ∞ number of capacitor banks, with $\lambda = 0$.

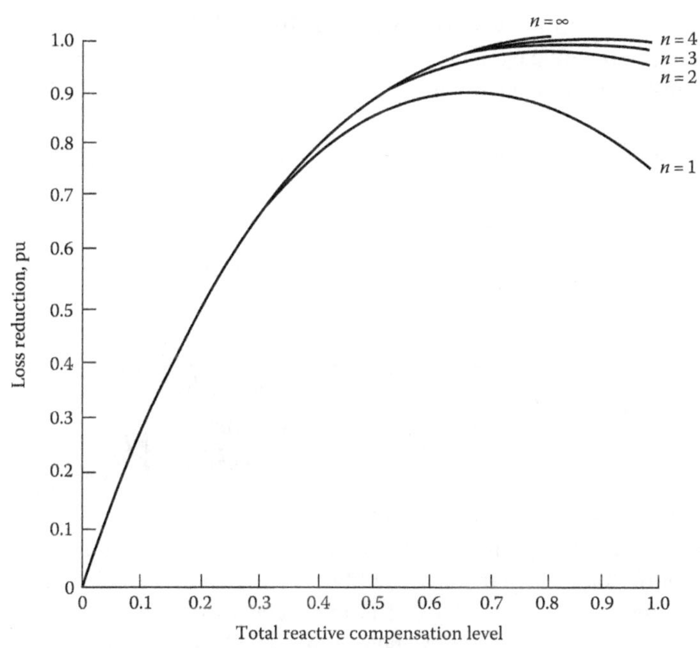

Figure 8.32 Comparison of loss reduction obtainable from $n = 1, 2, 3, 4$, and ∞ number of capacitor banks, with $\lambda = 1/4$.

The optimum locations for the fixed shunt capacitors for the maximum energy loss reduction can be found by differentiating Equation 8.91 with respect to x_i and setting the result equal to zero. Therefore,

$$\frac{\partial(\Delta \mathrm{EL})}{\partial x_i} = 3\alpha c[2F'_{\mathrm{LD}}(\lambda - 1)x_i + 2F'_{\mathrm{LD}} - (2i-1)c] \tag{8.91}$$

$$\frac{\partial^2(\Delta \mathrm{EL})}{\partial x_i^2} = -2F'_{\mathrm{LD}}(\lambda - 1) < 0 \tag{8.92}$$

The optimum capacitor location for the maximum energy loss reduction can be found by setting Equation 8.92 to zero, so that

$$x_{i,\mathrm{opt}} = \frac{1}{1-\lambda} - \frac{(2i-1)c}{2(1-\lambda)F'_{\mathrm{LD}}} \tag{8.93}$$

Similarly, the optimum total capacitor rating can be found as

$$C_T = \frac{2n}{2n+1}F'_{\mathrm{LD}} \tag{8.94}$$

From Equation 8.95, it can be observed that if the total number of capacitor banks approaches infinity, then the optimum total capacitor rating becomes equal to the reactive load factor.

If only one capacitor bank is used, the optimum capacitor rating to provide for the maximum energy loss reduction is

$$C_T = \frac{2}{3}F'_{\mathrm{LD}} \tag{8.95}$$

This equation gives the well-known *two-thirds rule for fixed shunt capacitors*. Figure 8.33 shows the relationship between the total capacitor compensation ratio and the reactive load factor, in order to achieve maximum energy loss reduction, for a line segment with uniformly distributed load where $\lambda = 0$ and $\alpha = 1$.

By substituting Equation 8.94 into Equation 8.95, the optimum energy loss reduction can be found as

$$\begin{aligned} \Delta \mathrm{EL}_{\mathrm{opt}} &= \frac{3\alpha c}{1-\lambda}\left[nF'_{\mathrm{LD}} - cn^2 + \frac{c^2 n(4n^2-1)}{12F'_{\mathrm{FD}}}\right]T \\ &= \frac{3\alpha cn}{1-\lambda}\left[F'_{\mathrm{LD}} - cn + \frac{c^2 n^2(4n^2-1)}{12n^2F'_{\mathrm{FD}}}\right]T \\ &= \frac{3\alpha C_T}{1-\lambda}\left[F'_{\mathrm{LD}} - C_T + \frac{C_T^2(4n^2-1)}{12n^2F'_{\mathrm{FD}}}\right]T \end{aligned} \tag{8.96}$$

where C_T is the total reactive compensation level = cn.

Based on Equation 8.97, the optimum energy loss reductions with any size capacitor bank located at the optimum location for various reactive load factors have been calculated, and the results have been plotted in Figures 8.34 through 8.40. It is important to note the fact that, for all values of λ,when reactive load factors are 0.2 or 0.4, the use of a fixed capacitor bank with corrective ratios of 0.4 and 0.8, respectively, gives a zero energy loss reduction.

Figures 8.39 through 8.43 show the effects of various reactive load factors on the maximum energy loss reductions for a feeder with different load patterns.

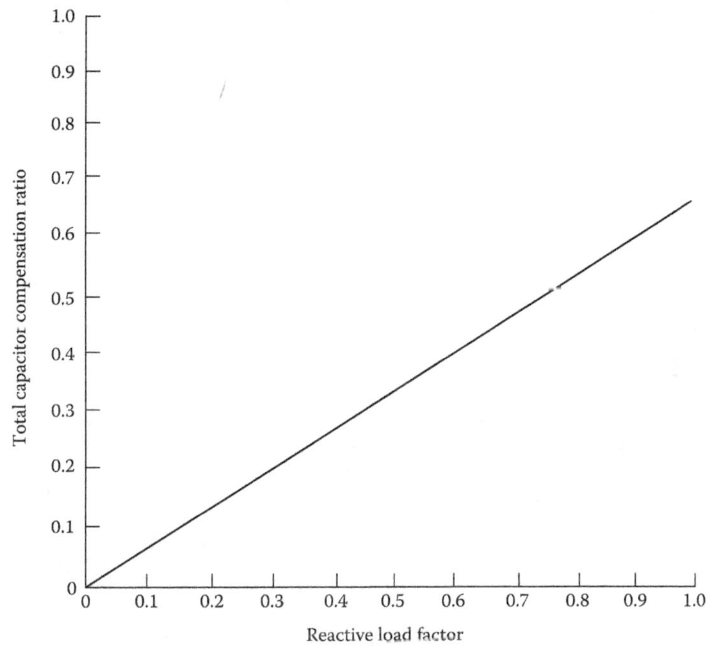

Figure 8.33 Relationship between the total capacitor compensation ratio and the reactive load factor for uniformly distributed load ($\lambda = 0$ and $\alpha = 1$).

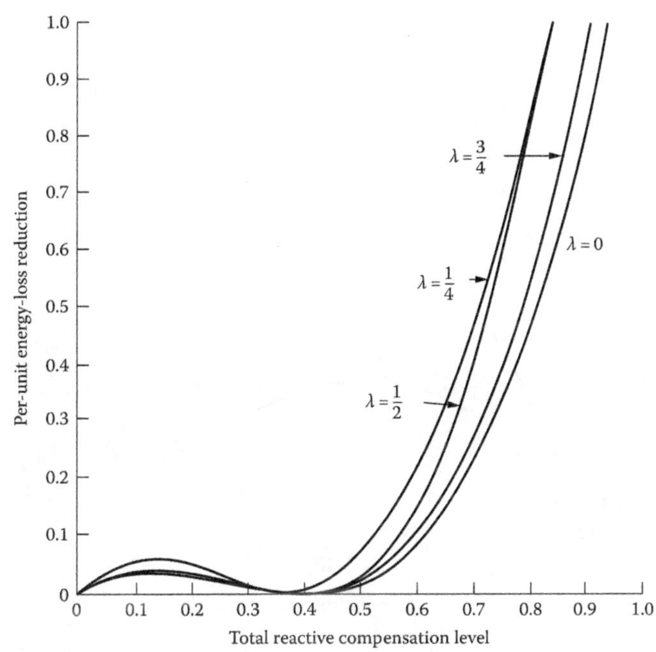

Figure 8.34 Energy loss reduction with any capacitor-bank size, located at optimum location ($F_{LD} = 0.2$).

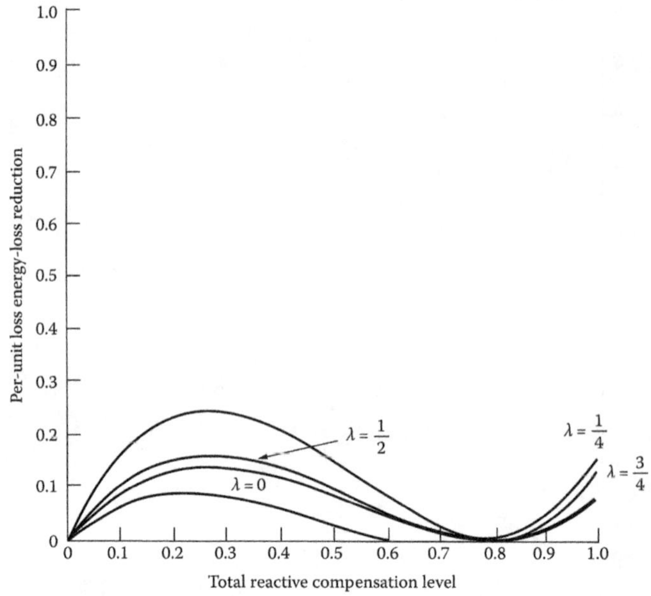

Figure 8.35 Energy loss reduction with any capacitor-bank size, located at the optimum location (F_{LD}= 0.4).

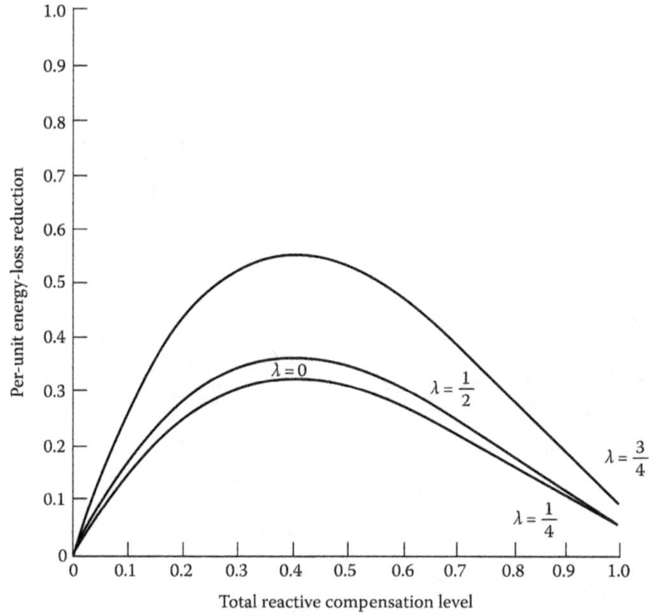

Figure 8.36 Energy loss reduction with any capacitor-bank size, located at the optimum location (F_{LD}= 0.6).

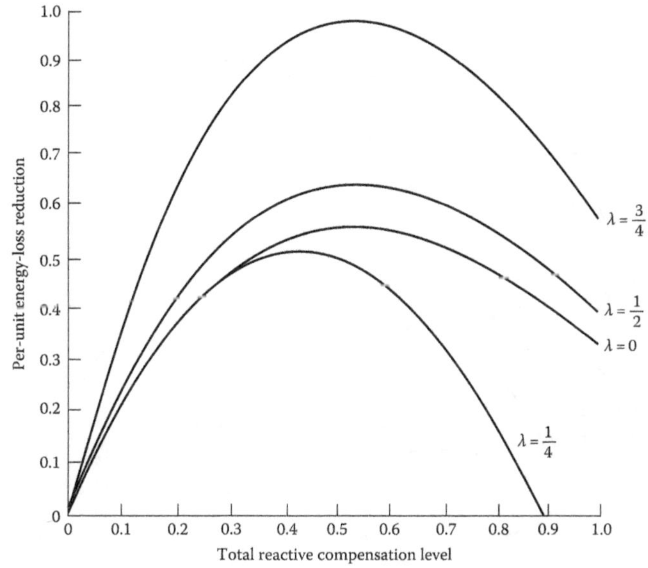

Figure 8.37 Energy loss reduction with any capacitor-bank size, located at the optimum location ($F_{LD} = 0.8$).

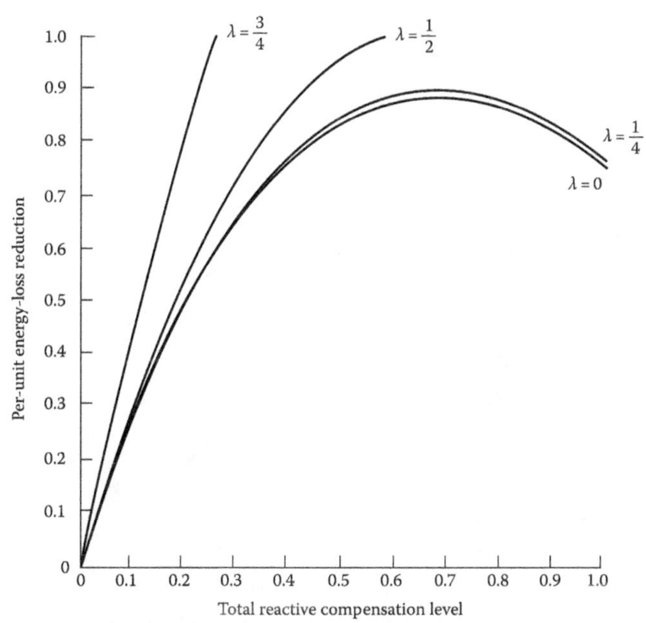

Figure 8.38 Energy loss reduction with any capacitor-bank size, located at the optimum location ($F'_{LD} = 1.0$).

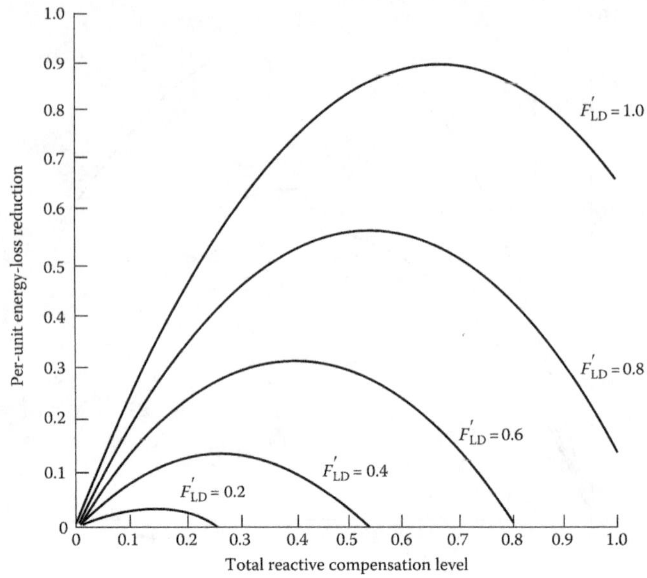

Figure 8.39 Effects of reactive load factors on energy loss reduction due to capacitor-bank installation on a line segment with uniformly distributed load ($\lambda = 0$).

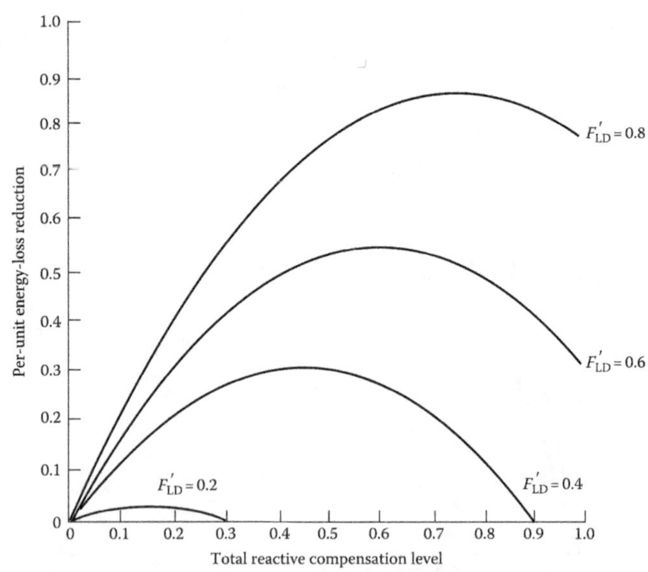

Figure 8.40 Effects of reactive load factors on energy loss reduction due to capacitor-bank installation on a line segment with a combination of concentrated and uniformly distributed loads (λ = 1/4).

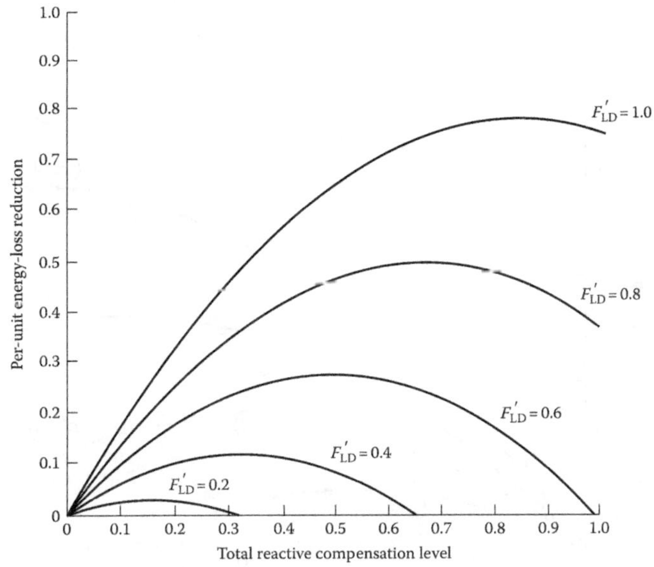

Figure 8.41 Effects of reactive load factors on energy loss reduction due to capacitor-bank installation on a line segment with a combination of concentrated and uniformly distributed loads (λ = 1/2).

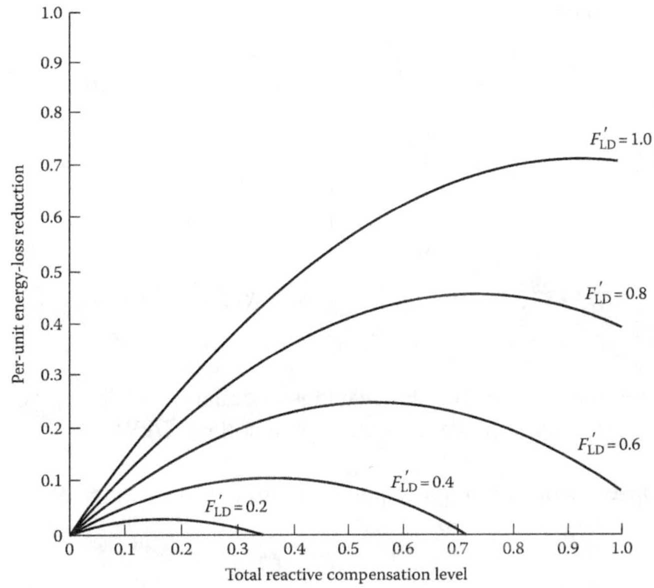

Figure 8.42 Effects of reactive load factors on loss reduction due to capacitor-bank installation on a line segment with a combination of concentrated and uniformly distributed loads (λ = 3/4).

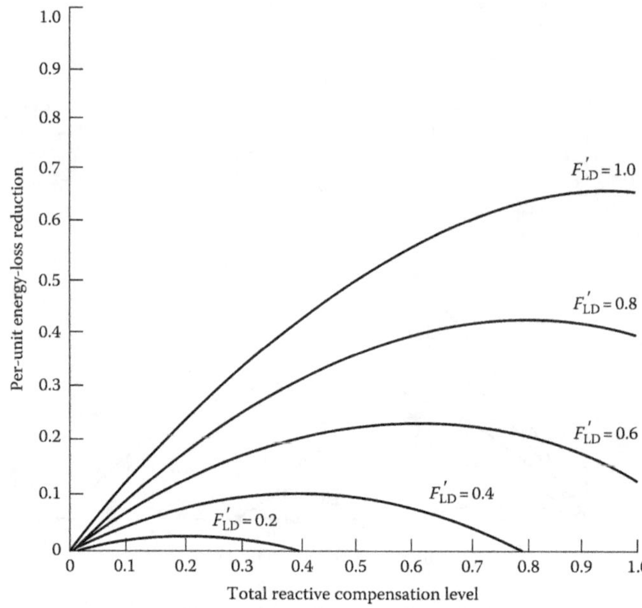

Figure 8.43 Effects of reactive load factors on energy loss reduction due to capacitor-bank installation on a line segment with a concentrated load (λ = 1).

8.9.4 RELATIVE RATINGS OF MULTIPLE FIXED CAPACITORS

The total savings due to having two fixed-shunt-capacitor banks located on a feeder with uniformly distributed load can be found as

$$\sum \$ = 3c_1\left(1 - c_1 + \frac{c_1^2}{4}\right)K_2 + 3c_2\left(1 - c_2 + \frac{c_2^2}{4}\right)K_2 + 3c_1\left(F_{LD'} - c_1 + \frac{c_1^2}{4F'_{LD}}\right)K_1T + 3c_2\left(F_{LD'} - c_2 + \frac{c_2^2}{4F'_{LD}}\right)K_1T \tag{8.97}$$

or

$$\sum \$ = 3[(c_1+c_2)(K_1+K_2TF_{LD'}) - (c_1^2+c_2^2)(K_2+K_1T)] + \frac{1}{4}(c_1^3+c_2^3)\left(K_2+\frac{K_1T}{F'_{LD}}\right) \tag{8.98}$$

where

K_1 is the a constant to convert energy loss savings to dollars, \$/kWh
K_2 is the a constant to convert power loss savings to dollars, \$/kWh

Since the total capacitor-bank rating is equal to the sum of the ratings of the capacitor banks,

$$C_T = c_1 + c_2 \tag{8.99}$$

or

$$c_1 = C_T - c_2 \tag{8.100}$$

By substituting Equation 8.101 into Equation 8.99,

$$\sum \$ = 3[C_T(K_1+K_2TF_{LD'}) - (c_T^2 + 2c_2^2 - 2c_2C_T)(K_1T+K_2)]\frac{1}{4}(c_T^3 + 3c_2c_T^2 - 2c_2^2C_T)\left(K_2 + \frac{K_1T}{F_{LD'}}\right) \tag{8.101}$$

The optimum rating of the second fixed capacitor bank as a function of the total capacitor-bank rating can be found by differentiating Equation 8.106 with respect to c_2, so that

$$\frac{\partial(\sum\$)}{\partial c_2} = -3(4c_2 - 2C_T)(K_2 + K_1 T) + \frac{3}{4}(-3C_T^2 + 6c_2 C_T)\left(K_2 + \frac{K_1 T}{F'_{\text{LD}}}\right) \tag{8.102}$$

and setting the resultant equation equal to zero,

$$2c_2 = C_T \tag{8.103}$$

and since

$$C_T = c_1 + c_2 \tag{8.104}$$

then

$$c_1 = c_2 \tag{8.105}$$

The outcome reveals that for a feeder with uniformly distributed loads and multiple fixed-shunt-capacitor banks, maximum savings can be achieved by ensuring that all capacitor banks have the same rating.

8.9.5 GENERAL SAVINGS EQUATION FOR ANY NUMBER OF FIXED CAPACITORS

By utilizing Equations 8.76 and 8.92, we can derive the equation for total savings in a three-phase primary feeder with a mix of concentrated and uniformly distributed loads as follows:

$$\begin{aligned}\sum\$ = {} & 3K_1\alpha c\sum_{i=1}^{n} x_i[(2-x_i)F_{\text{LD}'} + x_i\lambda F_{\text{LD}'} - (2i-1)c]T \\ & + 3K_2\alpha c\sum_{i=1}^{n} x_i[(2-x_i) + x_i\lambda - (2i-1)c] - K_3 C_T\end{aligned} \tag{8.106}$$

where

K_1 is the constant to convert energy loss savings to dollars, \$/kWh
K_2 is the constant to convert power loss savings to dollars, \$/kWh
K_3 is the constant to convert total fixed capacitor ratings to dollars, \$/kVAr
x_i is the ith capacitor location, pu length
n is the total number of capacitor banks
F'_{LD} is the reactive load factor
C_T is the total reactive compensation level
c is the capacitor compensation ratio at each location
λ is the ratio of reactive current at the end of the line segment to the reactive load current at the beginning of the line segment

$$\alpha = \frac{1}{(1+\lambda+\lambda^2)}$$

T is the total time period during which fixed-shunt-capacitor banks are connected

By taking the first- and second-order partial derivatives of Equation 8.107 with respect to x_i,

$$\frac{\partial(\sum\$)}{\partial x_i} = 3\alpha c[2x_i(K_2 + K_1 T F_{\text{LD}'})(\lambda - 1) + 2(K_2 + K_1 T F_{\text{LD}'})(2i-1)c(K_2 + K_1 T)] \tag{8.107}$$

and

$$\frac{\partial^2(\Sigma\$)}{\partial x_i^2} = -6\alpha c(1-\lambda)(K_2 + K_1 T F_{\mathrm{LD}'}) < 0 \tag{8.108}$$

Setting Equation 8.108 equal to zero, the optimum location for any fixed capacitor bank with any rating can be found as

$$x_i = \frac{1}{1-\lambda} - \frac{(2i-1)c}{1-\lambda}\frac{K_2 + K_1 T}{K_2 + K_1 T F_{\mathrm{LD}'}} \tag{8.109}$$

where $0 \le x_t \le 1.0$ pu length. Setting the capacitor bank anywhere else on the feeder would decrease rather than increase the savings from loss reduction.

Some of the *cardinal rules* that can be derived for the application of capacitor banks include the following:

1. The location of fixed shunt capacitors should be based on the average reactive load.
2. There is only one location for each size of capacitor bank that produces maximum loss reduction.
3. One large capacitor bank can provide almost as much savings as two or more capacitor banks of equal size.
4. When multiple locations are used for fixed-shunt-capacitor banks, the banks should have the same rating to be economical.
5. For a feeder with a uniformly distributed load, a fixed capacitor bank rated at two-thirds of the total reactive load and located at two-thirds of the distance out on the feeder from the source gives an 89% loss reduction.
6. The result of the two-thirds rule is particularly useful when the reactive load factor is high. It can be applied only when fixed shunt capacitors are used.
7. In general, particularly at low reactive load factors, some combination of fixed and switched capacitors gives the greatest energy loss reduction.
8. In actual situations, it may be difficult, if not impossible, to locate a capacitor bank at the optimum location; in such cases the permanent location of the capacitor bank ends up being suboptimum.

8.10 FURTHER THOUGHTS ON CAPACITORS AND IMPROVING POWER FACTORS

In a given circuit, reducing the relative current results in a reduction in the total current. In cases where the real current remains unchanged, which is typically true, the power factor improves as the reactive current decreases. As the reactive current approaches zero, all the current becomes real current, leading to a power factor of 1.0, indicating unity or 100% power factor. For example, in Figure 8.46a, if a capacitor is installed to supply the total reactive power of 60 kVAr, the line power factor can be enhanced by fulfilling the reactive power requirements of the load with the capacitor.

In the case of an induction load or equipment such as an induction motor, real power is supplied and subsequently lost, while reactive power is supplied without loss. In this scenario, reactive power is utilized to store energy in the motor's magnetic field. Since the current is alternating, the magnetic field experiences cycles of buildup and breakdown. When the current is in the breakdown phase, the reactive current flows out of the motor back to the power supply, as depicted in Figure 8.44a. It is important to note that real power flow is unidirectional, flowing from the source to the motor. However, it is possible to provide only real power from the power source and fulfill the reactive power requirements through a capacitor.

Unlike the unidirectional flow of real power, the flow of reactive power is cyclic in nature. It flows out of the capacitor to the motor and then returns back to the capacitor, as illustrated in Figure 8.44b.

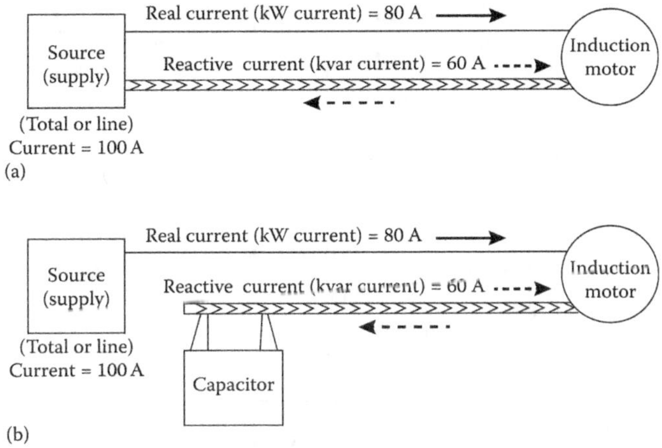

Figure 8.44 Schematic diagram for the use of capacitors to reduce total line current by supplying reactive power locally: (a) without the use of capacitors and (b) with the use of capacitors.

In this case, the capacitor acts as a temporary storage area for reactive power by storing energy units in the electric field.

Figure 8.44 demonstrates how capacitors can be used to provide reactive power locally. In the illustration, it is evident that the operating load necessitates 80 A, but due to the motor's magnetic requirements of 60 A, the supply circuit must accommodate 100 A. However, with the installation of the capacitor, the magnetic requirements of the motor can be fulfilled by the capacitor itself. As a result, the supply circuit only needs to deliver 80 A to perform the same work. The supply circuit now carries only kW, eliminating the need to transport the so-called "non-working current" and preventing wastage of system capacity.

Moreover, the power triangle highlights that subtracting real power from total apparent power does not yield reactive power, except in the case of unity power factor.

However, in practical scenarios, it is generally unnecessary and uneconomical to improve the power factor to 100%. For this reason, utility companies typically provide a power factor of around 0.95 at their power plants. The remaining reactive power is supplied by utilizing overexcited synchronous motors operating without loads or by employing capacitor banks.

8.11 CAPACITOR TANK-RUPTURE CONSIDERATIONS

When the energy input to a capacitor exceeds the strength of its tank enclosure, the tank may rupture. This energy input can occur under various current-time conditions. To address this concern, capacitor manufacturers have conducted extensive testing and developed tank-rupture curves that correlate with available fault currents.

These tank-rupture curves serve as a basis for coordinating fuse selection to provide effective protection against tank rupture. Figure 8.45 displays the results of tank-rupture tests conducted on all-film capacitors. Additionally, Figure 8.46 illustrates the reliability cycle of a capacitor throughout its lifespan. The longer it takes for the dielectric material to degrade due to the combined forces of electric stress and temperature, the higher the reliability of the capacitor. In other words, the wear-out process or time to failure serves as a measure of its life and reliability. Currently, various methods exist to detect capacitor tank ruptures, as categorized by Burrage [220], including:

1. Detection of sound produced by the rupture
2. Observation of smoke and/or vapor from the capacitor tank upon rupture
3. Observation of ultraviolet light generated by the arc getting outside the capacitor tank

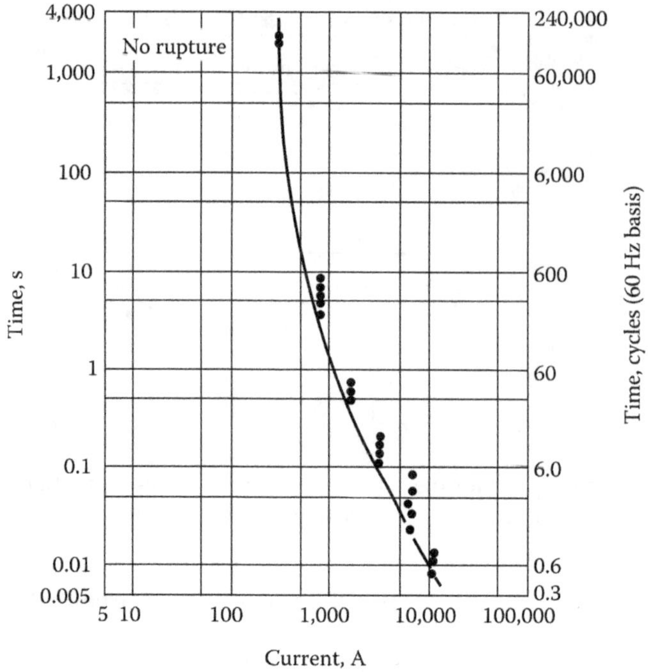

Figure 8.45 Time-to-rupture characteristics for 200 kVAr 7.2 kV all-film capacitors. (From McGraw-Edison Company, *The ABC of Capacitors*, Bulletin R230-90-1, 1968.)

4. Measurement of the change in arc voltage when the capacitor tank is breached
5. Detection of a sudden reduction in internal pressure
6. Measurement of the distortion generated by gas pressure within the capacitor tank

8.12 DYNAMIC BEHAVIOR OF DISTRIBUTION SYSTEMS

The characteristics of the distribution system dynamic behavior include (1) fault effects and transient recovery voltage (TRV), (2) switching and lightning surges, (3) inrush and cold-load current transients, (4) ferroresonance, and (5) harmonics.

In the event of a fault on a distribution system, there will be a substantial change in current magnitude on the faulted phase. It is also possible for the current to have a dc offset that is a function of the voltage wave at the time of the fault and the X/R ratio of the circuit. This may cause a voltage rise on the unfaulted phases due to neutral shift that results in saturation of transformers and increased load current magnitudes.

On the other hand, it may cause a reduction in voltage and load current on the faulted phase. When a circuit recloser clears the fault at a current zero, a higher-frequency transient voltage issuperimposed on the power frequency recovery voltage. The resultant voltage is called the TRV. It is possible to have its crest magnitudes be two or three times the nominal system voltage. This may cause failure to clear or restrike that may produce substantial switching surges.

Switching surges are generated when loads, station capacitor banks, or feeders are energized or reenergized; or when faults are initiated, cleared, and reinitiated. The factors affecting the magnitude and duration of the resultant voltage transients include (1) the system impedance characteristics, (2) the amount of capacitive kVAr connected at the time of switching, (3) the location of the capacitor bank on the system, (4) the type of breaker, and (5) the breaker pole-closing angles. In general, the switching surges on distribution systems have not been taken seriously so far.

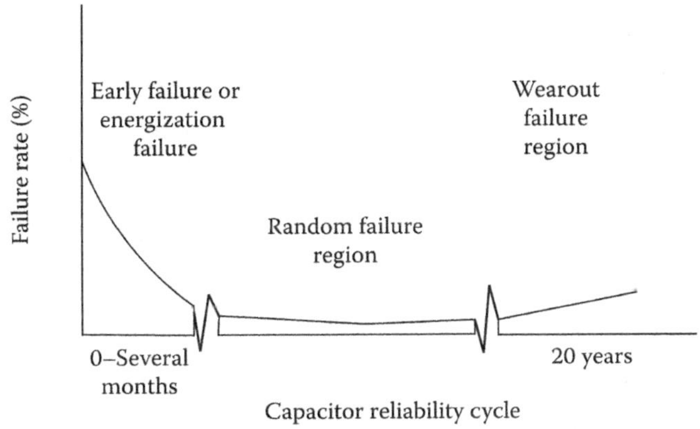

Figure 8.46 Capacitor reliability cycle. (From McGraw-Edison Company, *The ABC of Capacitors*,Bulletin R230-90-1, 1968.)

However, if the current trend toward higher-distribution voltage levels and reduced insulation levels continues, this may change. But voltage surges resulting from lightning strokes to the distribution line will usually require the most severe design requirements. The factors affecting the lightning surge include (1) the system configuration and the system grounding and shielding, (2) the stroke characteristics and stroke location, (3) the sparkover of arresters remote from the converters, (4) the amount of the connected capacitive kVAr in the surge path, and (5) the loss mechanism (corona, skin effect) in the surge path.

Current transients are generated during the energization of motors, transformers, capacitors, feeders, and loads. For instance, when motors and other loads draw high starting currents, capacitors experience a high-frequency inrush current determined by the instantaneous voltage, circuit inductance, and capacitance. Similarly, in transformers, the magnitude of the inrush current depends on the voltage waveform during energization and the residual flux in the core.

It is crucial to acknowledge that low voltage levels during inrush can damage equipment and prevent the circuit from recovering without sectionalizing. Additionally, protective devices may operate incorrectly or fail to operate due to the high magnitude and high-frequency nature of these transient currents.

8.12.1 FERRORESONANCE

The phenomenon known as *ferroresonance* occurs when the capacitance of a system interacts with the nonlinear inductance of a transformer, leading to an oscillatory behavior. This combination of capacitive and inductive elements forms a series-resonant circuit capable of generating high transient or sustained overvoltages that can cause damage to system equipment.

These overvoltages are more likely to occur in situations where a significant length of cable is connected to an overloaded three-phase transformer (or bank), and single-phase switching takes place at a remote point from the transformer, such as a riser pole. It is essential to identify the conditions that increase the likelihood of these overvoltages and implement appropriate preventive measures to prevent significant equipment damage.

Signs of severe overvoltages may include flashovers or damage to lightning arresters, transformer humming when only one phase is closed, equipment damage, three-phase motor reversals, and high secondary voltages.

While the phenomenon of ferroresonance has been recognized for some time, it was not considered a significant operational issue on electric distribution systems until recently. Changes in

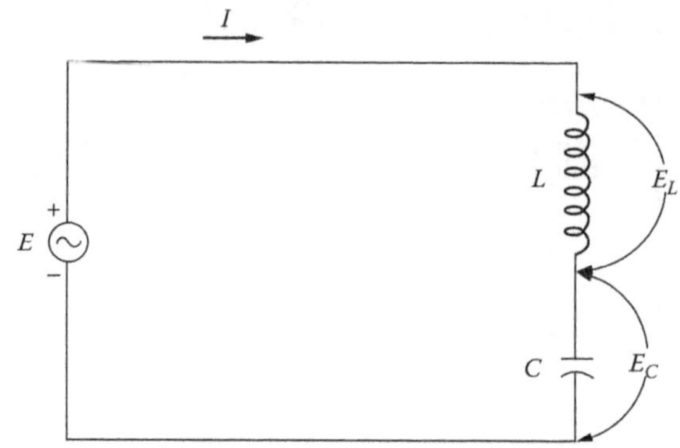

Figure 8.47 The *LC* circuit for ferroresonance.

distribution system characteristics and transformer designs have increased the probability of ferroresonant overvoltages when switching three-phase transformer installations.

As an illustration, underground cables have a higher capacitance (and thus lower capacitive reactance) compared to open wires. This is significant because current trends favor the increased use of underground cables due to aesthetic reasons. Additionally, operating distribution systems at voltages higher than the nominal level, as well as recent transformer design trends, have resulted in transformers operating with higher saturation levels.

Moreover, the adoption of higher-distribution voltages has led to the use of distribution transformers with increased magnetizing reactance. Simultaneously, the capacitance of underground systems becomes more pronounced, leading to lower capacitive reactance.

Let us examine the *LC* circuit depicted in Figure 8.47. In the interest of simplicity, we will neglect the presence of resistance. When the magnitude of the inductive reactance X_L of the inductor is equal to the magnitude of the capacitive reactance X_C of the capacitor, the circuit is said to be in resonance. At this state, the voltage E across the inductor is out of phase with the voltage E_C across the capacitor by 180°. Mathematically, the voltages E_L and E_C can be represented as follows:

$$\begin{aligned} E_L &= \frac{E}{jX_L - jX_c}(jX_L) \\ &= \frac{E}{1 - X_c/X_L} \end{aligned} \tag{8.110}$$

and

$$\begin{aligned} E_c &= \frac{E}{jX_L - jX_c}(jX_c) \\ &= \frac{E}{1 - X_L/X_c} \end{aligned} \tag{8.111}$$

Example 8.20. Consider a downtown 34.5 kV underground cable, and for the purpose of illustration, assume that $X_L/X_C = 0.9$ in the *LC* circuit given by Figure 8.47, and therefore, $X_C/X_L = 1.1111$. If the line-to-line voltage is 34.5 kV, determine

a The voltages E_L
b The voltages E_C

a.

$$E_L = -9 \times \left(\frac{34.5\text{ kV}}{\sqrt{3}}\right) \cong 179.3\text{ kV}$$

$$E_L = \frac{E}{1 - X_c/X_L} = \frac{E}{1 - 1.1111} = -9E$$

In practice, such situation often takes place in downtown underground networks when a 34.5 kV underground cable feeds a transformer. As a result of the ferroresonance (because of the X_L of the transformer and X_C of the cable), the 34.5 kV underground cable breaks down due to such high voltages, causing a short-circuit condition. The fault usually takes place at the entrance terminal to the transformer. It has often puzzled some of the protection engineers. Thus, in such situations, it is a good idea to check for possible ferroresonance conditions.

b.

$$E_c = \frac{E}{1 - X_L/X_c} = \frac{E}{1 - 0.9} = 10E$$

or

$$\begin{aligned} E_c &= 10 \times \left(\frac{34.5\text{ kV}}{\sqrt{3}}\right) \\ &= \frac{345\text{ kV}}{\sqrt{3}} \\ &\cong 199.2\text{ kV} \end{aligned}$$

Hence, in this scenario, the voltage across the capacitor is ten times greater than the voltage of the source. The closer the circuit is to the actual resonance, the higher the overvoltage will be.

While the previous example provides a relatively straightforward illustration of a resonant circuit, the underlying concept is similar to ferroresonance with one significant distinction. In a ferroresonant circuit, the capacitor is connected in series with a nonlinear inductor, typically an iron-core inductor. The volt-ampere or impedance characteristic of an iron-core reactor follows a pattern akin to the BH curve of the iron core. When the iron-core reactor operates near saturation, even a small increase in voltage can lead to a substantial reduction in the effective inductive reactance of the reactor. Consequently, the value of inductive reactance can vary considerably, and resonance can occur across a range of capacitance values. To mitigate the effects of ferroresonance, various measures can be implemented, including:

1. Using grounded-wye–grounded-wye transformer connection
2. Using open-wye–open-delta transformer connection
3. Using switches rather than fuses at the riser pole
4. Using single-pole devices only at the transformer location and three-pole devices for remote switching
5. Avoiding switching an unloaded transformer bank at a point remote from the transformers
6. Keeping X_C/X_M ratios high (10 or more)
7. Installing neutral resistance
8. Using dummy loads to suppress ferroresonant overvoltages
9. Assuring load is present during switching
10. Using larger transformers
11. Limiting the length of cable serving the three-phase installation
12. Using only three-phase switching and sectionalizing devices at the terminal pole
13. Temporarily grounding the neutral of a floating-wye primary during switching operations

8.12.2 HARMONICS ON DISTRIBUTION SYSTEMS

The power industry has recognized the issue of power system harmonics since the 1920s, when distorted voltage and current waveforms were observed on power lines. However, the levels of harmonics on distribution systems were generally considered insignificant in the past [221–224].

It is evident that the levels of harmonic voltages and currents on distribution systems are becoming a significant problem. Several crucial power system operational issues caused by harmonics have been reported, including the following [225]:

1. Capacitor-bank failure from dielectric breakdown or reactive power overload
2. Interference with ripple control and power-line carrier systems, causing misoperation of systems that accomplish remote switching, load control, and metering
3. Excessive losses in—and heating of—induction and synchronous machines
4. Overvoltages and excessive currents on the system from resonance to harmonic voltages or currents on the network
5. Dielectric instability of insulated cables resulting from harmonic overvoltages on the system
6. Inductive interference with telecommunication systems
7. Errors in induction watthour meters
8. Signal interference and relay malfunction, particularly in solid-state and microprocessor-controlled systems
9. Interference with large motor controllers and power plant excitation systems (reported to cause motor problems as well as nonuniform output [226])

These effects depend, of course, on the harmonic source, its location in the power system, and the network characteristics that facilitate the propagation of harmonics. There are numerous sources of harmonics. In general, the harmonic sources can be classified as (1) known harmonic sources and (2) new harmonic sources. The known harmonic sources include the following:

1. Tooth ripples or ripples in the voltage waveform of rotating machines
2. Variations in air-gap reluctance over synchronous machine pole pitch
3. Flux distortion in the synchronous machine from sudden load changes
4. Nonsinusoidal distribution of the flux in the air gap of synchronous machines
5. Transformer magnetizing currents
6. Network nonlinearities from loads such as rectifiers, inverters, welders, arc furnaces, voltage controllers, frequency converters

While the established sources of harmonics are still present on the system, the power network is also subjected to new harmonic sources:

1. Energy conservation measures, such as those for improved motor efficiency and load matching, which employ power semiconductor devices and switching for their operation. These devices often produce irregular voltage and current waveforms that are rich in harmonics.
2. Motor control devices such as speed controls for traction.
3. High-voltage dc power conversion and transmission.
4. Interconnection of wind and solar power converters with distribution systems.
5. Static-var compensators that have largely replaced synchronous condensors as continuously variable-var sources.
6. The development and potentially wide use of electric vehicles that require a significant amount of power rectification for battery charging.
7. The potential use of direct energy conversion devices, such as magnetohydrodynamics, storage batteries, and fuel cells, that require DC/AC power converters.

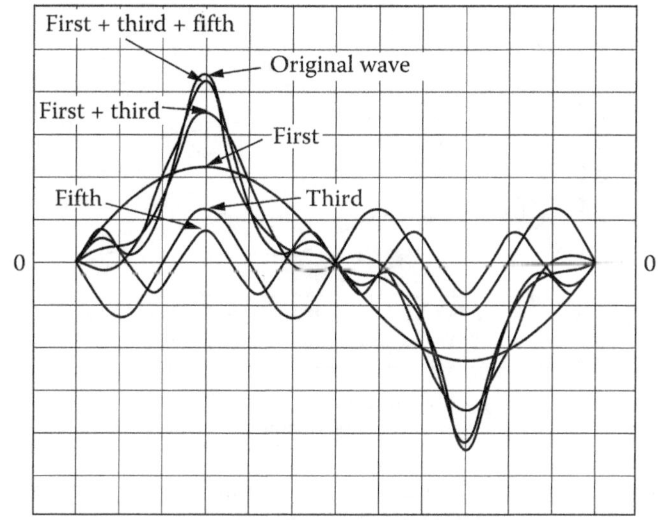

Figure 8.48 Harmonic analysis of peaked no-load current.

The presence of harmonics causes the distortion of the voltage or current waves. The distortions are measured in terms of the voltage or current harmonic factors. The IEEE Standard 519–1981 [227] defines the harmonic factors as the ratio of the root-mean-square value of all the harmonics to the root-mean-square value of the fundamental. Therefore, the voltage harmonic factor HF_V can be expressed as

$$HF_V = \frac{(E_3^2 + E_5^2 + E_7^2 + \cdots)^{1/2}}{E_1} \tag{8.112}$$

and the current harmonic factor HF_I can be expressed as

$$HF_I = \frac{(I_3^2 + I_5^2 + I_7^2 + \cdots)^{1/2}}{I_1} \tag{8.113}$$

The presence of voltage distortion leads to the generation of harmonic currents. Figure 8.48 illustrates the harmonic analysis of a peaked no-load current waveform.

The characteristics of harmonics in a distribution system depend on both the harmonic source and the system's response. For instance, utilities are currently installing larger and more numerous transformers to meet growing power demands. Each transformer itself becomes a source of harmonics in the distribution system.

Moreover, these transformers are operated closer to their saturation point. When a sinusoidal voltage is applied to a transformer in a saturated state, it results in a non-sinusoidal exciting current in the iron core. The degree of transformer saturation is influenced by the magnitude of the applied voltage.

When the applied voltage exceeds the rated voltage, the harmonic components of the exciting current experience a significant increase. Owen [228, 229] has demonstrated this phenomenon for a typical substation power transformer, as depicted in Figure 8.49. Additionally, some utility companies deliberately overexcite distribution transformers as part of their policy and practice, exacerbating the harmonic issue.

The significant current harmonics produced by transformers are typically of the third, fifth, and seventh orders. Table 8.6 provides a summary of the conditions associated with the occurrence of third harmonics in different configurations of double-wound three-phase transformers. The table presents information on third harmonics in double-wound single-phase core-type and shell-type

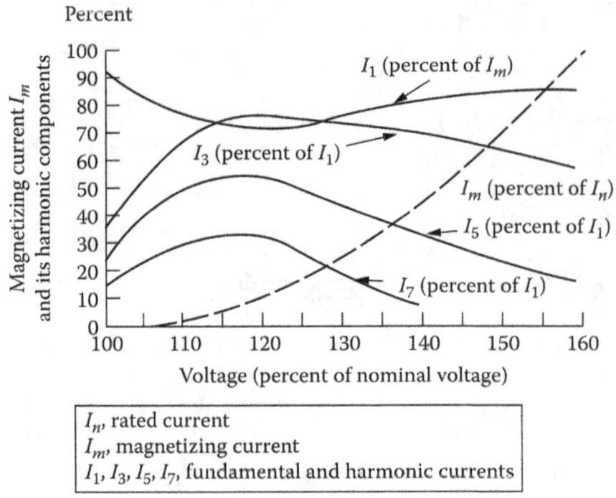

Figure 8.49 Harmonic components of transformer exciting current. (From Owen, R.E., Distribution system harmonics: Effects on equipment and operation, *Pacific Coast Electrical Association Engineering and Operating Conference*, Los Angeles, CA, March 15–16, 1979.)

transformers, as well as three-phase shell-type transformers for three-phase service. Figure 8.50 illustrates the waveforms resulting from the combination of the fundamental and third harmonic, with different phase relationships.

It is important to note that at harmonic frequencies, the phase angles can vary between 0° and 360° due to the different harmonic impedances of each load. Furthermore, as the harmonic order increases, the power-line impedance becomes a significant factor, causing the harmonic current to have varying phase angles at different locations.

There are several effects of harmonics on transformers. Voltage harmonics lead to increased iron losses, while current harmonics cause higher copper losses and stray flux losses. These losses can result in overheating of the transformer.

Harmonics can also create insulation stresses and resonances between transformer windings and line capacitances at the harmonic frequencies. The total eddy-current losses are directly proportional to the harmonic frequencies and can be mathematically expressed as follows:

$$\text{TECL} = \text{ECL}_1 \sum_{h=1}^{\infty} \left(\frac{h \times I_h}{I_1} \right)^2 \tag{8.114}$$

where

TECL is the total eddy-current loss

ECL_1 is the eddy-current loss at rated fundamental current

h is the harmonic order

I_1 is the rated fundamental current

I_h is the harmonic current

Capacitor-bank sizes and locations are critical factors in a distribution system's response to harmonic sources. The combination of capacitors and the system reactance causes both series- and parallel-resonant frequencies for the circuit.

The possibility of resonance between a shunt capacitor bank and the rest of the system, at a harmonic frequency, may be determined by calculating equal order of harmonic h at which resonance may take place [230, 231]. This equal order of harmonic is found from

$$h = \left(\frac{S_{SC}}{Q_{cap}} \right)^{1/2} \tag{8.115}$$

Table 8.6
The Influence of Three-Phase Transformer Connections on Third Harmonics

	Primary					Secondary			
	Currents		Voltages			Currents		Voltages	
Connections[a]	No-Load	Line	Line	Phase	Flux	No-Load	Line	Line	Phase
1. Wye I.N./wye I.N.	Sine	Sine	Sine	Contains 3d h(P)	Contains 3d h(FT)		Sine	Sine	Contains 3d h(P)
2. Wye N. to G./wye I.N.	Contains 3d h(P)[b]	Contains 3d h(P)[b]	Sine	Contains 3d h(P)[b]	Contains 3d h(FT)[b]		Sine	Sine	Contains 3d h(P)[b]
3. Wye I.N./wye, four-wire	Sine	Sine	Sine	Contains 3d h(P)[b]	Contains 3d h(FT)[b]	Contains 3d h(P)[b]	Contains 3d h(P)[b]	Sine	Contains 3d h(P)[b]
4. Wye I.N. tertiary delta/wye I.N.	Sine in star, 3d h in delta (P)	Sine	Sine	Sine	Sine		Sine	Sine	Sine
5. Wye I.N./delta	Sine	Sine	Sine	Sine	Sine	Contains 3d h(P)	Sine	Sine	Sine
6. Wye N. to G./delta	Contains 3d h(P)[b]	Contains 3d h(P)[b]	Sine	Sine	Sine	Contains 3d h(P)[b]	Sine	Sine	Sine
7. Wye I.N./interconnected wye I.N.	Sine	Sine	Sine	Contains 3d h(P)	Contains 3d h(FT)		Sine	Sine	Sine
8. Wye I.N./interconnected wye, four-wire	Sine	Sine	Sine	Contains 3d h(P)	Contains 3d h(FT)		Sine	Sine	Sine
9. Delta/wye I.N.	Contains 3d h(P)	Sine	Sine	Sine	Sine		Sine	Sine	Sine
10. Delta/wye, four-wire	Contains 3d h(P)	Sine	Sine	Sine	Sine	Contains 3d h(P)	Contains 3d h(P)	Sine	Sine
11. Delta/delta	Contains 3d h(P)	Sine	Sine	Sine	Sine	Contains 3d h(P)[b]	Sine	Sine	Sine

Source: Stigant, S.A. and Franklin, A.C., *The J&P Transformer Book,* Butterworth, London, U.K., 1973.

[a] I.N., isolated neutral; N. to G., transformer primary neutral connected to generator neutral; (P), peaked wave; (FT), flat-top wave.

[b] In all these cases the third-harmonic component is less than it otherwise would be if (1) the circulating third-harmonic current flowed through a closed delta winding only or (2) the neutral point was isolated.

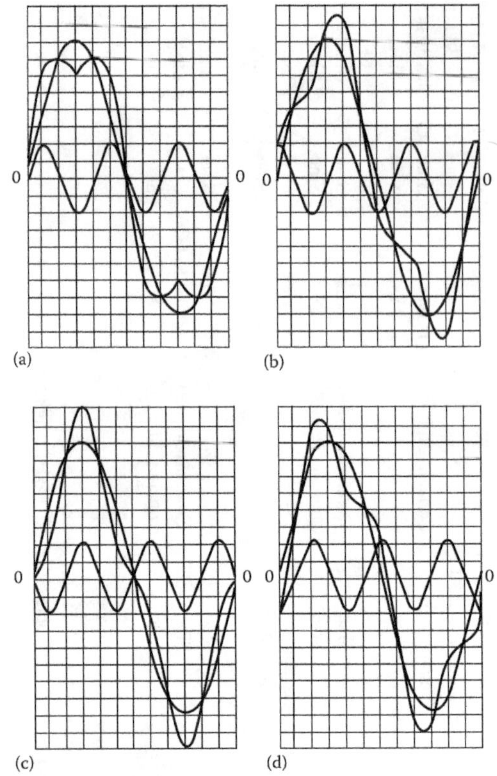

Figure 8.50 Combinations of fundamental and third-harmonic waves: (a) harmonic in phase, (b) harmonic 90° leading, (c) harmonic in opposition, and (d) harmonic 90° lagging. (From Stigant, S.A. and Franklin, A.C., *The J&P Transformer Book*, Butterworth, London, U.K., 1973 [75].)

where
S_{sc} is the short-circuit power of a system at the point of application, MVA
Q_{cap} is the capacitor-bank size, Mvar

The parallel-resonant frequency f_p can be expressed as

$$f_p = f_1 \times h \tag{8.116}$$

Substituting Equation 8.115 into Equation 8.116,

$$f_p = f_1 \left(\frac{S_{SC}}{Q_{cap}} \right)^{1/2} \tag{8.117}$$

or

$$f_p = f_1 \left(\frac{X_{cap}}{X_{sc}} \right)^{1/2} \tag{8.118}$$

or

$$f_p = \frac{1}{2\pi} \left(\frac{1}{L_{sc} \times C} \right)^{1/2} \tag{8.119}$$

where

f_1 is the fundamental frequency, Hz
X_{cap} is the reactance of capacitor bank, pu or Ω
X_{sc} is the reactance of power system, pu or Ω
L_{sc} is the inductance of power system, H
C is the capacitance of capacitor bank, F

The effects of the harmonics on the capacitor bank include (1) overheating of the capacitors, (2) overvoltage at the capacitor bank, (3) changed dielectric stress, and (4) losses in capacitors. According to Kimbark [230], the increase of losses in capacitors due to harmonics can be expressed as

$$\text{LCDH} = \sum C(\tan\delta)_h w_h V_h^2 \tag{8.120}$$

where

LCDH is the losses in capacitors due to harmonics
C is the capacitance
$(\tan\delta)_h$ is the loss factor at frequency of the hth harmonic
$w_h = 2\pi$ times frequency of the hth harmonic
V_h is the rms voltage of the hth harmonic

The harmonic control techniques include (1) locating the capacitor banks strategically, (2) selecting capacitor-bank sizes properly, (3) ungrounding or deleting the capacitor bank, (4) using shielded cables, (5) controlling grounds properly, and (6) using harmonic filters.

PROBLEMS

8.1 Assume that a feeder supplies an industrial consumer with a cumulative load of (1) induction motors totaling 300 hp that run at an average efficiency of 89% and a lagging average power factor of 0.85, (2) synchronous motors totaling 100 hp with an average efficiency of 86%, and (3) a heating load of 100 kW. The industrial consumer plans to use the synchronous motors to correct its overall power factor. Determine the required power factor of the synchronous motors to correct the overall power factor at peak load to

a. Unity
b. 0.96 lagging

8.2 A 2.4/4.16 kV wye-connected feeder serves a peak load of 300 A at a lagging power factor of 0.7 connected at the end of the feeder. The minimum daily load is approximately 135 A at a power factor of 0.62. If the total impedance of the feeder is 0.50 + j1.35Ω, determine the following

a. The necessary kVAr rating of the shunt capacitors located at the load to improve the peak-load power factor to 0.96
b. The reduction in kVA and line current due to the capacitors
c. The effects of the capacitors on the voltage regulation and voltage drop in the feeder
d. The power factor at minimum daily load level

8.3 Assume that a locked-rotor starting current of 90 A at a lagging load factor of 0.30 is supplied to a motor that is operated discontinuously. A normal operating current of 15 A, at a lagging power factor of 0.80, is drawn by the motor from the 2.4/4.16 kV feeder of Problem 8.2. Assume that a series capacitor is desired to be installed in the feeder to improve the voltage regulation and limit lamp flicker from the intermittent motor starting, and determine the following:

Table 8.7
Summary of Load Flows

Comment	At 90% PF	At 99% PF
Total loss reduction due to capacitors applied to substation buses, kW	496	488
Additional loss reduction due to capacitors applied to feeders, kW	84	72
Total demand reduction due to capacitors applied to substation buses and feeders, kVA	21,824	19,743
Total required capacitor additions at buses and feeders, kVAr	9,512	2,785

a. The voltage dip due to the motor starting, before the installation of the series capacitor
b. The necessary size of the capacitor to restrict the voltage dip at motor start to not more than 3%

8.4 Assume that a three-phase distribution substation transformer has a nameplate rating of 7250 kVA and a thermal capability of 120%. If the connected load is 8816 kVA with a 0.85 lagging power factor, determine the following:

a. The kVAr rating of the shunt capacitor bank required to decrease the kVA load on the transformer to its capability level
b. The power factor of the corrected load
c. The kVAr rating of the shunt capacitor bank required to correct the load power factor to unity
d. The corrected kVA load at this unity power factor

8.5 Assume that the NP&NL Utility Company is presently operating at 90% power factor. It is desired to improve the power factor to 98%. To study the power factor improvement, a number of load flow runs have been made and the results are summarized in the following table. Using the relevant additional information given in Table 8.7, repeat Example 8.19.

8.6 Assume that a manufacturing plant has a three-phase in-plant generator to supply only three-phase induction motors totaling 1200 hp at 2.4 kV with a lagging power factor and efficiency of 0.82 and 0.93, respectively. Using the given information, determine the following:

a. Find the required line current to serve the 1200 hp load and the required capacity of the generator.
b. Assume that 500 hp of the 1200 hp load is produced by an overexcited synchronous motor operating with a leading power factor and efficiency of 0.90 and 0.93, respectively. Find the required new total line current and the overall power factor.
c. Find the required size of shunt capacitors to be installed to achieve the same overall power factor as found in part b by replacing the overexcited synchronous motor.

8.7 Verify that the loss reduction with two capacitor banks is

$$\Delta L = 3\alpha c\{x_1[(2-x_1)+\lambda x_1 - 3c] + x_2[(2-x_2)+\lambda x_2 - c]\}$$

8.8 Derive Equation 8.96 from Equation 8.95.

8.9 Verify that the optimum loss reduction is

$$\Delta L_{\text{opt}} = 3\alpha c \sum_{i=2}^{n}\left[\frac{1}{1-\lambda} - \frac{(2i-1)c}{1-\lambda} + \frac{i^2c^2}{1-\lambda} - \frac{c^2}{4(1-\lambda)} - \frac{ic^2}{1-\lambda}\right]$$

8.10 Derive Equation 8.105 from Equation 8.96

8.11 Verify Equation 8.106.

8.12 If a power system has a 15,000 kVA capacity and is operating at a power factor of 0.65 lagging and the cost of synchronous capacitors is \$15/kVA, find the investment required to correct the power factor to

a. 0.85 lagging power factor
b. Unity power factor

8.13 If a power system has a 20,000 kVA capacity and is operating at a power factor of 0.6 lagging and the cost of synchronous capacitors is \$12.50/kVA, find the investment required to correct the power factor to

a. 0.85 lagging power factor
b. Unity power factor

8.14 If a power system has a 20,000 kVA capacity and is operating at a power factor of 0.6 lagging and the cost of synchronous capacitors is \$17.50/kVA, develop a table showing the required (leading) reactive power to correct the power factor to

a. 0.85 lagging power factor
b. 0.95 lagging power factor
c. Unity power factor

8.15 If a power system has a 25,000 kVA capacity and is operating at a power factor of 0.7 lagging and the cost of synchronous capacitors is \$12.50/kVA, develop a table showing the required (leading) reactive power to correct the power factor to

a. 0.85 lagging power factor
b. 0.95 lagging power factor
c. Unity power factor

8.16 If a power system has an 8000 kVA capacity and is operating at a power factor of 0.7 lagging and the cost of synchronous capacitors is \$15/kVA, find the investment required to correct the power factor to

a. Unity power factor
b. 0.85 lagging power factor

8.17 Assume that a feeder supplies an industrial consumer with a cumulative load of (1) induction motors totaling 200 hp that run at an average efficiency of 90% and a lagging average power factor of 0.80, (2) synchronous motors totaling 200 hp with an average efficiency of 80%, and a lagging average power factor of 0.80 and (3) a heating load of 50 kW. The industrial consumer plans to use the synchronous motors to correct its overall power factor. Determine the required power factor of synchronous motors to correct the overall power factor at peak load to unity power factor.

9 Distribution System Voltage Regulation

Nothing is so firmly believed as what we least know.

M.E. De Montaigne, Essays, 1580

Talk sense to a fool and he calls you foolish.

Euripides, TheBacchae, 407 B. C.

But talk nonsense to a fool and he calls you a genius.

Turan Gönen

9.1 STATE-OF-THE-ART REVIEW

As introduced in the previous chapter, capacitance introduces additional reactive power, resulting in power factor correction. This chapter, however, provides a voltage improvement along the feeder without the use of reactive power injection. A paper was introduced in the early millennium with an advanced method to regulate voltage along a feeder with coordination of interconnected energy storage and distributed generation [232]. Alternatively, some regulation of voltage can come from embedded generation, which applies techniques to restrict steady-state voltage rise [233]. Control of the voltage regulator has been modeled for the backward-forward sweeping technique of unbalanced power flow [234]. Development has extended to multi-objective locations using a micro-genetic algorithm for rural distribution feeders [235]. The optimality of voltage control can be coordinated with distributed generation [236]. The load tap changing (LTC) transformer at the distribution substation provides a root node of the spanning tree, i.e., the entire distribution feeder, for regulating the feeder load center with deadband control. Fuzzy logic control has been introduced to address the challenging issues encountered, including normalization required for fuzzy strategies with averaging active power supported by data acquired in the substation control system [237]. In recent years, studies have demonstrated that photovoltaic generation in a radial distribution feeder could assist in limiting voltage rise, especially on circuits that are overloaded [238]. As the distribution feeder has been gradually utilized with renewable energy resources (RES), the uncertainty related to intermittency of wind, solar irradiance forecast, as well as load variation has been a major challenge. Paper [239] provides a stochastic framework based on multiple scenarios of adaptation. Strategically, the distributed control architecture could be two-stage [240]. Many control applications can relate to EV charging, avoiding DT overload [241], a coordinated approach among LTC, voltage regulator, and DG [242, 243], load voltage control [244, 245], reconfiguration strategies [246, 247]. More extensively, applications with algorithms to manage decentralization of voltage control for EV charging and cost-effectiveness for PV generation and mobile storage have been extensively researched [248, 249].

9.2 BASIC DEFINITIONS

Voltage regulation: The percent voltage drop of a line (e.g., a feeder) with respect to the receiving-end voltage. Therefore,

$$\%\text{regulation} = \frac{|V_s| - |V_r|}{|V_r|} \times 100 \tag{9.1}$$

DOI: 10.1201/9781003129721-9

Voltage drop: The difference between the sending-end and the receiving-end voltages of a line
Nominal voltage: The nominal value assigned to a line or apparatus or a system of a given voltage class
Rated voltage: The voltage at which performance and operating characteristics of apparatus are referred
Service voltage: The voltage measured at the ends of the service-entrance apparatus
Utilization voltage: The voltage measured at the ends of an apparatus
Base voltage: The reference voltage, usually 120 V
Maximum voltage: The largest 5-min average voltage
Minimum voltage: The smallest 5-min voltage
Voltage spread: The difference between the maximum and minimum voltages, without voltage dips due to motor starting [77]

9.3 QUALITY OF SERVICE AND VOLTAGE STANDARDS

In general, the performance of distribution systems and the quality of service provided are assessed based on the absence of interruptions and the maintenance of satisfactory voltage levels at customers' premises, within appropriate limits for this type of service. Due to economic considerations, electric utility companies are unable to provide each customer with a constant voltage that exactly matches the nameplate voltage on their utilization apparatus.

As a result, a common practice among utilities is to adhere to preferred voltage levels and ranges of variation that ensure satisfactory operation of apparatus, as outlined by the American National Standards Institute (ANSI) Standard. In many states, the ANSI standard serves as the foundation for regulatory commission rulings that establish voltage requirements and limits for different classes of electric service.

In general, excessive steady-state voltage has been observed to cause reduced light bulb life, decreased lifespan of electronic devices, and premature failure of certain types of apparatus. Conversely, insufficient steady-state voltage leads to reduced illumination levels, diminished TV picture size, slow heating of heating devices, difficulties in motor starting, and motor overheating and/or burnout. However, most equipment and appliances can function satisfactorily within a certain range of voltage, allowing for reasonable tolerance.

The nominal voltage standards followed by the majority of electric utilities in the United States for residential and commercial customers are as follows: [Please provide the specific voltage standards you intended to mention.

1. 120/240 V three-wire single phase
2. 240/120 V four-wire three-phase delta
3. 208Y/120 V four-wire three-phase wye
4. 480Y/277 V four-wire three-phase wye

As shown in Figure 9.1, *the voltage on a distribution circuit varies from a maximum value at the customer nearest to the source (first customer) to a minimum value at the end of the circuit (last customer)*. For illustrative purposes, Table 9.1 presents typical secondary voltage standards applicable to residential and commercial customers. These voltage limits are often established by the state regulatory commission as guidelines for utilities to follow.

As indicated in Table 9.1, for any given nominal voltage level, the actual operating values can vary over a wide range. This range has been divided into three zones: (1) the *favorable zone* or *preferred zone*, (2) the *tolerable zone*, and (3) the *extreme zone*. The favorable zone encompasses the majority of the existing operating voltages, and voltages within this zone (range A) ensure satisfactory operation of customers' equipment.

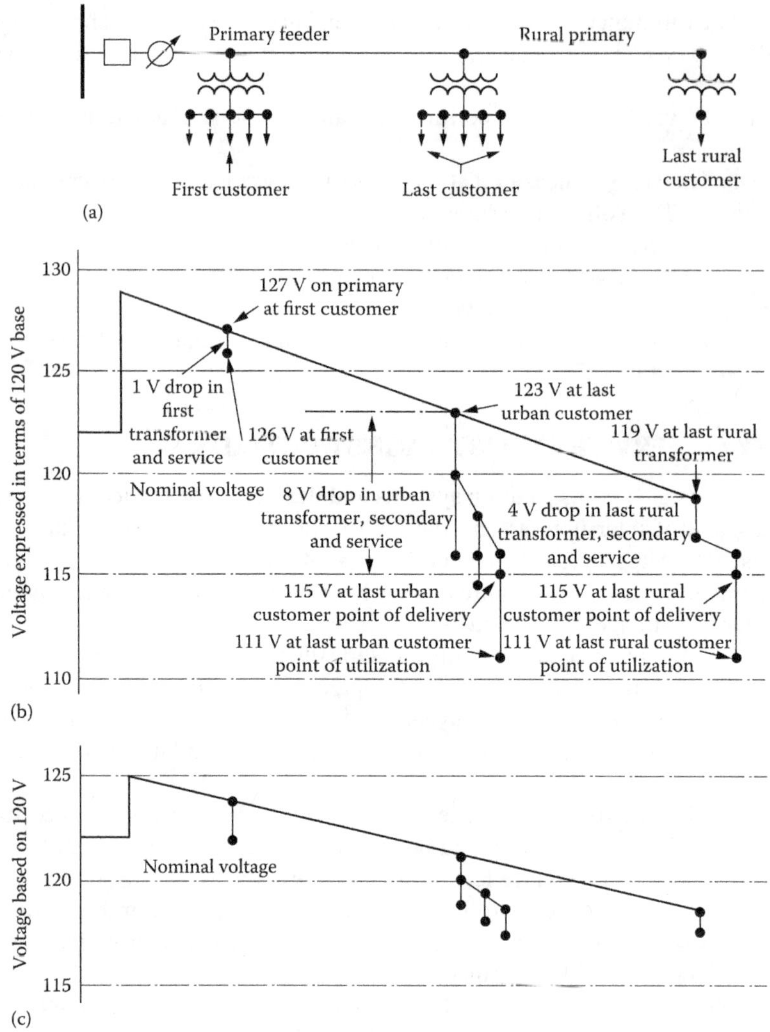

Figure 9.1 Illustration of voltage spread on a radial primary feeder: (a) one-line diagram of a feeder circuit, (b) voltage profile at peak-load conditions, and (c) voltage profile at light-load conditions.

Distribution engineers strive to maintain the voltage for each customer on a given distribution circuit within the favorable zone. Figure 9.1 illustrates the outcomes of such efforts for urban and rural circuits. The tolerable zone includes a range of operating voltages slightly above and below the favorable zone. The operating voltages in the tolerable zone (range B) are generally acceptable for most purposes. In this zone, customers' equipment is expected to function satisfactorily, although its performance may be slightly below the manufacturer's specifications.

However, if the voltage within the tolerable zone leads to unsatisfactory performance of customers' equipment, steps should be taken to improve the voltage. The extreme or emergency zone comprises voltages at the edges of the tolerable zone, usually within 2% or 3% above or below the tolerable zone. Acceptability of voltages in this zone depends on the specific application. Occasionally, under extraordinary conditions such as failure of the main supply line or non-operational voltage regulators, the voltages may reach the emergency limits.

Table 9.1
Typical Secondary Voltage Standards Applicable to Residential and Commercial Customers

	Voltage Limits		
	At Point of Delivery		At Point of Utilization
Nominal Voltage Class	**Maximum**	**Minimum**	**Minimum**
	120/240 V 1ϕ and 240/120 V 3ϕ		
Favorable zone, range A	126/252	114/228	110/220
Tolerable zone, range B	127/254	110/220	106/212
Extreme zone, emergency	130/260	108/216	104/208
	208Y/120 V 3ϕ		
Favorable zone, range A	218Y/126	197Y/114	191Y/110
Tolerable zone, range B	220Y/127	191Y/110	184Y/106
Extreme zone, emergency	225Y/130	187Y/108	180Y/104
	408Y/277 V 3ϕ		
Favorable zone, range A	504Y/291	456Y/263	440Y/254
Tolerable zone, range B	508Y/293	440Y/254	424Y/245
Extreme zone, emergency	520Y/300	432Y/249	416Y/240

Even in these conditions, if the operating voltage remains within the extreme zone, customers' equipment is expected to function reliably, albeit with a deviation from standard performance. However, voltages outside the extreme zone should never be tolerated and must be promptly improved. Typically, *a maximum voltage drop of* 4 V *is considered acceptable in the customer's wiring between the point of delivery and the point of utilization, based on a* 120 V *system.*

9.4 VOLTAGE CONTROL

In order to maintain distribution circuit voltages within acceptable limits, measures need to be implemented to control the voltage. This involves increasing the circuit voltage when it is too low and reducing it when it is too high. There are various methods available to enhance the overall voltage regulation of the distribution system. A comprehensive list of these methods is provided by Lokay [50, 250] as follows:

1. Use of generator voltage regulators
2. Application of voltage-regulating equipment in the distribution substations
3. Application of capacitors in the distribution substation
4. Balancing of the loads on the primary feeders
5. Increasing of feeder conductor size
6. Changing of feeder sections from single phase to multiphase
7. Transferring of loads to new feeders
8. Installing of new substations and primary feeders
9. Increasing of primary voltage level
10. Application of voltage regulators out on the primary feeders
11. Application of shunt capacitors on the primary feeders
12. Application of series capacitors on the primary feeders

The choice of a specific technique or techniques depends on the specific requirements of the system. However, *automatic voltage regulation is always achieved through (1) bus regulation at the substation, (2) individual feeder regulation in the substation, and (3) supplementary regulation along the main by regulators mounted on poles.* Distribution substations are equipped with *load-tap-changing (LTC) transformers* that automatically operate under load or with separate voltage regulators to ensure bus regulation [251].

Voltage-regulating equipment is designed to automatically maintain a predetermined voltage level that would otherwise fluctuate with the load. As the load increases, the regulating equipment boosts the voltage at the substation to compensate for the increased voltage drop in the distribution feeder. In cases where customers are located at long distances from the substation or where there is excessive voltage drop along the primary circuit, additional regulators or capacitors installed at selected points on the feeder provide supplementary regulation. Many utilities have found that the most cost-effective approach to regulating the voltage within the required limits is to use both step voltage regulators and shunt capacitors [252].

Capacitors are installed along the feeders and at the substation bus in sufficient quantities to achieve optimal power factor and economic benefits. Many of these installations feature sophisticated controls that enable automatic switching. While a fixed capacitor is not a voltage regulator and cannot be directly compared to regulators, in some cases, automatically switched capacitors can replace conventional step-type voltage regulators for voltage control on distribution feeders [253].

9.5 FEEDER VOLTAGE REGULATORS

Feeder voltage regulators are extensively employed to regulate the voltage of each feeder individually *in order to maintain a reasonably constant voltage at the point of utilization.* These regulators can be of the induction type or the step type. However, since modern step-type voltage regulators have largely replaced induction-type regulators, this chapter will focus solely on step-type voltage regulators.

Step-type voltage regulators can be classified as either (1) *station type*, which can be single-phase or three-phase and used for bus voltage regulation (BVR) at substations or individual feeder voltage regulation, or (2) *distribution type*, which are single-phase and mounted on poles along overhead primary feeders. Single-phase step-type voltage regulators are available in sizes ranging from 25 to 833 kVA, while three-phase step-type voltage regulators come in sizes ranging from 500 to 2000 kVA [254,255].

For certain units, the standard capacity ratings can be increased by 25% to 33% through forced-air cooling. Standard voltage ratings range from 2,400 to 19,920 V, enabling regulators to be used on distribution circuits with voltages from 2,400 to 34,500 V in grounded-wye/19,920 V in multigrounded-wye configurations. Station-type step voltage regulators for BVR can go up to 69 kV [256].

Fundamentally, a step-type voltage regulator is an autotransformer with numerous taps (or steps) in the series winding. Most regulators are designed to adjust the line voltage within a range of -10% to +10% in 32 steps, with each step corresponding to a voltage change of 5/8%.

It is important to note that the total voltage regulation range is 20%. By dividing this range by the 32 steps, the percentage regulation per step can be determined. When two internal coils of a regulator are connected in series, the regulator can achieve a +10% regulation; when they are connected in parallel, the current rating of the regulator would increase to 160%, but the regulation range would decrease to +5% [257,258]. Figure 9.2 illustrates the application of a step-type voltage regulator on a feeder, highlighting its essential components.

In addition to its autotransformer component, a step-type regulator consists of two other major components: the tap-changing mechanism and the control mechanism, as depicted in Figure 9.2. Each voltage regulator is typically equipped with the necessary controls and accessories to enable automatic tap changes under load. These tap changes are carried out by a tap changer that responds to a voltage-sensing control, ensuring a predetermined output voltage is maintained. The control

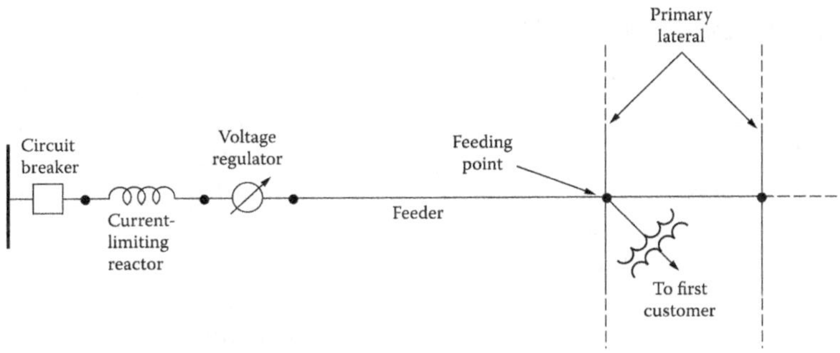

Figure 9.2 One-line diagram of a feeder, indicating the sequence of essential components.

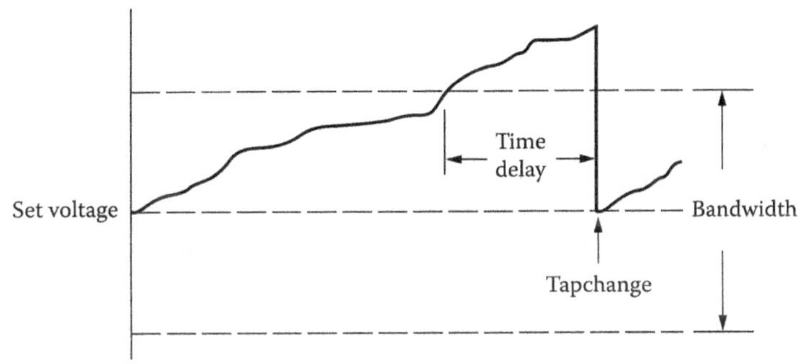

Figure 9.3 Regulator tap controls based on the set voltage, bandwidth, and time delay.

mechanism, which receives inputs from potential transformers (PT) and current transformers (CT), is responsible for regulating the voltage level and bandwidth (BW).

One example of such a control mechanism is a *voltage regulating relay* (VRR), which is responsible for controlling the tap changes. As illustrated in Figure 9.3, this relay has three fundamental settings that govern the tap changes:

1. *Set voltage*: It is the desired output of the regulator. It is also called the set point or band center.
2. *Bandwidth*: Voltage regulator controls monitor the difference between the measured voltage and the set voltage. Only when the difference exceeds one half of the BW will a tap change start.
3. *Time delay (TD)*: It is the waiting time between the time when the voltage goes out of the band and when the controller initiates the tap change. Longer TDs reduce the number of tap changes. Typical TDs are 10–120 s.

Moreover, the control mechanism also offers the capability to adjust line-drop compensation by selecting the resistance and reactance settings, as depicted in Figure 9.7. A standard direct-drive tap changer is illustrated in Figure 9.8. Figure 9.9 showcases four-step auto-booster regulators. Auto-boosters essentially function as single-phase regulating autotransformers, providing four-step regulation of feeder voltage without the high level of complexity found in 32-step regulators. They can be employed on circuits with ratings ranging from 2.4 to 12 kV in delta configuration and 2.4/4.16 to 19.92/34.5 kV in multigrounded-wye configuration. The auto-booster unit can have a continuous current rating of either 50 or 100 A. Each step represents a voltage change of either 1½% or 2½%, depending on whether the unit has a regulation range of 6% or 10%, respectively. Auto-boosters are considerably more cost-effective compared to standard voltage regulators.

9.6 LINE-DROP COMPENSATION

Voltage regulators positioned at the substation or along a feeder are utilized to maintain a constant voltage at a designated point called the regulation point (RP), regardless of the load's magnitude or power factor. Typically, the regulation point is selected somewhere between the regulator and the end of the feeder. This automatic voltage control is achieved by adjusting the resistance and reactance elements of a component known as the line-drop compensator (LDC), located on the control panel of the voltage regulator. The LDC's dial settings determine the voltage maintenance. Figure 9.4 presents a simplified schematic and phasor diagram of the control circuit and LDC circuit of a step or induction voltage regulator. The appropriate dial settings depend on whether there are any loads tapped off the feeder between the regulator and the regulation point.

If there are no loads tapped off the feeder between the regulator and the regulation point, the R dial setting of the LDC can be determined by the following method:

$$R_{\text{set}} = \frac{\text{CT}_P}{\text{PT}_N} \times R_{\text{eff}} \Omega \tag{9.2}$$

where

CT_P is the rating of the current transformer's primary

PT_N is the potential transformer's turns ratio = $V_{\text{pri}}/V_{\text{sec}}$

R_{eff} is the effective resistance of a feeder conductor from regulator station to regulation point, Ω

$$R_{\text{eff}} = r_a \times \frac{l - S_1}{2} \Omega \tag{9.3}$$

where

r_a is the resistance of a feeder conductor from regulator station to regulation point, Ω/mi per conductor

s_1 is the length of three-phase feeder between regulator station and substation, mi (multiply length by 2 if feeder is in single phase)

l is the primary feeder length, mi

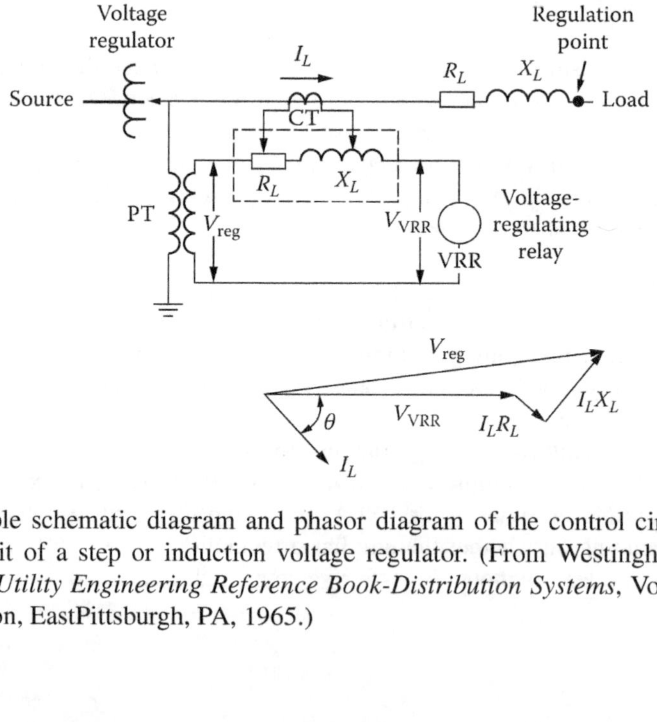

Figure 9.4 Simple schematic diagram and phasor diagram of the control circuit and line-drop compensator circuit of a step or induction voltage regulator. (From Westinghouse Electric Corporation, *Electric Utility Engineering Reference Book-Distribution Systems*, Vol. 3, Westinghouse Electric Corporation, EastPittsburgh, PA, 1965.)

Also, the *X dial setting of the LDC* can be determined from

$$X_{\text{set}} = \frac{\text{CT}_P}{\text{PT}_N} \times X_{\text{eff}} \Omega \tag{9.4}$$

where
X_{eff} is the effective reactance of a feeder conductor from regulator to regulation point, Ω

$$X_{\text{eff}} = X_L \times \frac{l - S_1}{2} \Omega \tag{9.5}$$

and

$$X_L = X_a + X_d \Omega/mi \tag{9.6}$$

where
x_a is the inductive reactance of individual phase conductor of feeder at 12-in spacing, Ω/mi
x_d is the inductive-reactance spacing factor, Ω/mi
x_L is the inductive reactance of feeder conductor, Ω/mi
It is important to note that when determining the *R* and *X* settings, they are based on the total connected load rather than a small group of customers. Therefore, the resistance and reactance values of the transformers are not taken into account in the calculations for effective resistance and reactance.

If load is tapped off the feeder between the regulator station and the regulation point, the *R* dial setting of the LDC can still be determined from Equation 9.2, but the determination of the R_{eff} is somewhat more involved. Lokay [50] gives the following equations to calculate the effective resistance:

$$R_{\text{eff}} = \frac{\sum_{i=1}^{n} |\text{VD}_R|i}{|I_L|} \Omega \tag{9.7}$$

and

$$\sum_{i=1}^{n} |\text{VD}_R|_i = |I_{L,1}| \times r_{a,1} \times l_1 + |I_{L,2}| \times r_{a,2} \times l_2 + \cdots + |I_{L,n}| \times r_{a,n} \times l_n \tag{9.8}$$

where
$|\text{VD}_R|_i$ is the voltage drop due to line resistance of *i*-th section of feeder between regulator station and regulation point, V/section
$\sum_{i=1}^{n} |\text{VD}_R|_i$ is the total voltage drop due to line resistance of feeder between regulator station and regulation point, V
$|I_L|$ is the magnitude of load current at regulator location, A
$|I_{Li}|$ is the magnitude of load current in ith feeder section, A
$r_{a,i}$ is the resistance of a feeder conductor in *i*th section of feeder, Ω/mi
l_i is the length of *i*th feeder section, mi

Also, the *X* dial setting of the LDC can still be determined from Equation 9.4, but the determination of the X_{eff} is again somewhat more involved. Lokay [50] gives the following equations to calculate the effective reactance:

$$X_{\text{eff}} = \frac{\sum_{i=1}^{n} |VD_X|_i}{|I_L|} \Omega \tag{9.9}$$

and

$$\sum_{i=1}^{n} |\text{VD}_X|_i = |I_{L,1}| \times X_{L,1} \times I_1 + |I_{L,2}| \times X_{L,2} \times I_2 + \cdots + |I_{L,n}| \times X_{L,n} \times I_n \tag{9.10}$$

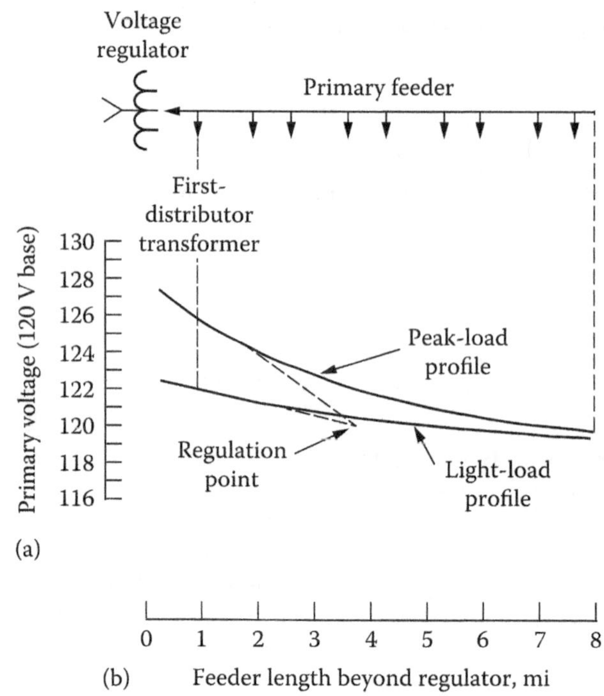

Figure 9.5 Determination of the voltage profiles for (a) peak loads and (b) light loads.

where

$|\mathrm{VD}_R|_i$ is the voltage drop due to line reactance of i-th section of feeder between regulator station and regulation point, V/section

$\sum_{i=1}^{n} |\mathrm{VD}_X|_i$ is the total voltage drop due to line reactance of feeder between regulator station and regulation point, V

$X_{L,1}$ is the inductive reactance [as defined in Equation 9.6] of ith section of feeder, Ω/mi

Given that the methods described above for determining the effective R and X values are complex, Lokay [50] proposes an alternative and practical approach. This method involves measuring the current (I_L) and voltage at the regulator location, as well as the voltage at the regulation point (RP). The discrepancy between the two voltage measurements represents the total voltage drop from the regulator to the regulation point. This voltage drop can be expressed as follows:

$$\mathrm{VD} = |I_L| \times R_{\mathrm{eff}} \times \cos\theta + |I_L| \times X_{\mathrm{eff}} \times \sin\theta \tag{9.11}$$

With the known load power factor of the feeder and the average R/X ratio of the feeder conductors between the regulator and the regulation point, the values of R_{eff} and X_{eff} can be easily determined.

Figure 9.5 illustrates an example that demonstrates the voltage profiles for peak and light loads. It presents a one-line diagram and voltage profiles of a feeder with distributed load beyond a voltage regulator location. The primary-feeder voltage values are based on a 120 V base.

The example assumes that the conductor size between the regulator and the first distribution transformer is #2/0 copper conductor with a 44-inch flat spacing, having resistance and reactance values of 0.481 and 0.718 Ω/mi, respectively. The voltage regulator has PT and CT ratios of 7960:120 and 200:5, respectively. The distance to the fictitious regulation point (RP) is 3.9 miles.

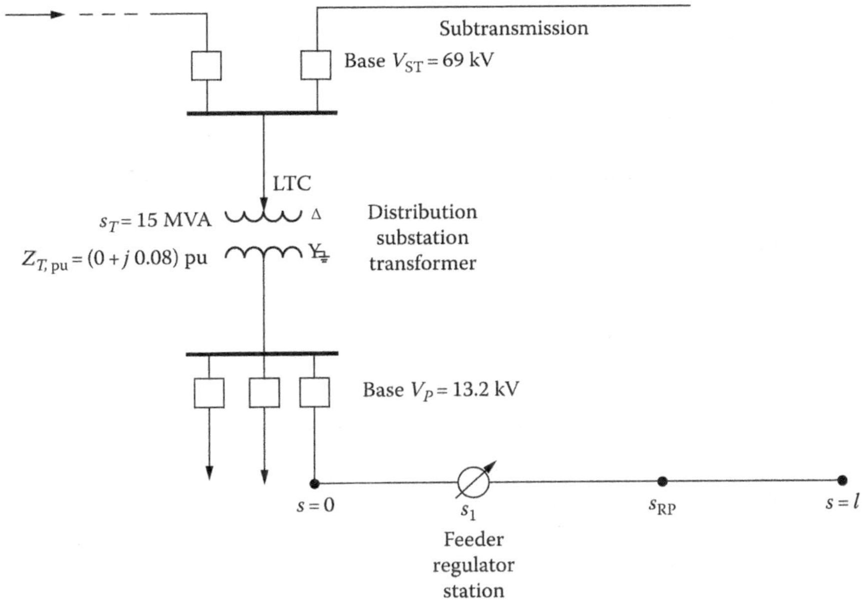

Figure 9.6 The elements of a distribution substation for Example 9.1.

The settings of the load distribution control (LDC) are as follows:

$$R_{set} = 200 \times \frac{120}{7960} \times 0.481 \times 3.9 = 5.656$$

$$R_{set} = 200 \times \frac{120}{7960} \times 0.718 \times 3.9 = 8.4428$$

The voltage-regulating relay is set to 120.1 V (adapted from [50]).

Example 9.1. This example explores the utilization of step-type voltage regulation/control to enhance the voltage profile in distribution systems. Figure 9.6 illustrates the components of a distribution substation that receives power from a subtransmission loop and supplies multiple radial primary feeders.

The substation's Load Tap Changer (LTC) transformer is employed to regulate the voltage at the primary distribution bus (V_P), ensuring a constant V_P despite variations in both the subtransmission voltage (V_{ST}) and the voltage drop IZ_T across the substation transformer due to load changes. By strategically employing a feeder voltage regulator bank, if the primary feeder's main is limited by voltage drop, it can be extended further or loaded more heavily. In Figure 9.6, the feeder voltage regulator, depicted by a symbol resembling a zero with an arrow passing through it, is located at point $s = s_1$ and automatically adjusts its boost and buck to maintain a predefined voltage at the regulation point (RP), denoted as $s = S_{RP}$.

Typical LTC and feeder regulator data: The abbreviation Voltage-Regulating Relay (VRR) represents a voltage-regulating relay or its solid-state equivalent, adjustable within the approximate range of 110 to 125 V. The VRR measures the voltage at the RP, denoted as V_{RP}, using the Load Distribution Control (LDC).

The LDC includes adjustable *R* and *X* settings, both within the approximate range of 0 to 24 Ω (often referred to as "volts" since regulators typically utilize CTs with 1A secondaries).

The bandwidth (BW) of the voltage-regulating relay can be adjusted within an approximate range of ± ¾ to ± 1½ V, based on a 120 V reference. The time delay (TD) is adjustable between approximately 10 and 120 seconds.

Table 9.2
Overloading of Step-Type Feeder Regulators

Reduced Range of Regulation (%)	Percent of Normal Load Current
±10.00	100
±8.75	110
±7.50	120
±6.25	135
±5.00	160

Table 9.3
Some Typical Single-Phase Regulator Sizes

Single Phase kVA	Volts	Amps	CTP[a]	PTN[b]
25	2500	100	100	20
125	2500	500	500	20
38.1	7620	50	50	63.5
57.2	7620	75	75	63.5
76.2	7620	100	100	63.5
114.3	7620	150	150	63.5
167	7620	219	250	63.5
250	7620	328	400	63.5

[a] Ratio of the current transformer contained within the regulator. (Here, the ratio is the high-voltage-side ampere rating becausethe low-voltage rating is 1.0 A.)
[b] Ratio of the potential transformer contained within the regulator. (All potential transformer secondaries are 120 V.)

The regulation point (RP) location is controlled by the R and X settings of the Load Distribution Control (LDC). When the R and X settings are set to zero, the regulator regulates the voltage at its local terminal to the VRR setting ± BW. In this example, SRP is equal to $s1$.

Regarding the overloading of step-type feeder regulators, ANSI standards specify the overload capacity of regulators, as shown in Table 9.2, in case the full 10% range of regulation is not required. Modern regulators are equipped with adjustments to limit the range within which the motor can operate the tap-changer switching mechanism.

The limited-range operation can be advantageous in certain situations. However, if there is load growth, a larger range of regulation and a larger regulator size (kVA or current) may be necessary. Table 9.3 provides some typical sizes of single-phase regulators.

Substation data Make the following assumptions:

Base MVA3_ϕ = 15 MVA
Subtransmission base $V_{L-L} = 69$ kV
Primary base $V_{L-L} = 13.2$ kV

The substation transformer is rated 15 MVA, 69–7.62/13.2 kV grounded-wye and has a per unit impedance ($Z_{T,pu}$) of 0 + j0.08 based on its ratings. Its three-phase LTC can regulate ± 10% voltage in 32 steps of 5/8% each.

Load flow data: Assume that the maximum subtransmission voltage (max V_{ST}) is 72.45 kV or 1.05 pu, which occurs during the off-peak period at which the off-peak kVA is 0.25 pu with a leading power factor of 0.95. The minimum subtransmission voltage (min V_{ST}) is 69 kV or 1.00 pu, which occurs during the peak period at which the peak kVA is 1.00 pu with a lagging power factor of 0.85.

Voltage data and voltage criteria: Assume that the maximum secondary voltage is 125 V or 1.0417 pu V (based on 120 V) and the minimum secondary voltage is 116 V or 0.9667 pu V, and that the maximum voltage drop in secondaries is 0.035 pu V.

Assume that the maximum primary voltage (max V_P) is 1.0417 pu V at zero load, and that at annual peak load, the maximum primary voltage is 1.0767 pu V (1.0417 + 0.035) considering the nearest secondary to the regulator and the minimum primary voltage is 1.0017 pu V (0.9667 + 0.035) considering the most remote secondary.

Feeder data: Assume that the annual peak load is 4000 kVA, at a lagging power factor of 0.85, and is distributed uniformly along the 10-mi-long feeder main. The main has 266.8 kcmil all-aluminum conductors (AACs) with 37 strands and 53-in geometric mean spacing. Use 3.88×10^{-6} pu VD/ (kVA · mi) at 0.85 lagging power factor as the constant K.

Assume that the substation transformer LTC is used for BVR. Use a BW of ± 1.0 V or (1 V/120 V) = 0.0083 pu V. Also use rounded figures of 1.075 and 1.000 pu V for the maximum and minimum primary voltages at peak load, respectively.

a. Specify the setting of the VRR for the highest allowable primary voltage (V_P), BW being considered, then round the setting to a convenient number.
b. Find the maximum number of steps of buck and boost that will be required.
c. Sketch voltage profiles of the feeder being considered for zero load and for the annual peak load. Label the significant voltage values on the curves.

Solution. a. Since the LDC of the regulator is not used,

$$R_{set} = 0 \text{ and } X_{set} = 0$$

Therefore, the setting of the VRR for the highest allowable primary voltage, BW being considered, occurs at the zero load and is

$$\begin{aligned} VRR &= (V_P)_{max} - \text{BW} \\ &= 1.0417 - 0.0083 \\ &= 1.0334 \text{ pu V} \\ &\cong 1.035 \text{ pu V} \\ &= 124.2 \text{ V} \end{aligned}$$

b. To find the maximum number of buck and boost that will be required, the highest allowable primary voltages at off-peak and on-peak have to be found. Therefore, *at off-peak*,

$$\overline{V}_{P,\text{pu}} = \overline{V}_{\text{ST,pu}} - \overline{I}_{P,\text{pu}} \times \overline{Z}_{T,\text{pu}} \tag{9.12}$$

where

$V_{\text{ST,pu}}$ is the per unit subtransmission voltage at primary side of the substation transformer = 1.05 $\angle 0°$ pu V

$I_{P,\text{pu}}$ is the per unit no-load primary current at substation (transformer) = 0.2381 puA = per unit impedance of substation transformer

$$Z_{T,\text{pu}} = 0 + j0.08 \text{ pu } \Omega$$

Therefore,

$$\begin{aligned} V_{P\text{pu}} &= 1.05 - (0.2381)(\cos\theta + j\sin\theta)(0 + j0.08) \\ &= 1.05 - (0.2381)(0.95 + j0.3118)(0 + j0.08) \\ &= 1.0589 \text{ pu V} \end{aligned}$$

whereas *at on-peak*,

$$\begin{aligned} V_{P\text{pu}} &= 1.0 - (1.00)(0.85 - j0.53)(0 + j0.08) \\ &= 0.9602 \text{ pu V} \end{aligned}$$

Since the LTC of the substation can regulate ± 10% voltage in 32 steps of 5/8% volts (or 0.00625 pu V) each, the maximum number of steps of buck required, *at off-peak*, is

$$\begin{aligned} \text{No. of steps} &= \frac{V_{P,\text{pu}} - \text{VRR}_{\text{pu}}}{0.00625} \\ &= \frac{1.0589 - 1.05}{0.00625} \\ &\cong 3 \text{ or } 4 \text{ steps} \end{aligned} \tag{9.13}$$

and the maximum number of steps of boost required, *at peak*, is

$$\begin{aligned} \text{No. of steps} &= \frac{V_{P,\text{pu}} - \text{VRR}_{\text{pu}}}{0.00625} \\ &= \frac{1.035 - 0.9602}{0.00625} \\ &\cong 12 \text{ steps} \end{aligned} \tag{9.14}$$

c. To sketch voltage profiles of the primary feeder for the annual peak load, the total voltage drop of the feeder has to be known. Therefore,

$$\begin{aligned} \sum \text{VD}_{\text{pu}} &= K \times S \times \frac{l}{2} \\ &= (3.88 \times 10^{-6})(4000 \text{ kVA})\left(\frac{10 \text{ mi}}{2}\right) \\ &= 0.077 \text{ pu V} \end{aligned} \tag{9.15}$$

and thus the minimum primary-feeder voltage at the end of the 10 mi feeder, as shown in Figure 9.7, is

$$\begin{aligned} \text{Min}V_{P,\text{pu}} &= \text{VRR}_{\text{pu}} - \sum \text{VD}_{\text{pu}} \\ &= 1.035 - 0.0776 \\ &= 0.9574 \text{ pu V} \end{aligned} \tag{9.16}$$

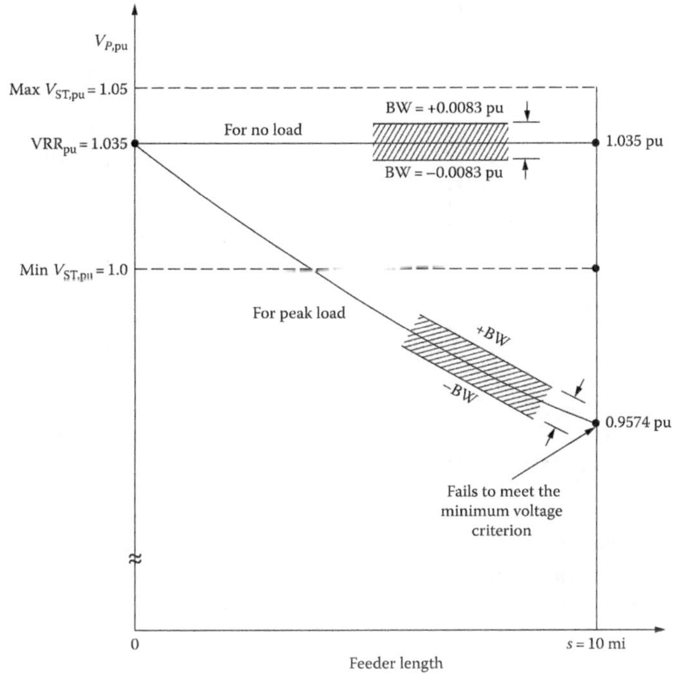

Figure 9.7 Feeder voltage profile.

At the annual peak load, the rounded voltage criteria are

$$\begin{aligned} \text{Max}V_{P,\text{pu}} &= 1.075 - \text{BW} \\ &= 1.075 - 0.0083 \\ &= 1.0667 \text{ puV} \end{aligned}$$

and

$$\begin{aligned} \text{Min}V_{P,\text{pu}} &= 1.00 + \text{BW} \\ &= 1.00 - 0.0083 \\ &= 1.0083 \text{ puV} \end{aligned}$$

At no-load, the rounded voltage criteria are

$$\begin{aligned} \text{Max}V_{P,\text{pu}} &= 1.0417 - \text{BW} \\ &= 1.0417 - 0.0083 \\ &= 1.035 \text{ puV} \end{aligned}$$

and

$$\text{Min}V_{P,\text{pu}} = 1.0083 \text{ pu V}$$

As can be seen from Figure 9.7, the minimum primary-feeder voltage at the end of the 10 mi feeder *fails to meet the minimum voltage criterion* at the annual peak load. Therefore, a voltage regulator has to be used.

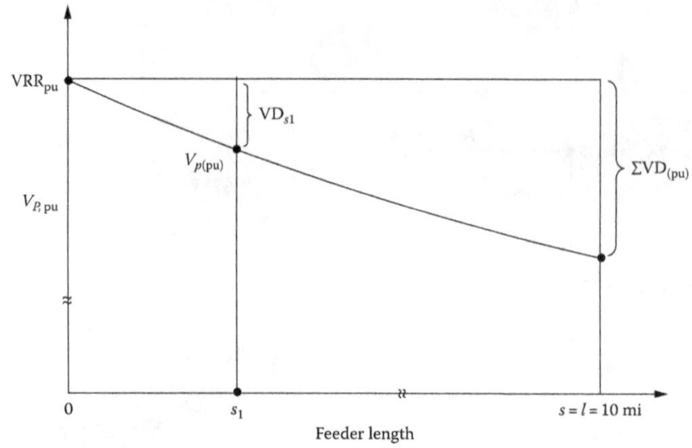

Figure 9.8 Voltage profile for Example 9.2.

Example 9.2. Use the information and data given in Example 9.1 and locate the voltage regulator, that is, determine the s_1 distance at which the regulator must be located as shown in Figure 9.6, for the following two cases, where the peak-load primary-feeder voltage $(V_{P,\text{pu}})$ at the input to the regulator is

a. $V_{P,\text{pu}} = 1.010$ pu V.
b. $V_{P,\text{pu}} = 1.000$ pu V.
c. What is the advantage of part *a* over part *b*, or vice versa?

Solution. a. When $V_{P,\text{pu}} = 1.010$ pu V, the associated voltage drop at the distance s_1, as shown in Figure 9.14, is

$$\begin{aligned} \text{VD}_{s1} &= \text{VRR}_{\text{pu}} - V_{p,\text{pu}} \\ &= 1.035 - 1.01 \\ &= 0.025 \text{ pu V} \end{aligned} \tag{9.17}$$

From Example 9.1, the total voltage drop of the feeder is

$$\sum \text{VD}_{\text{pu}} = 0.0776 \text{ pu V}$$

Therefore, the distance s_1 can be found from the following parabolic formula for the uniformly distributed load:

$$\frac{VD_{S_1}}{\sum \text{VD}_{\text{pu}}} = \frac{S_1}{l}\left(2 - \frac{S_1}{l}\right) \tag{9.18}$$

or

$$\frac{0.025}{0.0776} = \frac{s_1}{10}\left(2 - \frac{S_1}{10}\right)$$

from which the following quadratic equation can be obtained:

$$s_1^2 - 20s_1 + 32.2165 = 0$$

which has two solutions, namely, 1.75 and 18.23 mi. Therefore, the distance s_1, taking the acceptable answer, is 1.75 mi.

b. When $V_{P,\text{pu}} = 1.00$ pu V, the associated voltage drop at the distance s_l is

$$\begin{aligned} \text{VD}_{s_1} &= \text{VRR}_{\text{pu}} - \text{V}_{\text{P,pu}} \\ &= 1.035 - 1.00 \\ &= 0.035 \text{ pu V} \end{aligned}$$

Therefore, from Equation 9.18,

$$\frac{0.035}{0.0776} = \frac{S_1}{10}\left(2 - \frac{S_1}{10}\right)$$

or

$$s_1^2 - 20s_1 + 45.1031 = 0$$

which has two solutions, namely, 2.6 and 17.4 mi. Thus, taking the acceptable answer, the distance s1 is 2.6 mi.

c. The advantage of part (a) over part (b) is that it can compensate for future growth. Otherwise, the VP, pu might be less than 1.00 pu V in the future.

Example 9.3. Assume that the peak-load primary-feeder voltage at the input to the regulator is 1.010 pu V as given in Example 9.2. Determine the necessary minimum kVA size of each of three single-phase feeder regulators.

Solution. From Example 9.2, the distance s1 is found to be 1.75 mi. Previously, the annual peak load and the standard regulation range have been given as 4000 kVA and ±10%, respectively.

The *uniformly distributed three-phase load* at s_1 is

$$\begin{aligned} S_{3\phi}\left(1 - \frac{S_1}{l}\right) &= 4000\left(1 - \frac{1.75}{10.00}\right) \\ &= 3300 \text{ kVA} \end{aligned}$$

Therefore, the single-phase load at s_1 is

$$\frac{3300 \text{ kVA}}{3} = 1100 \text{ kVA}$$

Since the single-phase regulator kVA rating is given by

$$S_{\text{reg}} = \frac{(\%R_{\max})S_{\text{ckt}}}{100} \tag{9.19}$$

where S_{ckt} is the circuit kVA, then

$$S_{\text{reg}} = \frac{10 \times 1100 \text{ kVA}}{100} = 110 \text{ kVA}$$

Thus, from Table 9.3, the corresponding minimum kVA size of the regulator size can be found as 114.3 kVA.

Example 9.4. Use the distance of $s_1 = 1.75$ mi found in Example 9.2 and assume that the distance of the RP is equal to s_1, that is, $s_{RP} = s_1$, or, in other words, *the RP is located at the regulator station*, and determine the following:

a. Specify the best settings for the LDC's R and X, and for the VRR.
b. Sketch voltage profiles for zero load and for the annual peak load. Label significant voltage values on the curves.
c. Are the primary-feeder voltage ($V_{P,\text{pu}}$) criteria met?

Solution. a. The $x_{RP} = s_1$ means that the *RP is located at the feeder regulator station.* Therefore, the *best settings for the* LDC *of the regulator are when settings for both R and X are zero* and

$$\text{VRR}_{\text{pu}} = V_{\text{RP,PU}} = 1.035 \text{ pu V}$$

b. The voltage drop occurring in the feeder portion *between the RP and the end of the feeder* is

$$\begin{aligned}\text{VD}_{\text{pu}} &= K \times S \times \frac{l}{2} \\ &= (3.88 \times 10^{-6})(3300)\left(\frac{8.25}{2}\right) \\ &= 0.0528 \text{ puV}\end{aligned}$$

Thus the *primary-feeder voltage at the end of the feeder* for the annual peak load is

$$\begin{aligned}V_{Pi10mi} &= 1.035 - 0.0528 \\ &= 0.9809 \text{ puV}\end{aligned}$$

Note that the $V_{P,\text{pu}}$ used at the regulator point is the *no-load value* rather than the annual peak-load value. If, instead, the 1.0667 pu value is used, then, for example, television sets of those customers located at the vicinity of the RP might be damaged during the off-peak periods because of the too-high V_{RP} value.

As can be seen from Figure 9.9, the peak-load voltage profile is not in linear but in parabolic shape. *The voltage-drop value for any given point s between the substation and the regulator station* can be calculated from

$$\text{VD}_s = K\left(S_{3\phi} - \frac{S_{3\phi} \times S}{l}\right)s + K\left(\frac{S_{3\phi} \times S}{l}\right)\frac{s}{2} \text{ pu V} \tag{9.20}$$

where

K is the percent voltage drop per kVA-mile characteristic of feeder
$S_{3\phi}$ is the uniformly distributed three-phase annual peak load, kVA
l is the primary feeder length, mi
s is the distance from substation, mi

Therefore, from Equation 9.20,

$$\text{VD}_s = 3.88 \times 10^{-6}\left(4000 - \frac{4000s}{10}\right)s + 3.88 \times 10^{-6}\left(\frac{4000s}{10}\right)\frac{s}{2}\text{puV} \tag{9.21}$$

For various values of s, the associated values of the voltage drops and $V_{P,\text{pu}}$ can be found, as given in Table 9.4.

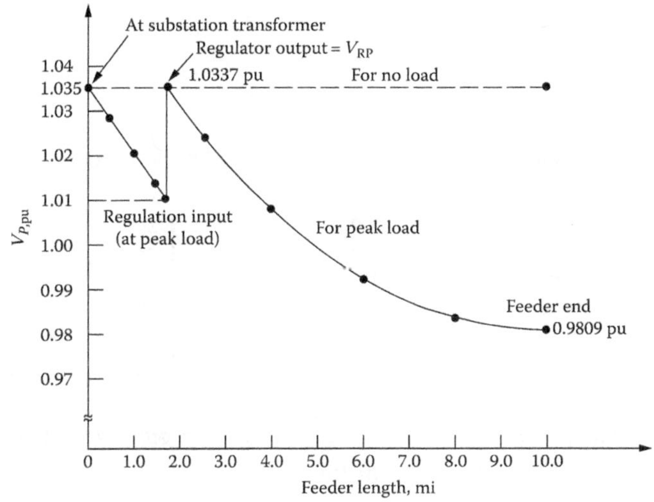

Figure 9.9 Feeder voltage profiles for zero load and for the annual peak load.

The *voltage-drop value for any given point s between the substation and the regulator station* can also be calculated from

$$\text{VD}_s = I(r \times \cos\theta + X \times \sin\theta)s\left(1 - \frac{s}{2l}\right) \tag{9.22}$$

where

I_L is the load current in feeder at substation end

$$= \frac{S_{3\phi}}{\sqrt{3} \times V_{L-L}} \tag{9.23}$$

r is the resistance of feeder main, Ω/mi per phase

x is the reactance of feeder main, Ω/mi per phase

Therefore, the voltage drop in per units can be found as

$$\text{VD}_s = \frac{\text{VD}_s}{V_{L-N}}\text{pu V} \tag{9.24}$$

Table 9.4
For Annual Peak Load

s (mi)	VD_s (pu V)	$V_{P,\,pu}$ (pu V)
0	0	1.035
0.5	0.0076	1.0274
1	0.0071	1.0203
1.5	0.0068	1.0135
1.75	0.025	1.01

Table 9.5
For Annual Peak Load

s (mi)	VD_s (pu V)	Vp, pu (pu V)
0	0	1.0337
0.75	0.0092	1.0245
2.25	0.0157	1.0088
4.25	0.0155	0.9933
6.25	0.0093	0.984
8.25	0.0031	0.9809

The *voltage-drop value for any given point s between the regulator station and the end of the feeder* can be calculated from the following equation:

$$VD_s = K\left(S'_{3\phi} - \frac{S'_{3\phi}\times s}{l-s}\right)s + \left(\frac{S'_{3\phi}\times s}{l-s}\right)\frac{s}{2}\text{pu V} \tag{9.25}$$

where $S'_{3\phi}$ = uniformly distributed three-phase annual peak load at distance s_1, kVA

$$= S_{3\phi} = \left(1 - \frac{S_1}{l}\right)\text{kVA}$$

s_1 is the distance of feeder regulator station from substation, mi Therefore, from Equation 9.25,

$$VD_s = 3.88\times10^{-6}\left(3300 - \frac{3300s}{8.25}\right)s + 3.88\times10^{-6}\left(\frac{3300s}{8.25}\right)\frac{s}{2}\text{pu V} \tag{9.26}$$

For various values of s, the corresponding values of the voltage drops and $V_{P,\text{pu}}$ can be found, as given in Table 9.5.

The voltage profiles for the annual peak load can be obtained by plotting the $V_{P,\text{pu}}$ values from Tables 9.4 and 9.5. Since there is no voltage drop at zero load, the $V_{P,\text{pu}}$ remains constant at 1.035 pu. Therefore, the voltage profile for the zero load is a horizontal line (with zero slope).

c. The minimum $V_{P,\text{pu}}$ criterion of 1.0083 pu V is not met even though the regulator voltage has been set as high as possible without exceeding the maximum voltage criterion of 1.035 pu V.

Example 9.5. Assume that the regulator station is located at a distance s_1 as found in part (a) of Example 9.2, but the RP has been moved to the end of the feeder so that $s_{RP} = l = 10$ mi.

a. Determine good settings for the values of VRR, R, and X so that all $V_{P,\text{pu}}$ voltage criteria will be met, if possible.
b. Sketch voltage profiles and label the values of significant voltages, in per unit volts.

Solution. a. From Table A.4 of Appendix A, the resistance at 50° C and the reactance of the 266.8 kcmil AAC with 37 strands are 0.386 and 0.4809 Ω/mi, respectively. From Table A.10, the inductive-reactance spacing factor for the 53-in geometric mean spacing is 0.1802 Ω/mi. Therefore, from Equation 9.6, the inductive reactance of the feeder conductor is

$$\begin{aligned} X_L &= X_a + X_d \\ &= 0.4809 + 0.1802 \\ &= 0.6611\Omega/\text{mile} \end{aligned}$$

From Equations 9.3 and 9.5,

$$\begin{aligned} R_{\text{eff}} &= r_a \times \frac{l - s_1}{2} \\ &= 0.386 \times \frac{8.25}{2} \\ &= 1.5923\ \Omega \end{aligned}$$

and

$$\begin{aligned} X_{\text{eff}} &= X_L \times \frac{l - s_1}{2} \\ &= 0.6611 \times \frac{8.25}{2} \\ &= 2.7270\ \Omega \end{aligned}$$

From Table 9.3, for the regulator size of 114.3 kVA found in Example 9.3, the primary rating of the CT and the PT ratio are 150 and 63.5, respectively. Therefore, from Equations 9.2 and 9.4, the R and X dial settings can be found as

$$\begin{aligned} R_{\text{set}} &= \frac{\text{CT}_P}{\text{PT}_N} \times R_{\text{eff}}\Omega \\ &= \frac{150}{63.5} \times 1.5923 \\ &= 3.761\ \text{V or} 0.0313\ \text{pu V} \end{aligned}$$

based on 120 V and

$$\begin{aligned} X_{\text{set}} &= \frac{\text{CT}_P}{\text{PT}_N} \times X_{\text{eff}}\Omega \\ &= \frac{150}{63.5} \times 2.727 \\ &= 6.442\ \text{V or } 0.0537\ \text{pu V} \end{aligned}$$

Assume that the voltage at the RP (V_{RP}) is arbitrarily set to be 1.0138 pu V using the R and X settings of the LDC of the regulator so that the V_{RP} is always the same for zero load or for the annual peak load. Therefore, the output voltage of the regulator for the annual peak load can be found from

$$\begin{aligned} V_{\text{reg}} &= V_{\text{RP}} + \frac{\frac{S_{1\phi}}{V_{L-N}}(R_{\text{set}} \times \cos\theta + X_{\text{set}} \times \sin\theta)}{\text{CT}_P \times V_B} \\ &= 1.0138 + \frac{(1100/7.62)(3.761 \times 0.85 + 6.442 \times 0.527)}{150 \times 120} \\ &= 1.0666\ \text{pu V} \end{aligned} \tag{9.27}$$

Here, note that the regulator regulates the regulator output voltage automatically according to the load at any given time in order to maintain the RP voltage at the predetermined voltage value.

Table 9.6
Actual Voltages vs. Voltage Criteria at Peak and Zero Loads

	Actual Voltage (pu V)		Voltage Criteria (pu V)	
Voltage	At Peak Load	At Zero Load	At Peak Load	At Zero Load
Max $V_{P,pu}$	1.0666	1.0138	1.0667	1.0337
Min Vp,pu	1.0138	1.0138	1.0083	1.0083

Table 9.7
Values Obtained

s (mi)	VD_s (pu V)	$V_{P,\,pu}$ (pu V)
0	0	1.0666
0.75	0.0092	1.0574
2.25	0.0157	1.0417
4.25	0.0155	1.0262
6.25	0.0093	1.0169
8.25	0.0031	1.0138

Table 9.6 gives the $V_{P,\text{pu}}$ values for the purpose of comparing the actual voltage values against the established voltage criteria for the annual peak and for zero load.

As can be observed from Table 9.6, the primary voltage criteria are met by using the R and X settings.

b. The voltage profiles for the annual peak load and zero load can be obtained by plotting the $V_{P,\text{pu}}$ values from Tables 9.6 and 9.7 (based on Equation 9.26), as shown in Figure 9.10.

Problem 9.6. Consider the results of Examples 9.4 and 9.5 and determine the following:

a. The number of steps of buck and boost the regulators will achieve in Example 9.4.
b. The number of steps of buck and boost the regulators will achieve in Example 9.5.

Solution. a. For Example 9.4, the number of steps of buck is

$$\text{No. of Steps} = \frac{1.035 - 1.0337}{0.00625}$$
$$= 0.208$$

thus it is either zero or one step. The number of boost is

$$\text{No. of Steps} = \frac{1.0337 - 1.010}{0.00625}$$
$$= 3.79$$

therefore it is either three or four steps.

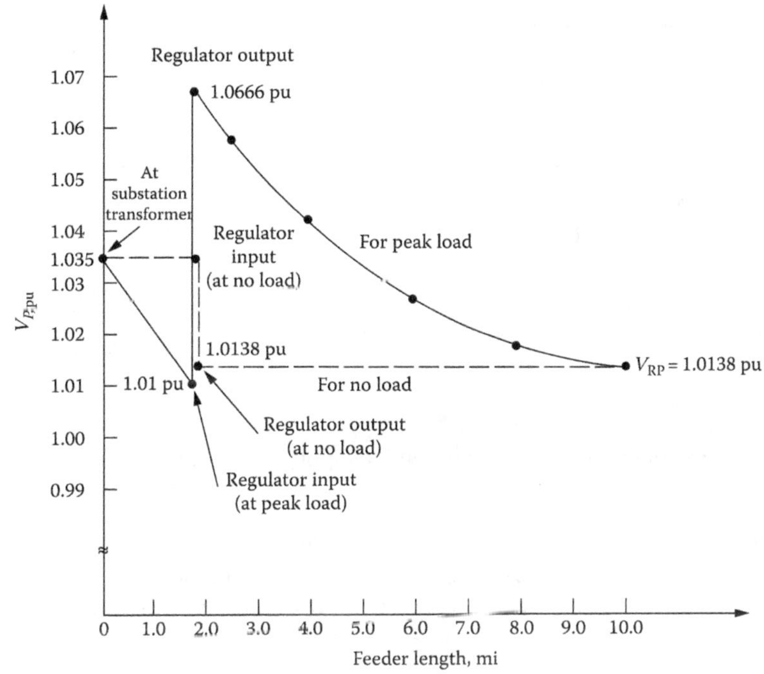

Figure 9.10 Voltage profiles.

b. For Example 9.5, the number of steps of buck is

$$\text{No. of Steps} = \frac{1.035 - 1.0138}{0.00625}$$
$$= 3.39$$

hence, it is either three or four steps. The number of steps of boost is

$$\text{No. of Steps} = \frac{1.0666 - 1.010}{0.00625}$$
$$= 9.06$$

therefore, it is either 9 or 10 steps.

Problem 9.7. Consider the results of Examples 9.4 and 9.5 and answer the following:

a. Can reduced range of regulation be used gainfully in Example 9.4? Explain.
b. Can reduced range of regulation be used gainfully in Example 9.5? Explain.

Solution. a. Yes, the reduced range of regulation can be used gainfully in Example 9.4 since the next-smaller-size regulator, that is, 76.2 kVA, at ± 5% regulation range can be selected. This ± 5% regulation range would allow the capacity of the regulator to be increased to 160% (see Table 9.2) so that

$$1.6 \times 76.2 \text{ kVA} = 121.92 \text{ kVA}$$

which is much larger than the required capacity of 110 kVA. It would allow the user ± 8 steps of buck and boost, which is more than the required one step of buck and four steps of boost.

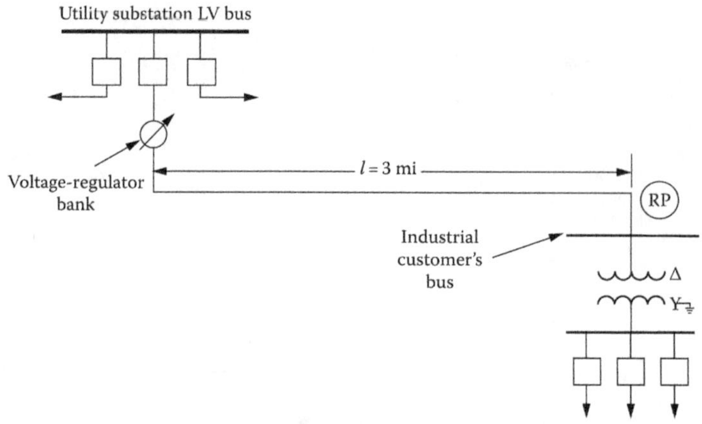

Figure 9.11 One-line diagram of a primary feeder supplying an industrial customer.

b. No, the reduced range of regulation cannot be used gainfully in Example 9.5 since the required steps of buck and boost are 4 and 10, respectively. The reduced range of regulation at $\pm$ 6.25% would provide the $\pm$ 10 steps of buck and boost, but it would allow the capacity of the regulator to be increased only up to 135% (see Table 9.2) so that

$$1.35 \times 76.2 \text{ kVA} = 102.87 \text{ kVA}$$

which is smaller than the required capacity of 110 kVA.

Problem 9.8. Figure 9.11 shows a one-line diagram of a primary feeder supplying an industrial customer. The nominal voltage at the utility substation low-voltage bus is 7.2/13.2 kV three-phase wye-grounded. The voltage regulator bank is made up of three single-phase step-type voltage regulators with a PT ratio of 63.5 (7620:120).

The industrial customer's bus is located at the end of a 3 mi primary line with a resistance of 0.30 Ω/mi and an inductive reactance of 0.80 Ω/mi.

The customer's transformer is rated 5000 kVA in three phase with a 12,800 V primary connected in delta (taps in use) and a 2400/4160 V secondary connected in grounded-wye. The transformer impedance is 0 + j0.05 puΩ based on the rated kVA and tap voltages in use. Assume that the bases to be used are 5000 kVA, 2400/4160 V, and 7390/12,800 V.

Assume that the customer asks that the low-voltage bus be regulated to 2450/4244 V and determine the following:

a. Find the necessary setting of the voltage-setting dial of the VRR of each single-phase regulator in use.
b. Assume that the ratio of the CT in each regulator is 250:1 A and find the necessary R and X dial settings of LDCs.

Solution. a. The voltage at the RP, *which is located at the customer's bus*, is

$$V_{RP} = \frac{2450 \text{ V}}{2400 \text{ V}}$$
$$= 1.02083 \text{ pu V}$$

Therefore,

$$\begin{aligned}\text{VRR} &= \frac{7390}{7620} \times 1.02083 \\ &= 0.99 \text{ pu V or } 7620 \text{ V}\end{aligned}$$

Thus,

$$\begin{aligned}\text{VRR} &= \frac{7620}{\text{PT}_N} \times 0.99 \\ &= \frac{7620}{63.5} \times 0.99 \\ &= 120 \times 0.99 \\ &= 118.8 \text{ V}\end{aligned}$$

or, alternatively,

$$\begin{aligned}\text{VRR}_{\text{set}} &= V_{RP} \times V_{B,sec} \\ &= 1.02083 \times \frac{12,800}{\sqrt{3} \times 63.5} \\ &\cong 118.8 \text{ V}\end{aligned} \tag{9.28}$$

b. The applicable impedance base is

$$\begin{aligned}Z_B &= \frac{(KV_{L-L})^2}{\text{MVA}} \\ &= \frac{(12.8 \text{ kV})^2}{5 \text{ MVA}} \\ &= 32.768 \ \Omega\end{aligned}$$

therefore, the transformer impedance is

$$\begin{aligned}Z_T &= Z_{T,\text{pu}} \times Z_B \\ &= (0 + j0.05) \times 32.768 \\ &= 0 + j1.6384 \ \Omega\end{aligned} \tag{9.29}$$

Since here the R and X settings are determined for only one customer, the resistance and reactance values of the customer's transformer have to be included in the effective resistance and reactance calculations. Therefore,

$$\begin{aligned}R_{\text{eff}} &= r \times I + R_T \\ &= (0.3 \ \Omega/\text{mi})(3 \text{ mi}) + 0 \\ &= 0.9 \ \Omega\end{aligned} \tag{9.30}$$

and

$$\begin{aligned}X_{\text{eff}} &= x \times I + X_T \\ &= (0.8 \ \Omega/\text{mi})(3 \text{ mi}) + 1.6384 \ \Omega \\ &= 4.0384 \ \Omega\end{aligned} \tag{9.31}$$

Thus, the R dial setting of the LDC is

$$\begin{aligned} R_{\text{set}} &= \frac{\text{CT}_P}{\text{PT}_N} \times R_{\text{eff}} \\ &= \frac{250}{63.5} \times 0.9 \\ &= 3.5433\ \Omega \end{aligned}$$

and the X dial setting is

$$\begin{aligned} X_{\text{set}} &= \frac{\text{CT}_P}{\text{PT}_N} \times X_{\text{eff}} \\ &= \frac{250}{63.5} \times 4.0384 \\ &= 15.8992\ \Omega \end{aligned}$$

Problem 9.9. Consider the 10 mi feeder of Example 9.1. Assume that the substation has a bus voltage regulator (BVR) with transformer LTC and that the primary feeder voltage (V_P) has been set on the VRR to be 1.035 pu V.

Assume that the main feeder is made up of 266.8 kcmil with 37 strands and 53-in geometric mean spacing. It has been found in Example 9.5 that the main feeder has an inductive reactance of 0.661 Ω/mi per conductor. Assume that the annual peak load is 4000 kVA at a lagging power factor of 0.85 and distributed uniformly along the main or, in other words, the uniformly distributed load is

$$S_{3\phi} = P_L + jQ_L = 3400 + j2100\ \text{kVA}$$

Assume that, as found in Example 9.1, the total voltage drop of the feeder is

$$\begin{aligned} \sum \text{VD}_{\text{pu}} &= \text{VD}_{1,\text{pu}} \\ &= 0.0776\ \text{pu V} \end{aligned}$$

and the reactive load factor is 0.40. Use the given data and determine the following:

a. Design a fixed, that is, nonswitched (NSW)-capacitor bank for the maximum loss reduction.
b. Sketch the voltage profiles when there is no-capacitor (N/C) bank installed and when there is a fixed-capacitor bank, that is, Q_{NSW} installed.
c. Add a switched-capacitor bank for voltage control on the feeder. Locate the switched-capacitor bank, that is, Q_{sw}, at the end of the feeder for the feeder at the annual peak load is 1.000 pu V. Sketch the associated voltage profiles.

Solution. a. From Figure 8.31, the corrective ratio (CR) for the given reactive load factor of 0.40 is found to be 0.27. Therefore, the required size of the NSW-capacitor bank is

$$\begin{aligned} Q_{\text{NSW}} &= CR \times Q_L \\ &= (0.27)(2100\ \text{kVAr}) \qquad (9.32) \\ &= 567\ \text{kVAr per three phase} \end{aligned}$$

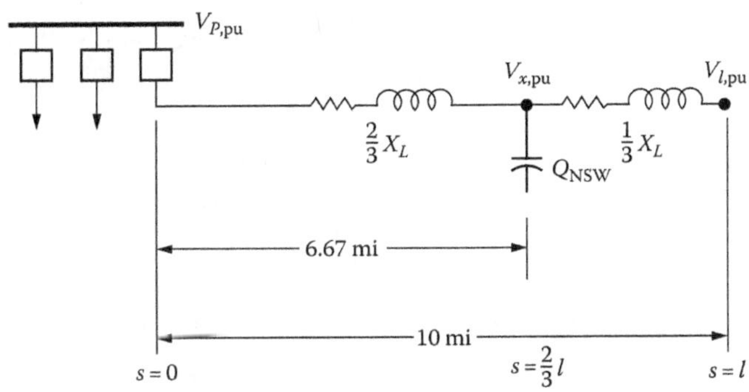

Figure 9.12 Optimum location of a capacitor bank.

Thus, two single-phase standard 100-kVAr-size capacitor units are required to be used on each phase and located on the feeder at a distance of

$$\begin{aligned} S &= \frac{2}{3} \times I \\ &= \frac{2}{3} \times 10 \\ &= 6.67 \text{ mi} \end{aligned}$$

for the optimum result, as shown in Figure 9.12.

Therefore, the per unit voltage rise (VRP_{pu}) due to the *NSW-capacitor bank* is

$$\begin{aligned} \text{VR}_{\text{pu}} &= \frac{Q_{\text{NSW}} \times \frac{2}{3}X_L}{1000(\text{kV}_{L-L})^2} \\ &= \frac{600 \times 4.41}{1000 \times 13.2^2} \\ &= 0.0152 \text{ pu V} \end{aligned} \tag{9.33}$$

b. When there is *N/C bank installed*, the voltage drop for the uniformly distributed load at a distance of $s = 2/3\ l$ can be found from Equation 9.18 as

$$\frac{\text{VD}_{Spu}}{\sum \text{VD}_{\text{pu}}} = \frac{S}{l}\left(2 - \frac{S}{l}\right) \tag{9.34}$$

or

$$\frac{\text{VD}_{S\text{pu}}}{0.0776} = \frac{6.67}{10}\left(2 - \frac{6.67}{10}\right)$$

from which

$$\text{VD}_{\text{spu}} = 0.069 \text{ pu V}$$

Therefore, the feeder voltage at the 2/3 l distance is

$$\begin{aligned} V_{S,\text{pu}} &= V_{P,\text{pu}} - \text{VD}_{S,\text{pu}} \\ &= 1.035 - 0.069 \\ &= 0.966 \text{ pu V} \end{aligned}$$

When there is a *fixed-capacitor bank installed*, the new voltage at the 2/3 l distance due to the voltage rise is

$$\begin{aligned} \text{New } V_{s,\text{PU}} &= V_{s,\text{PU}} + \text{VR}_{\text{pu}} \\ &= 0.966 + 0.0152 \\ &= 0.9812 \text{ pu V} \end{aligned} \tag{9.35}$$

When there is *N/C bank installed*, the voltage at the end of the feeder is

$$\begin{aligned} V_{I,\text{pu}} &= V_{P,\text{pu}} + \sum \text{VD}_{\text{pu}} \\ &= 1.035 - 0.0776 \\ &= 0.9574 \text{ pu V} \end{aligned} \tag{9.36}$$

When there is a *fixed-capacitor bank installed*, the new voltage at the end of the feeder due to the voltage rise is

$$\begin{aligned} \text{New } V_{I,\text{PU}} &= V_{I,\text{PU}} + \text{VR}_{\text{pu}} \\ &= 0.9574 + 0.0152 \\ &= 0.9726 \text{ pu V} \end{aligned} \tag{9.37}$$

The associated voltage profiles are shown in Figure 9.13.

c. Since the new voltage at the end of the feeder due to the Q_{sw} installation is 1.000 pu V at the annual peak load, the required voltage rise is

$$\begin{aligned} \text{VR}_{\text{pu}} &= V_{I,\text{pu}} - \text{new}V_{I,\text{PU}} \\ &= 1.000 - 0.9726 \\ &= 0.0274 \text{ pu V} \end{aligned} \tag{9.38}$$

Therefore, the *required size of the switched-capacitor bank* can be found from

$$\text{VR}_{\text{pu}} = \frac{Q_{3\phi,\text{SW}} \times X_L}{1000(\text{kV}_{L-L})} \text{pu V} \tag{9.39}$$

or

$$\begin{aligned} Q_{3\phi,\text{SW}} &= \frac{1000(\text{kV}_{L-L})^2 \text{VR}_{\text{pu}}}{X_L} \\ &= \frac{1000 \times 13.2^2 \times 0.0274}{6.611} \\ &= 722.2 \text{ kVAr} \end{aligned} \tag{9.40}$$

Hence, the possible combinations of the single-phase standard-size capacitor units to make up the capacitor bank are

i. Fifteen single-phase standard 50 kVAr capacitor units, for a total of 750 kVAr
ii. Six single-phase standard 100 kVAr capacitor units, for a total of 600 kVAr

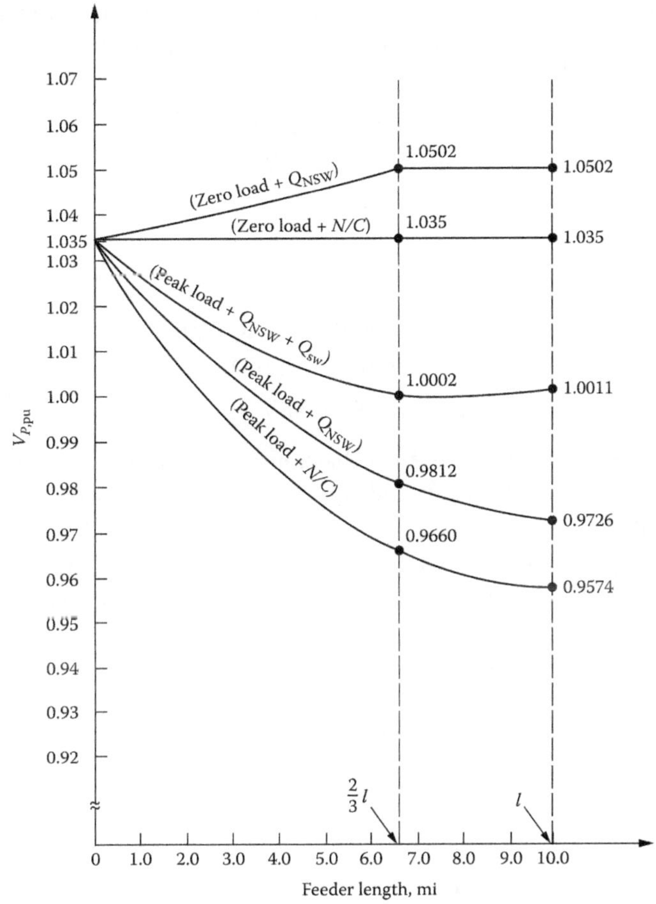

Figure 9.13 Voltage profiles.

iii. Nine single-phase standard 100 kVAr capacitor units, for a total of 900 kVAr For example, assume that the first combination, that is,

$$Q_{3\phi,\text{sw}} = 750 \text{ kVAr}$$

is selected. The *resultant new voltage rises* at the distance of l and $s = s = 2/3\ l$ are

$$\begin{aligned} \text{VR}_{l,\text{pu}} &= \text{VR}_{\text{pu}} \times \frac{\text{Selected } Q_{\text{sw}}}{\text{Required } Q_{\text{sw}}} \text{puV} \\ &= 0.0274 \times \frac{750 \text{ kVAr}}{722.2 \text{ kVAr}} \\ &\cong 0.0285 \text{ pu V} \end{aligned} \tag{9.41}$$

and

$$\begin{aligned} \text{VR}_{s,\text{pu}} &= \frac{2}{3}\text{VR}_{l,\text{pu}}\text{puV} \\ &= \frac{2}{3} \times 0.0285 \\ &= 0.0190 \text{ pu V} \end{aligned} \tag{9.42}$$

Therefore, *at the peak load when both the NSW-* (i.e., fixed) and the *switched-capacitor banks* are on, the voltage at two-thirds of the line and at the end of the line are

$$\begin{aligned} \mathrm{V}_{s,\mathrm{pu}} &= \mathrm{new}V_{s,\mathrm{pu}} + V_{s,\mathrm{pu}}\mathrm{pu\ V} \\ &= 0.9812 + 0.0190 \\ &= 1.0002\ \mathrm{pu\ V} \end{aligned} \tag{9.43}$$

and

$$\begin{aligned} \mathrm{V}_{s,\mathrm{pu}} &= \mathrm{new}V_{I,\mathrm{pu}} + \mathrm{VR}_{I,\mathrm{pu}}\mathrm{pu\ V} \\ &= 0.9726 + 0.0285 \\ &= 1.0011\ \mathrm{pu\ V} \end{aligned} \tag{9.44}$$

respectively. At zero load when there is N/C bank installed, the voltages at two-thirds of the line and at the end of the line are the same and equal to 1.035 pu V. The associated voltage profiles are shown in Figure 9.13.

Problem 9.10. Consider that the industrial load at the annual peak is 5000 kVA at 80% lagging power factor. Assume that the customer wishes to add some additional load, is currently paying a monthly power-factor penalty, and the single-phase voltage regulators are approaching full boost. Select a proper three-phase capacitor-bank size (in terms of the multiples of three-phase 150 kVAr capacitor units) to be connected to the 4 kV bus that will (1) produce a voltage rise of at least 0.020 pu V on the 4 kV bus and (2) raise the on-peak power factor of the present load to at least 88% lagging power factor.

Solution. The presently existing load is

$$S_{3\phi} = 5000\angle 36.87^\circ\ \mathrm{kVA}$$

or

$$S_{3\phi} = 4000 + j3000\ \mathrm{kVA}$$

at 80% lagging power factor. When a properly sized capacitor bank is connected to the bus to improve the on-peak power factor to 88%, the real power portion will be the same but the reactive power portion will be different. In other words,

$$|S|\angle 28.36^\circ = 4000 + jQ_{L,\mathrm{new}}$$

from which

$$\tan 28.36^\circ = \frac{Q_{L,\mathrm{new}}}{4000}$$

therefore

$$\begin{aligned} Q_{L,\mathrm{new}} &= 4000 \times \tan 28.36^\circ \\ &= 4000 \times 0.5397 \\ &= 2158.97\ \mathrm{kVAr} \end{aligned}$$

and hence the magnitude of the new apparent power is

$$|S| = 4545.45 \text{ kVA}$$

Therefore, the minimum size of the capacitor bank required to raise the load power factor to 0.88 is

$$\begin{aligned} Q_{3\phi} &= 3000 - 2158.97 \\ &= 841.03 \text{ kVAr} \end{aligned}$$

Thus, if a 900-kVAr-capacity bank is used, the resultant voltage rise from Equation 9.39 is

$$\text{VR}_{\text{pu}} = \frac{Q_{3\phi} \times X_L}{1000(kV_{B,L-L})^2}$$

where

$$\begin{aligned} X_L &= X_{\text{line}} + X_T \\ &= (0.83\ \Omega/\text{mi})(3\text{mi}) + 1.6384\ \Omega \\ &= 4.0384\ \Omega \end{aligned}$$

hence

$$\begin{aligned} \text{VR}_{\text{pu}} &= \frac{900 \times 4.0384}{1000 \times 12.8^2} \\ &= 0.0222 \text{ pu V} \end{aligned}$$

which is larger than the given voltage-rise criterion of 0.020 pu V. Therefore, it is proper to install six 150 kVAr three-phase units as the capacitor bank to meet the criteria.

9.7 DISTRIBUTION CAPACITOR AUTOMATION

Intelligent customer meters have the capability to monitor voltage levels at key customer sites and transmit this information to the utility company. This enables the fine-tuning of system information based on actual measured values at the endpoint, rather than relying on projected values. Additionally, var information is integrated into the control scheme, enhancing its effectiveness.

Distributed capacitor automation leverages the distributed-processing capabilities of electronic meters, capacitor controllers, radios, and substation processors. It utilizes an algorithm to remotely switch field and substation capacitors on and off. This switching is based on voltage information obtained from meters located at important customer sites, along with var information from the substation. Figure 9.14 provides an illustration of this process.

In the past, capacitors on distribution systems were primarily controlled by standalone controllers that monitored the circuit voltage at the capacitor. Different control strategies, such as temperature and/or time bias settings on the capacitor controllers, were implemented to ensure proper operation during predicted peak load conditions. Although this system adequately provided peak voltage/var support, it often involved overcompensation to ensure that all customers received satisfactory voltage service. Additionally, capacitors operated independently and were not integrated into a comprehensive system-wide control scheme [211,214,215,259].

Distribution capacitor automation integrates field and substation capacitors into a closed-loop control scheme, operating within the following structure:

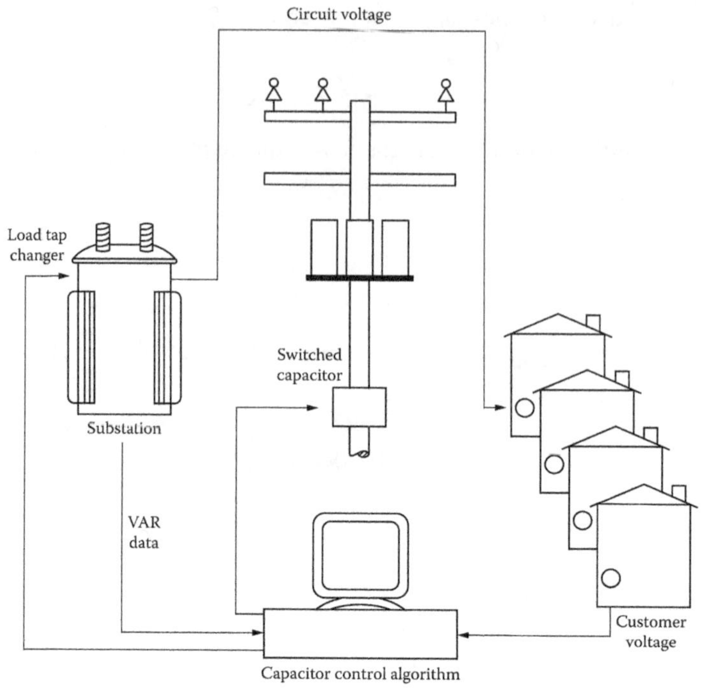

Figure 9.14 The distribution capacitor automation algorithm enables remote and automatic switching of capacitors based on voltage information obtained from customer meters and var information from the substation.

1. Intelligent customer meters provide exception reporting on voltages out of set BWs. They also report 5-min average voltages when polled.
2. Meters communicate via power-line carrier to the nearest packed radio, and via radio-frequency packet communication, to the designed capacitor controller.
3. Each capacitor controller automation programmed to receive meter voltages (received from several meters) to a substation processor.
4. The distribution capacitor automation program algorithm, running on an industrial-grade processor at the substation, determines the optimal capacitor-switching pattern and communicates control instructions to capacitor controllers.

Strategically positioned customer meters provide a representative sample of lower voltage customers to ensure consistency. The system's objective is to maintain the voltage of each customer within a narrower bandwidth, targeting a minimum of 114 V.

Optimization of substation reactive power flow is achieved by utilizing control bandwidth set points in the processor. For instance, the operator can specify a desired power factor measured at the substation transformer, and the algorithm will respond accordingly, selecting the appropriate capacitor switching pattern to maintain minimum customer voltage while meeting the substation's var requirements. In scenarios where distribution substations are equipped with load-tap changing (LTC) transformers, the control algorithm calculates the optimal bus voltage necessary to achieve unity power factor. The processor then issues commands to the LTC controller to maintain this optimal voltage level [260].

To regulate reactive power flow in the subtransmission system, the processor at the transmission substation interfaces with distribution substation processors to determine the subtransmission voltage level that minimizes var flow and ensures customer voltages are within desired limits.

9.8 VOLTAGE FLUCTUATIONS

Voltage fluctuations and lamp flicker on distribution systems are typically caused by the utilization apparatus of customers. Lamp flicker refers to sudden changes in illumination intensity due to abrupt voltage fluctuations across the lamp. Most flickering incidents occur during motor start-ups, where the large inrush of starting current results in a temporary dip in illumination for incandescent and fluorescent lamps, as their brightness is directly influenced by voltage.

Therefore, utility companies aim to avoid jeopardizing the quality of service for existing customers when serving a new customer who may generate excessive flicker according to the company's standards. Consequently, distribution circuits are evaluated to determine whether the flicker caused by the new customer's load, in addition to the existing flicker-generating loads, meets the company's voltage fluctuation criteria.

The decision to provide service to such a customer depends on factors such as the location of the load, load type, service voltage, frequency of motor starts, motor horsepower rating, and adherence to motor's NEMA code considerations. Momentary or pulsating loads are evaluated based on their starting requirements and the rate of change in power requirements over time. Flickers caused by the operation of pulsating loads during their continuous operation often have a more pronounced effect compared to starting loads. Special attention is usually given to pulsating loads like grinders, hammer mills, rock crushers, reciprocating pumps, and arc welders.

The annoyance caused by lamp flicker is subjective and varies from person to person. However, in certain cases, flickering can be highly objectionable and cause significant discomfort. The degree of objection to lamp flicker generally depends on the frequency of occurrence and the rate of voltage dip. Voltage changes that lead to lamp flicker can be cyclic or non-cyclic in nature, with cyclic flicker typically being more objectionable.

Figure 9.15 illustrates a typical curve utilized by utility companies to determine the acceptable amount of voltage flicker in their systems. Flicker values located above the curve are likely to be considered objectionable by lighting customers. For instance, based on the figure, it can be observed that 5 V dips, in relation to a 120 V system, are acceptable to lighting customers as long as the number of dips does not exceed 3 per hour. Consequently, more frequent dips of this magnitude fall within the objectionable flicker zone, indicating that they are problematic for lighting customers. The sinusoidal flicker curve should be used for voltage changes caused by pump compressors and equipment with similar characteristics. Each utility company develops its own voltage flicker limit curve based on their experiences and previous customer complaints.

Distribution engineers work diligently to ensure that voltage flickers remain within the satisfactory zone by ensuring compliance with the company's flicker standards and requirements.

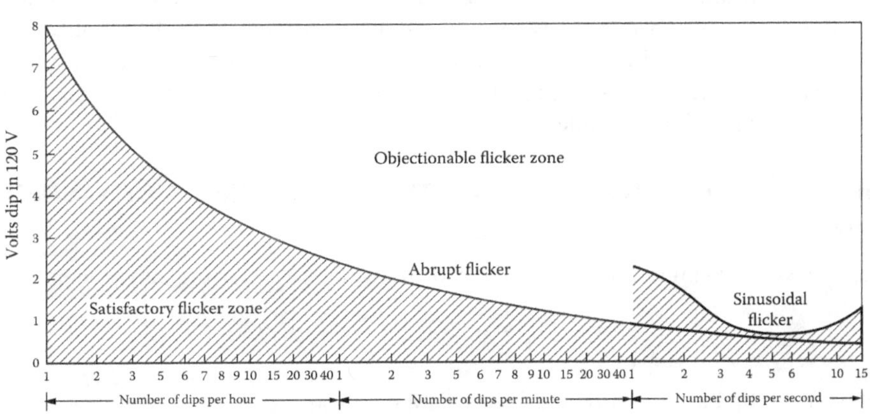

Figure 9.15 Permissible voltage-flicker-limit curve.

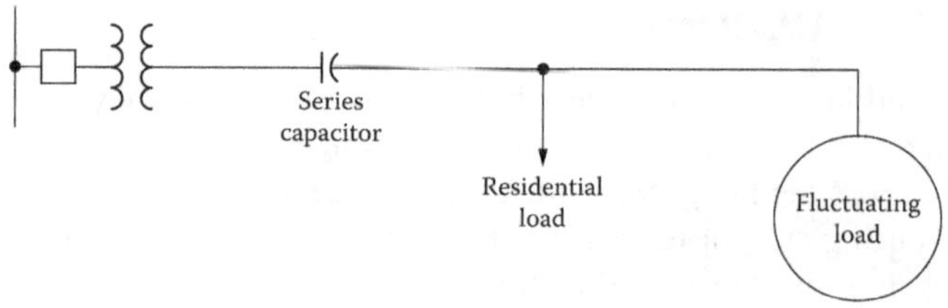

Figure 9.16 Installation of series capacitor to reduce the flicker voltage caused by a fluctuating load.

Additionally, when designing new extensions or rebuilds, they aim to provide service that falls within the satisfactory flicker zone. To mitigate flicker resulting from motor starting, the following remedies can be employed:

1. Using a motor that requires less kVA per horsepower to start
2. Choosing a low-starting-torque motor if the motor starts under light load
3. Replacing the large-size motor with a smaller-size motor or motors
4. Employing motor starters to reduce the motor inrush current at the start
5. Using shunt or series capacitors to correct the power factor

As previously mentioned in this section, distribution engineers make every reasonable effort to address the requirement of motor-start flicker. When other alternatives have been exhausted, they may opt to fulfill the flicker condition by installing shunt or series capacitors.

Shunt capacitors are utilized to compensate for the low power factor of the motor during start-up. These capacitors are disconnected from the circuit once the motor reaches its nominal running speed. During the start-up phase, shunt capacitors rated at line-to-neutral voltage are often connected line-to-line, meaning they are subjected to a voltage higher than their rating by a factor of $\sqrt{3}$. As a result, the momentary effective kVAr rating of the capacitors becomes three times their rated kVAr value, since $\left(\frac{V_{L-L}}{V_{L-N}}\right)^2 = 3$.

If series capacitors are employed, they should be installed between the substation transformer and the residential or lighting tap, as depicted in Figure 9.16. Placing the capacitor between the residential load and the fluctuating load would not mitigate the flicker voltage, as it would not reduce the impedance between the source and the lighting bus.

Since series capacitors are permanently integrated into the primary feeder, they require special protective devices to safeguard against overvoltage and resonance conditions. Consequently, a typical installation of series capacitors costs three to four times more than shunt capacitors used for motor start-ups. Series capacitors primarily correct the power factor of the system rather than the motor itself. The placement of series capacitors within the circuit is particularly crucial if there are other customers located along the line.

9.8.1 SHORTCUT METHOD TO CALCULATE THE VOLTAGE DIPS DUE TO A SINGLE-PHASE MOTOR START

If the starting kVA of a single-phase motor is known, the motor's starting current can be found as

$$I_{\text{start}} = \frac{S_{\text{start}}}{V_{L-N}}\,\text{A} \tag{9.45}$$

and the voltage dip based on 120 V can be calculated from

$$\text{VDIP} = \frac{120 \times I_{\text{start}} \times Z_G}{V_{L-N}}\,\text{V} \tag{9.46}$$

where

$$Z_G = \frac{V_{L-N}}{I_{f,L-G}} \Omega \tag{9.47}$$

Substituting Equations 9.45 and 9.47 into Equation 9.46, the voltage dip can be expressed as

$$\text{VDIP} = \frac{120 \times S_{\text{start}}}{I_{f,L-G} \times V_{L-N}} \text{V} \tag{9.48}$$

where
VDIP is the voltage dip due to single-phase motor start expressed in terms of 120 V base, V
S_{start} is the starting kVA of single-phase motor, kVA
$I_{f,L-G}$ is the line-to-ground fault current available at point of installation and obtained from fuse coordination, A
V_{L-N} is the line-to-neutral voltage, kV

Problem 9.11. Assume that a 10-hp single-phase 7.2 kV motor with NEMA code letter "G" starting 15 times per hour is to be served at a certain location. If the starting kVA per horsepower for this motor is given by the manufacturer as 6.3, and the line-to-ground fault at the installation location calculated to be 1438 A, determine the following:

a. The voltage dip due to the motor start, in volts
b. Whether or not the resultant voltage dip is objectionable

Solution. a. Since the starting kVA per horsepower is given as 6.3, the starting kVA can be found as

$$\begin{aligned} S_{\text{start}} &= (\text{kVA/hp})_{\text{start}} \times \text{hp}_{\text{motor}} \text{ kVA} \\ &= 6.3 \text{ kVA/hp} \times 10 \times \text{ hp} \\ &= 63 \text{ kVA} \end{aligned} \tag{9.49}$$

Therefore, the voltage dip due to the motor start, from Equation 9.48, can be calculated as

$$\begin{aligned} \text{VDIP} &= \frac{120 \times S_{\text{start}}}{I_{f,L-G} \times V_{L-N}} \\ &= \frac{120 \times 63 \text{ kVA}}{1438 \text{ A} \times 7.2 \text{ kV}} \\ &= 0.73 \text{ V} \end{aligned}$$

b. From Figure 9.15, it can be found that the voltage dip of 0.73 V with a frequency of 15 times/h is in the satisfactory-flicker zone and therefore is not objectionable to the immediate customers.

9.8.2 SHORTCUT METHOD TO CALCULATE THE VOLTAGE DIPS DUE TO A THREE-PHASE MOTOR START

If the starting kVA of a three-phase motor is known, its starting current can be found as

$$I_{\text{start}} = \frac{S_{\text{start}}}{\sqrt{3} \times V_{L-L}} \text{A} \tag{9.50}$$

and the voltage dip based on 120 V can be calculated from

$$\text{VDIP} = \frac{120 \times I_{\text{start}} \times Z_l}{V_{L-N}} \text{V} \tag{9.51}$$

where

$$Z_l = \frac{V_{L-N}}{I_{3\phi}} \Omega \tag{9.52}$$

Substituting Equations 9.50 and 9.52 into Equation 9.51, the voltage dip can be expressed as

$$\text{VDIP} = \frac{69.36 \times S_{\text{start}}}{I_{3\phi} \times V_{L-L}} \text{V} \tag{9.53}$$

where
VDIP is the voltage dip due to three-phase motor start expressed in terms of 120 V base, V
S_{start} is the starting kVA of three-phase motor, kVA
$I_{3\phi}$ is the three-phase fault current available at the point of installation and obtained from fusecoordination, A
V_{L-L} is the line-to-line voltage, kV

Problem 9.12. Assume that a 100-hp three-phase 12.47 kV motor with NEMA code letter "F" starting three times per hour is to be served at a certain location. If the starting kVA per horsepower for this motor is given by the manufacturer as 5.6, and the three-phase fault current at the installation location calculated to be 1765 A, determine the following:

a. The voltage dip due to the motor start
b. Whether or not the resultant voltage dip is objectionable

Solution. a. Since the starting kVA per horsepower is given as 5.6, the starting kVA can be found from Equation 9.49 as

$$\begin{aligned} S_{\text{start}} &= (\text{kVA/hp})_{\text{start}} \times \text{hp}_{\text{motor}} \\ &= 5.6 \text{ kVA/hp} \times 100 \times \text{ hp} \\ &= 560 \text{ kVA} \end{aligned}$$

Therefore, the voltage dip due to the motor start, from Equation 9.53, can be calculated as

$$\begin{aligned} \text{VDIP} &= \frac{69.36 \times S_{\text{start}}}{I_{3\phi} \times V_{L-L}} \\ &= \frac{69.36 \times 560 \text{ kVA}}{1765 \text{ A} \times 12.47 \text{ kV}} \\ &= 1.76 \text{ V} \end{aligned}$$

b. From Figure 9.15, it can be found that the voltage dip of 1.72 V with a frequency of three times per hour is in the satisfactory-flicker zone and therefore is not objectionable to the immediate customers.

PROBLEMS

9.1 Derive, or prove, Equation 9.18.

9.2 Derive, or prove, Equation 9.20.

9.3 Repeat Example 9.1, assuming 336.4 kcmil ACSR conductors and annual peak load of 5000 kVA at a lagging-load power factor of 0.90.

9.4 Repeat Example 9.2, assuming 336.4 kcmil ACSR conductors and annual peak load of 5000 kVA at a lagging-load power factor of 0.90.

9.5 Repeat Example 9.3, assuming 336.4 kcmil ACSR conductors and annual peak load of 5000 kVA at a lagging-load power factor of 0.90.

9.6 Repeat Example 9.4, assuming 336.4 kcmil ACSR conductors and annual peak load of 5000 kVA at a lagging-load power factor of 0.90.

9.7 Repeat Example 9.5, assuming 336.4 kcmil ACSR conductors and annual peak load of 5000 kVA at a lagging-load power factor of 0.90.

9.8 Repeat Example 9.6, assuming 336.4 kcmil ACSR conductors and annual peak load of 5000 kVA at a lagging-load power factor of 0.90.

9.9 Assume that a subtransmission line is required to be designed to carry a contingency peak load of 2 × SIL. A 60% series compensation is to be used, that is, the capacitive reactance (X_c) of the capacitor bank required to be installed is equal to 60% of the total series inductive reactance per phase of the transmission line. Assume that each phase of the series-capacitor bank is to be made up of series and parallel groups of two-bushing 12 kV 150 kVAr shunt power-factor-correction capacitors. Assume that the three-phase SIL of the line is 416.5 MVA and its inductive line reactance is 117.6 Ω/phase. Specify the necessary series–parallel arrangement of capacitors for each phase.

9.10 In this problem, design improvements of the designer choice to correct the undervoltage conditions are investigated on the radial system shown in Figure 9.17.

The voltage at the distribution substation low-voltage bus is kept at 1.04 pu V with bus-voltage regulation. The per unit voltages at annual peak-load values at the points *a*, *b*, *c*, *d*, *e*, and *f* are 1.0049, 0.9815, 0.9605, 0.8793, 0.8793, and 0.8793, respectively. Use the nominal operating voltage of 7,200/12,470 V of the three-phase four-wire wye-grounded system as the base voltage. Assume that all given kVA loads are annual peak values at 85% lagging-load power factor. The load between the substation bus and point *a* is a uniformly distributed load of 2000 kVA. The loads on the laterals *c*, *d*, $c - e$, and $c - f$ are also uniformly distributed, each with 400 kVA. There is a lumped load of 800 kVA at point *b*.

The line data for the #4/0 and 4 ACSR conductors are given in Table 9.8.

To improve voltage conditions, consider any or all combinations of the following design remedies:

a. Installation of shunt capacitor bank(s)
b. Installation of 32-step voltage regulators with a maximum regulation range of ± 10%
c. Addition of new phase conductors

Table 9.8
Table for Problem P9.10

Conductor Size	R, Ω/(phase · mi)	X, Ω/(phase · mi)	K,pu VD/(kVA · mi)
4/0	0.592	0.761	5.85×10^{-6}
4	2.55	0.835	1.69×10^{-5}

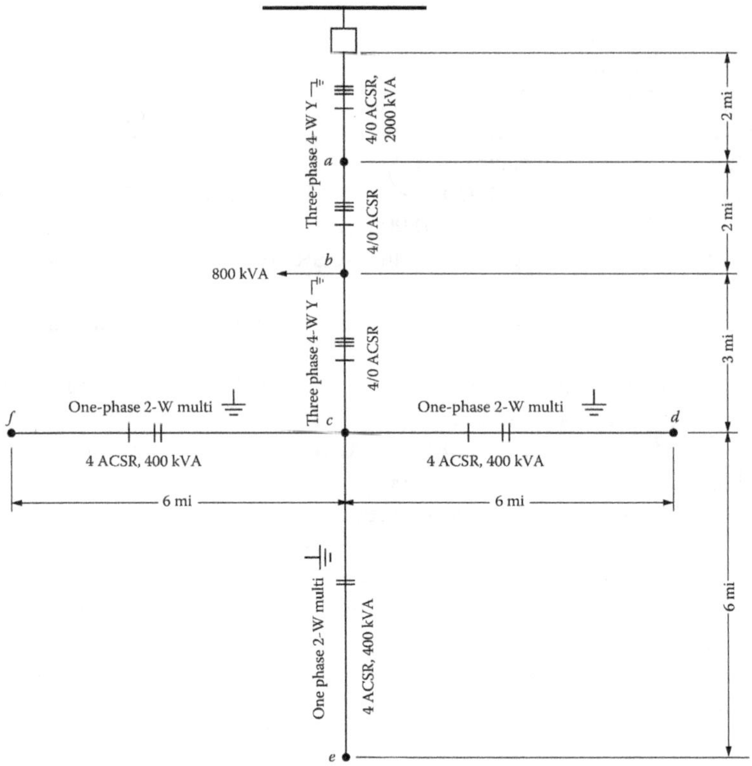

Figure 9.17 Figure for Problem 9.10.

Using these remedies attempt to meet the following primary-voltage criteria:

a. Maximum primary voltage must be 1.040 pu V at zero load.
b. Maximum primary voltage must be 1.07 pu V at peak load.
c. Minimum primary voltage must be 1.00 pu V at peak load.

If the installation of the capacitor alternative is chosen, determine the following:

a. The rating of the capacitor bank(s) in three-phase kVAr.
b. The location of the capacitor bank(s) on the given system.
c. Whether or not voltage-controlled automatic switching is required.

If the installation of the voltage-regulator alternative is chosen, determine the following:

a. The location of the regulator bank(s)
b. The standard kVA rating of each single-phase regulator
c. The location of the RP on the system
d. The setting of the voltage-regulating relay (VRR)
e. The *R* and *X* settings of the LDC

9.11 Assume that the annual peak load is 5000 kVA at a lagging power factor of 0.80 and that the reactive load factor is 0.60.

9.12 Figure 9.18 shows an open-wire primary line with many laterals and uniformly distributed load. The voltage at the distribution substation low-voltage bus is held at 1.03 pu V with BVR. When there is N/C bank installed on the feeder, that is, $Q_{NSW} = 0$, the per unit voltage at the end of the line at annual peak load is 0.97. Use the nominal operating voltage of 7.97/13.8 kV of the three-phase four-wire wye-grounded system as the base voltage. Assume that the off-peak load of the system is about 25% of the on-peak load. Also assume that the line reactance is 0.80 Ω/(phase · mi), but the line resistance is not given and determine the following:

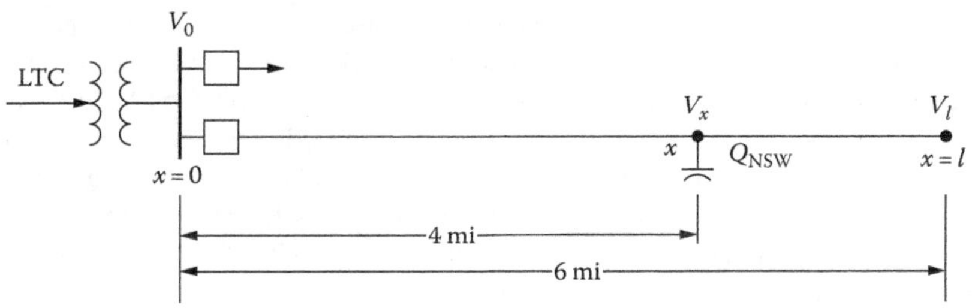

Figure 9.18 Figure for Problem 9.12.

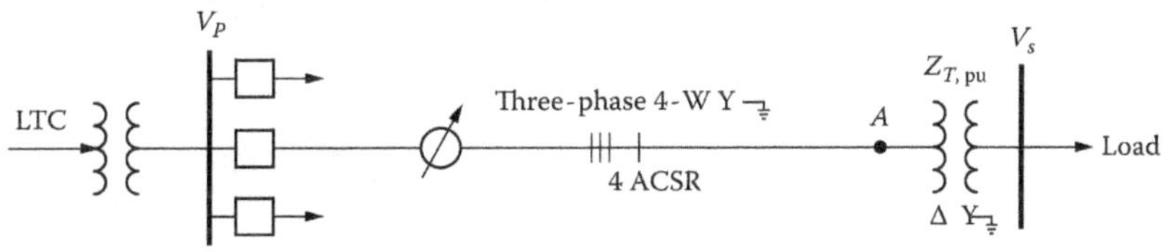

Figure 9.19 Figure for Problem 9.13.

a. When the shunt-capacitor bank is not used, find the V_x voltages at the times of peak load and off-peak load.
b. Apply an NSW-capacitor bank and locate it at the point of $x = 4$ mi on the line, and size the capacitor bank to yield the per unit voltage of 1.05 at point x at the time of zero load. Find the size of the capacitor (Q_{NSW}) in three-phase kVAr. Also find the per unit voltage of V_x and V_l at the time of peak load.

9.13 Figure 9.19 shows a system that has a load connected at the end of a 3.5 mi #4 ACSR open-wire primary line. The load belongs to an important scientific equipment installation, and it varies from nearly zero to 1000 kVA. The load requires a closely regulated voltage at the V_S bus. The consumer requests the voltage V_s to be equal to 1.000 ± 0.010 pu and offers to compensate properly the supplying utility company for such high-quality voltage-regulated service.

A junior engineer proposes to build the 3.5 mi #4 ACSR line to a nearby distribution substation and to place the feeder voltage regulators there in order to render the service requested. His wire size is generous for ampacity. He proposes a BW setting of ± 1.0 V based on 120 V.

There is BVR at the substation, but at times the BVR equipment is disconnected and bypassed for maintenance and repair. Therefore, the substation bus voltage $V_{P,pu}$ is as follows. When BVR is in use,

$$V_{P,pu} = 1.030 \text{ pu V}$$

When BVR is out,

$$\text{Max}V_{P,pu} = 1.060 \text{ pu V}$$
$$\text{Min}V_{P,pu} = 0.970 \text{ pu V}$$

The nominal and base voltage at the distribution substation low-voltage bus is 7,200/12,470 V for the three-phase four-wire wye-grounded system. The nominal and base voltage at the consumer's bus is 277/480 V for the three-phase four-wire wye-grounded service. Assume that the regulator bank is made up of three single-phase 32-step feeder voltage regulators with $\pm$ 10% regulation range. The feeder impedance is given as 2.55 + j0.835 Ω/(mi · phase). Assume that the precalculated K constant of the line is 1.69×10^{-5} pu VD/(kVA · mi) at 85% power factor. The consumer's transformer is rated as 1000 kVA, three phase, with 12,470 V high-voltage rating. It has an impedance of 0 + j0.055 pu based on the transformer ratings. Using the given information and data, determine and state whether or not the young engineer's proposed design will meet the consumer's requirements. (Check for both cases, that is, when BVR is in use and BVR is not in use.)

9.14 Assuming 20 starts per hour and a line-to-ground current of 350 A.

9.15 Assuming 10 starts per hour and a three-phase fault current of 750 A.

9.16 Resolve parts (a) and (b) of Example 9.1 by using MATLAB. Assume that all the quantities remain the same.

A Impedance Tables for Lines, Transformers, and Underground Cables

DOI: 10.1201/9781003129721-A

Table A.1
Characteristics of Copper Conductors, Hard-Drawn, 97.3% Conductivity

Size of Conductor Circular Mils	AWG or B & S	Number of Strands	Diameter of Individual Strands (in.)	Outside Diameter (in.)	Breaking Strength (lb)	Weight (lb/mi)	Approx. Current Carrying Capacity[a] (amps)	Geometric Mean Radius at 60 Cycles (ft)
1,000,000	—	37	0.1644	1.151	43,830	16,300	1300	0.0368
900,000	—	37	0.156	1.092	39,610	14,670	1220	0.0349
800,000	—	37	0.147	1.029	35,120	13,040	1130	0.0329
750,000	—	37	0.1424	0.997	33,400	12,230	1090	0.0319
700,000	—	37	0.1375	0.963	31,170	11,410	1040	0.0306
500,000	—	37	0.1273	0.891	27,020	9,781	940	0.0285
500,000	—	37	0.1162	0.814	22,610	8,161	840	0.026
500,000	—	19	0.1622	0.811	21,590	8,161	840	0.0256
450,000	—	19	0.1539	0.77	19,750	7,336	780	0.0243
400,000	—	19	0.1451	0.726	17,560	6,521	730	0.0229
350,000	—	19	0.1357	0.679	16,890	5,706	670	0.0214
350,000	—	12	0.1708	0.71	16,140	5,706	670	0.0225
300,000	—	19	0.1257	0.629	13,510	4,891	610	0.01987
300,000	—	12	0.1581	0.657	13,170	4,891	610	0.0208
250,000	—	19	0.1147	0.574	11,360	4,076	540	0.01813
250,000	—	12	0.1443	0.6	11,130	4,076	540	0.01902
211,600	4/0	19	0.1055	0.528	9,617	3,450	480	0.01668
211,600	4/0	12	0.1328	0.552	9,483	3,450	490	0.0175
211,600	4/0	7	0.1739	0.522	9,154	3,450	480	0.01579
167,800	3/0	12	0.1183	0.492	7,556	2,736	420	0.01569
167,800	3/0	7	0.1548	0.464	7,366	2,736	420	0.01404
133,100	2/0	7	0.1379	0.414	5,926	2,170	360	0.01252
106,600	1/0	7	0.1228	0.368	4,752	1,720	310	0.01113
83,690	1	7	0.1093	0.328	3,804	1,364	270	0.00992
63,690	1	3	0.167	0.36	3,620	1,351	270	0.01016
66,370	2	7	0.0974	0.292	3,045	1,082	230	0.00883
66,370	2	3	0.1487	0.32	2,913	1,071	240	0.00903
66,370	2	1	—	0.258	3,003	1,061	220	0.00836
52,630	3	7	0.0867	0.26	2,433	858	200	0.00787
52,630	3	3	0.1325	0.286	2,359	850	200	0.00805
52,630	3	1	—	0.229	2,439	841	190	0.00745
41,740	4	3	0.118	0.254	1,879	674	180	0.00717
41,740	4	1	—	0.204	1,970	667	170	0.00663
33,100	5	3	0.105	0.226	1,605	534	180	0.00638
33,100	5	1	—	0.1819	1,591	529	140	0.0059
26,250	6	3	0.0935	0.201	1,205	424	130	0.00568
26,250	6	1	—	0.162	1,280	420	120	0.00526
20,820	7	1	—	0.1443	1,030	333	110	0.00468
16,510	8	1	—	0.1286	826	264	90	0.00417

Source: Westinghouse Electric Corporation, *Electric Utility Engineering Reference Book—Distribution Systems*, East Pittsburgh, PA, 1965.

[a] For conductor at 75°C, air at 25°C, wind 1.4 mi/h (2 ft/s), frequency = 60 cycles.

r_a Resistance (Ω/Conductor/mi)								X_a Inductive Reactance (Ω/Conductor/mi) at 1 ft Spacing			X'_a Shunt Capacitive Reactance 1561 (mi/Conductor) at 1 ft Spacing		
	25°C (77°F)				50°C (122°F)								
DC	25 Cycles	50 Cycles	60 Cycles	DC	25 Cycles	50 Cycles	60 Cycles	25 Cycles	50 Cycles	60 Cycles	25 Cycles	50 Cycles	60 Cycles
0.0585	0.0594	0.062	0.0634	0.064	0.0648	0.0672	0.0685	0.1666	0.333	0.4	0.216	0.1081	0.0901
0.065	0.0658	0.0682	0.0695	0.0711	0.0718	0.074	0.0752	0.1693	0.339	0.406	0.22	0.11	0.0916
0.0731	0.0739	0.076	0.0772	0.08	0.0808	0.0826	0.0837	0.1722	0.344	0.413	0.224	0.1121	0.0934
0.078	0.0787	0.0807	0.6818	0.0853	0.0859	0.0878	0.0888	0.1739	0.348	0.417	10.225	0.1132	0.0943
0.0836	0.0842	0.0661	0.0671	0.0914	0.092	0.0937	0.0947	0.1789	0.352	0.422	0.229	0.1145	0.0954
0.0975	0.0981	0.0997	0.1006	0.1066	0.1071	0.1086	0.1095	0.1799	0.36	0.432	0.235	0.1173	0.0977
0.117	0.1175	0.1188	0.1196	0.128	0.1283	0.1296	0.1303	0.1845	0.369	0.443	0.241	0.1206	0.1004
0.117	0.1175	0.1188	0.1196	0.128	0.1283	0.1296	0.1303	0.1853	0.371	0.445	0.241	0.1206	0.1006
0.13	0.1304	0.1316	0.1323	0.1422	0.1426	0.1437	0.1443	0.1879	0.376	0.451	0.245	0.1224	0.102
0.1462	0.1466	0.1477	0.1484	0.16	0.1603	0.1613	0.1519	0.1909	0.382	0.458	0.249	0.1245	0.1038
0.1671	0.1675	0.1684	0.169	0.1828	0.1831	0.184	0.1845	0.1943	0.389	0.466	0.254	0.1269	0.1058
0.1671	0.1675	0.1684	0.169	0.1828	0.1831	0.184	0.1845	0.1918	0.384	0.46	0.251	0.1253	0.1044
0.195	0.1953	0.1961	0.1966	0.213	0.214	0.214	0.215	0.1982	0.396	0.476	0.259	0.1296	0.106
0.195	0.1953	0.1961	0.1966	0.213	0.214	0.214	0.215	0.1957	0.392	0.47	0.256	0.1281	0.1068
0.234	0.234	0.235	0.236	0.256	0.256	0.257	0.257	0.203	0.406	0.487	0.266	0.1329	0.1108
0.234	0.234	0.235	0.236	0.256	0.256	0.257	0.257	0.2	0.401	0.481	0.263	0.1313	0.1094
0.276	0.277	0.277	0.278	0.302	0.303	0.303	0.303	0.207	0.414	0.497	0.272	0.1359	0.1132
0.276	0.277	0.277	0.278	0.302	0.303	0.303	0.303	0.208	0.409	0.491	0.269	0.1343	0.1119
0.276	0.277	0.277	0.278	0.302	0.303	0.303	0.303	0.21	0.42	0.603	0.273	0.1363	0.1136
0.349	0.349	0.349	0.35	0.381	0.381	0.382	0.382	0.21	0.421	0.606	0.277	0.1384	0.1153
0.349	0.349	0.349	0.35	0.381	0.381	0.382	0.382	0.216	0.431	0.518	0.281	0.1405	0.1171
0.44	0.44	0.44	0.44	0.481	0.481	0.481	0.481	0.222	0.443	0.532	0.289	0.1445	0.1205
0.555	0.555	0.555	0.555	0.606	0.607	0.607	0.607	0.227	0.455	0.546	0.298	0.1488	0.124
0.599	0.699	0.699	0.699	0.766				0.233	0.467	0.56	0.306	0.1528	0.1274
0.692	0.692	0.692	0.692	0.757				0.232	0.464	0.557	0.299	0.1495	0.1246
0.881	0.882	0.882	0.882	0.964				0.239	0.478	0.574	0.314	0.157	0.1308
0.873				0.956				0.238	0.476	0.571	0.307	0.1637	0.1281
0.884				0.946				0.242	0.484	0.581	0.323	0.1614	0.1346
1.112				1.216				0.245	0.49	0.588	0.322	0.1611	0.1343
1.101				1.204				0.244	0.488	0.585	0.316	0.1578	0.1315
1.09		Same as DC		1.192		Same as DC		0.248	0.496	0.595	0.331	0.1656	0.138
1.388				1.518				0.25	0.499	0.599	0.324	0.1619	0.1349
1.374				1.503				0.264	0.507	0.609	0.339	0.1697	0.1416
1.75				1.914				0.256	0.511	0.613	0.332	0.1661	0.1384
1.733				1.895				0.26	0.519	0.623	0.348	0.1738	0.1449
2.21				2.41				0.262	0.523	0.628	0.341	0.1703	0.1419
2.18				2.39				0.265	0.531	0.637	0.356	0.1779	0.1483
2.75				3.01				0.271	0.542	0.651	0.364	0.1821	0.1517
3.47				3.8				0.277	0.564	0.665	0.372	0.1862	0.1652

Table A.2
Characteristics of Anaconda Hollow Copper Conductors

Design Number	Size of Conductor Circular Mils or AWG	Wires Number	Diameter (in.)	Outside Diameter (in.)	Breaking Strain (lb)	Weight (lb/mi)	Geometric Mean Radius at 60 Cycles (ft)	Approx. Current Carrying Capacity (amps)[a]
966	890,500	28	0.161	1.65	36,000	15,085	0.0612	1395
96R 1	750,000	42	0.1296	1.155	34,200	12,345	0.0408	1160
939	650,000	50	0.1097	1.126	29,500	10,761	0.0406	1060
360R 1	600,000	50	0.1053	1.007	27,500	9,905	0.0387	1020
938	550,000	50	0.1009	1.036	25,200	9,103	0.0373	960
4R 5	510,000	50	0.097	1	22,700	8,485	0.036	910
892R 3	500,000	18	0.1558	1.08	21,400	8,263	0.0394	900
933	450,000	21	0.1353	1.074	19,300	7,476	0.0398	850
924	400,000	21	0.1227	1.014	17,200	6,642	0.0376	810
925R 1	380,500	22	0.1211	1.003	16,300	6,331	0.0373	780
565R 1	350,000	21	0.1196	0.95	15,100	5,813	0.0353	750
936	350,000	15	0.1444	0.86	15,400	5,776	0.0311	740
378R 1	350,000	30	0.1059	0.736	16,100	5,739	0.0253	700
954	321,000	22	0.1113	0.92	13,850	5,343	0.034	700
935	300,000	18	0.1205	0.839	13,100	4,984	0.0307	670
903R 1	300,000	15	0.1338	0.797	13,200	4,953	0.0289	660
178R 2	300,000	12	0.1507	0.75	13,050	4,937	0.0266	650
926	250,000	18	0.11	0.766	10,950	4,155	0.0279	600
915R 1	250,000	15	0.1214	0.725	11,000	4,148	0.0266	590
24R 1	250,000	12	0.1368	0.683	11,000	4,133	0.0245	580
923	4/0	18	0.1005	0.7	9,300	3,521	0.0255	530
922	4/0	15	0.1109	0.663	9,300	3,510	0.0238	520
50R 2	4/0	14	0.1152	0.65	9,300	3,510	0.0234	520
158R 1	3/0	16	0.0961	0.606	7,500	2,785	0.0221	460
495R 1	3/0	15	0.0996	0.595	7,600	2,785	0.0214	460
570R 2	3/0	12	0.1123	0.56	7,600	2,772	0.0201	450
909R 2	2/0	15	0.088	0.53	5,950	2,213	0.0191	370
412R 2	2/0	14	0.0913	0.515	6,000	2,207	0.0184	370
937	2/0	13	0.095	0.505	6,000	2,203	0.0181	370
930	125,600	14	0.0885	0.5	5,650	2,083	0.018	360
934	121,300	15	0.0836	0.5	5,400	2,015	0.0179	350
901	119,400	12	0.0936	0.47	5,300	1,979	0.0165	340

Source: Westinghouse Electric Corporation, *Electric Utility Engineering Reference Book—Distribution Systems*, East Pittsburgh, PA, 1965.

[a] For conductor at 75°C, air at 25°C, wind 1.4 mi/h (2 ft/s), frequency = 60 cycles, average tarnished surface.

B Standard Device Numbers Used in Protection Systems

Some of the frequently used device numbers are listed in the following. A complete list and definitions are given in ANSI/IEEE Standard C37.2-1079.

1. Master element: normally used for hand-operated devices
2. Time-delay starting or closing relay
3. Checking or interlocking relay
4. Master contactor
5. Stopping device
6. Starting circuit breaker
7. Anode circuit breaker
8. Control power-disconnecting device
9. Reversing device
10. Unit sequence switch
12. Synchronous-speed device
14. Underspeed device
15. Speed- or frequency-matching device
17. Shunting or discharge switch
18. Accelerating or decelerating device
20. Electrically operated valve
21. Distance relay
23. Temperature control device
25. Synchronizing or synchronism-check device
26. Apparatus thermal device
27. Undervoltage relay
29. Isolating contactor
30. Annunciator relay
32. Directional power relay
37. Undercurrent or underpower relay
46. Reverse-phase or phase-balance relay
47. Phase-sequence voltage relay
48. Incomplete-sequence relay
49. Machine or transformer thermal relay
50. Instantaneous overcurrent or rate-of-rise relay
51. Ac time overcurrent relay
52. Ac circuit breaker: the mechanism-operated contacts are
    (a). 52a, 52aa: open when breaker, closed when breaker contacts closed
    (b). 52b, 52bb: operates just as the mechanism motion start; known as high-speed contacts
55. Power factor relay
57. Short-circuiting or grounding device
59. Overvoltage relay

DOI: 10.1201/9781003129721-B

60. Voltage or current balance relay
62. Time-delay stopping or opening relay
64. Ground detector relay
67. AC directional overcurrent relay
68. Blocking relay
69. Permissive control device
72. AC circuit breaker
74. Alarm relay
76. DC overcurrent relay
78. Phase-angle measuring or out-of-step protective relay
79. AC reclosing relay
80. Flow switch
81. Frequency relay
82. DC reclosing relay
83. Automatic selective control or transfer relay
84. Operating mechanism
85. Carrier or pilot-wire receiver relay
86. Lockout relay
87. Differential protective relay
89. Line switch
90. Regulating device
91. Voltage directional relay
92. Voltage and power directional relay
93. Field-changing contactor
94. Tripping or trip-free relay

C The per-Unit System

C.1 INTRODUCTION

Due to the various benefits involved, it is customary in power system analysis calculations to use per-unit values for impedances, currents, voltages, and powers. Per-unit values are scaled or normalized values that allow for easier comparison, as they eliminate the need to work with physical values in ohms, amperes, kV, and megavoltamperes (or megavars, or megawatts).

A per-unit system provides a convenient way to express quantities. In this system, the per-unit value of any quantity is defined as the ratio of that quantity to an arbitrarily chosen base (or reference) value with the same dimensions. As a result, the per-unit value of any quantity can be described as a dimensionless quantity.

$$\text{Quantity in unit} = \frac{\text{Physical quantity}}{\text{Base value of quantity}}. \tag{C.1}$$

In this context, the term "physical quantity" refers to a value expressed in units such as ohms, amperes, volts, and so on. The "base value" is also known as the unit value, as it corresponds to a value of 1 or unity in the per-unit system. Thus, a base current can also be referred to as a unit current.

Since both the physical quantity and the base quantity share the same dimensions, the resulting per-unit value, expressed as a decimal, becomes dimensionless and is denoted by the subscript "pu." The base quantity is denoted by the subscript "B." The symbol used to represent per unit is "pu" or "0/1." Alternatively, the percent system is obtained by multiplying the per-unit value by 100. Therefore,

$$\text{Quantity in percent} = \frac{\text{Physical quantity}}{\text{Base value of quantity}} \times 100 \tag{C.2}$$

However, the percent system is somewhat more difficult to work with and more subject to possible error since it must always be remembered that the quantities have been multiplied by 100.

Thus, the factor 100 has to be continually inserted or removed for reasons that may not be obvious at time. For example, 40% reactance times 100% current is equal to 4000% voltage, which, of course, must be corrected to 40% voltage. Hence, the per-unit system is preferred in power system calculations. The advantages of using the per-unit system include the following:

1. Network analysis is greatly simplified since all impedances of a given equivalent circuit can directly be added together regardless of the system voltages.
2. It eliminates the $\sqrt{3}$ multiplications and divisions that are required when balanced three-phase systems are represented by per-phase systems. Therefore, the factors $\sqrt{3}$ and 3 are associated with delta and wye quantities in a balanced three-phase system are directly taken into account by the base quantities.
3. Usually, the impedance of an electrical apparatus is given in percent or per-unit by its manufacturer based on its nameplate ratings (e.g. its rated voltamperes and rated voltage).
4. Differences in operating characteristics of many electrical parameters can be estimated by a comparison of their constants expressed in per units.

DOI: 10.1201/9781003129721-C

5. Average machine constants can easily be obtained since the parameters of similar equipment tend to fall in a relatively narrow range and, therefore, arc comparable when expressed as per units based on rated capacity.
6. The use of per-unit quantities is more convenient in calculations involving digital computers.

C.2 SINGLE-PHASE SYSTEM

In the event that any two of the four base quantities (i.e. base voltage, base current, base voltamperes, and base impedance) are "*arbitrarily*" specified, the other two can be determined immediately.

Here, the term *arbitrarily* is slightly misleading since in practice the base values are selected so as to force the results to fall into specified ranges. For example, the base voltage is selected such that the system voltage is normally close to unity.

Similarly, the base voltampere is usually selected as the kVA or megavoltampere rating of one of the machines or transformers in the system, or a convenient round number such as 1, 10, 100, or 1000 MVA, depending on system size. As aforementioned, on determining the base voltamperes and base voltages, the other base values are fixed. For example, current base can be determined as

$$I_B = \frac{S_B}{V_B} = \frac{VA_B}{V_B} \tag{C.3}$$

where

I_B is the current base in amperes
S_B is the selected voltampere base in voltamperes
V_B is the selected voltage base in volts

Note that

$$S_B = VA_B = P_B = Q_B = V_B I_B \tag{C.4}$$

Similarly, the impedance base[1*] can be determined as

$$Z_B = \frac{V_B}{I_B} \tag{C.5}$$

where

$$Z_B = X_B = R_B \tag{C.6}$$

Similarly,

$$Y_B = B_B = G_B = \frac{I_B}{V_B} \tag{C.7}$$

Note that by substituting Equation C.3 into Equation C.5, the impedance base can be expressed as

$$Z_B = \frac{V_B}{VA_B/V_B} = \frac{V_B^2}{VA_B} \tag{C.8}$$

or

$$Z_B = \frac{(kV_B)^2}{MVA_B} \tag{C.9}$$

where

kV_B is the voltage base in kV
MVA_B is the voltampere base in megavoltamperes

The per-unit value of any quantity can be found by the *normalization process*, that is, by dividing the physical quantity by the base quantity of the same dimension. For example, the per-unit impedance can be expressed as

$$Z_{pu} = \frac{Z_{physical}}{Z_B} \tag{C.10}$$

or

$$Z_{pu} = \frac{Z_{physical}}{V_B^2/(kVA_B \times 1000)} \tag{C.11}$$

or

$$Z_{pu} = \frac{(Z_{physical})(kVA_B)(1000)}{V_B^2} \tag{C.12}$$

or

$$Z_{pu} = \frac{(Z_{physical})(kVA_B)}{(kV_B)^2(1000)} \tag{C.13}$$

or

$$Z_{pu} = \frac{(Z_{physical})}{(kV_B)^2/MVA_B} \tag{C.14}$$

or

$$Z_{pu} = \frac{(Z_{physical})(MVA_B)}{(kV_B)^2(1000)} \tag{C.15}$$

Similarly, the others can be expressed as

$$I_{pu} = \frac{I_{physical}}{I_B} \tag{C.16}$$

or

$$I_{pu} = \frac{I_{physical}}{V_B} \tag{C.17}$$

or

$$kV_{pu} = \frac{kV_{physical}}{kV_B} \tag{C.18}$$

or

$$VA_{pu} = \frac{VA_{physical}}{VA_B} \tag{C.19}$$

or

$$kVA_{pu} = \frac{kVA_{physical}}{kVA_B} \tag{C.20}$$

or

$$MVA_{pu} = \frac{MVA_{physical}}{MVA_B} \tag{C.21}$$

Note that the base quantity is always a real number, whereas the physical quantity can be a complex number. For example, if the actual impedance quantity is given as $Z\angle\theta\ \Omega$, it can be expressed in the per-unit system as

$$Z_{pu} = \frac{Z\angle\theta}{Z_B} = Z_{pu}\angle\theta \tag{C.22}$$

that is, it is the magnitude expressed in per-unit terms.

Alternatively, if the impedance has been given in rectangular form as

$$Z = R + jX \tag{C.23}$$

then

$$Z_{pu} = R_{pu} + jX_{pu} \tag{C.24}$$

where

$$R_{pu} = \frac{R_{physical}}{Z_B} \tag{C.25}$$

and

$$X_{pu} = \frac{X_{physical}}{Z_B} \tag{C.26}$$

Similarly, if the complex power has been given as

$$S = P + jQ \tag{C.27}$$

then

$$S_{pu} = P_{pu} + jQ_{pu} \tag{C.28}$$

where

$$P_{pu} = \frac{P_{physical}}{S_B} \tag{C.29}$$

and

$$Q_{pu} = \frac{Q_{physical}}{S_B} \tag{C.30}$$

If the actual voltage and current values are given as

$$V = V\angle\theta_V \tag{C.31}$$

and

$$I = I\angle\theta_I \tag{C.32}$$

the complex power can be expressed as

$$S = VI^* \tag{C.33}$$

or

$$S\angle\theta = (V\angle\theta_V)(I\angle-\theta_I) \tag{C.34}$$

Therefore, dividing through by S_B,

$$\frac{S\angle\phi}{S_B} = \frac{(V\angle\theta_V)(I\angle-\theta_I)}{S_B} \tag{C.35}$$

However,

$$S_B = V_B I_B \tag{C.36}$$

Thus,

$$\frac{S\angle\theta}{S_B} = \frac{(V\angle\theta_V)(I\angle-\theta_I)}{V_B I_B} \tag{C.37}$$

or

$$S_{pu}\angle\theta = (V_{pu}\angle\theta_V)(I_{pu}\angle-\theta_I) \tag{C.38}$$

or

$$S_{pu} = V_{pu} I^*_{pu} \tag{C.39}$$

Example C.1. A 240/120 V single-phase transformer rated 5 kVA has a high-voltage winding impedance of 0.3603 Ω. Use 240 V and 5 kVA as the base quantities and determine the following:

1. The high-voltage side base current.
2. The high-voltage side base impedance in ohms.
3. The transformer impedance referred to the high-voltage side in per unit.
4. The transformer impedance referred to the high-voltage side in percent.
5. The turns ratio of the transformer windings.
6. The low-voltage side base current.
7. The low-voltage side base impedance.
8. The transformer impedance referred to the low-voltage side in per unit.

Solution. 1. The high-voltage side base current is

$$I_{B(HV)} = \frac{S_B}{V_{B(HV)}} = \frac{5000\ V}{240\ A} = 20.8333\ \text{A}$$

2. The high-voltage side base impedance is

$$Z_{B(HV)} = \frac{V_{B(HV)}}{V_{B(HV)}} = \frac{5000\ VA}{20.8333\ A} = 11.52\ \Omega$$

3. The transformer impedance referred to the high-voltage side is

$$Z_{B(HV)} = \frac{V_{HV}}{V_{B(HV)}} = \frac{0.3603\Omega}{11.51\ \Omega} = 0.0313\ \text{pu}$$

4. The transformer impedance referred to the high-voltage side is %

$$\%Z_{HV} = Z_{pu(HV)} \times 100 = (0.0313\ \text{pu})100 = 3.13\%$$

5. The turns ratio of the transformer windings is

$$n = \frac{V_{HV}}{V_{LV}} = \frac{240 \text{ V}}{120 \text{ V}} = 2$$

6. The low-voltage side base current is

$$I_{B(LV)} = \frac{S_B}{V_{B(LV)}} = \frac{5000 \text{ VA}}{120 \text{ V}} = 41.6667 \text{ A}$$

or

$$I_{B(LV)} = nI_{B(LV)} = 2(20.8333 \text{ A}) = 41.6667 \text{ A}$$

7. The low-voltage side base impedance is

$$I_{B(LV)} = \frac{nI_{B(LV)}}{I_{B(LV)}} = \frac{120 \text{ V}}{41.667 \text{ A}} = 2.88\ \Omega$$

or

$$Z_{B(HV)} = \frac{Z_{B(LV)}}{n^2} = 2.88\ \Omega$$

8. The transformer impedance referred to the low-voltage side is

$$Z_{LV} = \frac{V_{HV}}{n^2} = \frac{0.3603\ \Omega}{2^2} = 0.0901\ \Omega$$

Therefore,

$$Z_{pu(LV)} = \frac{V_{LV}}{Z_{B(LV)}} = \frac{0.0901\ \Omega}{0.88\ \Omega} = 0.0313 \text{ pu}$$

or

$$Z_{pu(LV)} = Z_{pu(HV)} = 0.0313 \text{ pu}$$

Notice that in terms of per units the impedance of the transformer is the same whether it is referred to the high-voltage side or the low-voltage side.

C.3 CONVERTING FROM PER-UNIT VALUES TO PHYSICAL VALUES

The physical values (or system values) and per-unit values are related by the following relationships:

$$I = I_{pu} \times I_B \tag{C.40}$$

$$V = V_{pu} \times V_B \tag{C.41}$$

$$Z = Z_{pu} \times Z_B \tag{C.42}$$

$$R = R_{pu} \times Z_B \tag{C.43}$$

$$X = X_{pu} \times Z_B \tag{C.44}$$

$$VA = VA_{pu} \times VA_B \tag{C.45}$$

$$P = P_{pu} \times VA_B \tag{C.46}$$

$$Q = Q_{pu} \times VA_B \tag{C.47}$$

C.4 CHANGE OF BASE

In general, the per-unit impedance of a power apparatus is given based on its own voltampere and voltage ratings and consequently based on its own impedance base. When such an apparatus is used in a system that has its own bases, it becomes necessary to refer all the given per-unit values to the system base values. Assume that the per-unit impedance of the apparatus is given based on its nameplate ratings as

$$Z_{pu(given)} = Z_{physical} \frac{MVA_{B(given)}}{[kV_{B(given)}]^2} \tag{C.48}$$

and that it is necessary to refer the very same physical impedance to a new set of voltage and voltampere bases such that

$$Z_{pu(new)} = Z_{physical} \frac{MVA_{B(new)}}{[kV_{B(new)}]^2} \tag{C.49}$$

By dividing Equation C.48 by Equation C.49 side by side,

$$Z_{pu(new)} = Z_{pu(old)} \left[\frac{MVA_{B(old)}}{MVA_{B(given)}}\right] \left[\frac{kV_{B(given)}}{kV_{B(old)}}\right]^2 \tag{C.50}$$

In certain situations, it is more convenient to use subscripts 1 and 2 instead of subscripts "given" and "new," respectively. Then Equation C.50 can be expressed as

$$Z_{pu(2)} = Z_{pu(1} \left[\frac{MVA_{B(2)}}{MVA_{B(1)}}\right] \left[\frac{kV_{B(1)}}{kV_{B(2)}}\right]^2 \tag{C.51}$$

In the event that the kV bases are the same bu the MVA bases are different, from Equation C.50,

$$Z_{pu(new)} = Z_{pu(given)} \frac{MVA_{B(new)}}{MVA_{B(given)}} \tag{C.52}$$

Similarly, if the megavoltampere bases are the same but the kV bases are different, from Equation C.50,

$$Z_{pu(new)} = Z_{pu(given)} \left[\frac{kV_{B(given)}}{kV_{B(new)}}\right]^2 \tag{C.53}$$

Equations C.49 through C.52 must only be used to convert the given per-unit impedance from the base to another but not for referring the physical value of an impedance from one side of the transformer to another.

Example C.2. Consider Example B.1 and select 300/150 V as the base voltages for the high-voltage and the low-voltage windings, respectively. Use a new base power of 10 kVA and determine the new per-unit, base, and physical impedances of the transformer referred to the high-voltage side.

Solution. By using Equation C.50, the new per-unit impedance can be found as

$$Z_{pu(new)} = Z_{pu(old)}\left[\frac{MVA_{B(old)}}{MVA_{B(given)}}\right]\left[\frac{kV_{B(given)}}{kV_{B(old)}}\right]^2 = (0.0313\text{ pu})\left(\frac{10{,}000\text{ VA}}{300\text{ V}}\right)\left(\frac{240\text{ V}}{300\text{ V}}\right)^2$$
$$= 33.334\text{ A}$$

The new current base is

$$I_{B(HV)new} = \frac{S_B}{V_{B(HV)new}} = \frac{10{,}000\text{ VA}}{300\text{ V}} = 33.334\text{ A}$$

Thus,

$$Z_{B(HV)new} = \frac{V_{B(HV)new}}{I_{B(HV)new}} = \frac{300\text{ V}}{33.334\text{ A}} = 9\ \Omega$$

Therefore, the physical impedance of the transformer is still

$$Z_{HV} = Z_{pu,new} \times V_{B(HV)new} = (0.0401\text{ pu})(9\Omega) = 0.3609\ \Omega$$

C.5 THREE-PHASE SYSTEMS

The three-phase problems involving balanced systems can be solved on a per-phase basis. In that case. the equations that are developed for single-phase systems can be used for three-phase systems as long as per-phase values are used consistently. Therefore,

$$I_B = \frac{S_{B(1\phi)}}{V_{B(L-N)}} \tag{C.54}$$

or

$$I_B = \frac{VA_{B(1\phi)}}{V_{B(L-N)}} \tag{C.55}$$

and

$$Z_B = \frac{V_{B(L-N)}}{I_B} \tag{C.56}$$

or

$$Z_B = \frac{[kV_{B(L-N)}]^2(1000)}{kVA_{B(1\phi)}} \tag{C.57}$$

or

$$Z_B = \frac{[kV_{B(L-N)}]^2}{MVA_{B(1\phi)}} \tag{C.58}$$

where the subscripts 1 ϕ and L-N denote per-phase and line to neutral, respectively. Note that, for a balanced system,

$$V_{B(L-N)} = \frac{V_{B(L-L)}}{\sqrt{3}} \tag{C.59}$$

and

$$S_{B(1\phi)} = \frac{S_{B(3\phi)}}{3} \tag{C.60}$$

However, it has been customary in three-phase system analysis to use line-to-line voltage and three-phase voltamperes as the base values. Therefore,

$$I_B = \frac{S_{B(3\phi)}}{\sqrt{3}V_{B(L-L)}} \tag{C.61}$$

or

$$I_B = \frac{kVA_{B(3\phi)}}{\sqrt{3}V_{B(L-L)}} \tag{C.62}$$

and

$$Z_B = \frac{V_{B(L-L)}}{\sqrt{3}I_B} \tag{C.63}$$

$$Z_B = \frac{[kV_{B(L-L)}]^2(1000)}{kVA_{B(3\phi)}} \tag{C.64}$$

or

$$Z_B = \frac{[kV_{B(L-L)}]^2}{MVA_{B(3\phi)}} \tag{C.65}$$

where the subscripts 3ϕ and L-L denote per three phase and line, respectively. Furthermore, base admittance can be expressed as

$$Y_B = \frac{1}{Z_B} \tag{C.66}$$

or

$$. \tag{C.67}$$

where

$$Y_B = B_B = G_B \tag{C.68}$$

The data for transmission lines are usually given in terms of the line resistance R in ohms per mile at a given temperature, the line inductive reactance X_L in ohms per mile at 60 Hz, and the line shunt capacitive reactance X_C in megaohms per mile at 60 Hz. Therefore, the line impedance and shunt susceptance in per units for 1 i of line can be expressed as

$$Z_{pu} = (Z, \Omega/mi)\frac{MVA_{B,(3\phi)}}{[kV_{B(L-L)}]^2} \text{ pu} \tag{C.69}$$

where

$$Z = R + jX_L = Z\angle\theta \ \Omega/mi$$

and

$$B_{pu} = \frac{[kV_{B(L-L)}]^2 \times 10^{-6}}{[MVA_{B(3\phi)}][X_c, M\Omega/mi]} \tag{C.70}$$

In the event that the admittance for a transmission line is given in microsiemens per mile, the per-unit admittance can be expressed as

$$B_{pu} = \frac{[kV_{B(L-L)}]^2 \times (Y, \mu S)}{[MVA_{B(3\phi)}] \times 10^{-6}} \tag{C.71}$$

Similarly, if it is given as reciprocal admittance in megohms per mile, the per-unit admittance can be found as

$$Y_{pu} = \frac{[kV_{B(L-L)}]^2 \times 10^{-6}}{[MVA_{B(3\phi)}][Z, M\Omega/mi]} \tag{C.72}$$

Figure 4.16 shows conventional three-phase transformer connections and associated relationships between the high-voltage and low-voltage side voltages and currents. The given relationships are correct for a three-phase transformer as well as for a three-phase bank of single-phase transformers. Note that in the figure, n is the turns ratio, that is,

$$n = \frac{N_1}{N_2} = \frac{V_1}{V_2} = \frac{I_2}{I_1} \tag{C.73}$$

where the subscripts 1 and 2 are used for the primary and secondary sides. Therefore, an impedance Z_2 in the secondary circuit can be referred to the primary circuit provided that

$$Z_1 = n^2 Z_2 \tag{C.74}$$

Thus, it can be observed that in an ideal transformer, voltages are transformed in the direct ratio of turns, currents in the inverse ratio, and impedances in the direct ratio squared, and power and voltamperes are, of course, unchanged. Note that a balanced delta-connected circuit of Z_Δ Ω/phase is equivalent to a balanced wye-connected circuit of Z_Y Ω/phase as long as

$$Z_Y = \frac{1}{3} Z_\Delta \tag{C.75}$$

The per-unit impedance of a transformer remains unchanged, regardless of whether it is calculated based on the high-voltage side or the low-voltage side of the transformer. This can be achieved by selecting appropriate bases for each side of the transformer, irrespective of whether it is connected in wye-wye, delta-delta, delta-wye, or wye-delta configuration. The voltage transformation performed by the transformer is equivalent to that of a wye-wye transformer when the same line-to-line voltage ratings are used.

In other words, the designated per-unit impedance values for transformers are determined based on the coil ratings. Since the coil ratings cannot be altered by a simple change in connection (e.g., from wye-wye to delta-wye), the per-unit impedance remains consistent regardless of the three-phase connection. However, the line-to-line voltage for the transformer may vary.

Due to the method of selecting the base values in different sections of the three-phase system, the per-unit impedances calculated in various sections can be combined on a single impedance diagram without considering whether the transformers are connected in wye-wye or delta-wye configuration.

Since the ratings of coils cannot be altered by a simple change in connection (e.g., from wye–wye to delta–wye), the per-unit impedance remains the same regardless of the three-phase connection. The line-to-line voltage for the transformer will differ. Because of the method of choosing the base in various sections of the three-phase system, the per-unit impedances calculated in various sections can be put together on one impedance diagram without paying any attention to whether the transformers are connected in wye–wye or delta–wye.

Example C.3. Assume that a 19.5 kV 120 MVA three-phase generator has a synchronous reactance of 1.5/Ω and is connected to a 150 MVA 18/230 kV delta–wye connected three-phase transformer with a 0.1/Ω reactance. The transformer is connected to a transmission line at the 230 kV side. Use the new MVA base of 100 MVA and 240 kV base for the line and determine the following:

1. The new reactance value for the generator in per unit ohms.
2. The new reactance value for the transformer in per unit ohms.

Solution. 1. Using Equation B.49, the new per unit impedance of the generator is

$$Z_{pu(new)} = Z_{pu(old)} \left[\frac{MVA_{B(new)}}{MVA_{B(old)}}\right] \left[\frac{kV_{B(old)}}{kV_{B(old)}}\right]^2$$

But, first determining the new kV base for the generator,

$$kV^{gen}_{B(new)} = (240kV)\left(\frac{18 \text{ kV}}{230 \text{ kV}}\right) = 18.783 \text{ kV}$$

Thus, the new and adjusted synchronous reactance of the generator is

$$Z^{gen}_{pu(new)} = Z^{gen}_{pu(old)} \left[\frac{MVA_{B(new)}}{MVA_{B(old)}}\right] \left[\frac{kV_{B(old)}}{kV_{B(old)}}\right]^2 = (1.5 \text{ pu})\left(\frac{100 \text{ MVA}}{120 \text{ MVA}}\right)\left(\frac{19.5 \text{ kV}}{300 \text{ kV}}\right)^2$$
$$= 1.347 \text{ pu}$$

2. The new reactance value for the transformer in per unit ohms, referred to high-voltage side is

$$Z^{trf}_{pu(new)} = (0.1 \text{ pu})\left(\frac{100 \text{ MVA}}{150 \text{ MVA}}\right)\left(\frac{230 \text{ kV}}{240 \text{ kV}}\right)^2 = 0.061 \text{ pu}$$

And referred to the low-voltage side is

$$Z^{trf}_{pu(new)} = (0.1 \text{ pu})\left(\frac{100 \text{ MVA}}{150 \text{ MVA}}\right)\left(\frac{18 \text{ kV}}{18.783 \text{ kV}}\right)^2 = 0.061 \text{ pu}$$

Note that the transformer reactance referred to the high-voltage side or the low-voltage side is the same, as it should be!

Example C.4. A three-phase transformer has a nameplate ratings of 20 MVA, 345Y/34.5Y kV with a leakage reactance of 12% and the transformer connection is wye–wye. Select a base of 20 MVA and 345 kV on the high-voltage side and determine the following:

1. Reactance of transformer in per units.
2. High-voltage side base impedance.
3. Low-voltage side base impedance.
4. Transformer reactance referred to high-voltage side in ohms.
5. Transformer reactance referred to low-voltage side in ohms.

Solution. 1. The reactance of the transformer in per units is 12/100, or 0.12 pu. Note that it is the same whether it is referred to the high-voltage or the low-voltage sides.

2. The high-voltage side base impedance is

$$Z_{B(HV)} = \frac{[kV_{B(HV)}]^2}{MVA_{B(3\phi)}} = \frac{345^2}{20} = 5951.25\ \Omega$$

3. The low-voltage side base impedance is

$$Z_{B(LV)} = \frac{[kV_{B(LV)}]^2}{MVA_{B(3\phi)}} = \frac{34.5^2}{20} = 59.5125\ \Omega$$

4. The reactance referred to the high-voltage side is

$$X_{(HV)} = X_{pu} \times X_{B(HV)} = (0.12)(5951.25) = 714.15\ \Omega$$

5. The reactance referred to the low-voltage side is

$$X_{(LV)} = X_{pu} \times X_{B(LV)} = (0.12)(59.5125) = 7.1415\ \Omega$$

or from

$$X_{(LV)} = \frac{X(HV)}{n^2} = \frac{714,15\ \Omega}{\left(\frac{345/\sqrt{3}}{34.5\sqrt{3}}\right)} = 7.1415\ \Omega$$

where n is defined as the turns ratio of the windings.

Example C.5. A three-phase transformer has a nameplate ratings of 20 MVA, and the voltage ratings of 345Y/34.5Δ kV with a leakage reactance of 12% and the transformer connection is wye–delta. Select a base of 20 MVA and 345 kV on the high-voltage side and determine the following:

1. Turns ratio of windings.
2. Transformer reactance referred to low-voltage side in ohms.
3. Transformer reactance referred to low-voltage side in per units.

Solution. 1. The turns ratio of the windings is

$$n = \frac{345/\sqrt{3}}{34.5} = 5.7735$$

2. Since the high-voltage side impedance base is

$$Z_{B(HV)} = \frac{[kV_{B)(HV)}]^2}{MVA_{B(3\phi)}} = \frac{345^2}{20} = 5951.25\ \Omega$$

and

$$X_{(HV)} = X_{pu} \times X_{B(HV)} = (0.12)(5951.25) = 714.15\ \Omega$$

Thus, the transformer reactance referred to the delta-connected low-voltage side is

$$X_{(LV)} = \frac{X_{(HV)}}{n^2} = \frac{714.14\ \Omega}{5.7735^2} = 21.4245\ \Omega$$

3. The reactance of the equivalent wye connection is

$$Z_Y = \frac{Z_\Delta}{3} = \frac{21.4245\ \Omega}{3} = 7.1415\ \Omega$$

Similarly,

$$Z_{B(LV)} = \frac{[kV_{B(LV)}]^2}{MVA_{B(3\phi)}} = \frac{34.5^2}{20} = 59.5125\ \Omega$$

Thus,

$$X_{pu} = \frac{7.1415\ \Omega}{Z_{B(LV)}} = \frac{7.1415\ \Omega}{59.5125\ \Omega} = 0.12\ \text{pu}$$

Alternatively, if the line-to-line voltages are used,

$$X_{pu} = \frac{X_{(HV)}}{n_2} = \frac{714.14\ \Omega}{(345/34.5)^2} = 7.1415\ \Omega$$

and therefore,

$$X_{pu} = \frac{X_{(LV)}}{Z_{B(LV)}} = \frac{7.1415\ \Omega}{59.5125\ \Omega} = 0.12\ \Omega$$

as before.

Example C.6. Consider a three-phase system that has a generator connected to a 2.4/24 kV, wye–wye connected, three-phase step-up transformer T_1. Suppose that the transformer is connected to three-phase power line. The receiving end of the line is connected to a second, wye–wye connected, three-phase 24/12 kV step-down transformer T_2. Assume that the line length between the two transformers is negligible and the three-phase generator is rated 4160 kVA, 2.4 kV, and 1000 A and that it supplies a purely inductive load of $I_{pu} = 2.08\angle{-90^\circ}$ pu. The three-phase transformer T_1 is rated 6000 kVA, 2.4Y–24Y kV, with leakage reactance of 0.04 pu. Transformer T_2 is made up of three single-phase transformers and is rated 4000 kVA, 24Y–12Y kV, with leakage reactance of 0.04 pu. Determine the following for all three circuits, 2.4, 24, and 12 kV circuits:

1. Base kVA values.
2. Base line-to-line kV values.
3. Base impedance values.
4. Base current values.
5. Physical current values (neglect magnetizing currents in transformers and charging currents in lines).
6. Per-unit current values.
7. New transformer reactances based on their new bases.
8. Per-unit voltage values at buses 1, 2, and 4.
9. Per-unit apparent power values at buses 1, 2, and 4.
10. Summarize results in a table.

Solution. 1. The kVA base for all three circuits is arbitrarily selected as 2080 kVA

2. The base voltage for the 2.4 kV circuit is arbitrarily selected as 2.5 kV. Since the turns ratios for transformers T_1 and T_2 are

$$\frac{N_1}{N_2} = 10 \quad \text{or} \quad \frac{N_2}{N_1} = 0.10$$

and

$$\frac{N_1'}{N_2'} = 2$$

the base voltages for the 24 and 12 kV circuits are determined to be 25 and 12.5 kV, respectively.

3. The base impedance values can be found as

$$Z_B = \frac{[kV_{B(L-L)}]^2(1000)}{kVA_{B(3\phi)}} = \frac{[2.5kV]^2 1000}{kVA} = 3.005\ \Omega$$

and

$$Z_B = \frac{[25kV]^2 1000}{2080kVA} = 300.5\ \Omega$$

and

$$Z_B = \frac{[12.5kV]^2 1000}{2080kVA} = 75.1\ \Omega$$

4. The base current values can be determined as

$$I_B = \frac{kVA_{B(3\phi)}}{\sqrt{3}kVA_{B(L-L)}} = \frac{2080\ \text{kVA}}{\sqrt{3}(2.5kV)} = 480\ \text{A}$$

and

$$I_B = \frac{2080\ \text{kVA}}{\sqrt{3}(25kV)} = 48\ \text{A}$$

and

$$I_B = \frac{2080\ \text{kVA}}{\sqrt{3}(12.5kV)} = 96\ \text{A}$$

5. The physical current values can be found based on the turns ratios as

$$I = 1000\ \text{A}\ I = \left(\frac{N_2}{N_1}\right)(1000\ \text{A}) = 100\text{A}\ I = \left(\frac{N_{\phi_1}}{N_{\phi_2}}\right)(100\text{A}) = 200\ \text{A}$$

6. The per-unit current values are the same, 2.08 pu, for all three circuits.

$$Z_{pu(new)} = Z_{pu(given)}\left[\frac{kVA_{B(new)}}{kVA_{B(given)}}\right]\left[\frac{kV_{B(given)}}{kV_{B(new)}}\right]^2$$

7. The given transformer reactances can be converted based on their new bases using

$$Z_{pu(T_1)} = j0.04\left[\frac{2080\ \text{kVA}}{6000\ \text{kVA}}\right]\left[\frac{2.4\ \text{kV}}{2.5\ \text{kV}}\right]^2 = j0.0128\ \text{pu}$$

Therefore, the new reactances of the two transformers can be found as

$$Z_{pu(T_2)} = j0.04\left[\frac{2080\text{ kVA}}{4000\text{ kVA}}\right]\left[\frac{12\text{ kV}}{12.5\text{ kV}}\right]^2 = j0.0192\text{ pu}$$

and

$$V_1 = \frac{2.4\text{ kV}\angle 0^\circ}{2.5\text{ kV}} = 0.96\angle 0^\circ\text{ pu } V_2$$
$$= V_1 - I_{pu}Z_{pu(T_1)} = 0.96\angle 0^\circ - (2.08\angle -90^\circ)(0.0128\angle 90^\circ) = 0.9334\angle 0^\circ\text{ pu}$$

8. Therefore, the per-unit voltage values at buses 1, 2, and 4 can be calculated as

$$S_1 = 2.00\text{ pu } S_2 = V_2 I_{pu} = (0.9334)(2.08) = 1.9415\text{ pu}$$
$$S_4 = V_4 I_{pu} = (0.8935)(2.08) = 1.8585\text{ pu}$$

9. Thus, the per-unit apparent power values at buses 1, 2, and 4 are

$$S_1 = 2{:}00\text{ pu}$$
$$S_2 = V_2 I_{pu} = (0{:}9334)(2{:}08) = 1{:}9415\text{ pu}$$
$$S_4 = V_4 I_{pu} = (0{:}8935)(2{:}08) = 1{:}8585\text{ pu}$$

10. The results are summarized in Table C.1.

Table C.1
Results of Example C.6

Quantity	2.4 kV Circuit	24 kV Circuit	12 kV Circuit
$KVA_{B(3\phi)}$	2080 kVA	2080 kVA	2080 kVA
$kV_{B(L-L)}$	2.5 kV	25 kV	12.5 kV
Z_B	3005 Ω	300.5 Ω	75.1 Ω
I_B	480 A	48 A	96 A
I_{physical}	1000 A	100 A	200 A
I_{pu}	2.08 pu	2.08 pu	2.08 pu
V_{pu}	0.96 pu	0.9334 pu	0.8935 pu
S_{pu}	2.00 pu	1.9415 pu	1.8585 pu

PROBLEMS

C.1 Solve Example C.1 for a transformer rated 100 kVA and 2400/240 V that has a high-voltage winding impedance of 0.911.

C.2 Consider the results of Problem C.1 and use 3000/300 V as new base voltages for the high-voltage and low-voltage windings, respectively. Use a new base power of 200 kVA and determine the new per-unit, base, and physical impedances of the transformer referred to the high-voltage side.

C.3 A 240/120 V single-phase transformer rated 25 kVA has a high-voltage winding impedance of 0.65 Ω. If 240 V and 25 kVA are used as the base quantities, determine the following:

1. The high-voltage side base current.
2. The high-voltage side base impedance in Q.
3. The transformer impedance referred to the high-voltage side in per unit.
4. The transformer impedance referred to the high-voltage side in percent.
5. The turns ratio of the transformer windings.
6. The low-voltage side base current.
7. The low-voltage side base impedance.
8. The transformer impedance referred to the low-voltage side in per unit.

C.4 A 240/120 V single-phase transformer is rated 25 kVA and has a high-voltage winding impedance referred to its high-voltage side that is 0.2821 pu based on 240 V and 25 kVA. Select 230/115 V as the base voltages for the high-voltage and low-voltage windings, respectively. Use a new base power of 50 kVA and determine the new per-unit base, and physical impedances of the transformer referred to the high-voltage side.

C.5 After changing the S base from 5 to 10 MVA, redo the Example C.1 by using MATLAB

1. Write the MATLAB program script.
2. Give the MATLAB program output.

D Glossary for Distribution System Terminology

Some of the most commonly used terms, both in this book and in general usage, are defined later. Most of the definitions given in this glossary are based on Refs. [1–8].

AA: Abbreviation for all-aluminum conductors.

AAAC: Abbreviation for all-aluminum-alloy conductors. Aluminum-alloy conductors have higher strength than those of the ordinary electric-conductor grade of aluminum.

ACAR: Abbreviation for aluminum conductor alloy reinforced. It has a central core of higher-strength aluminum surrounded by layers of electric-conductor grade aluminum.

ACL cable: A cable with a lead sheath over the cable insulation that is suitable for wet locations. It is used in buildings at low voltage.

ACSR: Abbreviation for aluminum conductor steel reinforced. It consists of a central core of steel strands surrounded by layers of aluminum strands.

Active filter: Any of a number of sophisticated power electronic devices for eliminating harmonic distortion.

Admittance: The ratio of the phasor equivalent of the steady-state sine-wave current to the phasor equivalent of the corresponding voltage.

Adverse weather: Weather conditions that cause an abnormally high rate of forced outages for exposed components during the periods such conditions persist, but which do not qualify as major storm disasters. Adverse weather conditions can be defined for a particular system by selecting the proper values and combinations of conditions reported by the Weather Bureau: thunderstorms, tornadoes, wind velocities, precipitation, temperature, etc.

Aerial cable: An assembly of insulated conductors installed on a pole line or similar overhead structures; it may be self-supporting or installed on a supporting messenger cable.

Air-blast transformer: A transformer cooled by forced circulation of air through its core and coils.

Air circuit breaker: A circuit breaker in which the interruption occurs in air.

Air switch: A switch in which the interruptions of the circuit occur in air. Al: Symbol for aluminum.

Ampacity: Current rating in amperes, as of a conductor.

ANSI: Abbreviation for American National Standards Institute.

Apparent sag (at any point): The departure of the wire at the particular point in the span from the straight line between the two points of the span, at 60°F, with no wind loading.

Arcing time of fuse: The time elapsing from the severance of the fuse link to the final interruption of the circuit under specified conditions.

Arc-over of insulator: A discharge of power current in the form of an arc following a surface discharge over an insulator.

Armored cable: A cable provided with a wrapping of metal, usually steel wires, primarily for the purpose of mechanical protection.

Askarel: A generic term for a group of nonflammable synthetic chlorinated hydrocarbons used as electrical insulating media. Askarels of various compositional types are used. Under arcing conditions, the gases produced, while consisting predominantly of noncombustible hydrogen chloride,

DOI: 10.1201/9781003129721-D

can include varying amounts of combustible gases depending upon the askarel type. Because of environmental concerns, it is not used in new installations anymore.

Automatic substations: Those in which switching operations are so controlled by relays that transformers or converting equipment are brought into or taken out of service as variations in load may require, and feeder circuit breakers are closed and reclosed after being opened by overload relays.

Autotransformer: A transformer in which at least two windings have a common section.

AWG: Abbreviation for American Wire Gauge. It is also sometimes called the Brown and Sharpe Wire Gauge.

Base load: The minimum load over a given period of time.

Benchboard: A switchboard with a horizontal section for control switches, indicating lamps, and instrument switches; may also have a vertical instrument section.

BIL: Abbreviation for basic impulse insulation levels, which are reference levels expressed in impulse-crest voltage with a standard wave not longer than 1.5×50 μs. The impulse waves are defined by a combination of two numbers. The first number is the time from the start of the wave to the instant crest value; the second number is the time from the start to the instant of half-crest value on the tail of the wave.

Billing demand: The demand used to determine the demand charges in accordance with the provisions of a rate schedule or contract.

Branch circuit: A set of conductors that extend beyond the last overcurrent device in the low-voltage system of a given building. A branch circuit usually supplies a small portion of the total load.

Breakdown: Also termed puncture, denoting a disruptive discharge through insulation.

Breaker, primary-feeder: A breaker located at the supply end of a primary feeder that opens on a primary-feeder fault if the fault current is of sufficient magnitude.

Breaker-and-a-half scheme: A scheme that provides the facilities of a double main bus at a reduction in equipment cost by using three circuit breakers for each two circuits.

Bus: A conductor or group of conductors that serves as a common connection for two or more circuits in a switchgear assembly.

Bus, transfer: A bus to which one circuit at a time can be transferred from the main bus.

Bushing: An insulating structure including a through conductor, or providing a passageway for such a conductor, with provision for mounting on a barrier, conductor or otherwise, for the purpose of insulating the conductor from the barrier and conducting from one side of the barrier to the other.

BVR: Abbreviation for bus voltage regulator or regulation.

BW: Abbreviation for bandwidth.

BX cable: A cable with galvanized interlocked steel spiral armor. It is known as AC cable and used in a damp or wet location in buildings at low voltage.

Cable: Either a standard conductor (single-conductor cable) or a combination of conductors insulated from one another (multiple-conductor cable).

Cable fault: A partial or total load failure in the insulation or continuity of the conductor.

Capability: The maximum load-carrying ability expressed in kVA or kW of generating equipment or other electric apparatus under specified conditions for a given time interval.

Capability, net: The maximum generation expressed in kWh per hour that a generating unit, station, power source, or system can be expected to supply under optimum operating conditions.

Capacitor bank: An assembly at one location of capacitors and all necessary accessories (switching equipment, protective equipment, controls, etc.) required for a complete operating installation.

Capacity: The rated load-carrying ability expressed in kVA or kW of generating equipment or other electric apparatus.

Capacity factor: The ratio of the average load on a machine or equipment for the period of time considered to the capacity of the machine or equipment.

Charge: The amount paid for a service rendered or facilities used or made available for use.

Circuit, earth (ground) return: An electric circuit in which the earth serves to complete a path for current.

Circuit breaker: A device that interrupts a circuit without injury to itself so that it can be reset and reused over again.

Circuit-breaker mounting: Supporting structure for a circuit breaker.

Circular mil: A unit of area equal to 1/4 of a square mil (=0.7854 square mil). The cross-sectional area of a circle in circular mils is therefore equal to the square of its diameter in mils. A circular inch is equal to 1 million circular mils. A mil is one one-thousandth of an inch. There are 1974 circular mils in a square millimeter. Abbreviated cmil.

CL: Abbreviation for current limiting (fuse).

cmil: Abbreviation for circular mil.

Coincidence factor: The ratio of the maximum coincident total demand of a group of consumers to the sum of the maximum power demands of individual consumers comprising the group, both taken at the same point of supply at the same time.

Coincident demand: Any demand that occurs simultaneously with any other demands; also the sum of any set of coincident demands.

Component: A piece of equipment, a line, a section of a line, or a group of items that is viewed as an entity.

Condenser: Also termed capacitor; a device whose primary purpose is to introduce capacitance into an electric circuit. The term condenser is deprecated.

Conductor: A substance that has free electrons or other charge carriers that permit charge flow when an emf is applied across the substance.

Conductor tension, final unloaded: The longitudinal tension in a conductor after the conductor has been stretched by the application for an appreciable period, with subsequent release of the loadings of ice and wind, at the temperature decrease assumed for the loading district in which the conductor is strung (or equivalent loading).

Conduit: A structure containing one or more ducts; commonly formed from iron pipe or electrical metallic tubing, used in buildings at low voltage.

Connection charge: The amount paid by a customer for connecting the customer's facilities to the supplier's facilities.

Contactor: An electric power switch, not operated manually and designed for frequent operation.

Contract demand: The demand that the supplier of electric service agrees to have available for delivery.

Cress factor: A value that is displayed on many power quality monitoring instruments representing the ratio of the crest value of the measured waveform to the rms value of the waveform. For example, the cress factor of a sinusoidal wave is 1.414.

CT: Abbreviation for current transformers.

Cu: Symbol for copper.

Customer charge: The amount paid periodically by a customer without regard to demand or energy consumption.

Demand: The load at the receiving terminals averaged over a specified interval of time.

Demand charge: That portion of the charge for electric service based upon a customer's demand.

Demand factor: The ratio of the maximum coincident demand of a system, or part of a system, to the total connected load of the system, or part of the system, under consideration.

Demand, instantaneous: The load at any instant.

Demand, integrated: The demand integrated over a specified period.

Demand interval: The period of time during which the electric energy flow is integrated in determining demand.

Depreciation: The component that represents an approximation of the value of the portion of plant consumed or "used up" in a given period by a utility.

Disconnecting or isolating switch: A mechanical switching device used for changing the connections in a circuit or for isolating a circuit or equipment from the source of power.

Disconnector: A switch that is intended to open a circuit only after the load has been thrown off by other means. Manual switches designed for opening loaded circuits are usually installed in a circuit with disconnectors to provide a safe means for opening the circuit under load.

Displacement factor (DPF): The ratio of active power (watts) to apparent power (voltamperes).

Distribution center: A point of installation for automatic overload protective devices connected to buses where an electric supply is subdivided into feeders and/or branch circuits.

Distribution switchboard: A power switchboard used for the distribution of electric energy at the voltages common for such distribution within a building.

Distribution system: That portion of an electric system that delivers electric energy from transformation points in the transmission, or bulk power system, to the consumers.

Distribution transformer: A transformer for transferring electric energy from a primary distribution circuit to a secondary distribution circuit or consumer's service circuit; it is usually rated in the order of 5–500 kVA.

Diversity factor: The ratio of the sum of the individual maximum demands of the various subdivisions of a system to the maximum demand of the whole system.

Duplex cable: A cable composed of two insulated stranded conductors twisted together. They may or may not have a common insulating covering.

Effectively grounded: Grounded by means of a ground connection of sufficiently low impedance that fault grounds that may occur cannot build up voltages dangerous to connected equipment.

EHV: Abbreviation for extra high voltage.

Electric rate schedule: A statement of an electric rate and the terms and conditions governing its application.

Electric system loss: Total electric energy loss in the electric system. It consists of transmission, transformation, and distribution losses between sources of supply and points of delivery.

Electrical reserve: The capability in excess of that required to carry the system load.

Emergency rating: Capability of installed equipment for a short time interval.

EMT: Abbreviation for electrical metallic tubing. A raceway that has a thin wall that does not permit threading. Connectors and couplings are secured by either compression rings or setscrews. It is used in buildings at low voltage.

Energy: That which does work or is capable of doing work. As used by electric utilities, it is generally a reference to electric energy and is measured in kWh.

Energy charge: That portion of the charge for electric service based upon the electric energy consumed or billed.

Energy loss: The difference between energy input and output as a result of transfer of energy between two points.

Express feeder: A feeder that serves the most distant networks and must traverse the systems closest to the bulk power source.

Extra high voltage: A term applied to voltage levels higher than 230 kV. Abbreviated EHV.

Facilities charge: The amount paid by the customer as a lump sum or, periodically, as reimbursement for facilities furnished. The charge may include operation and maintenance as well as fixed costs.

FCN: Abbreviation for full-capacity neutral.

Feeder: A set of conductors originating at a main distribution center and supplying one or more secondary distribution centers, one or more branch-circuit distribution centers, or any combination of these two types of load.

Feeder, multiple: Two or more feeders connected in parallel.

Feeder, tie: A feeder that connects two or more independent sources of power and has no tapped load between the terminals. The source of power may be a generating system, substation, or feeding point.

First-contingency outage: The outage of one primary feeder.

Fixed-capacitor bank: A capacitor bank with fixed, not switchable, capacitors.

Flicker: Impression of unsteadiness of visual sensation induced by a light stimulus whose luminance or spectral distribution fluctuates with time.

Flicker factor: A factor used to quantify the load impact of electric arc furnaces on the power system.

Forced interruption: An interruption caused by a forced outage.

Forced outage: An outage that results from emergency conditions directly associated with a component, requiring that it be taken out of service immediately, either automatically or as soon as switching operations can be performed; or an outage caused by improper operation of equipment or by human error.

Frequency deviation: An increase or decrease in the power frequency. Its duration varies from few cycles to several hours.

Fuel adjustment clause: A clause in a rate schedule that provides for adjustment of the amount of the bill as the cost of fuel varies from a specified base amount per unit.

Fuse: An overcurrent protective device with a circuit-opening fusible part that is heated and severed by the passage of overcurrent through it.

Fuse cutout: An assembly consisting of a fuse support and holder; it may also include a fuse link.

Ground: Also termed earth; a conductor connected between a circuit and the soil; an accidental ground occurs due to cable insulation faults, an insulator defect, etc.

Ground wire: A conductor having grounding connections at intervals that is suspended usually above but not necessarily over the line conductor to provide a degree of protection against lightning discharges.

Harmonic distortion: Periodic distortion of the sign wave.

Harmonic resonance: A condition in which the power system is resonating near one of the major harmonics being produced by nonlinear elements in the system, hence increasing the harmonic distortion.

Harmonics: Sinusoidal voltages or currents having frequencies that are an integer multiple of the fundamental frequency at which the supply system is designed to operate.

HMWPE: Abbreviation for high-molecular-weight polyethylene (cable insulation).

HV: Abbreviation for high voltage.

Impedance: The ratio of the phasor equivalent of a steady-state sine-wave voltage to the phasor equivalent of a steady-state sine-wave current.

Impulsive transient: A sudden (nonpower) frequency change in the steady-state condition of the voltage or current that is unidirectional in polarity.

Incremental energy costs: The additional cost of producing or transmitting electric energy above some base cost.

Index of reliability: A ratio of cumulative customer minutes that service was available during a year to total customer minutes demanded; can be used by the utility for feeder reliability comparisons.

Indoor transformer: A transformer that must be protected from the weather.

Installed reserve: The reserve capability installed on a system.

Interruptible load: A load that can be interrupted as defined by contract.

Interruption: The loss of service to one or more consumers. An interruption is the result of one or more component outages.

Interruption duration: The period from the initiation of an interruption to a consumer until service has been restored to that consumer.

Investment-related charges: Those certain charges incurred by a utility that are directly related to the capital investment of the utility. kcmil: Abbreviation for a thousand circular mils.

Isolated ground: It originates at an isolated ground-type receptacle or equipment input terminal block and terminates at the point where neutral and ground are bonded at the power source. Its conductor is insulated from the metallic raceway and all ground points throughout its length.

K-factor: A factor used to quantify the load impact of electric arc furnaces on the power system.

Lag: Denotes that a given sine wave passes through its peak at a later time than a reference time wave.

Lambda: The incremental operating cost at the load center, commonly expressed in mils per kWh.

Lateral conductor: A wire or cable extending in a general horizontal direction or at an angle to the general direction of the line; service wires either overhead or underground are considered laterals from the street mains.

LDC: Abbreviation for line-drop compensator.

Lightning arrestor: A device that reduces the voltage of a surge applied to its terminals and restores itself to its original operating condition.

Limit switch: A switch that is operated by a moving part at the end of its travel typically to stop or reverse the motion.

Limiter: A device in which some characteristic of the output is automatically prevented from exceeding a predetermined value.

Line: A component part of a system extending between adjacent stations or from a station to an adjacent interconnection point. A line may consist of one or more circuits.

Line-drop compensator: A device that causes the voltage-regulating relay to increase the output voltage by an amount that compensates for the impedance drop in the circuit between the regulator and a predetermined location at the circuit.

Line loss: Energy loss on a transmission or distribution line.

L–L: Abbreviation for line to line.

L–N: Abbreviation for line to neutral.

Load center: A point at which the load of a given area is assumed to be concentrated.

Load diversity: The difference between the sum of the maxima of two or more individual loads and the coincident or combined maximum load, usually measured in kW over a specified period of time.

Load duration curve: A curve of loads, plotted in descending order of magnitude, against time intervals for a specified period.

Load factor: The ratio of the average load over a designated period of time to the peak load occurring in that period.

Load-interrupter switch: An interrupter switch designed to interrupt currents not in excess of the continuous-current rating of the switch.

Load, interruptible: A load that can be interrupted as defined by contract.

Load losses, transformer: Those losses that are incident to the carrying of a specified load. They include I2R loss in the winding due to load and eddy currents, stray loss due to leakage fluxes in the windings, etc., and the loss due to circulating currents in parallel windings.

Load tap changer: A selector switch device applied to power transformers to maintain a constant low-side or secondary voltage with a variable primary voltage supply, or to hold a constant voltage out along the feeders on the low-voltage side for varying load conditions on the low-voltage side. Abbreviated LTC.

Load-tap-changing transformer: A transformer used to vary the voltage, or phase angle, or both, of a regulated circuit in steps by means of a device that connects different taps of tapped winding(s) without interrupting the load.

Loop feeder: A number of tie feeders in series, forming a closed loop. There are two routes by which any point on a loop feeder can receive electric energy, so that the flow can be in either direction.

Loop service: Two services of substantially the same capacity and characteristics, supplied from adjacent sections of a loop feeder. The two sections of the loop feeder are normally tied together on the consumer's bus through switching devices.

Loss factor: The ratio of the average power loss to the peak-load power loss during a specified period of time.

Low-side surges: The current surge that appears to be injected into the transformer secondary terminals upon a lighting strike to grounded conductors in the vicinity.

LTC: Abbreviation for load tap changer.

LV: Abbreviation for low voltage.

Main distribution center: A distribution center supplied directly by mains.

Maintenance expenses: The expense required to keep the system or plant in proper operating repair.

Maximum demand: The largest of a particular type of demand occurring within a specified period.

MC: Abbreviation for metal clad (cable).

Messenger cable: A galvanized steel or Copperweld cable used in construction to support a suspended current-carrying cable.

Metal-clad switchgear, outdoor: A switchgear that can be mounted in suitable weatherproof enclosures for outdoor installations. The base units are the same for both indoor and outdoor applications. The weatherproof housing is constructed integrally with the basic structure and is not merely a steel enclosure. The basic structure, including the mounting details and withdrawal mechanisms for the circuit breakers, bus compartments, transformer compartments, etc., is the same as that of indoor metal-clad switchgear.

Metal-enclosed switchgear: Primarily indoor-type switchgear. It can, however, be furnished in weatherproof houses suitable for outdoor operation. The switchgear is suitable for 600 V maximum service.

Minimum demand: The smallest of a particular type of demand occurring within a specified period.

Momentary interruption: An interruption of duration limited to the period required to restore service by automatic or supervisory-controlled switching operations or by manual switching at locations where an operator is immediately available.

Monthly peak duration curve: A curve showing the total number of days within the month during which the net 60 min clock-hour integrated peak demand equals or exceeds the percent of monthly peak values shown.

NC: Abbreviation for normally closed.

NEC: Abbreviation for National Electric Code.

NESC: Abbreviation for National Electrical Safety Code.

Net system energy: Energy requirements of a system, including losses, defined as (1) net generation of the system, plus (2) energy received from others, less, and (3) energy delivered to other systems.

Network distribution system: A distribution system that has more than one simultaneous path of power flow to the load.

Network protector: An electrically operated low-voltage air circuit breaker with self-contained relays for controlling its operation. It provides automatic isolation of faults in the primary feeders or network transformers. Abbreviated NP.

NO: Abbreviation for normally open.

No-load current: The current demand of a transformer primary when no current demand is made on the secondary.

No-load loss: Energy losses in an electric facility when energized at rated voltage and frequency but not carrying load.

Noise: An unwanted electrical signal with a less than 200 kHz superimposed upon the power-system voltage or current in-phase conductors, or found on neutral conductors or signal lines. It is not a harmonic distortion or transient. It disturbs microcomputers and programmable controllers.

Noncoincident demand: The sum of the individual maximum demands regardless of time of occurrence within a specified period.

Nonlinear load: An electrical load that draws current discontinuously or whose impedances vary throughout the cycle of the input AC voltage waveform.

Normal rating: Capacity of installed equipment.

Normal weather: All weather not designated as adverse or major storm disaster.

Normally closed: Denotes the automatic closure of contacts in a relay when deenergized. Abbreviated NC.

Normally open: Denotes the automatic opening of contacts in a relay when deenergized. Abbreviated NO.

NP: Abbreviation for network protector.

NSW: Abbreviation for nonswitched.

Notch: A switching (or other) disturbance of the normal power voltage waveform, lasting less than a half cycle, which is initially of opposite polarity than the waveform. It includes complete loss of voltage for up to a 0.5 cycle.

Notching: A periodic disturbance caused by normal operation of a power electronic device, when its current is commutated from one phase to another.

NX: Abbreviation for nonexpulsion (fuse).

Off-peak energy: Energy supplied during designated periods of relatively low system demands.

On-peak energy: Energy supplied during designated periods of relatively high system demands.

OH: Abbreviation for overhead.

Operating expenses: The labor and material costs for operating the plant involved.

Oscillatory transient: A sudden and nonpower frequency change in the steady-state condition of voltage or current that includes both positive and negative polarity values.

Outage: The state of a component when it is not available to perform its intended function due to some event directly associated with that component. An outage may or may not cause an interruption of service to consumers depending upon the system configuration.

Outage duration: The period from the initiation of an outage until the affected component or its replacement once again becomes available to perform its intended function.

Outage rate: For a particular classification of outage and type of component, the mean number of outages per unit exposure time per component.

Overhead expenses: The costs that in addition to direct labor and material are incurred by all utilities.

Overload: Loading in excess of normal rating of equipment.

Overload protection: Interruption or reduction of current under conditions of excessive demand, provided by a protective device.

Overvoltage: A voltage that has a value at least 10% above the nominal voltage for a period of time greater than 1 min.

Pad-mounted: A general term describing equipment positioned on a surface mounted pad located outdoors. The equipment is usually enclosed with all exposed surfaces at ground potential.

Pad-mounted transformer: A transformer utilized as part of an underground distribution system, with enclosed compartment(s) for high-voltage and low-voltage cables entering from below, and mounted on a foundation pad.

Panelboard: A distribution point where an incoming set of wires branches into various other circuits.

Passive filter: A combination of inductors, capacitors, and resistors designed to eliminate one or more harmonics. The most common variety is simply an inductor in series with a shunt capacitor, which short circuits the major distorting harmonic component from the system.

PE: Abbreviation used for polyethylene (cable insulation).

Peak current: The maximum value (crest value) of an alternating current.

Peak voltage: The maximum value (crest value) of an alternating voltage.

Peaking station: A generating station that is normally operated to provide power only during maximum load periods.

Peak-to-peak value: The value of an AC waveform from its positive peak to its negative peak. In the case of a sine wave, the peak-to-peak value is double the peak value.

Pedestal: A bottom support or base of a pillar, statue, etc.

Percent regulation: See Percent voltage drop.

Percent voltage drop: The ratio of voltage drop in a circuit to voltage delivered by the circuit, multiplied by 100 to convert to percent.

Permanent forced outage: An outage whose cause is not immediately self-clearing but must be corrected by eliminating the hazard or by repairing or replacing the component before it can be returned to service. An example of a permanent forced outage is a lightning flashover that shatters an insulator, thereby disabling the component until repair or replacement can be made.

Permanent forced outage duration: The period from the initiation of the outage until the component is replaced or repaired.

Phase: The time of occurrence of the peak value of an AC waveform with respect to the time of occurrence of the peak value of a reference waveform.

Phase angle: An angular expression of phase difference.

Phase shift: The displacement in time of one voltage waveform relative to other voltage waveform(s).

Pole: A column of wood or steel, or some other material, supporting overhead conductors, usually by means of arms or brackets.

Pole fixture: A structure installed in lieu of a single pole to increase the strength of a pole line or to provide better support for attachments than would be provided by a single pole. Examples are A fixtures, H fixtures.

Primary disconnecting devices: Self-coupling separable contacts provided to connect and disconnect the main circuits between the removable element and the housing.

Primary distribution feeder: A feeder operating at primary voltage supplying a distribution circuit.

Primary distribution mains: The conductors that feed from the center of distribution to direct primary loads or to transformers that feed secondary circuits.

Primary distribution network: A network consisting of primary distribution mains.

Primary distribution system: A system of AC distribution for supplying the primaries of distribution transformers from the generating station or substation distribution buses.

Primary distribution trunk line: A line acting as a main source of supply to a distribution system.

Primary feeder: That portion of the primary conductors between the substation or point of supply and the center of distribution.

Primary lateral: That portion of a primary distribution feeder that is supplied by a main feeder or other laterals and extends through the load area with connections to distribution transformers or primary loads.

Primary main feeder: The higher-capacity portion of a primary distribution feeder that acts as a main source of supply to primary laterals or direct connected distribution transformers and primary loads.

Primary network: A network supplying the primaries of transformers whose secondaries may be independent or connected to a secondary network.

Primary open-loop service: A service that consists of a single distribution transformer with dual primary switching, supplied from a single primary circuit that is arranged in an open-loop configuration.

Primary selective service: A service that consists of a single distribution transformer with primary throw-over switching, supplied by two independent primary circuits.

Primary transmission feeder: A feeder connected to a primary transmission circuit.

Primary unit substation: A unit substation in which the low-voltage section is rated above 1000 V.

Protective relay: A device whose function is to detect defective lines or apparatus or other power-system conditions of an abnormal or dangerous nature and to initiate appropriate control circuit action.

Power: The rate (in kW) of generating, transferring, or using energy.

Power, active: The product of the rms value of the voltage and the rms value of the in-phase component of the current.

Power, apparent: The product of the rms value of the voltage and the rms value of the current. Power, instantaneous: The product of the instantaneous voltage multiplied by the instantaneous current.

Power factor: The ratio of active power to apparent power.

Power-factor adjustment clause: A clause in a rate schedule that provides for an adjustment in the billing if the customer's power factor varies from a specified reference.

Power pool: A group of power systems operating as an interconnected system and pooling their resources.

Power, reactive: The product of the rms value of the voltage and the rms value of the quadrature component of the current.

Power transformer: A transformer that transfers electric energy in any part of the circuit between the generator and the distribution primary circuits.

PT: Abbreviation for potential transformers.

pu: Abbreviation for per unit.

Raceway: A channel for holding wires, cables, or busbars. The channel may be in the form of a conduit, electrical metallic tubing, or a square sheet-metal duct. It is used in buildings at low voltage.

Radial distribution system: A distribution system that has a single simultaneous path of power flow to the load.

Radial service: A service that consists of a single distribution transformer supplied by a single primary circuit.

Radial system, complete: A radial system that consists of a radial subtransmission circuit, a single substation, and a radial primary feeder with several distribution transformers each supplying radial secondaries; has the lowest degrees of service continuity.

Ratchet demand: The maximum past or present demands that are taken into account to establish billings for previous or subsequent periods.

Ratchet demand clause: A clause in a rate schedule that provides that maximum past or present demands be taken into account to establish billings for previous or subsequent periods.

Rate base: The net plant investment or valuation base specified by a regulatory authority upon which a utility is permitted to earn a specified rate of return.

RCN: Abbreviation for reduced-capacity neutral.

Recloser: A dual-timing device that can be set to operate quickly to prevent downline fuses from blowing.

Reclosing device: A control device that initiates the reclosing of a circuit after it has been opened by a protective relay.

Reclosing fuse: A combination of two or more fuse holders, fuse units, or fuse links mounted on a fuse support(s), mechanically or electrically interlocked, so that one fuse can be connected into the circuit at a time and the functioning of that fuse automatically connects the next fuse into the circuit, thereby permitting one or more service restorations without replacement of fuse links, refill units, or fuse units.

Reclosing relay: A programming relay whose function is to initiate the automatic reclosing of a circuit breaker.

Reclosure: The automatic closing of a circuit-interrupting device following automatic tripping. Reclosing may be programmed for any combination of instantaneous, time-delay, single-shot, multiple-shot, synchronism-check, dead-line-live-bus, or dead-bus-live-line operation.

Recovery voltage: The voltage that occurs across the terminals of a pole of a circuit-interrupting device upon interruption of the current.

Required reserve: The system planned reserve capability needed to ensure a specified standard of service.

Resistance: The real part of impedance.

Return on capital: The requirement that is necessary to pay for the cost of investment funds used by the utility. RP: Abbreviation for regulating point.

Sag: The distance measured vertically from a conductor to the straight line joining its two points of support. Unless otherwise stated, the sag referred to is the sag at the midpoint of the span.

Sag: A decrease to between 0.1 and 0.9 pu in rms voltage and current at the power frequency for a duration of 0.5 cycles to 1 min.

Sag, final unloaded: The sag of a conductor after it has been subjected for an appreciable period to the loading prescribed for the loading district in which it is situated, or equivalent loading, and the loading removed. Final unloaded sag includes the effect of inelastic deformation.

Sag, initial unloaded: The sag of a conductor prior to the application of any external load.

Sag of a conductor (at any point in a span): The distance measured vertically from the particular point in the conductor to a straight line between its two points of support.

Sag section: The section of line between snub structures. More than one sag section may be required to properly sag the actual length of conductor that has been strung.

Sag span: A span selected within a sag section and used as a control to determine the proper sag of the conductor, thus establishing the proper conductor level and tension. A minimum of two, but normally three, sag spans are required within a sag section to sag properly. In mountainous terrain or where span lengths vary radically, more than three sag spans could be required within a sag section.

Scheduled interruption: An interruption caused by a scheduled outage.

Scheduled maintenance (generation): Capability that has been scheduled to be out of service for maintenance.

Scheduled outage: An outage that results when a component is deliberately taken out of service at a selected time, usually for purposes of construction, preventive maintenance, or repair.

Scheduled outage duration: The period from the initiation of the outage until construction, preventive maintenance, or repair work is completed.

SCV: Abbreviation for steam cured (cable insulation).

Seasonal diversity: Load diversity between two (or more) electric systems that occurs when their peak loads are in different seasons of the year.

Secondary, radial: A secondary supplied from either a conventional or completely self-protected (type CSP) distribution transformer.

Secondary current rating: The secondary current existing when the transformer is delivering rated kVA at rated secondary voltage.

Secondary disconnecting devices: Self-coupling separable contacts provided to connect and disconnect the auxiliary and control circuits between the removable element and the housing.

Secondary distributed network: A service consisting of a number of network-transformer units at a number of locations in an urban load area connected to an extensive secondary cable grid system.

Secondary distribution feeder: A feeder operating at secondary voltage supplying a distribution circuit.

Secondary distribution mains: The conductors connected to the secondaries of distribution transformers from which consumers' services are supplied.

Secondary distribution network: A network consisting of secondary distribution mains.

Secondary distribution system: A low-voltage AC system that connects the secondaries of distribution transformers to the consumers' services.

Secondary distribution trunk line: A line acting as a main source of supply to a secondary distribution system.

Secondary fuse: A fuse used on the secondary-side circuits, restricted for use on a low-voltage secondary distribution system that connects the secondaries of distribution transformers to consumers' services.

Secondary mains: Those that operate at utilization voltage and serve as the local distribution main. In radial systems, secondary mains that supply general lighting and small power are usually separate from mains that supply three-phase power because of the dip in voltage caused by starting motors. This dip in voltage, if sufficiently large, causes an objectionable lamp flicker.

Secondary network: It consists of two or more network-transformer units connected to a common secondary system and operating continuously in parallel.

Secondary network service: A service that consists of two or more network transformer units connected to a common secondary system and operating continuously in parallel.

Secondary selective service: A service that consists of two distribution transformers, each supplied by an independent primary circuit, and with secondary main and tie breakers.

Secondary spot network: A network that consists of at least two and as many as six network-transformer units located in the same vault and connected to a common secondary service bus. Each transformer is supplied by an independent primary circuit.

Secondary system, banked: A system that consists of several transformers supplied from a single primary feeder, with the low-voltage terminals connected together through the secondary mains.

Secondary unit substation: A unit substation whose low-voltage section is rated 1000 V and below.

Secondary voltage regulation: A voltage drop caused by the secondary system, it includes the drop in the transformer and in the secondary and service cables.

Second-contingency outage: The outage of a secondary primary feeder in addition to the first one.

Sectionalizer: A device that resembles an oil circuit recloser but lacks the interrupting capability.

Service area: Territory in which a utility system is required or has the right to supply or make available electric service to ultimate consumers.

Service availability index: See Index of reliability.

Service drop: The overhead conductors, through which electric service is supplied, between the last utility company pole and the point of their connection to the service facilities located at the building or other support used for the purpose.

Service entrance: All components between the point of termination of the overhead service drop or underground service lateral and the building main disconnecting device, with the exception of the utility company's metering equipment.

Service entrance conductors: The conductors between the point of termination of the overhead service drop or underground service lateral and the main disconnecting device in the building.

Service entrance equipment: Equipment located at the service entrance of a given building that provides overcurrent protection to the feeder and service conductors, provides a means of disconnecting the feeders from energized service conductors, and provides a means of measuring the energy used by the use of metering equipment.

Service lateral: The underground conductors, through which electric service is supplied, between the utility company's distribution facilities and the first point of their connection to the building or area service facilities located at the building or other support used for the purpose.

SF_6: Formula for sulfur hexafluoride (gas). St: Abbreviation for steel.

Strand: One of the wires, or groups of wires, of any stranded conductor.

Stranded conductor: A conductor composed of a group of wires, or of any combination of groups of wires. Usually, the wires are twisted together.

Submarine cable: A cable designed for service under water. It is usually a lead-covered cable with a steel armor applied between layers of jute.

Submersible transformer: A transformer so constructed as to be successfully operable when submerged in water under predetermined conditions of pressure and time.

Substation: An assemblage of equipment for purposes other than generation or utilization, through which electric energy in bulk is passed for the purpose of switching or modifying its characteristics.

Substation voltage regulation: The regulation of the substation voltage by means of the voltage regulation equipment that can be load-tap-changing (LTC) mechanisms in the substation transformer, a separate regulator between the transformer and low-voltage bus, switched capacitors at the low-voltage bus, or separate regulators located in each individual feeder in the substation.

Subtransmission: That part of the distribution system between bulk power source(s) (generating stations or power substations) and the distribution substation.

Susceptance: The imaginary part of admittance.

Swell: An increase to between 1.1 and 1.8 pu in rms voltage or current at the power frequency for durations from 0.5 cycle to 1 min.

Sustained interruption: The complete loss of voltage (¡0.1 pu) on one or more phase conductors for a time greater than 1 min.

Switch: A device for opening and closing or for changing connections in a circuit.

Switch, isolating: An auxiliary switch for isolating an electric circuit from its source of power; it is operated only after the circuit has been opened by other means.

Switch, limi: A switch that is operated by some part or motion of a power-driven machine or equipment to alter the electric circuit associated with the machine or equipment.

Switchboard: A large single panel, frame, or assembly of panels on which are mounted (on the face, or back, or both) switches, fuses, buses, and usually instruments.

Switched-capacitor bank: A capacitor bank with switchable capacitors.

Switchgear: A general term covering switching or interrupting devices and their combination with associated control, instrumentation, metering, protective, and regulating devices; also assemblies of these devices with associated interconnections, accessories, and supporting structures.

Switching time: The period from the time a switching operation is required due to a forced outage until that switching operation is performed.

System: A group of components connected together in some fashion to provide flow of power from one point or points to another point or points.

System interruption duration index: The ratio of the sum of all customer interruption durations per year to the number of customers served. It gives the number of minutes out per customer per year.

Total demand distortion (TDD): The ratio of the root-mean-square (rms) of the harmonic current to the rms value of the rated or maximum demand fundamental current, expressed as a percent.

Total harmonic distortion (THD): The ratio of the root-mean-square of the harmonic content to the root-mean-square value of the fundamental quantity, expressed as a percent of the fundamental.

Triplen harmonics: A term frequently used to refer to the odd multiples of the third harmonic, which deserve special attention because of their natural tendency to be zero sequence.

True power factor (TPF): The ratio of the active power of the fundamental wave, in watts, to the apparent power of the fundamental wave, in root-mean-square voltamperes (including the harmonic components).

Underground distribution system: That portion of a primary or secondary distribution system that is constructed below the earth's surface. Transformers and equipment enclosures for such a system may be located either above or below the surface as long as the served and serving conductors are located underground.

Undervoltage: A voltage that has a value at least 10% below the nominal voltage for a period of time greater than 1 min.

Unit substation: A substation consisting primarily of one or more transformers that are mechanically and electrically connected to and coordinated in design with one or more switchgear or motor control assemblies or combinations thereof.

URD: Abbreviation for underground residential distribution.

Utilization factor: The ratio of the maximum demand of a system to the rated capacity of the system.

VD: Abbreviation for voltage drop. **VDIP:** Abbreviation for voltage dip.

Voltage, base: A reference value that is a common denominator to the nominal voltage ratings of transmission and distribution lines, transmission and distribution equipment, and utilization equipment.

Voltage imbalance (or unbalance): The maximum deviation from the average of the three-phase voltages or currents, divided by the average of the three-phase voltages or currents, expressed in percent.

Voltage, maximum: The greatest 5 min average or mean voltage.

Voltage, minimum: The least 5 min average or mean voltage.

Voltage, nominal: A nominal value assigned to a circuit or system of a given voltage class for the purpose of convenient designation.

Voltage, rated: The voltage at which operating and performance characteristics of equipment are referred.

Voltage, service: Voltage measured at the terminals of the service entrance equipment.

Voltage, utilization: Voltage measured at the terminals of the machine or device.

Voltage dip: A voltage change resulting from a motor starting.

Voltage drop: The difference between the voltage at the transmitting and receiving ends of a feeder, main or service.

Voltage flicker: Voltage fluctuation caused by utilization equipment resulting in lamp flicker, that is, in a lamp illumination change.

Voltage fluctuation: A series of voltage changes or a cyclical variation of the voltage envelope.

Voltage interruption: Disappearance of the supply voltage on one or more phases. It can be momentary, temporary, or sustained.

Voltage magnification: The magnification of capacitor switching oscillatory transient voltage on the primary side by capacitors on the secondary side of a transformer.

Voltage regulation: The percent voltage drop of a line with reference to the receiving-end voltage.

$$\%regulation = \frac{|\bar{E}_S| - |\bar{E}_r|}{|\bar{E}_r|}$$

where

$|\bar{E}_s|$ is the magnitude of the sending-end voltage

$|\bar{E}_r|$ is the magnitude of the receiving-end voltage

Voltage regulator: An induction device having one or more windings in shunt with, and excited from, the primary circuit, and having one or more windings in series between the primary circuit and the regulated circuit, all suitably adapted and arranged for the control of the voltage, or of the phase angle, or of both, of the regulated circuit.

Voltage spread: The difference between maximum and minimum voltages.

VRR: Abbreviation for voltage-regulating relay.

Waveform distortion: A steady-state deviation from an ideal sine wave of power frequency principally characterized by the special content of the deviation.

XLPE: Abbreviation for cross-linked polyethylene (cable insulation).

References

1. Z. Wang, H. Liu, D. C. Yu, X. Wang, and H. Song, "A practical approach to the conductor size selection in planning radial distribution systems," *IEEE Transactions on Power Delivery*, vol. 15, no. 1, pp. 350–354, 2000.
2. P. M. S. Carvalho, L. A. F. M. Ferreira, and L. M. F. Barruncho, "On spanning-tree recombination in evolutionary large-scale network problems-application to electrical distribution planning," *IEEE Transactions on Evolutionary Computation*, vol. 5, no. 6, pp. 623–630, 2001.
3. N. G. Boulaxis and M. P. Papadopoulos, "Optimal feeder routing in distribution system planning using dynamic programming technique and GIS facilities," *IEEE Transactions on Power Delivery*, vol. 17, no. 1, pp. 242–247, 2002.
4. T. Asakura, T. Genji, T. Yura, N. Hayashi, and Y. Fukuyama, "Long-term distribution network expansion planning by network reconfiguration and generation of construction plans," *IEEE Transactions on Power Systems*, vol. 18, no. 3, pp. 1196–1204, 2003.
5. W. El-Khattam, K. Bhattacharya, Y. Hegazy, and M. Salama, "Optimal investment planning for distributed generation in a competitive electricity market," *IEEE Transactions on Power Systems*, vol. 19, no. 3, pp. 1674–1684, 2004.
6. C. Monteiro, I. J. Ramírez-Rosado, V. Miranda, P. J. Zorzano-Santamaría, E. García-Garrido, and L. A. Fernández-Jiménez, "Gis spatial analysis applied to electric line routing optimization," *IEEE Transactions on Power Delivery*, vol. 20, no. 2, pp. 934–942, 2005.
7. I. J. Ramírez-Rosado and J. A. Domínguez-Navarro, "New multiobjective tabu search algorithm for fuzzy optimal planning of power distribution systems," *IEEE Transactions on Power Systems*, vol. 21, no. 1, pp. 224–233, 2006.
8. D. Singh, R. K. Misra, and D. Singh, "Effect of load models in distributed generation planning," *IEEE Transactions on Power Systems*, vol. 22, no. 4, pp. 2204–2212, 2007.
9. S. Haffner, L. F. A. Pereira, L. A. Pereira, and L. S. Barreto, "Multistage model for distribution expansion planning with distributed generation—part i: Problem formulation," *IEEE Transactions on Power Delivery*, vol. 23, no. 2, pp. 915–923, 2008.
10. A. Navarro and H. Rudnick, "Large-scale distribution planning—part i: Simultaneous network and transformer optimization," *IEEE Transactions on Power Systems*, vol. 24, no. 2, pp. 744–751, 2009.
11. M. Lavorato, M. J. Rider, A. V. Garcia, and R. Romero, "A constructive heuristic algorithm for distribution system planning," *IEEE transactions on power systems*, vol. 25, no. 3, pp. 1734–1742, 2010.
12. C. M. Domingo, T. G. San Roman, A. Sánchez-Miralles, J. P. P. Gonzalez, and A. C. Martinez, "A reference network model for large-scale distribution planning with automatic street map generation," *IEEE Transactions on Power Systems*, vol. 26, no. 1, pp. 190–197, 2010.
13. R. J. Millar, S. Kazemi, M. Lehtonen, and E. Saarijarvi, "Impact of mv connected microgrids on mv distribution planning," *IEEE Transactions on Smart Grid*, vol. 3, no. 4, pp. 2100–2108, 2012.
14. C.-M. Chan, C.-N. Lu, and Y. L. Lo, "Optimal use of existing distribution feeders to accommodate transportation electrification," *IEEE Transactions on Intelligent Transportation Systems*, vol. 16, no. 4, pp. 1941–1950, 2015.
15. S. M. Mazhari, H. Monsef, and R. Romero, "A multi-objective distribution system expansion planning incorporating customer choices on reliability," *IEEE Transactions on Power Systems*, vol. 31, no. 2, pp. 1330–1340, 2015.
16. X. Shen, M. Shahidehpour, Y. Han, S. Zhu, and J. Zheng, "Expansion planning of active distribution networks with centralized and distributed energy storage systems," *IEEE Transactions on Sustainable Energy*, vol. 8, no. 1, pp. 126–134, 2016.
17. N. Amjady, A. Attarha, S. Dehghan, and A. J. Conejo, "Adaptive robust expansion planning for a distribution network with ders," *IEEE Transactions on Power Systems*, vol. 33, no. 2, pp. 1698–1715, 2017.
18. Z. Lin, Z. Hu, and Y. Song, "Distribution network expansion planning considering $n-1$ criterion," *IEEE Transactions on Power Systems*, vol. 34, no. 3, pp. 2476–2478, 2019.
19. S. Vahedi, M. Banejad, and M. Assili, "Gis-based substation expansion planning," *IEEE Systems Journal*, vol. 15, no. 1, pp. 959–970, 2020.

20. H. Liang, J. Ma, and J. Lin, "Robust distribution system expansion planning incorporating thermostatically-controlled-load demand response resource," *IEEE Transactions on Smart Grid*, vol. 13, no. 1, pp. 302–313, 2021.
21. US Department of Energy, "The national electric reliability study: Technical study reports," Office of Emergency Operations, Washington, DC, Tech. Rep., April 1981.
22. T. Gonen and J. Thompson, "Distribution system planning: The state-of-the-art and the future trends." IEEE, January 22–24, 1980, pp. 13–18.
23. H. Willis, H. Tram, M. Engel, and L. Finley, "Optimization applications to power distribution," *IEEE Computer Applications in Power*, vol. 8, no. 4, pp. 12–17, 1995.
24. T. Gonen and I. J. Ramirez-Rosado, "Optimal multi-stage planning of power distribution systems," *IEEE Transactions on Power Delivery*, vol. 2, no. 2, pp. 512–519, 1987.
25. R. Delgado, "Load management—a planner's view," *IEEE Transactions on Power Apparatus and Systems*, vol. 6, no. 102, p. 1812–1813, 1983.
26. T. Gonen and B. Foote, "Application of mixed-integer programming to reduce suboptimization in distribution systems planning," *Proceedings of the 1979 Modeling and Simulation Conference, University of Pittsburgh, Pittsburgh, PA*, vol. 10, no. 3, pp. 1133–1139, 1979.
27. T. Gonen and D. Yu, "Bibliography of distribution system planning." IEEE, March 17–18, 1980, pp. 23–34.
28. N. Balijepalli, S. S. Venkata, and R. D. Christie, "Modeling and analysis of distribution reliability indices," *IEEE Transactions on Power Delivery*, vol. 19, no. 4, pp. 1950–1955, 2004.
29. A.-H. Mohsenian-Rad, V. W. Wong, J. Jatskevich, R. Schober, and A. Leon-Garcia, "Autonomous demand-side management based on game-theoretic energy consumption scheduling for the future smart grid," *IEEE transactions on Smart Grid*, vol. 1, no. 3, pp. 320–331, 2010.
30. K. Qian, C. Zhou, M. Allan, and Y. Yuan, "Modeling of load demand due to EV battery charging in distribution systems," *IEEE Transactions on Power Systems*, vol. 26, no. 2, pp. 802–810, 2010.
31. P. Du and N. Lu, "Appliance commitment for household load scheduling," *IEEE Transactions on Smart Grid*, vol. 2, no. 2, pp. 411–419, 2011.
32. R. A. Jabr, R. Singh, and B. C. Pal, "Minimum loss network reconfiguration using mixed-integer convex programming," *IEEE Transactions on Power Systems*, vol. 27, no. 2, pp. 1106–1115, 2012.
33. T. Logenthiran, D. Srinivasan, and T. Z. Shun, "Demand side management in smart grid using heuristic optimization," *IEEE Transactions on Smart Grid*, vol. 3, no. 3, pp. 1244–1252, 2012.
34. S. Shafiee, M. Fotuhi-Firuzabad, and M. Rastegar, "Investigating the impacts of plug-in hybrid electric vehicles on power distribution systems," *IEEE Transactions on Smart Grid*, vol. 4, no. 3, pp. 1351–1360, 2013.
35. W. Liu, Q. Wu, F. Wen, and J. Østergaard, "Day-ahead congestion management in distribution systems through household demand response and distribution congestion prices," *IEEE Transactions on Smart Grid*, vol. 5, no. 6, pp. 2739–2747, 2014.
36. M. Pipattanasomporn, M. Kuzlu, S. Rahman, and Y. Teklu, "Load profiles of selected major household appliances and their demand response opportunities," *IEEE Transactions on Smart Grid*, vol. 5, no. 2, pp. 742–750, 2013.
37. Y. Zhang, L. Wang, Y. Xiang, and C.-W. Ten, "Power system reliability evaluation with scada cybersecurity considerations," *IEEE Transactions on Smart Grid*, vol. 6, no. 4, pp. 1707–1721, 2015.
38. S.-C. Huang, C.-N. Lu, and Y.-L. Lo, "Evaluation of ami and scada data synergy for distribution feeder modeling," *IEEE Transactions on Smart Grid*, vol. 6, no. 4, pp. 1639–1647, 2015.
39. W. Luan, J. Peng, M. Maras, J. Lo, and B. Harapnuk, "Smart meter data analytics for distribution network connectivity verification," *IEEE Transactions on Smart Grid*, vol. 6, no. 4, pp. 1964–1971, 2015.
40. X. Tang, K. Hasan, J. Milanovic, K. Bailey, and S. Stott, "Estimation and validation of characteristic load profile through smart grid trials in a medium voltage distribution network," *IEEE Transactions on Power Systems*, vol. 33, no. 2, pp. 1848–1859, 2018.
41. M. M. Albu, M. Sănduleac, and C. Stănescu, "Syncretic use of smart meters for power quality monitoring in emerging networks," *IEEE Transactions on Smart Grid*, vol. 8, no. 1, pp. 485–492, 2016.
42. W. Liu, Q. Gong, H. Han, Z. Wang, and L. Wang, "Reliability modeling and evaluation of active cyber physical distribution system," *IEEE Transactions on Power Systems*, vol. 33, no. 6, pp. 7096–7108, 2018.

43. W. Kong, Z. Y. Dong, J. Ma, D. J. Hill, J. Zhao, and F. Luo, “An extensible approach for non-intrusive load disaggregation with smart meter data,” *IEEE Transactions on Smart Grid*, vol. 9, no. 4, pp. 3362–3372, 2016.
44. A. Narimani, G. Nourbakhsh, A. Arefi, G. F. Ledwich, and G. R. Walker, “SAIDI constrained economic planning and utilization of central storage in rural distribution networks,” *IEEE Systems Journal*, vol. 13, no. 1, pp. 842–853, 2018.
45. J. Su, T. T. Lie, and R. Zamora, “Integration of electric vehicles in distribution network considering dynamic power imbalance issue,” *IEEE Transactions on Industry Applications*, vol. 56, no. 5, pp. 5913–5923, 2020.
46. V. C. Cunha, W. Freitas, F. C. Trindade, and S. Santoso, “Automated determination of topology and line parameters in low voltage systems using smart meters measurements,” *IEEE Transactions on Smart Grid*, vol. 11, no. 6, pp. 5028–5038, 2020.
47. H. Sheng, C. Wang, B. Li, J. Liang, M. Yang, and Y. Dong, “Multi-timescale active distribution network scheduling considering demand response and user comprehensive satisfaction,” *IEEE Transactions on Industry Applications*, vol. 57, no. 3, pp. 1995–2005, 2021.
48. A. A. S. D. of Electric Terms, “Generation, Transmission and Distribution,” 1957.
49. R. Sarikas and H. Thacker, “Distribution system load characteristics and their use in planning and design,” *Transactions of the American Institute of Electrical Engineers. Part III: Power Apparatus and Systems*, vol. 76, no. 3, pp. 564–572, 1957.
50. W. E. Corporation, *Electric Utility Engineering Reference Book: Distribution Systems*. The Corporation, East Pittsburgh, PA, 1965, vol. 3.
51. H. P. Seelye, *Electrical Distribution Engineering*. McGraw-Hill Book Company, Incorporated, 1930.
52. F. Buller and C. Woodrow, “Load factor-equivalent hour values compared,” *Electrical World*, vol. 92, no. 2, pp. 59–60, 1928.
53. C. Arvidson, “Diversified demand method of estimating residential distribution transformer loads,” *Edison Electrical Institute Bulletin*, vol. 8, pp. 469–479, 1940.
54. W. Fu, J. D. McCalley, and V. Vittal, “Risk assessment for transformer loading,” *IEEE Transactions on Power Systems*, vol. 16, no. 3, pp. 346–353, 2001.
55. H. Khodr, J. Molea, I. Garcia, C. Hidalgo, P. Paiva, J. Yusta, and A. J. Urdaneta, “Standard levels of energy losses in primary distribution circuits for scada application,” *IEEE Transactions on Power Systems*, vol. 17, no. 3, pp. 615–620, 2002.
56. Y.-T. Chao, S.-T. Lee, H.-C. Chang, and T.-H. Chen, “An improvement project for distribution transformer load management in taiwan,” *IEEE Transactions on Power Systems*, vol. 18, no. 2, pp. 875–881, 2003.
57. C. A. Dortolina and R. Nadira, “The loss that is unknown is no loss at all: A top-down/bottom-up approach for estimating distribution losses,” *IEEE Transactions on Power Systems*, vol. 20, no. 2, pp. 1119–1125, 2005.
58. M. Nagpal, F. Plumptre, R. Fulton, and T. G. Martinich, “Dispersed generation interconnection-utility perspective,” *IEEE Transactions on Industry Applications*, vol. 42, no. 3, pp. 864–872, 2006.
59. C. Zhang, Q. Chen, Y. Zhao, D. Li, and Y. Xiong, “A novel active power filter for high-voltage power distribution systems application,” *IEEE Transactions on Power Delivery*, vol. 22, no. 2, pp. 911–918, 2007.
60. W. S. Hopper and B. L. Etzel, “Distribution equipment modernization to reduce arc flash hazards,” *IEEE Transactions on Industry Applications*, vol. 44, no. 3, pp. 940–948, 2008.
61. W. H. Kersting, “Center-tapped transformer and 120-/240-v secondary models,” *IEEE transactions on industry applications*, vol. 45, no. 2, pp. 575–581, 2009.
62. A. Borghetti, M. Bosetti, S. Grillo, S. Massucco, C. A. Nucci, M. Paolone, and F. Silvestro, “Short-term scheduling and control of active distribution systems with high penetration of renewable resources,” *IEEE Systems Journal*, vol. 4, no. 3, pp. 313–322, 2010.
63. E. T. Jauch, “Possible effects of smart grid functions on ltc transformers,” *IEEE Transactions on Industry Applications*, vol. 47, no. 2, pp. 1013–1021, 2010.
64. C.-S. Chen, T.-T. Ku, and C.-H. Lin, “Design of phase identification system to support three-phase loading balance of distribution feeders,” *IEEE Transactions on Industry Applications*, vol. 48, no. 1, pp. 191–198, 2012.

65. J. M. Sexauer, K. D. McBee, and K. A. Bloch, "Applications of probability model to analyze the effects of electric vehicle chargers on distribution transformers," *IEEE Transactions on Power Systems*, vol. 28, no. 2, pp. 847–854, 2012.
66. P. Arboleya, C. González-Morán, and M. Coto, "Unbalanced power flow in distribution systems with embedded transformers using the complex theory in stationary reference frame," *IEEE Transactions on Power Systems*, vol. 29, no. 3, pp. 1012–1022, 2013.
67. N. Papanikolaou, A. Kyritsis, M. Loupis, C. Tzotzos, and E. Zoga, "Design considerations for single-phase line frequency transformers applied at photovoltaic systems," *IEEE Power and Energy Technology Systems Journal*, vol. 2, no. 3, pp. 82–93, 2015.
68. M. A. Awadallah, T. Xu, B. Venkatesh, and B. N. Singh, "On the effects of solar panels on distribution transformers," *IEEE Transactions on Power Delivery*, vol. 31, no. 3, pp. 1176–1185, 2015.
69. A. Parchure, S. J. Tyler, M. A. Peskin, K. Rahimi, R. P. Broadwater, and M. Dilek, "Investigating pv generation induced voltage volatility for customers sharing a distribution service transformer," *IEEE Transactions on Industry Applications*, vol. 53, no. 1, pp. 71–79, 2016.
70. M. Soleimani and M. Kezunovic, "Mitigating transformer loss of life and reducing the hazard of failure by the smart EV charging," *IEEE Transactions on Industry Applications*, vol. 56, no. 5, pp. 5974–5983, 2020.
71. O. Pereira, J. Quirós-Tortós, and G. Valverde, "Phase rebalancing of distribution circuits dominated by single-phase loads," *IEEE Transactions on Power Systems*, vol. 36, no. 6, pp. 5333–5344, 2021.
72. M. T. A. Khan, A. A. Milani, A. Chakrabortty, and I. Husain, "Dynamic modeling and feasibility analysis of a solid-state transformer-based power distribution system," *IEEE Transactions on Industry Applications*, vol. 54, no. 1, pp. 551–562, 2017.
73. S. A. Saleh, E. Ozkop, B. Alsayid, C. Richard, X. F. S. Onge, K. M. McDonald, and L. Chang, "Solid-state transformers for distribution systems–part ii: Deployment challenges," *IEEE Transactions on Industry Applications*, vol. 55, no. 6, pp. 5708–5716, 2019.
74. W. E. Corporation, *Electrical Transmission and Distribution Reference Book*. Westinghouse Electric Corporation, 1964.
75. S. Stigant and A. Franklin, *The J&P Transformer Book*, 1973.
76. A. N. S. Institute, "The guide for loading mineral oil-immersed overhead-type distribution transformers with 55° c and 65°c average winding rise," p. C57.91, 1969.
77. H. W. Beaty and D. G. Fink, *Standard Handbook for Electrical Engineers*, 11th ed. McGraw-Hill Education, 1978.
78. E. Clarke, *Circuit Analysis of AC Power Systems*. General Electric Series, Schenectady, NY, 1943, vol. 1.
79. G. E. Company, "Distribution transformer manual," 1975.
80. I. Roytelman and V. Ganesan, "Coordinated local and centralized control in distribution management systems," *IEEE Transactions on Power Delivery*, vol. 15, no. 2, pp. 718–724, 2000.
81. M. Dilek, R. P. Broadwater, J. C. Thompson, and R. Seqiun, "Simultaneous phase balancing at substations and switches with time-varying load patterns," *IEEE Transactions on Power Systems*, vol. 16, no. 4, pp. 922–928, 2001.
82. M. A. Yehia, E. E. Matar, N. Y. Hobeila, and A. D. Avedikian, "A heuristic algorithm for electric distribution networks optimal feeder configuration using geographic information system," *IEEE Transactions on Power Systems*, vol. 17, no. 4, pp. 1232–1237, 2002.
83. A. A. Chowdhury and D. O. Koval, "Current practices and customer value-based distribution system reliability planning," *IEEE Transactions on Industry Applications*, vol. 40, no. 5, pp. 1174–1182, 2004.
84. P. Paiva, H. Khodr, J. Dominguez-Navarro, J. Yusta, and A. Urdaneta, "Integral planning of primary-secondary distribution systems using mixed integer linear programming," *IEEE Transactions on Power Systems*, vol. 20, no. 2, pp. 1134–1143, 2005.
85. C.-L. Hor and P. A. Crossley, "Unsupervised event extraction within substations using rough classification," *IEEE Transactions on Power Delivery*, vol. 21, no. 4, pp. 1809–1816, 2006.
86. C.-L. Hor, P. A. Crossley, and S. J. Watson, "Building knowledge for substation-based decision support using rough sets," *IEEE Transactions on Power Delivery*, vol. 22, no. 3, pp. 1372–1379, 2007.
87. R. Walling, R. Saint, R. C. Dugan, J. Burke, and L. A. Kojovic, "Summary of distributed resources impact on power delivery systems," *IEEE Transactions on Power Delivery*, vol. 23, no. 3, pp. 1636–1644, 2008.

88. S. Najafi, S. H. Hosseinian, M. Abedi, A. Vahidnia, and S. Abachezadeh, "A framework for optimal planning in large distribution networks," *IEEE Transactions on Power Systems*, vol. 24, no. 2, pp. 1019–1028, 2009.
89. G. A. Jiménez-Estévez, L. S. Vargas, and V. Marianov, "Determination of feeder areas for the design of large distribution networks," *IEEE Transactions on Power Delivery*, vol. 25, no. 3, pp. 1912–1922, 2010.
90. C. A. Hill, M. C. Such, D. Chen, J. Gonzalez, and W. M. Grady, "Battery energy storage for enabling integration of distributed solar power generation," *IEEE Transactions on Smart Grid*, vol. 3, no. 2, pp. 850–857, 2012.
91. C. K. Gan, D. Pudjianto, P. Djapic, and G. Strbac, "Strategic assessment of alternative design options for multivoltage-level distribution networks," *IEEE Transactions on Power Systems*, vol. 29, no. 3, pp. 1261–1269, 2013.
92. G. H. Fox, "Design guidelines for safer low-voltage distribution systems," *IEEE Transactions on Industry Applications*, vol. 51, no. 3, pp. 2060–2070, 2015.
93. Q. Song, W. Sheng, L. Kou, D. Zhao, Z. Wu, H. Fang, and X. Zhao, "Smart substation integration technology and its application in distribution power grid," *CSEE Journal of Power and Energy Systems*, vol. 2, no. 4, pp. 31–36, 2016.
94. A. Dede, D. Della Giustina, G. Massa, and L. Cremaschini, "Toward a new standard for secondary substations: the viewpoint of a distribution utility," *IEEE Transactions on Power Delivery*, vol. 32, no. 2, pp. 1123–1132, 2016.
95. A. Dos Santos, B. Soares, C. Fan, M. Kuipers, S. Sabino, A. M. Grilo, P. R. B. Pereira, M. S. Nunes, and A. Casaca, "Characterization of substation process bus network delays," *IEEE Transactions on Industrial Informatics*, vol. 14, no. 5, pp. 2085–2094, 2017.
96. L. Zhang, J. Liang, W. Tang, G. Li, Y. Cai, and W. Sheng, "Converting AC distribution lines to dc to increase transfer capacities and DG penetration," *IEEE Transactions on Smart Grid*, vol. 10, no. 2, pp. 1477–1487, 2018.
97. A. Kotsonias, M. Asprou, L. Hadjidemetriou, and E. Kyriakides, "State estimation for distribution grids with a single-point grounded neutral conductor," *IEEE Transactions on Instrumentation and Measurement*, vol. 69, no. 10, pp. 8167–8177, 2020.
98. Y. Deng, W. Jiang, F. Hu, K. Sun, and J. Yu, "Resilience-oriented dynamic distribution network with considering recovery ability of distributed resources," *IEEE Journal on Emerging and Selected Topics in Circuits and Systems*, vol. 12, no. 1, pp. 149–160, 2022.
99. E. S. Pulido, L. A. Morán, J. J. Silva, I. González-Torres, J. A. Rohten, F. A. Villarroel, and J. R. Espinoza, "Design and implementation of a parallel-connected fault current attenuator for power distribution systems," *IEEE Journal of Emerging and Selected Topics in Power Electronics*, vol. 10, no. 1, pp. 402–412, 2022.
100. H. L. Willis and G. B. Rackliffe, "Introduction to integrated resource T&D planning," *ABB Power T&D Company, Inc*, 1994.
101. T. Gonen, *Power Distribution, in the Electrical Engineering Handbook*, 1st ed. Academic Press, New York, 2005, Chapter 6.
102. T. Gönen, "Toward automated distribution system planning," in *Proceedings of the IEEE Control of Power Systems Conference*, March 19–21, 1979, pp. 23–30.
103. F. Van Wormer, "Some aspects of distribution load area geometry [includes discussion]," *Transactions of the American Institute of Electrical Engineers. Part III: Power Apparatus and Systems*, vol. 73, no. 2, pp. 1343–1349, 1954.
104. W. Denton and D. Reps, "Distribution-substation and primary-feeder planning," *AIEE Trans*, vol. 74, no. 3, p. 484–499, 1955.
105. C. Morrison, "A linear approach to the problem of planning new feed points into a distribution system," *AIEE Trans*, vol. III, no. PAS, p. 819–832, December 1963.
106. T. Gönen, *Engineering Economy for Engineering Managers: With Computer Applications*. Wiley New York, 1990.
107. S. C. Sciacca and W. R. Block, "Advanced SCADA concepts," *IEEE Computer Applications in Power*, vol. 8, no. 1, pp. 23–28, 1995.

108. T.-H. Chen and J.-T. Cherng, "Optimal phase arrangement of distribution transformers connected to a primary feeder for system unbalance improvement and loss reduction using a genetic algorithm," *IEEE Transactions on Power Systems*, vol. 15, no. 3, pp. 994–1000, 2000.
109. C. Wang and M. H. Nehrir, "Analytical approaches for optimal placement of distributed generation sources in power systems," *IEEE Transactions on Power Systems*, vol. 19, no. 4, pp. 2068–2076, 2004.
110. C. Zhang, J. Li, Y. J. Zhang, and Z. Xu, "Optimal location planning of renewable distributed generation units in distribution networks: an analytical approach," *IEEE Transactions on Power Systems*, vol. 33, no. 3, pp. 2742–2753, 2018.
111. T.-H. Chen, W.-T. Huang, J.-C. Gu, G.-C. Pu, Y.-F. Hsu, and T.-Y. Guo, "Feasibility study of upgrading primary feeders from radial and open-loop to normally closed-loop arrangement," *IEEE Transactions on Power Systems*, vol. 19, no. 3, pp. 1308–1316, 2004.
112. E. C. Senger, G. Manassero, C. Goldemberg, and E. L. Pellini, "Automated fault location system for primary distribution networks," *IEEE Transactions on Power Delivery*, vol. 20, no. 2, pp. 1332–1340, 2005.
113. N. C. Hien, N. Mithulananthan, and R. C. Bansal, "Location and sizing of distributed generation units for loadabilty enhancement in primary feeder," *IEEE Systems Journal*, vol. 7, no. 4, pp. 797–806, 2013.
114. O. Gandhi, W. Zhang, C. D. Rodriguez-Gallegos, M. Bieri, T. Reindl, and D. Srinivasan, "Analytical approach to reactive power dispatch and energy arbitrage in distribution systems with ders," *IEEE Transactions on Power Systems*, vol. 33, no. 6, pp. 6522–6533, 2018.
115. R.-C. Leou, C.-L. Su, and C.-N. Lu, "Stochastic analyses of electric vehicle charging impacts on distribution network," *IEEE Transactions on Power Systems*, vol. 29, no. 3, pp. 1055–1063, 2014.
116. L. Sun and D. Lubkeman, "Agent-based modeling of feeder-level electric vehicle diffusion for distribution planning," *IEEE Transactions on Smart Grid*, vol. 12, no. 1, pp. 751–760, 2021.
117. S. Sharma and A. Abhyankar, "Loss allocation for weakly meshed distribution system using analytical formulation of shapley value," *IEEE Transactions on Power Systems*, vol. 32, no. 2, pp. 1369–1377, 2017.
118. K. Montano-Martinez, S. Thakar, S. Ma, Z. Soltani, V. Vittal, M. Khorsand, R. Ayyanar, and C. Rojas, "Detailed primary and secondary distribution system model enhancement using ami data," *IEEE Open Access Journal of Power and Energy*, vol. 9, pp. 2–15, 2022.
119. T. Gonen, B. L. Foote, and J. C. Thompson, "Development of advanced methods for planning electric energy distribution systems," Tech. Rep., October 1979.
120. T. Gönen and D. You, "A comparative analysis of distribution feeder costs," in *Proceeding of the Southwest Electrical Exposition and IEEE Conference*, no. 1980, January 22–24.
121. E. E. Institute, *Underground Systems Reference Book*, 2nd ed. Edison Electric Institute, 1957.
122. F. Andrews, "Residential underground distribution adaptable," pp. 107–113, December 12, 1955.
123. D. Smith and J. Barger, "Impedance and circulating current calculations for urd multi-wire concentric neutral circuits," *IEEE Transactions on Power Apparatus and Systems*, no. 3, pp. 992–1006, 1972.
124. T. Gönen, *Power Distribution*, 1st ed. Academic Press, 2005, Chapter 6.
125. EEI and NEMA, "Preferred voltage ratings for AC systems and equipment," no. R-6 and 117, 1949.
126. C. Morrison, "A linear approach to the problem or planning new feed points into a distribution system," *AIEE Transactions*, vol. 82, no. III, pp. 819–832, 1963.
127. T. Gönen, "High-temperature superconductors," pp. 127–129, 1992.
128. D. L. Stone, "Mathematical analysis of direct buried rural distribution cable impedance," *IEEE Transactions on Power Apparatus and Systems*, no. 3, pp. 1015–1022, 1972.
129. R. C. Company, *URD Technical Manual*, 4th ed. Rome Cable Company, 1962.
130. T. Gonen, *Engineering Economy for Engineering Managers: With Computer Applications*. Wiley New York, 1990.
131. J. Davey, "Practical application of weather sensitive load forecasting to system planning," in *Proceedings, the IEEE PES Summer Meeting*, July 9–14, 1972.
132. N. Chang, "Loading distribution transformers, transmission and distribution," pp. 58–59, August 1974.
133. N. Chang, "Determination and evaluation of distribution transformer losses of the electric system through transformer load monitoring," *IEEE Transactions on Power Apparatus and Systems*, pp. 1282–1284, 1970.

134. E. P. R. Institute, "Analysis of distribution r&d planning, EPRI report 329," Tech. Rep., October 1975.
135. J. Shao, N. D. Rao, and Y. Zhang, "An expert system for secondary distribution system design," *IEEE Transactions on Power Delivery*, vol. 6, no. 4, pp. 1607–1615, 1991.
136. J. Gomez, H. Khodr, P. De Oliveira, L. Ocque, J. Yusta, R. Villasana, and A. Urdaneta, "Ant colony system algorithm for the planning of primary distribution circuits," *IEEE Transactions on Power Systems*, vol. 19, no. 2, pp. 996–1004, 2004.
137. A. M. Cossi, R. Romero, and J. R. S. Mantovani, "Planning of secondary distribution circuits through evolutionary algorithms," *IEEE Transactions on Power Delivery*, vol. 20, no. 1, pp. 205–213, 2005.
138. K. Strunz and R. H. Fletcher, "Optimal distribution system horizon planning–part i: formulation," *IEEE Transactions on Power Systems*, vol. 22, no. 2, pp. 791–799, 2007.
139. R. H. Fletcher and K. Strunz, "Optimal distribution system horizon planning–part ii: application," *IEEE Transactions on Power Systems*, vol. 22, no. 2, pp. 862–870, 2007.
140. S. Haffner, L. F. A. Pereira, L. A. Pereira, and L. S. Barreto, "Multistage model for distribution expansion planning with distributed generation—part ii: Numerical results," *IEEE Transactions on Power Delivery*, vol. 23, no. 2, pp. 924–929, 2008.
141. A. M. Cossi, R. Romero, and J. R. S. Mantovani, "Planning and projects of secondary electric power distribution systems," *IEEE Transactions on Power Systems*, vol. 24, no. 3, pp. 1599–1608, 2009.
142. A. Navarro and H. Rudnick, "Large-scale distribution planning—part ii: Macro-optimization with voronoi's diagram and tabu search," *IEEE Transactions on Power Systems*, vol. 24, no. 2, pp. 752–758, 2009.
143. S. Agah and H. A. Abyaneh, "Distribution transformer loss-of-life reduction by increasing penetration of distributed generation," *IEEE Transactions on Power Delivery*, vol. 26, no. 2, pp. 1128–1136, 2011.
144. R. Tonkoski, D. Turcotte, and T. H. El-Fouly, "Impact of high pv penetration on voltage profiles in residential neighborhoods," *IEEE Transactions on Sustainable Energy*, vol. 3, no. 3, pp. 518–527, 2012.
145. E. Dall'Anese, S. V. Dhople, and G. B. Giannakis, "Optimal dispatch of photovoltaic inverters in residential distribution systems," *IEEE Transactions on Sustainable Energy*, vol. 5, no. 2, pp. 487–497, 2014.
146. S. A. El Batawy and W. G. Morsi, "Optimal secondary distribution system design considering rooftop solar photovoltaics," *IEEE Transactions on Sustainable Energy*, vol. 7, no. 4, pp. 1662–1671, 2016.
147. R. Gholizadeh-Roshanagh, S. Najafi-Ravadanegh, and S. H. Hosseinian, "A framework for optimal coordinated primary-secondary planning of distribution systems considering mv distributed generation," *IEEE Transactions on Smart Grid*, vol. 9, no. 2, pp. 1408–1415, 2016.
148. G. Lou, W. Gu, X. Lu, Y. Xu, and H. Hong, "Distributed secondary voltage control in islanded microgrids with consideration of communication network and time delays," *IEEE Transactions on Smart Grid*, vol. 11, no. 5, pp. 3702–3715, 2020.
149. S. Chang, "Economic design of secondary distribution system by computer," Tech. Rep., 1974.
150. E. E. I.-N. E. M. Association, *EEI-NEMA Standards for Secondary Network Transformers*. EEI-NEMA Publication, 1968.
151. R. Lawrence, D. Reps, and A. Patton, "Distribution system planning through optimal design, i-distribution transformers and secondaries," *Transactions of the American Institute of Electrical Engineers. Part III: Power Apparatus and Systems*, vol. 79, no. 3, pp. 199–204, 1960.
152. D. Robb, *ECDES Program User Manual. Power System Computer Service*, 1975.
153. C. W. Brice, "Voltage-drop calculations and power-flow studies for rural electric distribution lines," *IEEE Transactions on Industry Applications*, vol. 28, no. 4, pp. 774–781, 1992.
154. R. Taleski and D. Rajicic, "Energy summation method for energy loss computation in radial distribution networks," *IEEE Transactions on Power Systems*, vol. 11, no. 2, pp. 1104–1111, 1996.
155. D. L. Flaten, "Distribution system losses calculated by percent loading," *IEEE Transactions on Power Systems*, vol. 3, no. 3, pp. 1263–1269, 1988.
156. E. Díaz-Dorado, E. Miguez, and J. Cidrás, "Design of large rural low-voltage networks using dynamic programming optimization," *IEEE Transactions on Power Systems*, vol. 16, no. 4, pp. 898–903, 2001.
157. E. Diaz-Dorado, J. Cidrás, and E. Míguez, "Application of evolutionary algorithms for the planning of urban distribution networks of medium voltage," *IEEE Transactions on Power Systems*, vol. 17, no. 3, pp. 879–884, 2002.
158. P. M. Costa and M. A. Matos, "Loss allocation in distribution networks with embedded generation," *IEEE Transactions on Power Systems*, vol. 19, no. 1, pp. 384–389, 2004.

159. G. Celli, E. Ghiani, S. Mocci, and F. Pilo, "A multiobjective evolutionary algorithm for the sizing and siting of distributed generation," *IEEE Transactions on Power Systems*, vol. 20, no. 2, pp. 750–757, 2005.
160. V. M. Quezada, J. R. Abbad, and T. G. S. Roman, "Assessment of energy distribution losses for increasing penetration of distributed generation," *IEEE Transactions on Power Systems*, vol. 21, no. 2, pp. 533–540, 2006.
161. O. M. Mikic, "Variance-based energy loss computation in low voltage distribution networks," *IEEE Transactions on Power Systems*, vol. 22, no. 1, pp. 179–187, 2007.
162. V. F. Martins and C. L. Borges, "Active distribution network integrated planning incorporating distributed generation and load response uncertainties," *IEEE Transactions on Power Systems*, vol. 26, no. 4, pp. 2164–2172, 2011.
163. L. M. Queiroz, M. A. Roselli, C. Cavellucci, and C. Lyra, "Energy losses estimation in power distribution systems," *IEEE Transactions on Power Systems*, vol. 27, no. 4, pp. 1879–1887, 2012.
164. L. F. Ochoa and G. P. Harrison, "Minimizing energy losses: Optimal accommodation and smart operation of renewable distributed generation," *IEEE Transactions on Power Systems*, vol. 26, no. 1, pp. 198–205, 2011.
165. M. Alam, K. Muttaqi, and D. Sutanto, "Mitigation of rooftop solar pv impacts and evening peak support by managing available capacity of distributed energy storage systems," *IEEE Transactions on Power Systems*, vol. 28, no. 4, pp. 3874–3884, 2013.
166. M. Alam, D. Sutanto, and K. Muttaqi, "An approach for online assessment of rooftop solar pv impacts on low-voltage distribution networks," *IEEE Transactions on Sustainable Energy*, vol. 5, no. 2, pp. 663–672, 2014.
167. M. Ghofrani, A. Arabali, M. Etezadi-Amoli, and M. S. Fadali, "Smart scheduling and cost-benefit analysis of grid-enabled electric vehicles for wind power integration," *IEEE Transactions on Smart Grid*, vol. 5, no. 5, pp. 2306–2313, 2014.
168. A. Sharma, D. Srinivasan, and A. Trivedi, "A decentralized multiagent system approach for service restoration using DG islanding," *IEEE Transactions on Smart Grid*, vol. 6, no. 6, pp. 2784–2793, 2015.
169. P. Mohammadi and S. Mehraeen, "Challenges of pv integration in low-voltage secondary networks," *IEEE Transactions on Power Delivery*, vol. 32, no. 1, pp. 525–535, 2016.
170. Y. Xu, C.-C. Liu, Z. Wang, K. Mo, K. P. Schneider, F. K. Tuffner, and D. T. Ton, "Dgs for service restoration to critical loads in a secondary network," *IEEE Transactions on Smart Grid*, vol. 10, no. 1, pp. 435–447, 2017.
171. J. Light, "An economic approach to distribution conductor size selection," April 12–14, 1978.
172. M. Schlegel, "New selection method reduces conductor losses," Public Service Co., Tulsa, OK, Tech. Rep., 1977.
173. T. Gonen and D. Yu, "A distribution system planning model." IEEE, March 17–18, 1980, pp. 28–34.
174. U. D. of Agriculture, "Economic design of primary lines for rural distribution systems," pp. 60–69, May, 1960.
175. M. Delaney, "Economic analysis of electrical equipment losses," vol. 74, no. 4, pp. 7–8, 1974.
176. K. Klein, "Evaluation of distribution transformer losses and loss ratios," *Electric Light and Power*, pp. 56–61, July 15, 1960.
177. R. Smith and D. Ward, "Does early distribution transformer retirement make sense?" in *GE ELECTRIC FORUM Magazine*, vol. 6, no. 3, 1980, pp. 6–9.
178. I. Roytelman, B. Wee, and R. Lugtu, "Volt/var control algorithm for modern distribution management system," *IEEE Transactions on Power Systems*, vol. 10, no. 3, pp. 1454–1460, 1995.
179. I. Roytelman, B. Wee, R. Lugtu, T. Kulas, and T. Brossart, "Pilot project to estimate the centralized volt/var control effectiveness," *IEEE Transactions on Power Systems*, vol. 13, no. 3, pp. 864–869, 1998.
180. S. Miske, "Considerations for the application of series capacitors to radial power distribution circuits," *IEEE Transactions on Power Delivery*, vol. 16, no. 2, pp. 306–318, 2001.
181. K. Wada, H. Fujita, and H. Akagi, "Considerations of a shunt active filter based on voltage detection for installation on a long distribution feeder," *IEEE Transactions on Industry Applications*, vol. 38, no. 4, pp. 1123–1130, 2002.
182. S. Mekhamer, S. Soliman, M. Moustafa, and M. El-Hawary, "Application of fuzzy logic for reactive-power compensation of radial distribution feeders," *IEEE Transactions on Power Systems*, vol. 18, no. 1, pp. 206–213, 2003.

183. J.-P. Chiou, C.-F. Chang, and C.-T. Su, "Ant direction hybrid differential evolution for solving large capacitor placement problems," *IEEE Transactions on Power Systems*, vol. 19, no. 4, pp. 1794–1800, 2004.
184. R. H. Simpson, "Misapplication of power capacitors in distribution systems with nonlinear loads-three case histories," *IEEE Transactions on Industry Applications*, vol. 41, no. 1, pp. 134–143, 2005.
185. A. Ulinuha, M. A. Masoum, and S. M. Islam, "Optimal scheduling of ltc and shunt capacitors in large distorted distribution systems using evolutionary-based algorithms," *IEEE Transactions on Power Delivery*, vol. 23, no. 1, pp. 434–441, 2008.
186. D. T. Chessmore, W.-J. Lee, W. E. Muston, T. L. Anthony, F. Daniel, and L. Kohrmann, "Voltage-profile estimation and control of a distribution feeder," *IEEE Transactions on Industry Applications*, vol. 45, no. 4, pp. 1467–1474, 2009.
187. V. Farahani, B. Vahidi, and H. A. Abyaneh, "Reconfiguration and capacitor placement simultaneously for energy loss reduction based on an improved reconfiguration method," *IEEE Transactions on Power Systems*, vol. 27, no. 2, pp. 587–595, 2012.
188. T. Tanaka, T. Sekiya, H. Tanaka, M. Okamoto, and E. Hiraki, "Smart charger for electric vehicles with power-quality compensator on single-phase three-wire distribution feeders," *IEEE Transactions on Industry Applications*, vol. 49, no. 6, pp. 2628–2635, 2013.
189. M. R. Kleinberg, K. Miu, N. Segal, H. Lehmann, and T. R. Figura, "A partitioning method for distributed capacitor control of electric power distribution systems," *IEEE Transactions on Power Systems*, vol. 29, no. 2, pp. 637–644, 2014.
190. H. Ahmadi and J. R. Martí, "Distribution system optimization based on a linear power-flow formulation," *IEEE Transactions on Power Delivery*, vol. 30, no. 1, pp. 25–33, 2015.
191. F. Yang and Z. Li, "Improve distribution system energy efficiency with coordinated reactive power control," *IEEE Transactions on Power Systems*, vol. 31, no. 4, pp. 2518–2525, 2015.
192. A. Safayet, P. Fajri, and I. Husain, "Reactive power management for overvoltage prevention at high pv penetration in a low-voltage distribution system," *IEEE Transactions on Industry Applications*, vol. 53, no. 6, pp. 5786–5794, 2017.
193. A. Mosaddegh, C. A. Cañizares, K. Bhattacharya, and H. Fan, "Distributed computing architecture for optimal control of distribution feeders with smart loads," *IEEE Transactions on Smart Grid*, vol. 8, no. 3, pp. 1469–1478, 2017.
194. R. Anilkumar, G. Devriese, and A. K. Srivastava, "Voltage and reactive power control to maximize the energy savings in power distribution system with wind energy," *IEEE Transactions on Industry Applications*, vol. 54, no. 1, pp. 656–664, 2017.
195. H. V. Padullaparti, Q. Nguyen, and S. Santoso, "Optimal placement and dispatch of lv-svcs to improve distribution circuit performance," *IEEE Transactions on Power Systems*, vol. 34, no. 4, pp. 2892–2900, 2019.
196. S. Abdelhady, A. Osama, A. Shaban, and M. Elbayoumi, "A real-time optimization of reactive power for an intelligent system using genetic algorithm," *IEEE Access*, vol. 8, pp. 11 991–12 000, 2020.
197. A. R. Di Fazio, C. Risi, M. Russo, and M. De Santis, "Coordinated optimization for zone-based voltage control in distribution grids," *IEEE Transactions on Industry Applications*, vol. 58, no. 1, pp. 173–184, 2021.
198. J. Lapp, "The impact of technical developments on power capacitors," vol. 80, no. 2, pp. 19–24, July 1980.
199. *The ABC of Capacitors*. McGraw-Edison Company, 1968.
200. R. Hopkinson, "Economic power factor: key to kVAr supply," in *Electr. Forum;(United States)*, vol. 6, no. 3. Power Distribution Systems Engineering Operation, Schenectady, NY, 1980, pp. 20–22.
201. R. Zimmerman, "Economic merits of secondary capacitors," *Transactions of the American Institute of Electrical Engineers*, vol. 72, no. 4, pp. 694–697, 1953.
202. W. Baum and W. Frederick, "A method of applying switched and fixed capacitors for voltage control," *IEEE Transactions on Power Apparatus and Systems*, vol. 84, no. 1, pp. 42–48, 1965.
203. R. Wallace, "Capacitors reduce system investment and losses," vol. 76, no. 1, pp. 15–17, 1976.
204. *Oklahoma Gas and Electric Company: Engineering Guides*. Oklahoma Gas and Electric Company, January 1981.

205. N. Neagle and D. Samson, "Loss reduction from capacitors installed on primary feeders," *Transactions of the American Institute of Electrical Engineers. Part III: Power Apparatus and Systems*, vol. 75, no. 3, pp. 950–959, 1956.
206. R. Schmidt, "Dc circuit gives easy method of determining value of capacitors in reducing i^2r losses [includes discussion]," *Transactions of the American Institute of Electrical Engineers. Part III: Power Apparatus and Systems*, vol. 75, no. 3, pp. 840–848, 1956.
207. N. Maxwell, "The economic application of capacitors to distribution feeders," *Transactions of the American Institute of Electrical Engineers. Part III: Power Apparatus and Systems*, vol. 79, no. 3, pp. 353–358, 1960.
208. R. Cook, "Optimizing the application of shunt capacitors for reactive-volt-ampere control and loss reduction," *Transactions of the American Institute of Electrical Engineers. Part III: Power Apparatus and Systems*, vol. 80, no. 3, pp. 430–441, 1961.
209. J. Schmill, "Optimum size and location of shunt capacitors on distribution feeders," *IEEE Transactions on Power Apparatus and Systems*, vol. 84, no. 9, pp. 825–832, 1965.
210. N. E. Chang, "Determination of primary-feeder losses," *IEEE Transactions on Power Apparatus and Systems*, vol. 87, no. 12, pp. 1991–1994, 1968.
211. N. E. Chang, "Locating shunt capacitors on primary feeder for voltage control and loss reduction," *IEEE Transactions on Power Apparatus and Systems*, no. 10, pp. 1574–1577, 1969.
212. N. E. Chang, "Generalized equations on loss reduction with shunt capacitor," *IEEE Transactions on Power Apparatus and Systems*, no. 5, pp. 2189–2195, 1972.
213. Y. Bae, "Analytical method of capacitor allocation on distribution primary feeders," *IEEE Transactions on Power Apparatus and Systems*, no. 4, pp. 1232–1238, 1978.
214. T. Gönen and F. Djavashi, "Optimum loss reduction from capacitors installed on primary feeders," in *IEEE Midwest Power Symposium*, October 27–28, 1980.
215. T. Gonen and F. Djavashi, "Optimum shunt capacitor allocation on primary feeders," in *IEEE MEXICON-80 International Conference*, October 22–25, 1980.
216. J. J. Grainger and S. Lee, "Optimum size and location of shunt capacitors for reduction of losses on distribution feeders," *IEEE Transactions on Power Apparatus and Systems*, no. 3, pp. 1105–1118, 1981.
217. S. Lee and J. Grainger, "Optimum placement of fixed and switched capacitors on primary distribution feeders," *IEEE Transactions on Power Apparatus and Systems*, no. 1, pp. 345–352, 1981.
218. J. Grainger, A. El-Kib, and S. Lee, "Optimal capacitor placement on three-phase primary feeders: load and feeder unbalance effects," *IEEE Transactions on Power Apparatus and Systems*, no. 10, pp. 3296–3305, 1983.
219. J. Grainger, S. Civanlar, and S. Lee, "Optimal design and control scheme for continuous capacitive compensation of distribution feeders," *IEEE Transactions on Power Apparatus and Systems*, no. 10, pp. 3271–3278, 1983.
220. L. Burrage, "Capacitor tank ruptures studied," vol. 76, no. 2, pp. 2–5, 1976.
221. M. McGranaghan, R. Dugan, and W. Sponsler, "Digital simulation of distribution system frequency-response characteristics," *IEEE PES Summer Meeting*, July 13–18, 1980.
222. M. McGranaghan, J. Shaw, and R. Owen, "Measuring voltage and current harmonics on distribution systems," *IEEE PES Summer Meeting*, February 1–6, 1981.
223. T. Gönen and A. Mahmoud, "Bibliography of power system harmonics, part i," *IEEE Transactions on Power Apparatus and Systems*, vol. 103, no. 9, pp. 2460–2469, 1984.
224. T. Gönen and A. Mahmoud, "Bibliography of power system harmonics, part ii," *IEEE Transactions on Power Apparatus and Systems*, vol. 103, no. 9, pp. 2470–2479, 1984.
225. A. Mahmoud, R. Owen, and A. Emanuel, "Power system harmonics: an overview," *IEEE Transactions on Power Apparatus and Systems*, no. 8, pp. 2455–2460, 1983.
226. R. Owen, M. McGranaghan, and J. Vivirito, "Distribution system harmonics: controls for large power converters," *IEEE PES Summer Meeting*, July 26–31, 1981.
227. IEEE, "Ieee guide for harmonic control and reactive compensation of static power converters," *IEEE: Piscataway, NJ, USA*, 1981.
228. R. Owen and J. Vivirito, "Study of distribution system surge and harmonic characteristics. final report," GrElectric Power Research Institute, Tech. Rep., 1980.
229. R. Owen, "Distribution system harmonics: Effects on equipment and operation," *The Line*, vol. 2, March 15–16, 1979.

230. E. W. Kimbark, *Direct Current Transmission*. Wiley, 1971.
231. B. Szabados, E. J. Burgess, and W. Noble, "Harmonic interference corrected by shunt capacitors on distribution feeders," *IEEE Transactions on Power Apparatus and Systems*, vol. 96, no. 1, pp. 234–239, January/February 1977.
232. J.-H. Choi and J.-C. Kim, "Advanced voltage regulation method at the power distribution systems interconnected with dispersed storage and generation systems," *IEEE Transactions on Power Delivery*, vol. 15, no. 2, pp. 691–696, 2000.
233. N. C. Scott, D. J. Atkinson, and J. E. Morrell, "Use of load control to regulate voltage on distribution networks with embedded generation," *IEEE transactions on Power Systems*, vol. 17, no. 2, pp. 510–515, 2002.
234. W. H. Kersting, "Distribution feeder voltage regulation control," *IEEE Transactions on Industry Applications*, vol. 46, no. 2, pp. 620–626, 2010.
235. J. E. Mendoza, D. A. Morales, R. A. Lopez, E. A. Lopez, J.-C. Vannier, and C. A. C. Coello, "Multiobjective location of automatic voltage regulators in a radial distribution network using a micro genetic algorithm," *IEEE Transactions on Power Systems*, vol. 22, no. 1, pp. 404–412, 2007.
236. T. Senjyu, Y. Miyazato, A. Yona, N. Urasaki, and T. Funabashi, "Optimal distribution voltage control and coordination with distributed generation," *IEEE Transactions on Power Delivery*, vol. 23, no. 2, pp. 1236–1242, 2008.
237. D. H. Spatti, I. N. da Silva, W. F. Usida, and R. A. Flauzino, "Real-time voltage regulation in power distribution system using fuzzy control," *IEEE Transactions on Power Delivery*, vol. 25, no. 2, pp. 1112–1123, 2009.
238. R. A. Shayani and M. A. G. de Oliveira, "Photovoltaic generation penetration limits in radial distribution systems," *IEEE Transactions on Power Systems*, vol. 26, no. 3, pp. 1625–1631, 2010.
239. T. Niknam, M. Zare, and J. Aghaei, "Scenario-based multiobjective volt/var control in distribution networks including renewable energy sources," *IEEE Transactions on Power Delivery*, vol. 27, no. 4, pp. 2004–2019, 2012.
240. B. A. Robbins, C. N. Hadjicostis, and A. D. Domínguez-García, "A two-stage distributed architecture for voltage control in power distribution systems," *IEEE Transactions on Power Systems*, vol. 28, no. 2, pp. 1470–1482, 2013.
241. S. Weckx, R. D'Hulst, B. Claessens, and J. Driesensam, "Multiagent charging of electric vehicles respecting distribution transformer loading and voltage limits," *IEEE Transactions on Smart Grid*, vol. 5, no. 6, pp. 2857–2867, 2014.
242. K. M. Muttaqi, A. D. Le, M. Negnevitsky, and G. Ledwich, "A coordinated voltage control approach for coordination of oltc, voltage regulator, and DG to regulate voltage in a distribution feeder," *IEEE Transactions on Industry Applications*, vol. 51, no. 2, pp. 1239–1248, 2015.
243. A. Bokhari, A. Raza, M. Diaz-Aguiló, F. De Leon, D. Czarkowski, R. E. Uosef, and D. Wang, "Combined effect of cvr and DG penetration in the voltage profile of low-voltage secondary distribution networks," *IEEE Transactions on Power Delivery*, vol. 31, no. 1, pp. 286–293, 2016.
244. Z. Akhtar, B. Chaudhuri, and S. Y. R. Hui, "Smart loads for voltage control in distribution networks," *IEEE Transactions on Smart Grid*, vol. 8, no. 2, pp. 937–946, 2017.
245. G. Cavraro and R. Carli, "Local and distributed voltage control algorithms in distribution networks," *IEEE Transactions on Power Systems*, vol. 33, no. 2, pp. 1420–1430, 2018.
246. Y. Liu, J. Li, and L. Wu, "Coordinated optimal network reconfiguration and voltage regulator/der control for unbalanced distribution systems," *IEEE Transactions on Smart Grid*, vol. 10, no. 3, pp. 2912–2922, 2019.
247. A. Bedawy, N. Yorino, K. Mahmoud, Y. Zoka, and Y. Sasaki, "Optimal voltage control strategy for voltage regulators in active unbalanced distribution systems using multi-agents," *IEEE Transactions on Power Systems*, vol. 35, no. 2, pp. 1023–1035, 2020.
248. A. Kriukov, M. Gavrilaş, O. Ivanov, G. Grigoraş, B.-C. Neagu, and F. Scarlatache, "Novel decentralized voltage-centered EV charging control algorithm using dsrc system in low voltage distribution networks," *IEEE Access*, vol. 9, pp. 164 779–164 800, 2021.
249. X. Sun, J. Qiu, Y. Yi, and Y. Tao, "Cost-effective coordinated voltage control in active distribution networks with photovoltaics and mobile energy storage systems," *IEEE Transactions on Sustainable Energy*, vol. 13, no. 1, pp. 501–513, 2022.

250. H. Lokay and D. Reps, "Distribution system primary-feeder voltage control: Part i-a new approach using the digital computer," in *AIEE Trans*, vol. 77, no. 3, October 1958, pp. 845–855.
251. G. Kirk Jr and D. Reps, "Distribution system primary-feeder voltage control: Part ii-digital computer program," in *AIEE Trans*, vol. 77, no. 3, October 1958, pp. 856–865.
252. M. R. McCrary, "Regulating voltage on a major power system," vol. 81, no. 1, pp. 2–8, 1981.
253. R. Hopkinson, "Recap computer program aids voltage-regulation studies," in *Electr. Forum;(United States)*, vol. 4, no. 4, 1978, pp. 20–23.
254. H. Amchin, R. Bentzel, and D. Reps, "Distribution system primary-feeder voltage control: Part iii-computer program application," in *AIEE Trans*, vol. 77, no. 3, October 1958, pp. 865–879.
255. R. F. Cook and D. N. Reps, "Distribution system primary-feeder voltage control: Part iv-a supplementary computer program for main-circuit analysis," in *AIEE Trans*, vol. 78, no. 3, October 1958, pp. 904–913.
256. W. C. Sealey, "Increased current ratings for step regulators," *Transactions of the American Institute of Electrical Engineers. Part III: Power Apparatus and Systems*, vol. 74, no. 3, pp. 737–742, 1955.
257. W. Bovenizer, "New 'simplified' regulator for lower-cost distribution voltage regulation," vol. 75, no. 1, pp. 25–28, 1975.
258. W. bovenizer, "Paralleling voltage regulators," vol. 75, no. 2, pp. 6–9, 1975.
259. W. Ku, "Economic comparison of switched capacitors and voltage regulators for system voltage control," *Transactions of the American Institute of Electrical Engineers. Part III: Power Apparatus and Systems*, vol. 76, no. 3, pp. 891–906, 1957.
260. B. R. Williams and D. G. Walden, "Distribution automation strategy for the future," in *IEEE Comp. Appi. Power*, vol. 7, no. 3, July 1994, pp. 16–21.

Index

Note: Page numbers in **bold** refer to tables and those in *italic* refer to figures.